3rd EDITION

VOYAGES
TO THE PLANETS

Andrew Fraknoi
Foothill College

David Morrison
NASA Ames Research Center

Sidney Wolff
National Optical Astronomy Observatories

THOMSON
BROOKS/COLE

Australia • Canada • Mexico • Singapore • Spain • United Kingdom • United States

THOMSON
✳
BROOKS/COLE

Astronomy Editor: *Keith Dodson*
Development Editor: *Marie Carigma-Sambilay*
Assistant Editor: *Carol Benedict*
Editorial Assistant: *Melissa Newt*
Technology Project Manager: *Sam Subity*
Marketing Manager: *Kelley McAllister*
Marketing Assistant: *Leyla Jowza*
Advertising Project Manager: *Nathaniel Bergson-Michelson*
Project Manager, Editorial Production: *Hal Humphrey*
Print/Media Buyer: *Barbara Britton*
Permissions Editor: *Joohee Lee*

Production Service: *Nancy Shammas, New Leaf Publishing Service*
Cover and Text Designer: *Roy R. Neuhaus*
Art Editor: *Kathy Joneson*
Photo Researcher: *Jane Sanders*
Copy Editor: *Carol Reitz*
Illustrator: *Rolin Graphics*
Cover Photo: *NASA and The Hubble Heritage Team (STScI-AURA)*
Compositor: *G & S Typesetters, Inc.*
Cover Printer: *The Lehigh Press, Inc.*
Printer: *Quebecor World/Dubuque*

For more information about our products, contact us at:
Thomson Learning Academic Resource Center
1-800-423-0563
For permission to use material from this text, contact us by:
Phone: 1-800-730-2214 **Fax:** 1-800-730-2215
Web: http://www.thomsonrights.com

Library of Congress Control Number: 2003108533

ISBN 0-534-39567-8

Brooks/Cole—Thomson Learning
10 Davis Drive
Belmont, CA 94002-3098
USA

Asia
Thomson Learning
5 Shenton Way #01-01
UIC Building
Singapore 068808

Australia/New Zealand
Thomson Learning
102 Dodds Street
Southbank, Victoria 3006
Australia

Canada
Nelson
1120 Birchmount Road
Toronto, Ontario M1K 5G4
Canada

Europe/Middle East/Africa
Thomson Learning
High Holborn House
50/51 Bedford Row
London WC1R 4LR
United Kingdom

Latin America
Thomson Learning
Seneca, 53
Colonia Polanco
11560 Mexico D.F.
Mexico

Spain/Portugal
Paraninfo Thomson Learning
Calle/Magallanes, 25
28015 Madrid, Spain

About the Authors

Andrew Fraknoi is the Chair of the Astronomy Department at Foothill College near San Francisco, where his courses are taken by about 900 students per year. He is also Director of Project ASTRO at the Astronomical Society of the Pacific—a national program that forms partnerships between volunteer astronomers and school teachers in their communities and develops astronomy activity kits for families to use outside of school. From 1978 to 1992 he was Executive Director of the Society, as well as Editor of *Mercury* Magazine and the *Universe in the Classroom* Newsletter. He has taught astronomy and physics at San Francisco State University, Cañada College, and the University of California Extension Division. He is co-author and editor of *The Universe at Your Fingertips* and *More Universe at Your Fingertips,* two widely used collections of astronomy teaching activities and resources. For five years he was the lead author of a nationally syndicated newspaper column on astronomy, and he appears regularly on radio and television explaining astronomical developments. With Sidney Wolff, he is co-editor of *Astronomy Education Review,* a new on-line journal/magazine for those working in space science education (see: aer.noao.edu). In addition, he has organized three national symposia on teaching introductory astronomy at the college level, and over 20 national workshops on improving the way astronomy is taught in earlier grades. He has received the Annenberg Foundation Prize of the American Astronomical Society and the Klumpke-Roberts Prize of the Astronomical Society of the Pacific for his contributions to the public understanding of astronomy. Asteroid 4859 was named Asteroid Fraknoi in 1992 in recognition of his work in astronomy education.

David Morrison is the Senior Scientist at the NASA Astrobiology Institute, where he participates in a variety of research programs in astrobiology—the study of the living universe. From 1996–2001 he was the Director of Astrobiology and Space Research at NASA Ames Research Center, managing basic and applied research programs in the space, life, and Earth sciences. Dr. Morrison received his Ph.D. in astronomy from Harvard University, and until he joined NASA he was Professor of Astronomy at the University of Hawaii. Internationally known for his research on small bodies in the solar system, Dr. Morrison is the author of more than 130 technical papers and has published a dozen books. He chaired the official NASA study of impact hazards that recommended that a Spaceguard Survey be carried out to search for potentially threatening asteroids and comets and in 1995 received the NASA Outstanding Leadership Medal for this work. He is also the recipient of the Dryden Medal for research from the American Institute of Aeronautics and Astronautics, and of the Klumpke-Roberts award of the Astronomical Society of the Pacific for contributions to science education. He has served as President of the Astronomical Society of the Pacific, Chair of the Astronomy Section of the American Association for the Advancement of Science, and President of the Planetary Commission of the International Astronomical Union. Asteroid 2410 Morrison is named in his honor.

Sidney C. Wolff received her Ph.D. from the University of California at Berkeley and then joined the Institute for Astronomy at the University of Hawaii. During the 17 years she spent in Hawaii, the Institute for Astronomy developed Mauna Kea into the world's premier international observatory. She became Associate Director of the Institute for Astronomy in 1976 and Acting Director in 1983. During that period, she earned international recognition for her research, particularly on stellar atmospheres and how they can help us understand the evolution, formation, and composition of stars. She is currently working on problems relating to how stars form. In 1984, Wolff was named Director of the Kitt Peak National Observatory and in 1987 became Director of the National Optical Astronomy Observatories. She was the first woman to head a major observatory in the United States. As Director of NOAO, she and her staff oversaw facilities used annually by nearly 1000 visiting scientists. During its early phases, she was Director of the Gemini Project, which is an international program to build two state-of-the-art 8m telescopes. Wolff is currently involved with the design of a telescope that would scan the whole visible sky every week or so; conventional telescopes would take decades to complete a similar survey. Such a facility would open up the systematic study of time variable and moving objects. She has served as President of the American Astronomical Society, and is also a member of the Board of Trustees of Carleton College, a liberal arts school that excels in science education. The author of more than 70 professional articles, she has written a monograph, *The A-Type Stars: Problems and Perspectives,* as well as several astronomy textbooks. She is the editor of the new on-line journal, *Astronomy Education Review.*

Preface for the Student

IN COLLEGE TEXTBOOKS, there is a long tradition that the preface of the book is read by the instructor and the rest of the book by the student. Still, many students start reading the preface (it does come first) and then wonder why it doesn't say much to them.

So we begin our book with a preface specifically for student readers. It's not about the subject matter of astronomy, which is introduced in the Prologue. Rather, we want to tell you a little about the book and give you some hints for the effective study of astronomy. (Your professor will probably have other, more specific suggestions for doing well in your class.)

THE WORLD OF ASTRONOMY

Astronomy, the study of the universe beyond the borders of our planet, is one of the most exciting and rapidly changing branches of science. Even scientists from other fields often confess to having had a lifelong interest in astronomy, though they may now be doing something earth-bound like biology, chemistry, engineering, or writing software. There are fewer than 10,000 professional astronomers in the entire world. However, astronomy has a large group of amateur devotees who spend many an evening with a telescope under the stars observing the sky, and who occasionally make a discovery, such as a new comet or a developing storm on Saturn.

Many other people are "armchair astronomers"—fascinated just to read about bizarre worlds and processes that astronomers are uncovering. Some are intrigued by the scientific search for planets or life in other star systems. Others like to follow the challenges of space exploration, such as upgrading the Hubble Space Telescope by the shuttle astronauts or the results from the Cassini mission to probe the giant planet Saturn and its intriguing moon Titan. Hearing about an astronomical event in the news media may be what first sparked your interest in taking an astronomy course.

But some of the things that make astronomy so interesting also make it a challenge for the beginning student.

The universe is a big place, full of objects and processes that do not have familiar counterparts here on Earth. Like a visitor to a new country, you will need time to feel familiar with the territory or the local customs. Astronomy, like other sciences, has its own special vocabulary, some of which you will have to learn if you want to have a good voyage with us. (Since the title of this text includes the word *Voyages*, we will use voyaging metaphors from time to time. Humor your authors; we mean well.)

FEATURES OF THIS BOOK

To assist students taking their first college-level course in astronomy, we have built a number of special features into this book, and we invite you to make use of them:

- All technical terms are printed in **boldface** type the first time they are used and clearly defined in the text; their definitions are listed alphabetically in Appendix 3 (the Glossary), so you can refer to them at any time. The summaries at the end of each chapter also include these boldface terms as a review.

- The book begins with a historical summary of astronomy and then surveys our planetary system as well as planetary systems around other stars. But don't worry if your instructor doesn't assign all the chapters or doesn't assign the chapters in order. Throughout the book "directional signs" lead you to any earlier material you need to know before tackling the current section.

- We use tables to bring together numerical data for your convenience. For example, some tables summarize the important properties of each planet in the solar system (including the Earth). Students who want to see more of the data that astronomers use regularly can investigate the appendices at the back of the book.

- Figure captions clearly and fully describe what phenomena or objects students are looking at. We have tried to look at each figure with student eyes and ask what would help clarify the diagram or image for a reader just being introduced to astronomy.

- Each chapter ends with a summary of the essential points in the chapter, plus small-group activities (more about these in a minute), review questions, thought questions, and numerical problems to help you "process" what you have learned. We have put one or more worked out examples into a "Figuring for Yourself" box in most chapters to help you with these problems. Some simple multiple-choice quiz questions to help you review are also available on our Web site: http://info.brookscole.com/voyages.

- Suggestions for further reading in print and on the Web are included at the end of each chapter, in case you want to learn more about a particular topic. These books, articles, and Web sites are written at the same introductory level as this text. Appendices 1 and 2 contain a listing of general sources of astronomical information. Appendix 1 also includes Web sites that can guide you to the best new astronomical images (many of which we could not fit into the text).

- Each chapter features a series of small-group activities—mostly questions that take off in new directions from the material in the chapter and are meant to foster debate and discussion among students. Your instructor or discussion section leader may use these in class, or you may want to discuss questions that interest you with the other members of your study group. (What study group? you may be asking. See below.)

STUDYING ASTRONOMY

How can you increase your chances of doing well in your astronomy class? Here are a few suggestions for studying astronomy that come from good teachers and good students from around the country:

- First, the best advice we can give you is to be sure to leave enough time in your schedule to study the material in this class regularly. It sounds obvious, but it is not very easy to catch up with a subject like astronomy by trying to do everything just before an exam. Try to put aside some part of each day, or every other day, when you can have uninterrupted time for reading and studying astronomy.

- Try to read each assignment in the book twice, once before it is discussed in class and once afterward. Take notes or use a highlighter to outline ideas that you may want to review later. Also, take some time to coordinate the notes from your reading with the notes you take in class. Many students start college without good note-taking habits. If you are not a good note-taker, try to get some help. Many colleges and universities have student learning centers that offer short courses, workbooks, or videos on developing good study habits. Good note-taking skills will also be useful for many jobs or activities you may get involved with after college.

- Form a small astronomy study group with people in your class; get together as often as you can and discuss the topics that may be giving group members trouble. Make up sample exam questions and make sure everyone in the group can answer them confidently. If you have always studied alone, you may at first resist this idea, but don't be too hasty to say no. Study groups are a very effective way of digesting new information, learning a foreign language (which astronomy words sometimes resemble), or studying a broad field like law or astronomy.

- Before each exam, do a concise outline of the main ideas discussed in class and presented in your text. Compare your outline with those of other students as a check on your own study habits.

- If you find a topic in the text or in class especially difficult or interesting, don't hesitate to make use of the resources in your library or on the Web for additional study (each chapter has suggestions for you).

- Don't be too hard on yourself! If astronomy is new to you, many of the ideas and terms in this book may be unfamiliar. Astronomy is like any new language; it may take a while to become a good conversationalist. Practice as much as you can, but also realize that it is natural to be overwhelmed by the vastness of the universe and the variety of things that are going on in it.

We hope you enjoy reading this text as much as we enjoyed writing it. We are always glad to hear from students who have used the text and invite you to send us your reactions to the book and suggestions for how we can improve future editions. We promise you we will read and consider every serious letter we receive. You can send your comments to Andrew Fraknoi, Astronomy Department, Foothill College, 12345 El Monte Road, Los Altos Hills, CA 94022, USA. (Please note that we will not send you the answers to the chapter problems or do your homework for you, but all other thoughts are most welcome.)

Andrew Fraknoi
David Morrison
Sidney Wolff

Preface for the Instructor

WE'RE DELIGHTED THAT YOU ARE using or considering this book. *Voyages* is a planetary astronomy text written with today's students in mind—it is designed for non-science majors who may even be a little intimidated by science, and who approach this subject with more interest than experience. With features that we hope will appeal to everyone from university economics majors to first-time community college students, *Voyages* is written to draw in and engage readers while preserving the accuracy and timeliness that our colleagues expect of us.

In thinking through how best to present planetary astronomy to students whose science and math background may be limited, we have tried to make the language friendly and inviting and to use examples and analogies drawn from everyday experiences. Yet we have not hesitated to include the very latest research results and to explain the latest thinking in the many disciplines that make up our field. We very much hope our ways of drawing in readers—including vignettes from the lives of scientists, references to other fields that are influenced by astronomy and planetary science, and occasional touches of humor—can help make this a book that your students will actually enjoy reading. There is, we believe, nothing less useful than a textbook the instructor opens more often than the student!

■ ORGANIZATION AND SPECIAL FEATURES

We realize that some instructors will want to teach topics in a different order than we have presented or leave some out altogether. It's fair to say that in astronomy—much more than in any other introductory science class—there are as many approaches to the basic course as there are instructors teaching it. *Voyages to the Planets* is designed to serve a wide range of courses that cover planetary science or solar system astronomy. We have included chapters on the sky, on the history of astronomy, and on solar atmospheric phenomena for instructors who integrate this material into

their teaching, but the other chapters in the book will make sense to students even if you do not have time to include these subjects.

Since we do not expect students to remember every concept introduced in previous chapters, we have inserted unobtrusive verbal "sign posts" throughout to help students find where a concept was defined or explained in detail. We sometimes review a key idea that may have been introduced many chapters ago. Our aim is that your student should be able to use the book as an easy navigational tool, no matter how you teach your course. A complete glossary (even including terms not used in the book but often used by instructors) is supplied in Appendix 3.

Our book is not an encyclopedia of planetary astronomy (which is why students do not need to work out for six weeks in a gym before they can carry it around). We tried not to overwhelm the reader with detail, but instead to focus on the major threads and overarching ideas of planetary science. We probably have a bit less jargon than most textbooks, but we have worked hard not to sacrifice any of the key concepts that we believe students should learn in a basic course.

The third edition includes many of the latest ideas and discoveries in planetary science, not merely for their novelty but for their value in advancing the quest for a coherent understanding of our place in the universe. In each case, we have tried to fit the latest research results into a wider context and to explain clearly what they mean. Among the recent topics included in this edition are:

- a new chapter on astrobiology, with careful attention to recent ideas about the origin and early evolution of life on Earth;
- a full report on the NEAR-Shoemaker spacecraft and its exploration of the asteroid Eros;
- updated discussions of the discoveries of extrasolar planets and their significance;
- the latest on the relationship between solar activity and Earth climate.

We also portray science as a human endeavor and have tried to include descriptions and images of some of the key men and women who have created our science over the years. Illustrations also include many of the latest images from Galileo, Mars Global Surveyor, the Hubble Space

Telescope and other space instruments, as well as an up-to-date collection of color images from ground-based observatories around the world. Full-color diagrams are used as teaching tools, not as cosmetic devices. Figure captions contain full explanations of what the student should be seeing and understanding, not just the cryptic one-liners featured in so many other books.

Among the other features of the book that we hope will particularly help students and instructors are:

- suggestions for accurate, accessible Web sites and readings at the end of each chapter and in the appendices for those who want to explore further;
- three levels of end-of-chapter questions: review questions, thought-provoking exercises that ask students to apply what they have learned, and numerical problems;
- *Inter-Activity* questions in each chapter, consisting of a series of interactive group exercises (see box) that can be a change of pace during a lecture, form the basis of a discussion section, or serve as homework to teams of students.

At the request of a number of our adopters, our third edition features a new series of "Figuring for Yourself" boxes in each chapter. These demonstrate how to solve problems relating to the chapter and prepare students for doing the numerical questions. Our aim is to give instructors who use mathematics more extensively in their lectures both examples and questions to assign, and yet not to break up the main flow of our narrative for those who don't. We have significantly increased the number of numerical problems in this edition as well.

All our teaching materials are kept updated at the text Web site: info.brookscole.com/voyages.

The chapters in *Voyages* also feature highlighted boxes designed to help non-science students appreciate the breadth of astronomy without distracting them from the main narrative.

- *Making Connections:* These special boxes show how science connects to students' experiences with other fields of human activity, from poetry to engineering, from popular culture to natural disasters.
- *Voyagers in Astronomy:* These profiles of noted scientists focus not only on their work but also on their lives and times.
- *Astronomy Basics:* In these short sections we try to explain fundamental science ideas and terms that other texts often assume (incorrectly) that students know.
- *Seeing for Yourself:* Here, students become familiar with the sky and everyday astronomical phenomena through observations using simple equipment.

ANCILLARY MATERIALS

We are happy to offer instructors who adopt *Voyages* a growing series of learning and teaching aids to make their lives a little bit easier. These include:

1. *The* Voyages *Instructor's Manual/Test Bank*
   Our unique instructor's manual contains answers to thought questions and numerical problems in the textbook, hints for selected *Inter-Activity* exercises, various possible course outlines, and annotated guides to printed and Web resources for planetary astronomy instructors. It's especially aimed at instructors new to teaching but will, we think, still contain much of interest to veteran professors. The manual also includes a large number of multiple-choice test questions for use in a variety of classroom settings (also provided on disk). We're proud to say that, unlike most texts, our multiple-choice questions have five answers, and are far more than just regurgitation of factoids. Many of the questions require applying the principles students learned in new situations, or finding similarities or differences among disparate phenomena. There is also a liberal sprinkling of humor among the answers to cheer up the suffering students during the exam.

2. *ExamView® Computerized Test Bank for Windows and Macintosh*
   ExamView features all the questions from the printed test bank in a format that allows instructors to edit them, add questions, and compose many versions of the same test, in print or online.

3. *The* Voyages *Transparency Collection*
   This set contains 100 of the best figures and images in the book as full-page transparencies for your use in the classroom.

4. *The* Voyages Multimedia Manager
   This CD-ROM includes virtually all the diagrams, photos, and paintings from the text, together with a number of other images that we wanted to use in the book but couldn't find room for (especially recent images from planetary exploration and from large telescopes in orbit and on the ground). The software allows you to include them in your own presentations and add your own files or graphics in any order.

5. The Sky™ *Student Edition on CD-ROM (packaged free with every text)*
   See the sky from anywhere over a period of 12,000 years with this well-reviewed planetarium program. Comes with a database of 118,000 stars and 13,000 deep-sky objects and prints star charts for labs or observing sessions.

6. InfoTrac® *College Edition (free 4-month access code packaged with the book)*
   This online library of magazines and journals lets your students read or print complete articles from nearly 4,000 publications. Great for instructors who assign papers or require outside research but have students who whine about going to a library.

7. *Virtual Laboratories (free access code packaged with the book)*
   Michael Guidry, of the University of Tennessee, has created a series of online interactive lab exercises on 20 important concepts in astronomy. Students can print out or e-mail completed lab assignments to the instructor.

Collaborative Group Activities as a Teaching Tool

Recent research has shown that listening to an hour or hour-and-a-half lecture on some topic in astronomy is not always conducive to effective student learning. Students seem to learn better when they are active participants rather than passive listeners. Unfortunately, many of us teach huge classes, where it might seem that student participation could not possibly be practical. Yet, science instructors in many fields are beginning to come up with ways that even large-lecture classes can do small-group activities. (Readings and Web sites to assist you with trying such activities can be found in the *Voyages* Instructors' Manual.)

In the text, we focus on one approach, which involves small groups and collaborative group activities either during lecture or in discussion section. The instructor divides the class (or asks the students to divide themselves) into groups of three or four. The smaller you make the groups, the less likely it is that some students will be able to remain uninvolved, but groups of two are typically not sufficient for good discussions to evolve.

Each group is given a task and a time limit. Each chapter of *Voyages* presents a series of *Inter-Activity* exercises that are good for group discussion and typically do not require any outside materials or data. In addition, some instructors like to present each group with some simple data to analyze (on a handout or an overhead). After group members discuss or do the activity, they elect a spokesperson, who shares their results with the whole class. If your class is really large,

perhaps only a representative number of groups will report on any given day.

Based on what they hear in the reports, groups can then return to their discussion and change or refine their answers. Some instructors require each person or the whole group to write a short summary of their conclusions; others feel that the experience of interacting with their peers is sufficient.

This approach works well even with showing slides. Put up a high-resolution Galileo image of the surface of Europa, for example, and ask the groups to discuss what they are seeing, how large an area they think it is, and how such features might originate. The important thing is not for students to get a "right answer" but to get them thinking, talking, interacting, and participating in their own education. Students are so used to being passive in lecture classes that this approach may take a few tries before they get fully involved, but an instructor (and section leaders) walking around to facilitate discussion and get those who are congenitally uninvolved to participate can work wonders after a while.

We should warn you that many of the suggested group activities in *Voyages* are open-ended and may engender a great deal of discussion. You should probably have a strategy for coming to closure in a way that still leaves students feeling that their opinions are valued. A short paper (perhaps for extra credit), an interactive Web site, or a chance for further debate in a discussion section are some ways of capturing student enthusiasm while still being able to move on in the main lecture.

---

8. *WebTutor (with WebCT and Blackboard)*
   Software to help you set up course-specific and text-specific course management tools, with syllabi, office hours, student discussions, quizzes, and much more.

In addition, your Brooks/Cole–Thomson Learning representative can discuss with you many other materials available for adopters, including *RedShift*™ software, lab workbooks for both *The Sky* and *RedShift*, an *InfoTrac College Edition* student guide, a collection of short astronomy videos from CNN, and *My Course 2.0* Web software.

■ THE REGISTERED ADOPTERS PROGRAM

With many textbooks, the time you place your order is the last time you hear from the publisher or the authors. In fact, many authors and publishers have no idea who adopted their book and never communicate with the instructors except through sales reps. We hope to be a little

bit different; we'd like to know who our adopters are, stay in touch with them, and provide new materials, updated information, and so on. Over the years, the community of *Voyages* adopters has not only grown but has been very important to us in shaping the development of the texts.

Accordingly, we have created a "Registered Adopters" program for *Voyages*. To join, you merely send us an e-mail with your name, e-mail, college mailing address, the course you are teaching, and which text you are using. We will verify the adoption with your bookstore and then be in touch with you from time to time, mostly to let you know when we are publishing something (or giving away something free) that may help your teaching. There is no cost involved, and we promise not to release the names for any other purpose.

If you are not yet a registered adopter, we urge you to get on our list. You can do so by e-mailing the lead author of the book at:

fraknoiandrew@fhda.edu

(Note that the last name does not have an "n" before the "k"—that misspelling is the number one reason e-mails get returned.) If you prefer, you can send the information by particle mail to: Andrew Fraknoi, Astronomy Dept., Foothill College, 12345 El Monte Road, Los Altos Hills, CA 94022.

LET US HEAR FROM YOU

Unlike stars and planets, textbooks in astronomy do not exist in a vacuum. All three authors have benefited tremendously over the years from the advice of colleagues and stu-dents who teach astronomy. We want to be sure that we continue to make changes and updates in *Voyages* that will be the most useful to you. We welcome your comments and suggestions for improving the *Voyages* texts, our Web site, and our ancillary materials.

Please address your comments to Andrew Fraknoi (see addresses above), who will share them with the other two authors.

ACKNOWLEDGMENTS

We would like to thank the many colleagues and friends who have provided information, images, and encouragement, including: Charles Avis, Charles Bailyn, Bruce Balick, Roger Bell, Roy Bishop, Clark Chapman, Chip Clark, Grace Deming, David DesMarais, Richard Dreiser, Doug Duncan, George Djorgovski, Alex Filippenko, Debra Fischer, Dale Frail, Alex Fraknoi, Lola Fraknoi, Kathy Fransham, Michael Friedlander, Alan Friedman, Ian Gatley, Paul Geissler, Margaret Geller, Cheryl Gundy, Heidi Hammel, Alan Harris, Todd Henry, George Jacoby, Bruce Jakosky, William Keel, Jonathan Lunine, Geoff Marcy, Jeff McClintock, John McNeel, Michael Merrill, Jeff Moore, Janet Morrison, Jerome Orosz, Donald Osterbrock, Tobias Owen, Michael Perryman, Carle Pieters, Axel Quetz, Dennis Schatz, Rudy Schild, Maarten Schmidt, William Schopf, Joe Term, Ray Villard, Adrienne Wasserman, Richard Wolff, and Don Yeomans.

We are grateful to Lynette Cook, Bill Hartmann, John Spencer, Don Davis, and Don Dixon for permission to reproduce their wonderfully illustrative astronomical paintings, and to David Malin for his assistance and his superb astronomical photographs.

We appreciate the guidance of colleagues who reviewed all or part of the manuscript.

For the First Edition

Grady Blount
Texas A&M University

Michael Briley
University of Wisconsin, Oshkosh

David Buckley
East Strohdsburg University

John Burns
Mt. San Antonio College

Paul Campbell
Western Kentucky University

Eugene R. Capriotti
Michigan State University

George L. Cassiday
University of Utah

John Cunningham
Miami-Dale Community College

Grace Deming
University of Maryland

Miriam Dittman
DeKalb College

Gary J. Ferland
University of Kentucky

George Hamilton
Community College of Philadelphia

Ronald Kaitchuck
Ball State University

William C. Keel
University of Alabama

Steven L. Kipp
Mankato State University

Jim Lattimer
SUNY, Stonybrook

Robert Leacock
University of Florida

Terry Lemley
Heidelberg College

Bennett Link
Montana State University

Charles H. McGruder III
Western Kentucky University

Stephen A. Naftilan
Claremont Colleges

Anthony Pabon
DeAnza College

Cynthia W. Peterson
University of Connecticut

Andrew Pica
Salisbury State University

Terry Richardson
College of Charleston

Margaret Riedinger
University of Tennessee, Knoxville

Jim Rostirolla
Bellevue Community College

Michael L. Sitko
University of Cincinnati

John Stolar
West Chester University

Charles R. Tolbert
University of Virginia

Steve Velasquez
Heidelberg College

David Weinrich
Moorhead State University

David Weintraub
Vanderbilt University

Mary Lou West
Montclair State University

Dan Wilkins
University of Nebraska, Omaha

J. Wayne Wooten
Pensacola Junior College

For the Second Edition

Mitchell C. Begelman
University of Colorado

Stephen Danford
University of North Carolina, Greensboro

Richard French
Wellesley College

Catharine Garmany
University of Colorado, Boulder

Owen Gingerich
Harvard University

Edward Harrison
University of Massachusetts

Scott Johnson
Idaho State University

Mark Lane
Palomar College

John Patrick Lestrade
Mississippi State University

Michael C. LoPresto
Henry Ford Community College

Anthony Marston
Drake University

Ronald A. Schorn
Texas A&M University

Vernon Smith
University of Texas, El Paso

Ronald Stoner
Bowling Green State University

Jack W. Sulentic
University of Alabama

Stephen Walton
California State University, Northridge

Warren Young
Youngstown State University

For the Third Edition

Paul DeVries
Miami University

Steven Doty
Denison University

John F. Hawley
University of Virginia

Michael Kaufman
San Jose State University

Robbie F. Kouri
Our Lady of the Lake University

Michael C. Lopresto
Henry Ford Community College

Ed Oberhofer
Lake Sumter Community College

No book project of this complexity could succeed without the diligent efforts of many people at Brooks/Cole–Thomson Learning. We very much appreciate the assistance of:

Keith Dodson—Acquisitions Editor
Marie Carigma-Sambilay—Development Editor
Carol Benedict—Assistant Editor
Melissa Newt—Editorial Assistant
Samuel Subity—Technology Project Manager
Kelley McAllister—Marketing Manager
Sandra Perin—Marketing Assistant
Hal Humphrey—Production Project Editor

Nancy Shammas, New Leaf Publishing—Production Service
Nathaniel Bergson-Michelson—Executive Advertising Project Manager

But most of all, we appreciate the hundreds of instructors and many thousands of students around the United States, Canada, and the world, who have honored us by using this textbook on their own voyages through the universe.

Andrew Fraknoi
David Morrison
Sidney Wolff

Brief Contents

Contents

APPENDICES

Eight Planets and a Moon This montage of images shows off the kind of detail we can make out using instruments on board spacecraft that travel to the other worlds in our solar system. (Since spacecraft have not yet visited Pluto, only eight of the nine planets that orbit the Sun are shown.) The top four worlds—Mercury, Venus, Earth, and Mars—are called terrestrial planets (the Earth's moon is included as a bonus); the bottom four—Jupiter, Saturn, Uranus, and Neptune—are called giant or jovian planets. Note that the planets in each group are shown roughly to scale, but the two scales are different. Eleven Earths could fit side by side in the equator of Jupiter. The Venus image is a globe made from radar images and is artificially colored. From space Venus is shrouded with clouds and no surface details are visible.

JPL/NASA

Prologue: Science and the Universe: A Brief Tour

Everybody, at last, is getting nosy. I predict that in our time astronomy will become the gossip of the marketplace.

Galileo speaking in Bertolt Brecht's play Galileo

We invite you to come with us on a series of voyages to explore the realm of the planets as scientists understand it today. Along our journey, we will encounter:

- a grand canyon system so large that on Earth it would stretch from Los Angeles to Washington, D.C. (Figure P.1)
- a scar and other evidence on Earth that tell us that the dinosaurs (and many other creatures) died because of a cosmic collision
- a tiny moon whose gravity is so weak that one good throw from its surface could put a baseball into orbit
- a system of rings so vast that if we put them next to the Earth, they would stretch almost to the Moon (Figure P.2)

USGS

■ **FIGURE P.1**

Mars Mosaic This image of Mars was constructed at the U.S. Geological Survey in Flagstaff, Arizona, by combining many individual images taken by the Viking spacecraft starting in 1976. The view is centered on the Valles Marineris (Mariner Valley) complex of canyons, which is as long as the United States is wide.

It is these kinds of discoveries that make planetary science such an exciting field for both scientists and people in other walks of life. But we will explore more than just the individual objects in our solar system and the latest discoveries about them. We will pay equal attention to the *process* by which we have come to understand the realm beyond our Earth and the tools we need to increase that understanding.

1 THE NATURE OF ASTRONOMY

Astronomy is defined as the study of the objects that lie beyond our planet Earth, and of the processes by which these objects interact with one another. But, as we will see, it is

JPL/NASA

■ **FIGURE P.2**

Saturn and Its Rings The camera aboard the Voyager spacecraft took this image of the ringed planet in 1980, from a distance of 5.3 million km. The Sun is to the right, and Saturn casts a dramatic shadow on its own ring system.

much more than that. It is also humanity's attempt to organize what we learn into a clear history of the universe, from the instant of its birth in the Big Bang to the present moment in which you are finishing this sentence.

Putting a cosmic history together takes some nerve—especially for a little creature living on a rocky ball circling a nondescript star—and we are surely far from finished with it. Throughout the book, we emphasize how much science is a *progress report*—one that constantly changes as new techniques and instruments allow us to probe the universe more deeply.

In considering the history of the universe, we will see again and again that the cosmos *evolves;* it changes in profound ways over long periods of time. Although you may have read in the popular press that the concept of evolution is controversial in science, this is simply not true. Few ideas are better established than the notion that the universe is not the same today as it was long ago; it has existed and been evolving for billions of years.

For example, the universe could not have produced the readers of this book during the first generation of stars after the Big Bang. The ingredients in the recipe for an astronomy student simply did not exist then in sufficient numbers. The universe still had to make the carbon, the calcium, and the oxygen needed to construct something as interesting and complicated as you. Today, many billions of years later, the universe has evolved into a more hospitable place for life. Tracing the evolutionary processes that continue to shape the universe is one of the most important (and satisfying) parts of modern astronomy.

2 THE NATURE OF SCIENCE

Unlike religion or parts of philosophy, science accepts nothing on faith. The ultimate judge in science is always the experiment or observation: what nature itself reveals. Science, then, is not merely a body of knowledge, but a *method* by which we attempt to understand nature and how it behaves. This method begins with many observations over a period of time. From the trends in the observations, scientists may come up with a *model* of the particular phenomenon we want to understand. Such models are always approximations of nature itself, subject to further testing.

To take a concrete astronomical example, ancient astronomers constructed a model (partly from observations, partly from philosophical beliefs) that the Earth was the center of the universe and that everything moved around it. At first the available observations of the Sun, Moon, and planets could be fit to this model, but eventually better observations required the model to add circle after circle to the movements of the planets to keep the Earth at the center. As the centuries went by and improved instruments were developed for keeping track of objects in the sky, the old model (even with a huge number of circles)

could no longer explain all the observed facts. As we will see in Chapter 1, a new model, with the Sun at the center, fit the experimental evidence better. After a period of philosophical struggle, it became accepted as our view of the universe.

When they are first proposed, new models or ideas are sometimes called *hypotheses.* Many students today think that there can be no new hypotheses in a science such as astronomy, that everything important has already been learned. Nothing could be further from the truth. Throughout this book you will find discussions of recent, and occasionally still controversial, hypotheses in astronomy—concerning, for instance, the significance for life on our planet of huge chunks of rock and ice that hit the Earth. Such hypotheses are based on difficult observations done at the forefront of our technology, and all require further testing before we fully incorporate them into our standard scientific models.

This last point is crucial: If there is no possible way of *testing* a hypothesis, it does not belong in the realm of science. The most straightforward approach to such testing in science is to perform an experiment. If the experiment is properly conducted, either its results will agree with the predictions of the hypothesis or they will contradict it. If the experimental result is truly inconsistent with the hypothesis, a scientist must discard the hypothesis (no matter how fond of it she may have grown) and try to develop an alternative. If the experimental result agrees with predictions, that is satisfying but does not prove that the hypothesis is correct. Perhaps later experiments will contradict crucial parts of the hypothesis. But the more experiments a hypothesis survives, the more likely we are to accept it as a useful description of nature.

When you read about experiments, you probably have a mental picture of a scientist in a laboratory, performing some careful test or measurement. This is certainly the case for a biologist or a chemist, but what can we do when our laboratory is the universe? It's impossible to put a group of stars into a test tube or order another comet from a scientific supply company.

As a result, astronomy is sometimes called an *observational* science; we often make our tests by observing many samples of the kind of object we want to study and carefully noting how different samples vary. The invention of new instruments sometimes lets us look at astronomical objects from new perspectives or in greater detail. Our hypotheses are then judged in the light of this new information, and they pass or fail in the same way that we would evaluate the result of a laboratory experiment. Because the objects we want to study are far away and hard to see, many years may pass before the appropriate observational tests can be made in astronomy. We must simply wait patiently until either nature or better instruments make some deciding test possible.

Much of astronomy is also a *historical* science—meaning that what we observe has already happened in the universe, and we can do nothing to change it. In the same way, a geologist cannot alter what has happened to our planet, and a paleontologist cannot bring an ancient animal back to life. While this sometimes makes astronomy challenging, it also gives us fascinating opportunities to discover the secrets of our cosmic past.

You might compare an astronomer to a detective trying to solve a crime that occurred before the detective ever got to the scene. There is all sorts of evidence, but both the detective and the scientist must sift through it carefully and organize it to test various hypotheses about what really happened. And there is another way in which the scientist is like a detective: They both must prove their case. The detective must convince the district attorney, the judge, and perhaps ultimately the jury that his hypothesis is correct. Similarly the scientist must convince colleagues, editors of journals, and ultimately a broad cross section of other scientists that his hypothesis is provisionally correct. In both cases, one can only ask for evidence "beyond a reasonable doubt." And sometimes new evidence will force both the detective and the scientist to revise the most recent hypotheses.

It is this self-correcting aspect of science that sets it off from most human activities. Scientists spend a great deal of time questioning and criticizing one another. No project is funded and no report is published without *peer review*— that is, without careful examination by other scientists in the same field. In nonscience areas, young people are often taught to accept the authority of their elders without question, but in science (after proper training) everyone is encouraged to try new and better experiments and to challenge any and all hypotheses. Young scientists know that one of the best ways to advance their careers is to find a weakness in our current understanding and to correct it with a new or modified hypothesis.

This is one of the reasons science has made such dramatic progress. An undergraduate science major today knows more science and math than did Isaac Newton, one of the most brilliant scientists who ever lived. Even in this introductory astronomy course you will learn about objects and processes that a few generations ago no one even dreamed existed. While the domain of science is limited, within that domain its achievements have been glorious.

3 THE LAWS OF NATURE

Over the centuries, scientists have extracted from countless observations certain fundamental principles, called *scientific laws*. These are, in a sense, the rules of the game nature plays. One remarkable discovery about nature— one that underlies everything you will read in this book— is that the same laws apply everywhere in the universe. The rules that govern the behavior of gravity on Earth, for example, are the same rules that determine the motion of two

stars in a system so far away that your eye cannot find them in the sky.

Note that without the existence of such universal laws, we could not make much headway in astronomy. If each pocket of the universe had not only different objects but also different rules, we would have little chance of interpreting what happened in other "neighborhoods." But the consistency of the laws of nature gives us enormous power to understand distant objects without traveling to them and learning the local laws. In the same way, if every U.S. state or Canadian province had completely different laws, it would be very difficult to carry out commerce or even to understand the behavior of people in different regions. But a consistent set of laws allows us to apply what we learn or practice in one state to any other state.

You might ask whether the laws of the natural world could (under the right circumstances) be suspended. This is an enticing fantasy, but, despite many attempts, not a shred of scientific evidence has been found to support such an idea. While it would be nice if we could suspend the law of gravity and float away from our porch through effort of will alone, in real life such experiments generally result in broken bones.

This is not to say that our models or rules cannot change. New experiments and observations can lead to new or more sophisticated models—models that can include even new phenomena and new laws about their behavior. But wishing isn't going to bring such new models into existence; only the patient process of observing nature ever more finely can reap such rewards.

One important problem about describing scientific models has to do with the limitations of language. When we try to describe complex phenomena in everyday terms, the words themselves may not be adequate to do the job. For example, you may have heard the structure of the atom likened to a miniature solar system. While some aspects of our modern model of the atom do remind us of planetary orbits, many other of its aspects are fundamentally different.

This problem is the reason scientists often prefer to describe their theories using equations rather than words. In this book, designed to introduce the field of astronomy, we use mainly words to discuss what scientists have learned. We avoid math beyond basic algebra. But if this course piques your interest and you go on in science, more and more of your studies will involve the precise language of mathematics.

4 NUMBERS IN ASTRONOMY

You may have heard a television reporter refer to a large number (such as the national debt) as "astronomical." In astronomy, we do deal with distances on a scale you may never have thought about before, and with numbers larger than any you may have encountered. Most students take a while to learn to navigate among the millions and billions that astronomers tend to throw about in their everyday discussions. But with some practice, you can become just as good at it!

By the way, if you sometimes have trouble sorting out millions and billions, take heart. Even so distinguished a group as the Presidential Advisory Committee on the Future of the U.S. Space Program, in a 1990 report, listed the distance to the planet Uranus as 1.7 million miles, when in fact Uranus is 1.7 billion miles away. A million (1,000,000) is a thousand times less than a billion (1,000,000,000). If Uranus were suddenly that close, we would all have noticed it because it would be bigger and brighter than the Moon in our skies.

In this book we adopt two approaches that make dealing with astronomical numbers a little bit easier. First, we use a system for writing large and small numbers called *powers-of-ten notation* (or sometimes *scientific notation*). This system is very appealing because it does away with the huge number of zeros that can seem overwhelming to the reader.

In scientific notation, if you want to write a figure like $490,000 (which is definitely not the starting salary for an astronomer but might be for a television star), you write:

$$4.9 \times 10^5$$

The small raised number after the 10, called an *exponent*, keeps track of the number of tens you have to multiply together to get the number you want. In our example, five tens are multiplied together, so that $10 \times 10 \times 10 \times 10 \times 10$ gives 100,000. Multiply 100,000 by 4.9 and you get our astronomical starting salary. Another way to remember the basics of this notation is to note that 5 is the number of places you have to move the decimal point to the right to convert 4.9 to 490,000. If you are encountering this system for the first time or would like a refresher, we suggest you look at Appendix 4 for more information and examples.

Small numbers are written with negative exponents. Three millionths (0.000003) is expressed as:

$$3.0 \times 10^{-6}$$

One reason this notation is so popular among scientists—trust us, it is, even if you at first don't like it—is that it makes arithmetic a lot easier. To multiply two numbers in scientific notation, you need only add their exponents. Thus, you can multiply a thousand times a billion:

$$10^3 \times 10^9 = 10^{12}$$

To divide numbers, just subtract exponents.

The second way we try to keep numbers simple is to use a consistent set of units—the Metric International System of Units. Unlike the system used in the United States, in which a completely arbitrary number such as 5280 feet equals 1 mile, metric units are related by powers of ten; a kilometer, for example, equals a thousand (10^3) meters. If you are used to U.S. units, the kilometer is about $^6/_{10}$ of a mile. The metric system, which has been adopted by every major country in the world except the United States, is summarized in Appendix 5. You may want to make it part of your vocabulary so you are ready for the future (when everyone will be using metric units).

5 LIGHT YEARS

To give you a chance to practice scientific notation, and to set the scene for the tour of the universe in the next section, let's define a common unit astronomers use to describe distances in the universe. A *light year* (LY) is the distance that light travels during 1 year. Since light always travels at the same speed, and since its speed turns out to be the fastest possible speed in the universe, it makes a good standard for keeping track of distances. Some students complain about this name for a unit of distance; light years seem to imply that we are measuring time. But this mix-up of time and distance is common in everyday life as well—for example, when we tell a friend to meet us at a movie theater that's 20 minutes away.

So, how many kilometers are there in a light year? If you are new to scientific notation, we suggest you work through all the steps of the following example for yourself on a separate sheet of paper. Light travels at the amazing pace of 3×10^5 kilometers per second (km/s). Think about that: Light covers 300,000 km every single second. In 1 second, it can travel seven times around the circumference of the Earth; a commercial airplane, in contrast, would take about two days to go around once, not counting time to refuel.

Now that we know how far light goes in a second, we can calculate how far it goes in a year:

1. There are 60 (6×10^1) seconds in each minute and 6×10^1 minutes in every hour.
2. Multiply those numbers and you get that there are 3.6×10^3 seconds per hour.
3. Thus light covers 3×10^5 km/s $\times$ 3.6×10^3 s/h = 1.08×10^9 km/h.
4. There are 24 or 2.4×10^1 hours in a day and 365.24 (3.65×10^2) days in a year.
5. The product of those two numbers is 8.77×10^3 h/year.
6. Multiplying that product by 1.08×10^9 km/h gives 9.46×10^{12} km in a light year.

That's almost 10 trillion km that light covers in a year. A string 1 LY long could fit around the circumference of the Earth 236 million times!

You might think that such a long unit would more than reach to the nearest star. But the stars are far more remote than our imaginations (or episodes of *Star Trek*) might lead

C. R. O'Dell and NASA

■ FIGURE P.3

The Orion Nebula This beautiful cloud of cosmic raw material (gas and dust from which new stars and planets are being made) called the Orion Nebula is about 1500 LY away. That's a distance of roughly 1.4×10^{16} km—a pretty big number! The picture we see is actually a seamless mosaic of 15 smaller images taken with the Hubble Space Telescope. The field of view is about 2.5 LY wide and shows only a small part of a vast reservoir of mostly dark material. The gas and dust in this region are illuminated by the intense light from a few extremely energetic adolescent stars in the neighborhood.

us to believe. Even the nearest star is 4.3 LY away—more than 40 trillion km. Other stars visible to the unaided eye are hundreds or even thousands of light years away (Figure P.3). This is why astronomers are skeptical that UFOs are extraterrestrial spacecraft coming here across vast distances, briefly picking up two rural fishermen or loggers, and then going straight home. It seems like such a small reward for such a large investment.

6 CONSEQUENCES OF LIGHT TRAVEL TIME

There is another reason the speed of light is such a natural unit of distance for astronomers. Information about the universe comes to us almost exclusively through various forms of light, and all such light travels at the speed of light—that is, 1 LY every year. This sets a limit on how quickly we can learn about events in the universe. If a star is 100 LY away, the light we see from it tonight left that star 100 years ago and is just now arriving in our neighborhood. The soonest we can learn about any changes in that star— its blowing up, for example—is 100 years after the fact. For a star 500 LY away, the light we detect tonight left 500 years ago and is carrying 500-year-old news.

Some students, accustomed to CNN and other news media known for instant world coverage, at first find this frustrating. "You mean when I see that star up there," they ask, "I won't know what's actually happening there now for another 500 years?" But that's not really the right way to think about the situation. For astronomers, *now* is when the light reaches us here on Earth. There is no way for us to know anything about that star (or other object) until its light reaches us; despite the fondest dreams of science-fiction writers, instant communication through the universe is not possible.

But what at first may seem a great frustration is actually a tremendous benefit in disguise. If astronomers really want to piece together what has happened in the universe since its beginnings, they need to find evidence about each epoch of the past. Where can we find evidence today about cosmic events that occurred billions of years ago? Unfortunately, the universe doesn't leave written records (or even decent videotapes) of its main activities!

The delay in the arrival of light provides such evidence automatically. The farther out in space we look, the longer the light has taken to get here, and the longer ago it left its place of origin. By looking billions of light years out into space, astronomers are actually seeing billions of years into the past. In this way, we can reconstruct the history of the cosmos and get a sense of how it has evolved over time.

FIGURE P.4

Telescope in Orbit The Hubble Space Telescope, shown here being repaired aboard the Space Shuttle *Endeavour* in December 1993, is an example of the new generation of astronomical instruments in space.

This is one reason astronomers strive to build telescopes that can collect more and more of the faint light the universe sends us. The more light we collect, the fainter the objects we can make out. On average, fainter objects are farther away and can thus tell us about periods of time even deeper in the past. Recent instruments, such as the Hubble Space Telescope (Figure P.4) and Hawaii's Keck Telescope (see Chapter 5), are giving astronomers views of deep space and deep time better than any we have had before.

7 A TOUR OF THE UNIVERSE

Let us now take a brief introductory tour of the universe as astronomers understand it today, just to get acquainted with the sorts of objects and distances we will encounter throughout the text.[1] We begin at home with the Earth, a nearly spherical planet about 13,000 km in diameter (Figure P.5). A space traveler entering our planetary system

---

[1] This textbook focuses mostly on our "local neighborhood" in the cosmos, but we provide a tour of the entire universe here to give you the cosmic context for the discussions in the book. Don't worry; we won't charge you extra for this service!

would easily distinguish the Earth by the large amount of liquid water that covers some two-thirds of its crust. If the traveler had equipment to receive radio or television signals, or came close enough to see the lights of our cities at night, she would soon find signs that this water planet has intelligent life. (Of course, depending on what television channel the traveler tuned to, that conclusion might need to be changed to "semi-intelligent" life.)

Our nearest astronomical neighbor is the Earth's satellite, commonly called the Moon. Figure P.6 shows the Earth and the Moon drawn to scale on the same diagram. Notice how small we have to make these bodies to fit them on the page with the right scale. The Moon's distance from Earth is about 30 times the Earth's diameter, or approximately 384,000 km, and it takes about a month for the Moon to revolve around the Earth. The Moon's diameter is 3476 km, one-fourth the size of the Earth.

Light (or a radio wave) takes 1.3 seconds to travel between the Earth and the Moon. If you've seen videos of the Apollo flights to the Moon, you may recall that there was a delay of about 3 seconds between the time Mission Control asked a question and the time the astronauts replied. The reason is not that the astronauts were thinking slowly, but that it took the radio waves almost 3 seconds to make the round trip.

The Earth revolves around our star, the Sun, which is about 150 million km away—approximately 400 times as far as the Moon. We still call the average Earth–Sun distance an astronomical unit (AU) because in the early days of astronomy it was the most important measuring standard. Light takes a little more than 8 minutes to travel 1 AU, which means our latest news from the Sun is always 8 minutes old.

It takes the Earth 1 year (3×10^7 seconds) to go around the Sun at our distance; to make it around, we must travel at approximately 110,000 km/h. (If you, like many students in the United States, still prefer miles to kilometers, you might find the following trick helpful. To convert kilometers to miles, you can multiply kilometers by 0.6. Thus 110,000 km/h becomes 66,000 mi/h—a fast clip no matter what units you use.) Since gravity holds us firmly to the Earth and there is no resistance to the Earth's motion in the vacuum of space, we participate in this breakneck journey without being aware of it day by day.

The diameter of the Sun is about 1.5 million km; our Earth could fit comfortably inside one of the minor eruptions that occur on the surface of our star. If the Sun were reduced to the size of a basketball, the Earth would be a small apple seed some 30 m from the ball.

The Earth is only one of nine planets that revolve around the Sun. These planets, along with their satellites and swarms of smaller bodies, make up the solar system, what we might call the family of the Sun (Figure P.7).

FIGURE P.5

Humanity's Home Base Planet Earth as viewed from a perspective in space. This image was taken on May 3, 1990, by the Meteosat weather satellite high above the Earth's equator.

FIGURE P.6
The Earth and Moon, Drawn to Scale

Painting by Lynette Cook

Mercury

Venus

Earth

Mars

Jupiter

Saturn

Uranus

Neptune

Pluto

■ FIGURE F

A Stellar Nurse
called the Rosette
be seen forming ir

8   THE
     ON

In a very rough
as your house o
made up of mar
tury, astronome
is made up of n
up of enormous
verse as everytl
servations.) Ga
scopes can see,
modern instrur
some astronom
term is aptly d
stars in the vas

The neare
one that lies 75
constellation S

■ FIGURE P.7

Our Solar Family   The Sun and
the planets shown to scale. Notice the
size of the Earth compared to the
giant planets.

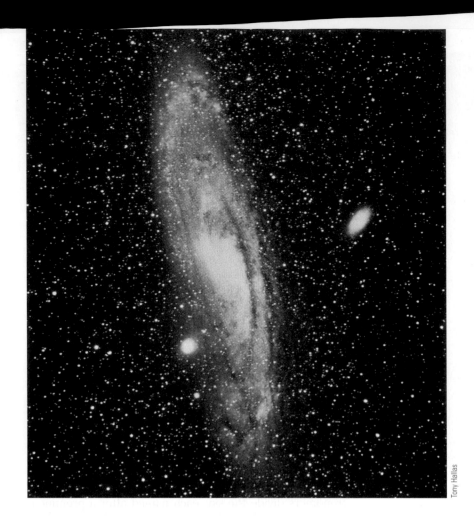

Closest Spiral Galaxy The Andromeda Galaxy (M31) is a spiral-shaped collection of stars similar to our own Milky
Way. Its flat disk of stars looks blue because its light is dominated by the vigorous
light output of hot blue stars. The bulge in
its center, on the other hand, contains older
stars, which are predominantly yellow and
red. Two smaller galaxies can be seen on
either side of M31.

Tony Hallas

FIGURE P.1

A Spiral Galaxy
number NGC 1232,
thought to be similar
see the giant wheel-sh
the colorful disk of sta
young, bluish stars.

penetrate the smog
a pretty good map o

Recent observ
rather surprising an
more—much more
the telescope). Fro
dence that much o
cannot observe di
with those that de
mentioned). We
Galaxy *dark matte*
the pull its gravity
can observe, but w
much of it exists r

By the way, n
does. Many are be
or three stars rev
numbers of partn
each other in suc

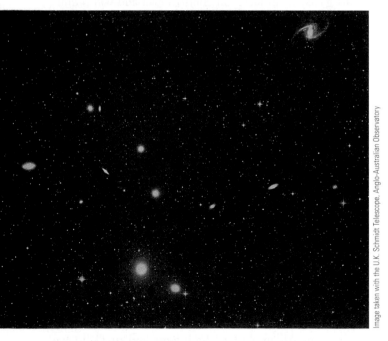

FIGURE P.16
Fornax Cluster of Galaxies On this image, you can
see a cluster of galaxies located about 60 million LY away, in
the constellation of Fornax. All the objects that are not pinpoints
of light on the picture are galaxies of billions of stars.

Image taken with the U.K. Schmidt Telescope, Anglo-Australian Observatory

At even greater distances, where many ordinary galaxies are too dim to see, we find the quasars. These are brilliant centers of galaxies, glowing with the light of some extraordinarily energetic process. Whatever the quasars are,
their brilliance makes them the most distant beacons we
can see in the dark oceans of space. They allow us to probe
the universe 10 billion or more LY away, and thus 10 billion
or more years in the past.

With the quasars we see a substantial way back to the
Big Bang explosion that marks the beginning of time. Beyond the quasars, we can detect only the feeble glow of the
explosion itself, filling the universe and thus coming to us
from all directions in space. The discovery of this "afterglow
of creation" was one of the most exciting events in 20th-
century science, and we are still exploring the many things
it has to tell us about the earliest times of the universe.

For now, we might just note that such ideas are far easier to state than to discover. Measurements of the properties of galaxies and quasars in remote locations require
large telescopes, sophisticated light-amplifying devices,
and painstaking labor. Every clear night, at observatories
around the world, astronomers and students are at work on
such mysteries as the birth of new stars and the large-scale
structure of the universe, mostly observing one star or one
galaxy at a time and fitting their results into the tapestry of
our understanding.

9 THE UNIVERSE OF THE VERY SMALL

The foregoing discussion should impress on you that the universe is extraordinarily large and extraordinarily empty. The universe on average is 10,000 times emptier than our Galaxy. Yet, as we have seen, even the Galaxy is mostly empty space. The air we breathe has about 10^{19} atoms in each cubic centimeter—and we think of air as pretty empty stuff. In the interstellar gas of the Galaxy, there is about one atom in every cubic centimeter. Intergalactic space is so sparsely filled that to find one atom, on average, we must search through a cubic meter of space! Most of the universe is fantastically empty; places that are dense, such as the bodies of our readers, are tremendously rare.

Yet even the familiar solids, such as this textbook, are mostly space. If we could take such a solid apart, piece by piece, we would eventually reach the molecules of which it is formed. Molecules are the smallest particles into which matter can be divided while still retaining its chemical properties. A molecule of water (H_2O), for example, consists of two hydrogen atoms and one oxygen atom, bonded together.

Molecules, in turn, are built up of atoms, which are the smallest particles of an element that can still be identified as that element. For example, an atom of gold is the smallest possible piece of gold (although one atom of gold won't impress your sweetheart very much). Nearly 100 different kinds of atoms (elements) exist in nature, but most of them are rare, and only a handful account for more than 99 percent of everything with which we come in contact. The most abundant elements in the cosmos today are listed in Table P.1; think of this table as the "greatest hits" of the universe when it comes to elements.

TABLE P.1 *The Cosmically Abundant Elements*

Element°	Symbol	Number of Atoms per Million Hydrogen Atoms
Hydrogen	H	1,000,000
Helium	He	80,000
Carbon	C	450
Nitrogen	N	92
Oxygen	O	740
Neon	Ne	130
Magnesium	Mg	40
Silicon	Si	37
Sulfur	S	19
Iron	Fe	32

° Our list of elements is arranged in order of the atomic number, which is the number of protons in each nucleus.

All atoms consist of a central, positively charged nucleus surrounded by negatively charged electrons. The bulk of the matter in each atom is found in the nucleus, which consists of positive protons and electrically neutral neutrons all tightly bound together in a very small space. Each element is defined by the number of protons in its atoms: Thus, any atom with 6 protons in its nucleus is called carbon, any with 50 protons is called tin, and any with 70 protons is called ytterbium. (Ytterbium, as you can probably guess, is not big on the cosmic hit parade of elements, but we like its name. For a list of the elements, see Appendix 13.)

The distance from an atomic nucleus to its electrons is typically 100,000 times the size of the nucleus itself. This is why we say that even solid matter is mostly space. The typical atom is far emptier than the solar system out to Pluto. (The distance from the Earth to the Sun, for example, is only 100 times the size of the Sun.) This is another reason atoms are not like miniature solar systems.

Remarkably, physicists have discovered that everything that happens in the universe, from the smallest atoms to the largest superclusters of galaxies, can be explained through the action of only four forces: gravity, electromagnetism (which combines the actions of electricity and magnetism), and two forces that act at the nuclear level. The fact that there are four forces (and not a million, or just one) has puzzled physicists and astronomers for many years, and has led to a quest for a unified picture of nature.

10 A CONCLUSION AND A BEGINNING

If you are typical of students new to astronomy, you have probably reached the end of our tour in this Prologue with mixed emotions. On the one hand, you may be fascinated by some of the new ideas you've read about, and eager to learn more. On the other, you may be feeling overwhelmed by the number of topics we have covered, and the number of new words and ideas we have introduced. Learning astronomy is a little like learning a new language: At first it seems there are so many new expressions that you'll never learn them all, but with practice, you soon develop facility with them.

At this point you may also feel a bit small and insignificant, dwarfed by the cosmic scales of distance and time. Such a feeling is not a bad thing from time to time (sometimes we wish more of our politicians and film stars felt this way). And just before a difficult exam, or when you've ended a treasured relationship, it can certainly help to see your problems in a cosmic perspective. But there is another way to look at what we've learned from our first glimpses of the cosmos.

Let us consider the history of the universe from the Big Bang to today and compress it, for easy reference, into a single year. (We have borrowed this idea from Carl Sagan's Pulitzer-Prize-winning book, *The Dragons of Eden*, published

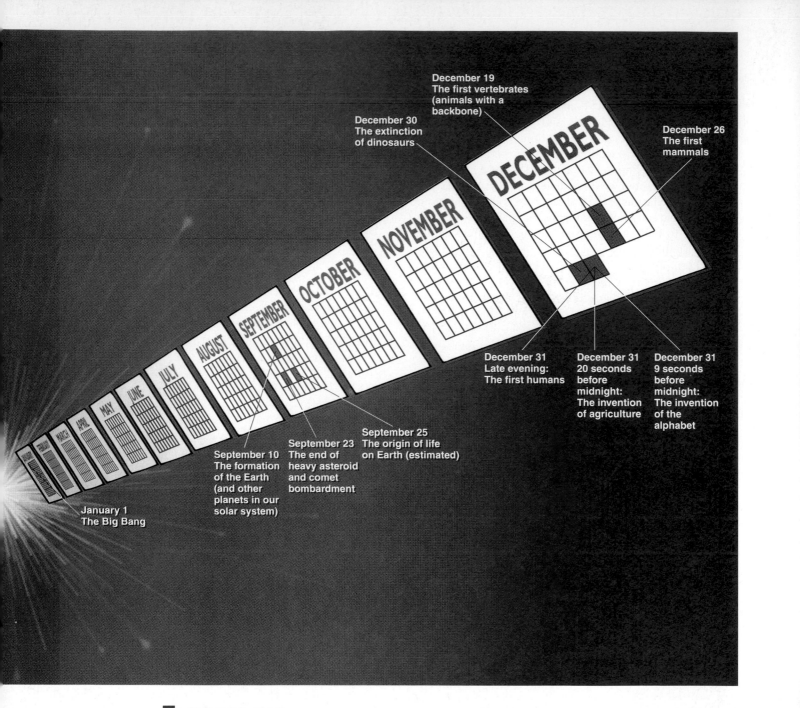

December 19
The first vertebrates
(animals with a
backbone)

December 30
The extinction
of dinosaurs

December 26
The first
mammals

December 31
Late evening:
The first humans

December 31
20 seconds
before
midnight:
The invention
of agriculture

December 31
9 seconds
before
midnight:
The invention
of the
alphabet

September 25
The origin of life
on Earth (estimated)

September 23
The end of
heavy asteroid
and comet
bombardment

September 10
The formation
of the Earth
(and other
planets in our
solar system)

January 1
The Big Bang

FIGURE P.17

Charting Cosmic Time   On a cosmic calendar, where the time since the Big Bang is compressed into 1 year, creatures we would call human do not emerge on the scene until the evening of December 31.

in 1977 by Random House.) On this scale, the Big Bang happened at the first moment of January 1, the solar system formed around September 10, and the oldest rocks we can date on Earth go back to the third week in September (Figure P.17).

Where in this "cosmic year" does the origin of human beings fall? The answer turns out to be the evening of De-

cember 31! The invention of the alphabet doesn't occur until the 50th second of 11:59 P.M. on December 31. And the beginnings of modern astronomy are a mere fraction of a second before the New Year. Seen in a cosmic context, the amount of time we have had to study the planets and the stars is minute, and our success at piecing together as much of the story as we have is remarkable.

Certainly, our attempts to understand the universe are not complete. As new instruments and new ideas allow us to gather even better data about the cosmos, our present picture of astronomy will very likely undergo many changes. In fact, we would not be surprised if your grandchildren's grandchildren found some of the contents of this book a bit primitive. Still, as you read our current progress report on the exploration of the universe, take a few minutes every once in a while just to savor how much we have already learned.

SURFING THE WEB

If you have enjoyed some of the spectacular photographs in this Prologue and have access to the World Wide Web, we want to recommend the following Web sites to you. They will allow you to check out some of the best images astronomers have of cosmic objects and to keep track of new images as they come in:

 Hubble Space Telescope Images:
hubblesite.org/newscenter/archive
All the magnificent Hubble Space Telescope images are here, with captions, background information, and new research results. Check the latest images, browse Hubble's "greatest hits," or search for objects of interest to you.

 National Optical Astronomy Observatories Image Gallery: www.noao.edu/image_gallery/
NOAO includes a number of major telescopes in the United States and the Southern Hemisphere. Some of the best images from NOAO instruments are collected at this site.

 Planetary Photojournal: photojournal.jpl.nasa.gov
This site features thousands of the best images from planetary exploration, with detailed captions and excellent indexing.

 Astronomy Picture of the Day:
antwrp.gsfc.nasa.gov/apod/astropix.html
Astronomers Robert Nemiroff and Jerry Bonell feature one relatively new celestial image each day with a brief non-technical caption. Over the years, some of the best astronomical images have been featured here.

 European Southern Observatory:
www.eso.org/outreach/gallery/
This growing album contains images from the large telescopes in the Southern Hemisphere run by a consortium of European countries.

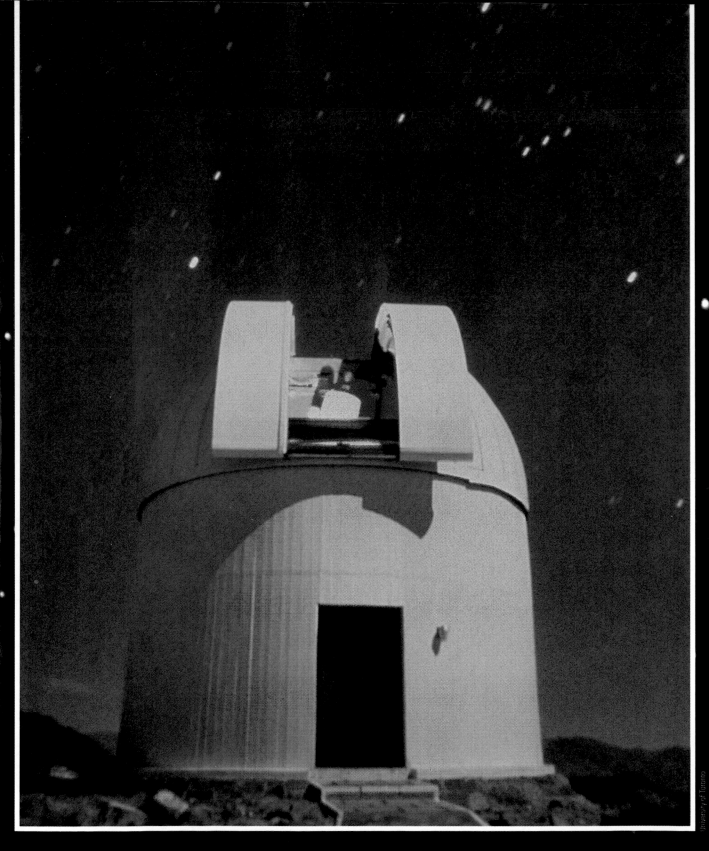

The constellation of Orion (a star grouping with a distinctive belt of three stars, part of the figure of the great hunter of ancient Greek mythology) can be seen above and to the right of the University of Toronto's telescope dome on Las Campanas mountain in Chile. To find our way among the stars, we rely on signposts like groups of stars that make a recognizable pattern we can point out to one another.

1

Observing the Sky: The Birth of Astronomy

If the Lord Almighty had consulted me before embarking upon Creation, I should have recommended something simpler.

Statement attributed to Alfonso the tenth, King of Castile, after having the Ptolemaic system of planet motions explained to him

THINKING AHEAD

Much to your surprise, a member of the Flat Earth Society moves in next door. He believes that the Earth is flat and that all the NASA images of a round Earth are either faked or simply show the round (but flat) disk of the Earth from above. How could you prove to this neighbor that the Earth is really a sphere? (You can find some of our answers in the chapter and a summary in a box later in the chapter, but try to figure out some ways by yourself before peeking at our suggestions! You might want to discuss ideas with other students in the class.)

Today, few people really spend much time looking at the night sky. When we see patterns in the dark, we're much more likely to be looking at a television or movie screen than at the heavens over our heads. But in ancient days, before electric lights and television robbed so many people of the beauty of the sky, the stars and planets were an important aspect of everyone's daily life. All the records that we have—on paper and in stone—show that ancient civilizations around the world noticed, worshipped, and tried to understand the lights in the sky and fit them into their own view of the world. These ancient observers found both majestic regularity and never-ending surprise in the motions of the heavens.

The Babylonians and Greeks, for example, believing the planets to be gods, studied their motions in hopes of understanding the presumed influences of those planet-gods on human affairs; thus developed the religion of astrology. Through their careful study of the planets, however, the ancient Greeks and later the Romans were also laying the foundations of the science of astronomy. In this chapter we take a look at the night sky as seen with the unaided eye and also examine some of the interesting history of how we came to understand the realm above our heads.

1.1 THE SKY ABOVE

Our senses suggest to us that the Earth is the center of the universe—the hub around which the heavens turn. This **geocentric** (Earth-centered) view was what almost everyone believed until the Renaissance. After all, it is simple, logical, and seemingly self-evident. Furthermore, the geocentric perspective reinforced those philosophical and religious systems that taught the unique role of human beings as the central focus of the cosmos. However, the geocentric view happens to be wrong. One of the great themes of our intellectual history is the overthrow of the geocentric perspective. Let us therefore take a look at the steps by which we reevaluated the place of our world in the cosmic order.

1.1.1 The Celestial Sphere

If you go on a camping trip or live far from city lights, your view of the sky on a clear night is pretty much identical to that seen by people all over the world before the invention of the telescope. Gazing up, you get the impression that the sky is a great hollow dome, with you at the center (Figure 1.1). The top of that dome, the point directly above your head, is called the **zenith,** and where the dome meets the Earth is called the **horizon.** It's easy to see the horizon as a circle around you from the sea or a flat prairie, but from most places where people live today, the horizon is hidden by mountains, trees, buildings, or smog.

If you lie back in an open field and observe the night sky for hours, as ancient shepherds and travelers regularly did, you will see stars rising on the eastern horizon (just as the Sun and Moon do), moving across the dome of heaven in the course of the night, and setting on the western horizon. Watching the sky turn like this night after night, you might eventually get the idea that the dome of the sky is really part of a great sphere that is turning around you, bringing different stars into view as it turns. The early Greeks regarded the sky as just such a **celestial sphere** (Figure 1.2). Some thought of it as an actual sphere of transparent crystalline material, with the stars embedded in it like tiny jewels.

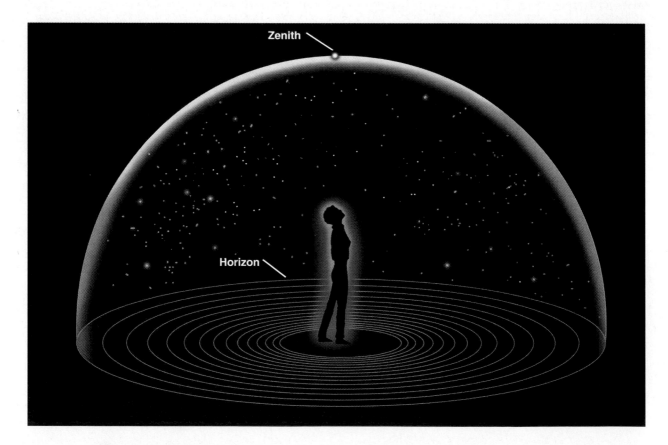

■ **FIGURE 1.1**

The Sky Around Us   The dome of the sky, as it appears to a naive observer. The horizon is where the sky meets the ground, and the observer's zenith is the point directly overhead.

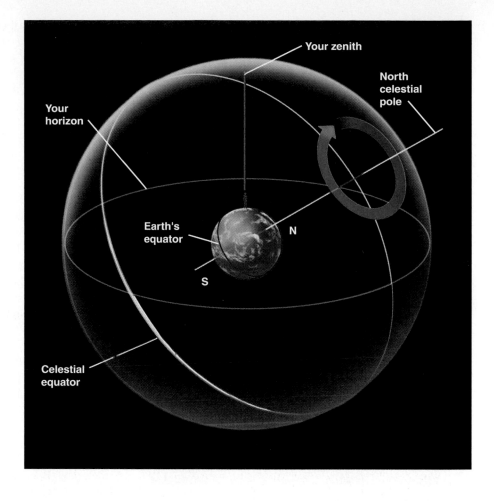

The image shows labels: Your zenith, North celestial pole, Your horizon, Earth's equator, N, S, Celestial equator

■ **FIGURE 1.2**

Circles on the Celestial Sphere
Here we show the (imaginary) celestial sphere around the Earth, on which objects in the sky can turn. Note the axis through the Earth. In reality, it is the Earth that turns around this axis, creating the illusion that the sky revolves around us. Note that the Earth in this picture has been tilted so that your location is at the top and the North Pole is where the N is.

Today, we know that it's not the celestial sphere that turns as the night and day proceed, but rather the planet on which we live. We can put an imaginary stick through the Earth's North and South Poles; we call this stick our planet's **axis.** It is because the Earth turns on this axis every 24 hours that we see the Sun, Moon, and stars rise and set with clockwork regularity. These celestial objects are not on a dome, but at greatly varying distances from us in space. Nevertheless, it is sometimes still convenient to talk about the celestial dome or sphere to help us keep track of objects in the sky. There is even a special theater, called a *planetarium,* in which we project a simulation of the stars and planets onto a white dome.

As the celestial sphere rotates, the objects on it maintain their positions with respect to one another. A grouping of stars like the Big Dipper has the same shape during the course of the night, although it turns with the sky. During a single night, even objects that we know to have significant motions of their own, such as the nearby planets, seem fixed relative to the stars. (Only the *meteors*—brief "shooting stars" that flash into view for just a few seconds—move appreciably with respect to the celestial sphere. This is because they are not stars at all. Rather, they are small pieces of cosmic dust, burning up as they hit the Earth's atmosphere.) We can use the fact that the entire celestial sphere seems to turn together to help us set up systems for keeping track of what things are visible in the sky and where they happen to be at a given time.

1.1.2 Celestial Poles and Celestial Equator

To help orient us in the turning sky, astronomers use a system that extends our Earth's special points into the sky. We extend the line of the Earth's axis outward: The points where the axis meets the celestial sphere are defined as the **north celestial pole** and the **south celestial pole.** As the Earth rotates about its axis, the sky appears to turn in the opposite direction about those celestial poles (Figure 1.3). We also (in our imaginations) throw the Earth's equator onto the sky and call this the **celestial equator.** It lies halfway between the celestial poles, just as the Earth's equator lies halfway between our planet's poles.

Now let's imagine how riding on different parts of the spinning Earth affects our view of the sky. The apparent motion of the celestial sphere depends on your latitude (position north or south of the equator). Bear in mind that the Earth's axis is pointing at the celestial poles, so those two points on the sky do not appear to turn.

If you stood at the North Pole of the Earth, for example, you would see the north celestial pole overhead, at your zenith. The celestial equator, 90° from the celestial

■ **FIGURE 1.3**

Circling the North Celestial Pole Time exposure showing trails left by stars as a result of the apparent rotation of the celestial sphere. (In reality, it is the Earth that rotates.) The bright trail at top center was made by Polaris (the North Star), which is very close to the north celestial pole.

National Optical Astronomy Observatories

poles, lies along your horizon. As you watched them during the course of the night, the stars would all circle around the celestial pole, with none rising or setting (Figure 1.4). Only that half of the sky that is north of the celestial equator is ever visible to an observer at the North Pole. Similarly, an observer at the South Pole would see only the southern half of the sky.

If you were at the Earth's equator, on the other hand, you would see the celestial equator (which after all is just an "extension" of the Earth's equator) pass overhead through your zenith. The celestial poles, being 90° from the celestial equator, must then be at the north and south points on your horizon. As the sky turns, all stars rise and set; they move straight up from the east side of the horizon and set straight down on the west side. During a 24-hour period, all stars are above the horizon exactly half the time. (Of course, during some of those hours, the Sun is too bright for us to see them.)

What would an observer in the latitudes of the United States or Europe see? Remember, we are neither at the Earth's pole nor at the equator, but in between them. In our case, the north celestial pole is neither overhead nor on the horizon, but in between. It appears above the northern horizon at an angular height, or altitude, equal to the observer's latitude (see Figure 1.4). In San Francisco, for example, where the latitude is 38° N, the north celestial pole is 38° above the northern horizon.

For an observer at 38° N latitude, the south celestial pole is 38° below the southern horizon and thus never visible. As the Earth turns, the whole sky seems to pivot about the north celestial pole. For this observer, stars within 38° of the North Pole can never set. They are always above the horizon, day and night. This part of the sky is called the **north circumpolar zone.** To observers in the continental United States, the Big and Little Dippers and Cassiopeia are examples of star groups in the north circumpolar zone. On the other hand, stars within 38° of the south celestial pole never rise. That part of the sky is the **south circumpolar zone.** To most U.S. observers, the Southern Cross is in that zone. (Don't worry if you are not familiar with the star groups we just mentioned. We will introduce them more formally later on.)

At this particular time in Earth history, there happens to be a star very close to the north celestial pole. It is called Polaris, the pole star, and has the distinction of being the star that moves the least amount as the northern sky turns each day. (Because it moved so little while the other stars moved much more, it played a special role in the mythology of several Native American tribes, for example. Some called it the "fastener of the sky.")

1.1.3 Rising and Setting of the Sun

We have described the movement of stars in the night sky. The stars continue to circle during the day, but the brilliance of the Sun makes them difficult to see. (The Moon is

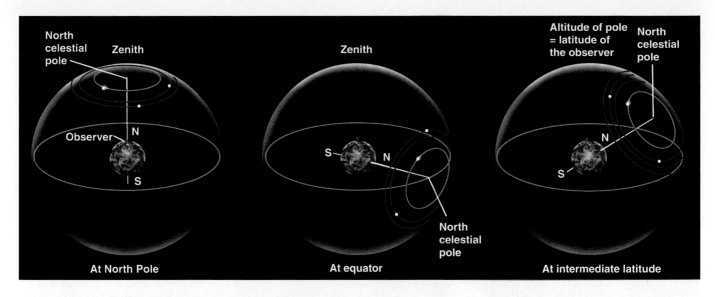

■ FIGURE 1.4

Star Circles at Different Latitudes The turning of the sky looks different depending on your latitude on Earth. (a) At the North Pole, the stars circle the zenith and do not rise and set. (b) At the equator, the celestial poles are on the horizon, and the stars rise straight up and set straight down. (c) At intermediate latitudes, the north celestial pole is at some position between overhead and the horizon. Its angle turns out to be equal to the observer's latitude. Stars rise and set at an angle to the horizon.

still easily seen in the daylight, however.) On any given day, we can think of the Sun as being located at some position on the hypothetical celestial sphere. When the Sun rises— that is, when the rotation of the Earth carries the Sun above the horizon—sunlight is scattered by the molecules of our atmosphere, filling our sky with light and hiding the stars that are above the horizon.

For thousands of years, astronomers have been aware that the Sun does more than just rise and set. It gradually changes position on the celestial sphere, moving each day about 1° to the east relative to the stars. Very reasonably, the ancients thought this meant that the Sun was slowly moving around the Earth, taking a period of time we call one **year** to make a full circle. Today, of course, we know that it is the Earth that is going around the Sun, but the effect is the same: The Sun's position in our sky changes day to day. We have a similar experience when we walk around a campfire at night; we see the flames appear in front of each person seated about the fire in turn.

The path the Sun appears to take around the celestial sphere each year is called the **ecliptic** (Figure 1.5). Because of its motion on the ecliptic, the Sun rises about 4 minutes later each day with respect to the stars. The Earth must make just a bit more than one complete rotation (with respect to the stars) to bring the Sun up again. As the months go by and we look at the Sun from different places in our orbit, we see it projected against different stars in the background, or we would, at least, if we could see the stars in the daytime. In practice, we must deduce what stars lie behind and beyond the Sun by observing the stars visible in the opposite direction at night. After a year, when the Earth has completed one trip around the Sun, the Sun will appear to have completed one circuit of the sky along the ecliptic.

The ecliptic does not lie along the celestial equator but is inclined to it at an angle of about 23°. In other words, the Sun's annual path in the sky is not lined up with the Earth's equator. This is because our planet's axis of rotation is tilted by about 23° from a vertical line sticking out of the plane of the ecliptic (Figure 1.6). Being tilted from "straight up" is not at all unusual among planets; Uranus and Pluto are actually tilted so much that they orbit the Sun "on their side." The inclination of the ecliptic is the reason the Sun moves north and south in the sky as the seasons change. In Chapter 3 we will discuss the progression of the seasons in more detail.

1.1.4 Fixed and Wandering Stars

The Sun is not the only object that moves among the fixed stars. The Moon and each of the five planets that are visible to the unaided eye—Mercury, Venus, Mars, Jupiter, and Saturn—also slowly change their positions from day to day. During a single day, the Moon and planets all rise and set as the Earth turns, just as the Sun and stars do. But like the Sun, they have independent motions among the stars, superimposed on the daily rotation of the celestial sphere. Noticing these motions, the Greeks of two thousand years ago distinguished between what they called the fixed stars, those that maintain fixed patterns among themselves

Constellation on the Ecliptic	Dates When the Sun Crosses It	Constellation on the Ecliptic	Dates When the Sun Crosses It
Capricornus	Jan 21–Feb 16	Leo	Aug 10–Sept 16
Aquarius	Feb 16–Mar 11	Virgo	Sept 16–Oct 31
Pisces	Mar 11–Apr 18	Libra	Oct 31–Nov 23
Aries	Apr 18–May 13	Scorpius	Nov 23–Nov 29
Taurus	May 13–June 22	Ophiuchus	Nov 29–Dec 18
Gemini	June 22–July 21	Sagittarius	Dec 18–Jan 21
Cancer	July 21–Aug 10		

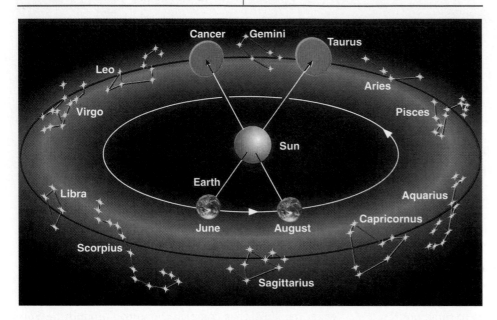

■ **FIGURE 1.5**

Constellations on the Ecliptic As the Earth revolves around the Sun, we sit on "platform Earth" and see the Sun moving around the sky. The circle in the sky that the Sun appears to make around us in the course of a year is called the ecliptic. This circle (like all circles in the sky) goes through a set of constellations; the ancients thought that these constellations, which the Sun (and the Moon and planets) visited, must be special and incorporated them into their system of astrology (see Section 1.3). Note that at any given time of the year, some of the constellations crossed by the ecliptic are visible in the night sky and others are in the day sky and thus hidden by the brilliance of the Sun. As we discuss later in the chapter, today the zodiac constellations in which we "find" the Sun each month are no longer lined up with the signs the astrologers use.

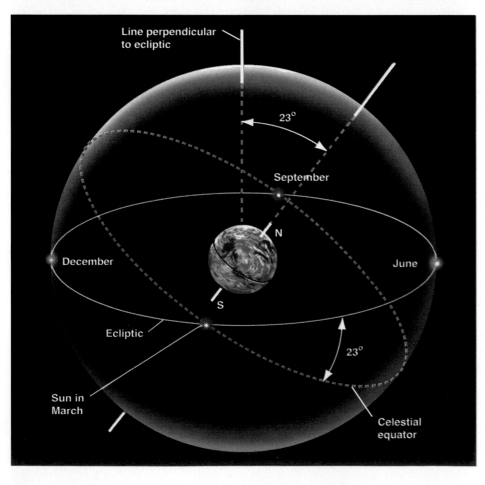

■ **FIGURE 1.6**

The Celestial Tilt The celestial equator is tilted by 23° to the ecliptic. As a result, North Americans and Europeans see the Sun north of the celestial equator and high in our sky in June, and south of the celestial equator and low in the sky in December.

through many generations, and the wandering stars, or **planets.** The word *planet,* in fact, means "wanderer" in Greek.

Today we do not regard the Sun and Moon as planets, but the ancients applied the term to all seven of the moving objects in the sky. Much of ancient astronomy was devoted to observing and predicting the motions of these celestial wanderers. The Moon, being the Earth's nearest celestial neighbor, has the fastest apparent motion; it completes a trip around the sky in about one month (*moonth*). To do this, the Moon moves about 12°, or 24 times its own apparent width on the sky, each day.

The individual paths of the Moon and planets in the sky all lie close to the ecliptic, although not exactly on it. This is because the paths of the planets about the Sun, and of the Moon about the Earth, are all in nearly the same plane, as if they were circles on a huge sheet of paper. The planets, the Sun, and the Moon are thus always found in the sky within a narrow 18°-wide belt called the **zodiac** that is centered on the ecliptic (see Figure 1.5). (The root of the term *zodiac* is the same as that of the word *zoo* and means a collection of animals; many of the patterns of stars within the zodiac belt reminded the ancients of animals, such as a fish or a goat.)

How the planets appear to move in the sky as the months pass is a combination of their actual motions plus the motion of the Earth about the Sun; consequently, their paths are somewhat complex. As we will see, this complexity has fascinated and challenged astronomers for centuries.

1.1.5 Constellations

The backdrop for the motions of the "wanderers" in the sky is the canopy of stars. If there were no clouds in the sky and we were on a flat plain with nothing to obstruct our view, we could see about 3000 stars with the unaided eye. To find their way around such a multitude, the ancients found groupings of stars that made some familiar geometric pattern or (more rarely) resembled something they knew. Each civilization found its own patterns in the stars, much like a modern Rorschach test in which you are asked to discover patterns or pictures in a set of inkblots. The ancient Chinese, Egyptians, and Greeks, among others, found their own groupings or **constellations** of stars helpful in navigating among them and in passing their star-lore on to their children.

You may be familiar with some of the star patterns we still use today, such as the Big and Little Dippers or Orion, the hunter, with his distinctive belt of three stars (Figure 1.7). However, many of the stars we see are not part of a distinctive star pattern at all, and a telescope can reveal millions of stars too faint for the eye to see. Therefore, in the early decades of the 20th century, astronomers from many countries decided to establish a more formal system for organizing the sky.

Today, we use the term *constellation* to mean one of 88 sectors into which we divide the sky, much as the United States has been divided into 50 states. The modern bound-

■ **FIGURE 1.7**

Orion The winter constellation of Orion, the hunter, as illustrated in the 17th-century atlas by Hevelius.

aries between the constellations are imaginary lines in the sky running north–south and east–west, so that each point in the sky falls in a specific constellation. All the constellations are listed in Appendix 14. Whenever possible, we have named each modern constellation after the Latin translation of one of the ancient Greek star patterns that lies within it. Thus, the constellation of Orion is a kind of box on the sky, which includes, among many other objects, the stars that made up the ancient picture of the hunter. Some people use the term *asterism* to denote an especially noticeable star pattern within a constellation (or sometimes spanning parts of several constellations). For example, the Big Dipper is an asterism within the constellation of Ursa Major, the Big Bear.

Students are sometimes puzzled because the constellations seldom resemble the people or animals for which they were named. In all likelihood, the Greeks themselves did not name groupings of stars because they looked like actual people or objects (any more than the outline of Washington State resembles George Washington). Rather, they named sections of the sky in honor of the characters in their mythology and then fit the star configurations to the animals and people as best they could.

1.2 ANCIENT ASTRONOMY

Let us now look briefly back into history. Much of modern Western civilization is derived in one way or another from the ideas of the ancient Greeks and Romans, and this is true in astronomy as well. However, many other ancient cultures also developed sophisticated systems for observing and interpreting the sky.

1.2.1 Astronomy Around the World

Ancient Babylonian, Assyrian, and Egyptian astronomers knew the approximate length of the year. The Egyptians of 3000 years ago, for example, adopted a calendar based on a 365-day year. They kept careful track of the rising time of the bright star Sirius in the predawn sky, whose yearly cycle corresponded with the flooding of the river Nile. The Chinese also had a working calendar and determined the length of the year at about the same time. They recorded comets, bright meteors, and dark spots on the Sun. (Many types of astronomical objects were introduced in the Prologue of this book. If you are not familiar with terms like *comets* and *meteors,* you may want to look at the Prologue.) Later, Chinese astronomers kept careful records of "guest stars"—those that are normally too faint to see but suddenly flare up to become visible to the naked eye for a few weeks or months. We still use some of these records in studying stars that exploded a long time ago.

The Mayan culture in Central America developed a sophisticated calendar based on the planet Venus, and they made astronomical observations from sites dedicated to this purpose a thousand years ago. The Polynesians learned to navigate by the stars over hundreds of kilometers of open ocean, a skill that enabled them to colonize new islands far away from where they began.

In the British Isles, before the widespread use of writing, ancient people used stones to keep track of the motions of the Sun and Moon. We still find some of the great stone circles they built for this purpose, dating from as far back as 2800 B.C.E. The best known of these is Stonehenge, which is discussed in Chapter 3.

1.2.2 Early Greek and Roman Cosmology

Our concept of the cosmos—its basic structure and origin—is called **cosmology,** a Greek word. Before the invention of telescopes, humans had to depend on the simple evidence of their senses for a picture of the universe. The ancients developed cosmologies that combined their direct view of the heavens with a rich variety of philosophical and religious symbolism.

At least 2000 years before Columbus, educated people in the eastern Mediterranean region knew the Earth was round. Belief in a spherical Earth may have stemmed from the time of Pythagoras, a philosopher and mathematician who lived 2500 years ago. He believed circles and spheres to be "perfect forms" and suggested that the Earth should therefore be a sphere. As evidence that the gods liked spheres, the Greeks cited the fact that the Moon is a sphere, using evidence we will describe later.

The writings of Aristotle (384–322 B.C.E.), the tutor of Alexander the Great, summarize many of the ideas of his day. They describe how the progression of the Moon's phases—its changing shape—results from our seeing different portions of the Moon's sunlit hemisphere during the month (see Chapter 3). Aristotle also knew that the Sun has

■ FIGURE 1.8

The Earth's Round Shadow A lunar eclipse, with the Moon moving into and out of the Earth's shadow. Note the curved shape of the shadow—evidence for a spherical Earth that has been recognized since antiquity.

to be farther away from the Earth than is the Moon, because occasionally the Moon passes exactly between the Earth and the Sun and temporarily hides the Sun from view. We call this a solar eclipse (see Chapter 3).

Aristotle cited two convincing arguments that the Earth must be round. First is the fact that as the Moon enters or emerges from the Earth's shadow during an eclipse of the Moon, the shape of the shadow seen on the Moon is always round (Figure 1.8). Only a spherical object always produces a round shadow. If the Earth were a disk, for example, there would be some occasions when the sunlight would be striking it edge-on and its shadow on the Moon would be a line.

As a second argument, Aristotle explained that travelers who go south a significant distance are able to observe stars that are not visible farther north. And the height of the North Star—the star nearest the north celestial pole—decreases as a traveler moves south. On a flat Earth, everyone would see the same stars overhead. The only possible explanation is that the traveler must have moved over a curved surface on the Earth, showing stars from a different angle. (See the box *How Do We Know the Earth Is Round?* for more ideas on proving the Earth is round.)

One brave Greek thinker, Aristarchus of Samos (310–230 B.C.E.), even suggested that the Earth was moving around the Sun, but Aristotle and most of the ancient Greek scholars rejected this idea. One of the reasons for

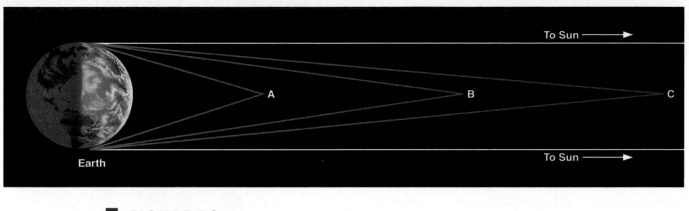

■ FIGURE 1.9

Light Rays from Space   The more distant an object, the more nearly parallel are the rays of light coming from it.

their conclusion was the thought that if the Earth moved about the Sun, they would be observing the stars from different places along Earth's orbit. As the Earth moved along, nearby stars should shift their positions in the sky relative to more distant stars. (In a similar way, we see foreground objects appear to move against a more distant background whenever we are in motion. When we ride on a train, the trees in the foreground appear to shift their position relative to distant hills as the train rolls by. Unconsciously, we use this phenomenon all of the time to estimate distances around us.)

The apparent shift in the direction of an object as a result of the motion of the observer is called *parallax*. We call the shift in the apparent direction of a star due to the Earth's orbital motion *stellar parallax*. The Greeks made dedicated efforts to observe stellar parallax, even enlisting the aid of Greek soldiers with the clearest vision, but to no avail. The brighter (and presumably nearer) stars just did not seem to shift as the Greeks observed them in the spring and then again in the fall.

This meant either that the Earth was not moving or that the stars had to be so tremendously far away that the parallax shift was immeasurably small. A cosmos of such enormous extent required a leap of imagination that most ancient philosophers were not prepared to make, so they retreated to the safety of the Earth-centered view, which would dominate Western thinking for nearly two millennia.

1.2.3   Measurement of the Earth by Eratosthenes

The Greeks not only knew the Earth was round, they were also able to measure its size. The first fairly accurate determination of the Earth's diameter was made in about 200 B.C.E. by Eratosthenes, a Greek living in Alexandria, Egypt. His method was a geometrical one, based on observations of the Sun.

The Sun is so distant from us that all the light rays that strike our planet approach us along essentially parallel lines. To see why, look at Figure 1.9. Take a source of light near

How Do We Know the Earth Is Round?

In addition to the two ways (from Aristotle's writings) discussed in this chapter, you might also reason as follows:

1. Let's watch a ship leave its port and sail into the distance on a clear day. On a flat Earth, we would just see the ship get smaller and smaller as it sailed away. But that's not what we actually observe. Instead, ships sink below the horizon, with the hull disappearing first and the mast remaining visible for a while longer. Eventually only the top of the mast can be seen, as the ship sails around the curvature of the Earth; then finally the ship disappears.

2. The Space Shuttle circles the Earth once every 90 minutes or so. Photographs taken from the shuttle and other satellites show that the Earth is round from every perspective.
3. Suppose you made a friend in each time zone of the Earth. You could call all of them in the same hour and ask, "Where is the Sun?" On a flat Earth, each caller would give you roughly the same answer. But on a round Earth you would find that for some, the Sun would be high in the sky, while for others it would be rising, or setting, or completely out of sight (and this last group of friends would be upset with you for waking them up).

the Earth—say, at position A. Its rays strike different parts of the Earth along diverging paths. From a light source at B, or at C still farther away, the angle between rays that strike opposite parts of the Earth is smaller. The more distant the source, the smaller is the angle between the rays. For a source infinitely distant, the rays travel along parallel lines.

Of course, the Sun is not infinitely far away, but light rays striking the Earth from a point on the Sun diverge from one another by an angle far too small to be observed with the unaided eye. As a consequence, if people all over the Earth who could see the Sun were to point at it, their fingers would all be essentially parallel to one another. (The same is also true for the planets and stars, an idea we will use in our discussion of how telescopes work.)

Eratosthenes noticed that on the first day of summer at Syene, Egypt (near modern Aswan), sunlight struck the bottom of a vertical well at noon. This indicated that the Sun was right over the well (that Syene was on a direct line from the center of the Earth to the Sun). At the corresponding time and date in Alexandria, he observed that the Sun was not directly overhead but was slightly south of the zenith, so that its rays made an angle with the vertical equal to about 1/50 of a circle (7°). Since the Sun's rays striking the two cities are parallel to one another, why would the two rays not make the same angle with the Earth's surface? Eratosthenes reasoned that the curvature of the round Earth meant that "straight up" was not the same in the two cities. And the measurement of the angle in Alexandria, he realized, allowed him to figure out the size of the Earth. Alexandria, he saw, must be 1/50 of the Earth's circumference north of Syene (Figure 1.10). Alexandria had been measured to be 5000 stadia north of Syene (the *stadium* was a Greek unit of length, derived from the length of the racetrack in a stadium). Eratosthenes thus found that the Earth's circumference must be 50 × 5000, or 250,000 stadia.

It is not possible to evaluate precisely the accuracy of Eratosthenes' solution because there is doubt about which of the various kinds of Greek stadia he used as his unit of distance. If it was the common Olympic stadium, his result was about 20 percent too large. According to another interpretation, he used a stadium equal to about 1/6 km, in which case his figure was within 1 percent of the correct value of 40,000 km. Even if his measurement was not exact, his success at measuring the size of our planet by using only shadows, sunlight, and the power of human thought was one of the greatest intellectual achievements in history.

1.2.4 Hipparchus and Precession

Perhaps the greatest astronomer of pre-Christian antiquity was Hipparchus, born in Nicaea in what is present-day Turkey. He erected an observatory on the island of Rhodes in the period around 150 B.C.E., when the Roman Republic was increasing its influence throughout the Mediterranean region. There he measured as accurately as possible the directions of objects in the sky, compiling a pioneering star

■ FIGURE 1.10

How Eratosthenes Measured the Size of the Earth
The Sun's rays come in parallel, but because the Earth's surface curves, a ray at Syene comes straight down, while a ray at Alexandria makes an angle of 7° with the vertical. That means, in effect, that at Alexandria the Earth's surface has curved away from Syene by 7° out of 360°, or 1/50 of a full circle. Thus the distance between the two cities must be 1/50 the circumference of the Earth.

catalog with about 850 entries. He designated celestial coordinates for each star, specifying its position in the sky, just as we can specify the position of a point on the Earth by giving its latitude and longitude. He also divided the stars into **magnitudes** according to their apparent brightnesses. He called the brightest ones "stars of the first magnitude," the next brightest group "stars of the second magnitude," and so forth. This system, in modified form, still remains in use today.

By observing the stars and comparing his data with older observations, Hipparchus made one of his most remarkable discoveries: The position in the sky of the north celestial pole had altered over the previous century and a half. Hipparchus correctly deduced that this had happened not only during the period covered by his observations, but was in fact happening all the time: The direction around which the sky appears to rotate changes slowly but continuously.

You will recall from Section 1.1.2 that the north celestial pole is just the projection of the Earth's North Pole into the sky. If the north celestial pole is wobbling around, then the Earth itself must be doing the wobbling. Today we understand that the direction in which the Earth's axis points does indeed change slowly but regularly, a motion we call **precession.** If you've ever watched a child's spinning top wobble, you have observed a similar kind of motion. The

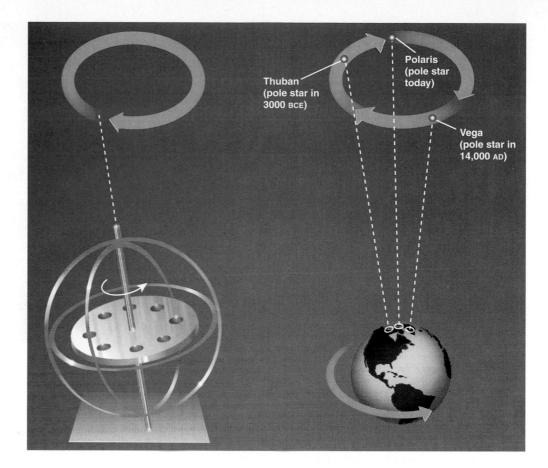

FIGURE 1.11

Precession Just as the axis of a rapidly spinning top slowly wobbles in a circle, so the axis of the Earth wobbles in a 26,000-year cycle. Today the north celestial pole is near the star Polaris, but about 5000 years ago it was close to a star called Thuban, and in 14,000 it will be closest to the star Vega.

top's axis describes a path in the shape of a cone, as the Earth's gravity tries to topple it over (Figure 1.11).

Because our planet is not an exact sphere, but bulges a bit at the equator, the pulls of the Sun and Moon cause it to wobble like a top. It takes about 26,000 years for the Earth's axis to complete one circle of precession. As a result of this motion, the point where our axis points in the sky changes as time goes on. While Polaris is the star closest to the north celestial pole today (it will reach its closest point around the year 2100), the star Vega in the constellation of Lyra will be the north star in the year 14,000.

1.2.5 Ptolemy's Model of the Solar System

The last great astronomer of the Roman era was Claudius Ptolemy (or Ptolemaeus), who flourished in Alexandria in about the year 140. He wrote a mammoth compilation of astronomical knowledge, which today is called by its Arabic name, *Almagest* (meaning "The Greatest"). *Almagest* does not deal exclusively with Ptolemy's own work; it includes a discussion of the astronomical achievements of the past, principally those of Hipparchus. Today it is our main source of information about the work of Hipparchus and other Greek astronomers.

Ptolemy's most important contribution was a geometrical representation of the solar system that predicted the positions of the planets for any desired date and time. Hipparchus, not having enough data on hand to solve the problem himself, had instead amassed observational material for posterity to use. Ptolemy supplemented this material with new observations of his own and produced a cosmological model that endured more than a thousand years, until the time of Copernicus.

The complicating factor in explaining the motions of the planets is that their apparent wanderings in the sky result from the combination of their own motions with the Earth's orbital revolution. As we watch the planets from our vantage point on the moving Earth, it's a little like watching a car race from inside one of the cars. Sometimes other cars pull ahead, while at other times we might pass opponents' cars, making them seem to move backward for a while with respect to our car.

Figure 1.12a shows the motion of the Earth and a planet farther from the Sun, such as Mars or Jupiter. The Earth travels around the Sun in the same direction as the other planet and in nearly the same plane, but its speed in orbit is faster. As a result, it periodically overtakes the planet, like a faster race car on the inside track. The figure shows where we see the planet in the sky at different times. In Figure 1.12b, we show what the path of the planet looks like among the stars.

Normally, planets move eastward in the sky over the weeks and months as they orbit the Sun, but from positions

(a)

(b)

■ **FIGURE 1.12**

Retrograde Motion of a Planet Beyond the Earth's Orbit   The letters on the diagram show where the Earth (blue) and the other planet (yellow) are at different times. (a) Actual positions of the planet and the Earth. (b) What the path of the planet looks like in the Earth's skies against the background of stars.

B to *D* in Figure 1.12, as the Earth passes the planet, it appears to drift backward, moving west in the sky. Even though it is actually moving to the east, the faster Earth has overtaken it and seems, from our perspective, to be leaving it behind. As the Earth rounds its orbit toward position *E,* the planet again takes up its apparent eastward motion in the sky. The temporary apparent westward motion of a planet as the Earth swings between it and the Sun is called **retrograde motion.** Such backward motion is much easier for us to understand today, now that we know the Earth is one of the moving planets and not the unmoving center of all creation. But Ptolemy was faced with the far more complex problem of explaining such motion while assuming a stationary Earth.

Furthermore, because the Greeks believed that celestial motions had to be circles, Ptolemy had to construct a model using circles alone. To do it, he needed dozens of circles, some moving around other circles, in a complex structure that makes a modern viewer dizzy. But we must not let our modern judgment cloud our admiration for Ptolemy's achievement. In his day, a complex universe centered on the Earth was perfectly reasonable and, in its own way, quite beautiful. (On the other hand, see the quote that starts this chapter.)

Ptolemy solved the problem of explaining the observed motions of planets by having each planet revolve in a small orbit called an **epicycle.** The center of the epicycle then revolved about the Earth on a circle called a *deferent* (Figure 1.13). When the planet is at position *x* in Figure 1.13 on the epicycle orbit, it is moving in the same direction as the center of the epicycle; from Earth the planet appears to be moving eastward. When the planet is at *y,* however, its motion is in the direction opposite to the motion of the epicycle's center around the Earth. By choosing the right combination of speeds and distances, Ptolemy succeeded in having the planet moving westward at *y* at the correct speed and for the correct interval of time, thus replicating retrograde motion with his model.

However, we shall see in the next chapter that the planets, like the Earth, travel about the Sun in orbits that are ellipses, not circles. Their actual behavior cannot be represented accurately by a scheme of uniform circular motions. In order to match the observed motions of the planets, Ptolemy had to center the deferent circles not on the Earth but at points some distance from the Earth. In addition, he introduced uniform circular motion around yet another axis, called the *equant point.* All of this considerably complicated his scheme.

It is a tribute to the genius of Ptolemy as a mathematician that he was able to develop such a complex system to account successfully for the observations of planets. It may be that Ptolemy did not intend his cosmological model to describe reality, but merely to serve as a mathematical representation that allowed him to predict the positions of the

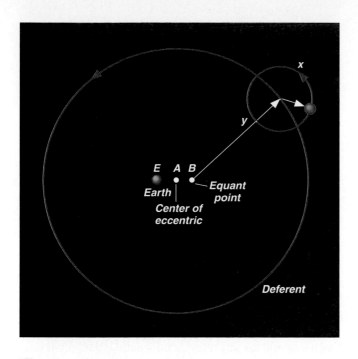

FIGURE 1.13

Ptolemy's Cosmological System   Each planet orbits around a small circle called an epicycle. The system is not exactly centered on the Earth but on a point called the equant. The Greeks needed all this complexity to explain the motions in the sky because they believed that the Earth was stationary.

planets at any time. Whatever his thinking, his model, with some modifications, was accepted as authoritative in the Muslim world and (later) in Christian Europe.

 ## 1.3   ASTROLOGY AND ASTRONOMY

Many ancient cultures regarded the planets and stars as representatives or symbols of the gods or other supernatural forces that controlled their lives. For them, the study of the heavens was not an abstract subject; it was directly connected to the life-and-death necessity of understanding the actions of the gods and currying favor with them. Before the time of our scientific perspective, everything that happened in nature—from the weather, to diseases and accidents, to celestial surprises like eclipses or new comets—was thought to be an expression of the whims or displeasures of the gods. Any signs that helped people understand what these gods had in mind were considered extremely important.

The movements of the seven objects that had the power to "wander" through the realm of the sky—the Sun, the Moon, and five planets visible to the unaided eye—clearly must have special significance in such a system of thinking.

Most ancient cultures associated these seven objects with various supernatural rulers in their pantheon and kept track of them for religious reasons. Even in the comparatively sophisticated Greece of antiquity, the planets had the names of gods and were credited with having the same powers and influences as the gods whose names they bore. From such ideas was born the ancient system called **astrology,** still practiced by some people today, in which the positions of these bodies among the stars of the zodiac are thought to hold the key to understanding what we can expect from life.

1.3.1   The Beginnings of Astrology

Astrology began in Babylonia about two and a half millennia ago. The Babylonians, believing that the planets and their motions influenced the fortunes of kings and nations, used their knowledge of astronomy to guide their rulers. When the Babylonian culture was absorbed by the Greeks, astrology gradually came to influence the entire Western world and eventually spread to the Orient as well.

By the 2nd century B.C.E. the Greeks democratized astrology by developing the idea that the planets influence every individual. In particular, they believed that the configuration of the Sun, Moon, and planets at the moment of birth affected a person's personality and fortune, the doctrine called *natal astrology.* Natal astrology reached its zenith with Ptolemy 400 years later. As famous for his astrology as for his astronomy, Ptolemy compiled the *Tetrabiblos,* a treatise on astrology that remains the "bible" of the subject. It is essentially this ancient religion, older than Christianity or Islam, that is still practiced by today's astrologers.

1.3.2   The Horoscope

The key to natal astrology is the **horoscope,** a chart showing the positions of the planets in the sky at the moment of an individual's birth. When a horoscope is charted, the planets (including the Sun and Moon, classed as wanderers or planets by the ancients) must first be located in the zodiac. At the time the system of astrology was set up, the circle of the zodiac was divided into 12 sectors called *signs,* each 30° long. Each sign was named after a constellation in the sky through which the Sun, Moon, and planets were seen to pass—the sign of Virgo after the constellation of Virgo, for example.

When someone today casually asks you your "sign," what they are asking for is your "sun-sign"—which zodiac sign the Sun was in at the moment you were born. But more than 2000 years have passed since the signs received their names from the constellations. Because of precession, the constellations of the zodiac slide westward along the ecliptic, going once around the sky in about 26,000 years. Thus today the real stars have slipped around by about 1/12 of the zodiac—about the width of one sign (see Figure 1.5).

In most forms of astrology, however, the signs have remained assigned to the dates of the year they had when astrology was first set up. This means that the astrological signs and the real constellations are out of step; the sign of Aries, for example, now occupies the constellation of Pisces. When you look up your sun-sign in a newspaper astrology column, the name of the sign associated with your birthday is no longer the name of the constellation in which the Sun was actually located when you were born. To know that constellation, you must look for the sign *before* the one that includes your birthday.

A complete horoscope shows the location of not only the Sun but also the Moon and each planet in the sky by indicating its position in the appropriate sign of the zodiac. However, as the celestial sphere turns (owing to the rotation of the Earth), the entire zodiac moves across the sky to the west, completing a circuit of the heavens each day. Thus the position in the sky (or "house" in astrology) must also be calculated. There are more or less standardized rules for the interpretation of the horoscope, most of which (at least in Western schools of astrology) are derived from the *Tetrabiblos* of Ptolemy. Each sign, each house, and each planet, the last acting as a center of force, is supposed to be associated with particular matters in a person's life.

The detailed interpretation of a horoscope is a very complicated business, and there are many schools of astrological thought on how it should be done. Although some of the rules may be standardized, how each rule is to be weighed and applied is a matter of judgment—and "art." It also means that it is very difficult to tie astrology down to specific predictions or to get the same predictions from different astrologers.

1.3.3 Astrology Today

Astrologers today use the same basic principles laid down by Ptolemy nearly 2000 years ago. They cast horoscopes (a process much simplified by the development of appropriate computer programs) and suggest interpretations. Sun-sign astrology (which you read in the newspapers and many magazines) is a recent simplified variant of natal astrology. Although even professional astrologers do not place much trust in such a limited scheme, which tries to fit everyone into just 12 groups, sun-sign astrology is taken seriously by many people (perhaps because it is so commonly discussed in the media). In a recent poll of teenagers in the United States, more than half said they "believed in astrology."

Today, we know much more about the nature of the planets as physical bodies, as well as about human genetics, than the ancients could. It is hard to imagine how the positions of the Sun, Moon, or planets in the sky at the moment of our birth could have anything to do with our personality or future. There are no known forces, not gravity or anything else, that could cause such effects. (For example, a simple calculation shows that the gravitational pull of the obstetrician delivering a newborn baby is greater than that of Mars.) Astrologers thus have to argue that there are unknown forces exerted by the planets that depend on their configurations with respect to one another and that do not vary according to the distance of the planet—forces for which there is no shred of evidence.

Another curious aspect of astrology is its emphasis on planet configurations at birth. What about the forces that might influence us at conception? Isn't our genetic makeup more important for determining our personality than the circumstances of our birth? Would we really be a different person if we had been born a few hours earlier or later, as astrology claims? (Back when astrology was first conceived, birth was thought of as a moment of magic significance, but today we understand a lot more about the long process that precedes it.)

Actually, very few thinking people today buy the claim that our entire lives are predetermined by astrological influences at birth, but many people apparently believe that astrology has validity as an indicator of affinities and personality. A surprising number of Americans make judgments about people—whom they will hire, associate with, and even marry—on the basis of astrological information. To be sure, these are difficult decisions, and you might argue that we should use any relevant information that might help us to make the right choices. But does astrology actually provide any useful information on human personality? This is the kind of question that can be tested using the scientific method (see Making Connections: *Testing Astrology*).

The results of hundreds of tests are all the same: There is no evidence that natal astrology has any predictive power, even in a statistical sense. Why then do people often seem to have anecdotes about how well their own astrologer advised them? Effective astrologers today use the language of the zodiac and the horoscope only as the outward trappings of their craft. Mostly they work as amateur therapists, offering simple truths that clients like or need to hear. (Recent studies have shown that just about any sort of short-term therapy makes people feel a little better. This is because the very act of talking about our problems with someone is in itself beneficial.)

The scheme of astrology has no basis in scientific fact, however. It is an interesting historical system, left over from prescientific days and best remembered for the impetus it gave people to learn the cycles and patterns of the sky. From it grew the science of astronomy, which is our main subject for discussion.

1.4 THE BIRTH OF MODERN ASTRONOMY

Astronomy made no major advances in strife-torn medieval Europe. The birth and expansion of Islam after the 7th century led to a flowering of Arabic and Jewish cultures that preserved, translated, and added to many of the astronomical ideas of the Greeks. (Many of the names of the brightest stars, for example, are today taken from the Arabic, as are such astronomical terms as *zenith*.)

Testing Astrology

In response to modern public interest in astrology, scientists have carried out a wide range of statistical tests to assess its predictive power. The simplest of these examine sun-sign astrology to determine whether—as astrologers assert—some signs are more likely than others to be associated with some objective measure of success, such as winning Olympic medals, earning high corporate salaries, or achieving elective office or high military rank. (You can make such a test yourself by looking up the birth dates of all members of Congress, for example, or all members of the U.S. Olympic Team.) Are our political leaders somehow selected at birth by their horoscopes and thus more likely to be Leos, say, than Scorpios?

You don't even need to be specific about your prediction in such tests. After all, many schools of astrology disagree about which signs go with which personality characteristics. To demonstrate the validity of the astrological hypothesis, it would be sufficient if the birthdays of all our leaders clustered in any one or two signs in some statistically significant way. Dozens of such tests have been performed, and all have come up completely negative: The birth dates of leaders in all fields tested have been found to be randomly distributed among *all* the signs. Sun-sign astrology does not predict anything about a person's future occupation or strong personality traits.

In a fine example of such a test, two statisticians examined the reenlistment records of the Marine Corps. We suspect our readers will agree that it takes a certain kind of personality to not only enlist but also reenlist in the Marines. If sun-signs can predict strong personality traits—as astrologers claim—then the reenlistees (with similar personalities) should have been distributed preferentially in those one or few signs that matched the personality of someone who loves being a Marine. However, the reenlistees were distributed randomly among all the signs.

More sophisticated studies have also been done, involving full horoscopes calculated for thousands of individuals. The results of all of these studies are also negative: None of the systems of astrology has been shown to be at all effective in connecting astrological aspects to personality, success, or finding the right person to love.

Other tests show that it hardly seems to matter what a horoscope interpretation says, as long as it is vague enough, and as long as each subject feels it was personally prepared just for him or her. The French statistician Michel Gauguelin, for example, sent the horoscope interpretation for one of the worst mass-murderers in history to 150 people, but told each recipient that it was a "reading" prepared exclusively for him or her. Ninety-four percent of the readers said they recognized themselves in the interpretation of the mass-murderer's horoscope.

Geoffrey Dean, an Australian researcher, reversed the astrological readings of 22 subjects, substituting phrases that were the opposite of what the horoscope actually said. Yet his subjects said that the resulting readings applied to them just as often (95 percent) as the people to whom the original phrases were given. Apparently, those who seek out astrologers just want guidance, and almost any guidance will do.

As European culture began to emerge from its long dark age, trading with Arab countries led to a rediscovery of ancient texts such as *Almagest* and to a reawakening of interest in astronomical questions. This time of rebirth (in French, *Renaissance*) in astronomy was embodied in the work of Copernicus (Figure 1.14).

1.4.1 Copernicus

One of the most important events of the Renaissance was the displacement of Earth from the center of the universe, an intellectual revolution initiated by a Polish cleric in the 16th century. Nicolaus Copernicus (1473–1543) was born in Torun, a mercantile town along the Vistula River. His training was in law and medicine, but his main interests were astronomy and mathematics. His great contribution to science was a critical reappraisal of the existing theories of planetary motion and the development of a new Sun-centered, or **heliocentric,** model of the solar system. Copernicus concluded that the Earth is a planet and that all the planets circle the Sun. Only the Moon orbits the Earth (Figure 1.15).

Copernicus described his ideas in detail in his book *De Revolutionibus Orbium Coelestium (On the Revolution of Celestial Orbs),* published in 1543, the year of his death. By this time, the old Ptolemaic system (like a cranky old machine) needed significant adjustments to predict the positions of the planets correctly. Copernicus wanted to develop an improved theory from which to calculate planetary positions, but in doing so, he was himself not free of all traditional prejudices.

He began with several assumptions that were common in his time, such as the idea that the motions of the heavenly bodies must be made up of combinations of uniform

Yerkes Observatory

■ **FIGURE 1.14**

Copernicus Nicolaus Copernicus (1473–1543), cleric and scientist, played a leading role in the emergence of modern science. While he could not prove that the Earth revolves about the Sun, he presented such compelling arguments for this idea that he turned the tide of cosmological thought and laid the foundations upon which Galileo and Kepler so effectively built in the following century.

Crawford Collection, Royal Observatory, Edinburgh

■ **FIGURE 1.15**

Copernicus' System Heliocentric plan of the solar system in the first edition of Copernicus' *De Revolutionibus*. Notice the word "Sol" for Sun in the middle.

circular motions. But he did not assume (as most people did) that the Earth had to be in the center of the universe, and he presented a defense of the heliocentric system that was elegant and persuasive. His ideas, although not widely accepted until more than a century after his death, were much discussed among scholars and ultimately had a profound influence on the course of world history.

One of the objections raised to the heliocentric theory was that if the Earth were moving, we would all sense or feel this motion. Solid objects would be ripped from the surface, a ball dropped from a great height would not strike the ground directly below it, and so forth. But a moving person is not necessarily aware of that motion. We have all experienced seeing an adjacent train, car, or ship appear to move, only to discover that it is we who are moving.

Copernicus argued that the apparent motion of the Sun about the Earth during the course of a year could be represented equally well by a motion of the Earth about the Sun. He also reasoned that the apparent rotation of the celestial sphere could be explained by assuming that the Earth rotates while the celestial sphere is stationary. To the ob-

jection that if the Earth rotated about an axis, it would fly into pieces, Copernicus answered that if such motion would tear the Earth apart, the still faster motion of the much larger celestial sphere required by the geocentric hypothesis would be even more devastating.

1.4.2 The Heliocentric Model

The most important idea in Copernicus' *De Revolutionibus* is that the Earth is one of six (then known) planets that revolve about the Sun. Using this concept, he was able to work out the correct general picture of the solar system. He placed the planets, starting nearest the Sun, in the correct order: Mercury, Venus, Earth, Mars, Jupiter, and Saturn. Further, he deduced that the nearer a planet is to the Sun, the greater is its orbital speed. With his theory, he was able to explain the complex retrograde motions of the planets without epicycles and to work out a roughly correct scale for the solar system.

Copernicus could not *prove* that the Earth revolves about the Sun. In fact, with some adjustments the old Ptolemaic system could have accounted as well for the motions of the planets in the sky. But Copernicus pointed out that the Ptolemaic cosmology was clumsy and lacking the beauty and symmetry of its successor.

In Copernicus' time, in fact, few people thought there were ways to prove whether the heliocentric or the older geocentric system was correct. A long philosophical tradition, going back to the Greeks and defended by the Catholic Church, held that pure human thought combined with divine revelation represented the path to truth. Nature, as revealed by our senses, was suspect. For example, Aristotle had reasoned that heavier objects (having more of the quality that made them heavy) must fall to Earth faster than lighter ones. This is absolutely incorrect, as any simple experiment dropping two balls of different weights will show. But in Copernicus' day, experiments did not carry much weight (if you will pardon the expression); Aristotle's brilliant reasoning was more convincing.

In this environment, there was little motivation to carry out observations or experiments to distinguish between competing cosmological theories (or anything else). It should not surprise us, therefore, that the heliocentric idea was debated for more than half a century without any tests being applied to determine its validity. (In fact, in the American colonies, the older geocentric system was still taught at Harvard University in the first years after it was founded in 1636.)

Contrast this with the situation today, when scientists rush to test each new hypothesis and do not accept any ideas until the results are in. For example, when two researchers at the University of Utah announced in 1989 that they had discovered a way to achieve nuclear fusion (the process that powers the stars) at room temperature, other scientists at more than 25 laboratories around the United States attempted to duplicate "cold fusion" within a few weeks—without success, as it turned out. The cold fusion theory soon went down in flames!

How would we look at Copernicus' model today? When a new hypothesis or theory is proposed in science, it must first be checked for consistency with what is already known. Copernicus' heliocentric idea passes this test, for it allows planetary positions to be calculated at least as well as does the geocentric theory. The next step is to see what predictions the new hypothesis makes that differ from those of competing ideas. In the case of Copernicus, one example is the prediction that, if Venus circles the Sun, the planet should go through the full range of phases just as the Moon does, whereas if it circles the Earth, it should not (Figure 1.16). But in those days, before the telescope, no one imagined testing this prediction.

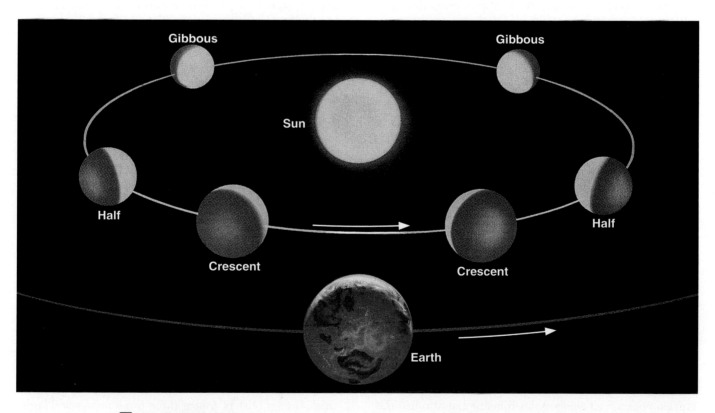

■ **FIGURE 1.16**

Venus Shows Phases The phases of Venus according to the heliocentric theory. As Venus moves around the Sun, we see changing illumination of its surface, just as we see the face of the Moon illuminated differently in the course of a month.

1.4.3 Galileo and the Beginning of Modern Science

Many of the modern scientific concepts of observation, experimentation, and the testing of hypotheses through careful quantitative measurements were pioneered by a man who lived nearly a century after Copernicus. Galileo Galilei (1564–1642; Figure 1.17), a contemporary of Shakespeare, was born in Pisa. Like Copernicus, he began training for a medical career, but he had little interest in the subject and later switched to mathematics. He held faculty positions at the Universities of Pisa and Padua, and eventually he became mathematician to the Grand Duke of Tuscany in Florence.

Galileo's greatest contributions were in the field of mechanics, the study of motion and the actions of forces on bodies. It was familiar to all persons then, as it is to us now, that if something is at rest, it tends to remain at rest and requires some outside influence to start it in motion. Rest was thus generally regarded as the natural state of matter. Galileo showed, however, that rest is no more natural than motion.

If an object is slid along a rough horizontal floor, it soon comes to rest, because friction between it and the floor acts as a retarding force. However, if the floor and the object are both highly polished, the object, given the same initial speed, will slide farther before stopping. On a smooth layer of ice, it will slide farther still. Galileo reasoned that if all resisting effects could be removed, the object would continue in a steady state of motion indefinitely. He argued that a force is required not only to start an object moving from rest but also to slow down, stop, speed up, or change the direction of a moving object. You will appreciate this if you have ever tried to stop a rolling car by leaning against it, or a moving boat by tugging on a line.

Galileo also studied the way objects **accelerate**—change their speed or direction of motion. Galileo watched objects as they fell freely or rolled down a ramp. He found that such objects accelerate uniformly; that is, in equal intervals of time they gain equal increments in speed. Galileo formulated these newly found laws in precise mathematical terms that enabled future experimenters to predict how far and how fast objects would move in various lengths of time.

Sometime in the 1590s Galileo adopted the Copernican hypothesis of a heliocentric solar system. In Roman Catholic Italy, this was not a popular philosophy, for the Church authorities still upheld the ideas of Aristotle and Ptolemy, and they had powerful political and economic reasons for insisting that the Earth was the center of creation. Galileo not only challenged this thinking but also had the audacity to write in Italian rather than scholarly Latin, and to lecture publicly on these topics. For him, there was no contradiction between the authority of the Church in matters of religion and morality, and the authority of nature

■ **FIGURE 1.17**

Galileo Galileo Galilei (1564–1642) advocated that we perform experiments or make observations to ask nature its ways. When Galileo turned the telescope to the sky, he found that things were not the way philosophers had supposed.

(revealed by experiments) in matters of science. It was primarily because of Galileo and his "dangerous" opinions that in 1616 the Church issued a prohibition decree stating that the Copernican doctrine was "false and absurd" and not to be held or defended.

1.4.4 Galileo's Astronomical Observations

It is not certain who first conceived of the idea of combining two or more pieces of glass to produce an instrument that enlarged images of distant objects, making them appear nearer. The first such telescopes that attracted much notice were made by the Dutch spectacle maker Hans Lippershey in 1608. Galileo heard of the discovery and, without ever having seen an assembled telescope, constructed one of his own with a three-power magnification, which made distant objects appear three times nearer and larger (Figure 1.18).

On August 25, 1609, Galileo demonstrated a telescope with a magnification of 9× to government officials of the city-state of Venice. By a magnification of 9×, we mean

■ **FIGURE 1.18**

Telescopes Used by Galileo The longer telescope has a wooden tube covered with paper and a lens 26 mm across.

that the linear dimensions of the object being viewed appeared nine times larger or, alternatively, that the objects appeared nine times closer than they really were. There were obvious military advantages associated with a device for seeing distant objects. For his invention, Galileo's salary was nearly doubled, and he was granted lifetime tenure as a professor. (His university colleagues were outraged, particularly because the invention was not even original.)

Others had used the telescope before Galileo to observe things on Earth. But, in a flash of insight that changed the history of astronomy, Galileo realized that he could turn the power of the telescope toward the heavens. Before using his telescope for astronomical observations, Galileo had to devise a stable mount, and he improved the optics to provide a magnification of 30×. Galileo also needed to acquire confidence in the telescope.

At that time, human eyes were believed to be the final arbiter of truth about sizes, shapes, and colors. Lenses, mirrors, and prisms were known to distort distant images by enlarging, reducing, or even inverting them. Galileo undertook repeated experiments to convince himself that what he saw through the telescope was identical to what he saw up close. Only then could he begin to believe that the miraculous phenomena the telescope revealed in the heavens were real.

Beginning his astronomical work late in 1609, Galileo found that many stars too faint to be seen with the unaided eye became visible with his telescope. In particular, he found that some nebulous blurs resolved into many stars, and that the Milky Way—the strip of whiteness across the night sky—was also made up of a multitude of individual stars.

Examining the planets, Galileo found four satellites, or moons, revolving about Jupiter in times ranging from just

under 2 days to about 17 days. This discovery was particularly important because it showed that not everything has to revolve around the Earth! Furthermore, it demonstrated that there could be centers of motion that are themselves in motion. Defenders of the geocentric view had argued that if the Earth were in motion, then the Moon would be left behind because it could hardly keep up with a rapidly moving planet. Yet here were Jupiter's satellites doing exactly that! (To recognize this discovery and honor his work, NASA named a spacecraft that explored the Jupiter system Galileo.)

With his telescope, Galileo was able to carry out the test of the Copernican theory mentioned earlier, based on the phases of Venus. Within a few months he had found that Venus goes through phases like the Moon, showing that it must revolve about the Sun, so that we see different parts of its daylight side at different times (see Figure 1.16). These observations could not be reconciled with any model in which Venus circled about the Earth.

Galileo also observed the Moon and saw craters, mountain ranges, valleys, and flat dark areas that he thought might be water. These discoveries showed that the Moon might be not so dissimilar to the Earth—suggesting that the Earth, too, could belong to the realm of celestial bodies.

After Galileo's work, it became increasingly difficult to deny the Copernican view, and slowly the Earth was dethroned from its central position in the universe and given its rightful place as one of the planets attending the Sun. Galileo himself had to appear before the Inquisition to answer charges that his work was heretical, and he was condemned to house arrest. His books were on the Church's forbidden list until 1836, although in countries where the Roman Catholic Church held less sway, they were widely read and discussed. Not until 1992 did the Catholic Church publicly admit that it had erred in the matter of censoring Galileo's ideas.

The new ideas of Copernicus and Galileo began a revolution in our conception of the cosmos. It eventually became evident that the universe is a vast place and that the Earth's role in it is relatively unimportant. The idea that the Earth moves around the Sun like the other planets raised the possibility that they might be worlds themselves, perhaps even supporting life. As the Earth was demoted from its position at the center of the universe, so too was humanity. The universe, despite what we may wish, does not revolve around us, and we must find a more modest place for ourselves in the scheme of the cosmos.

Most of us take these things for granted today, but four centuries ago such concepts were frightening and heretical for some, immensely stimulating for others. The pioneers of the Renaissance started the European world along the path toward science and technology that we still tread today. For them, nature was rational and ultimately knowable, and experiments and observations provided the means to reveal its secrets.

Observing the Planets

At almost any time of the night, and at any season, you can spot one or more bright planets in the sky. All five of the planets known to the ancients—Mercury, Venus, Mars, Jupiter, and Saturn—are more prominent than any but the brightest stars, and they can be seen even from urban locations if you know where and when to look. One way to tell planets from bright stars is that planets twinkle less.

Venus, which stays close to the Sun from our perspective, appears either as an "evening star" in the west after sunset or as a "morning star" in the east before sunrise. It is the brightest object in the sky after the Sun and Moon. It far outshines any real star, and under the most favorable circumstances it can even cast a visible shadow. Some young military recruits have tried to shoot Venus down as an approaching enemy craft or UFO!

■ ■

One way to tell planets from bright stars is that planets twinkle less.

■ ■

Mars, with its distinctive red color, can be nearly as bright as Venus is when close to the Earth, but normally it remains much less conspicuous. Jupiter is most often the second-brightest planet, approximately equaling in brilliance the brightest stars. Saturn is dimmer, and it varies considerably in brightness, depending on whether its large rings are seen nearly edge-on (faint) or more widely opened (bright).

Mercury is quite bright, but few people ever notice it because it never moves very far from the Sun (it's never more than 28° away in the sky) and is always seen against bright twilight skies.

True to their name, the planets "wander" against the background of the "fixed" stars. Although their apparent motions are complex, they reflect an underlying order, upon which the heliocentric model of the solar system, as described in this chapter, was based. The positions of the planets are often listed in newspapers (sometimes on the weather page), and clear maps and guides to their locations can be found each month in such magazines as *Sky & Telescope* and *Astronomy* (available at most libraries). There are also a number of computer programs (including shareware and commercial programs) that allow you to calculate and display where the planets are on any night.

Activities

1. If either Venus or Mercury is visible, plot its position with respect to the background stars and note the date. Use a map of the relevant portion of the sky, or sketch the location of the brighter stars. Wait a few days and repeat the observation. Estimate how many degrees the planet moves per day. (The width of your thumb, held straight out at arm's length, covers about 1° of sky; your fist at arm's length covers about 10°.) If you make enough observations, you can estimate each planet's maximum distance in degrees from the Sun.
2. Use binoculars to try to determine the shape of Venus. This experiment will be easier when Venus is on the same side of the Sun as the Earth.
3. Determine which of the outer planets—Mars, Jupiter, and Saturn—are visible. For each visible planet, draw its position with respect to the background stars and note the date. Repeat this experiment every few days for two months or as long as each planet is visible. Is the planet moving east or west? Did it appear to change direction? Did you observe retrograde motion for any of the planets?

SURFING THE WEB

The Constellations and Their Stars:
www.astro.wisc.edu/~dolan/constellations/
Chris Dolan has assembled handy information about all the constellations and the most important stars, nebulae, clusters, and galaxies that can be found in them.

The Galileo Project: es.rice.edu/ES/humsoc/Galileo/
Historian of astronomy Albert VanHelden of Rice University maintains this rich lode of information about Galileo's life, work, and times, with text, images, maps, timelines, and so on.

Ancient Astronomy Site:
www.astronomy.pomona.edu/archeo/INDEX.HTML
Brian Penprase of Pomona College and his students have assembled this introduction to the astronomical lore and knowledge of ancient civilizations, with an interactive atlas, timelines, tours, images, and links.

 The History of Astronomy Site:
www.astro.uni-bonn.de/~pbrosche/astoria.html
Professor Wolfgang Dick maintains this comprehensive Web site with many links to sites about all aspects of astronomical history.

 Activities About Astrology:
www.astrosociety.org/education/astro/act3/astrology.html
This site has suggested activities for "testing" the hypothesis of astrology, an article by *Voyages* co-author Andrew Fraknoi with embarrassing questions to ask astrology believers, and a bibliography of astrology-debunking resources.

SUMMARY

1.1 The direct evidence of our senses supports a **geocentric** perspective, with the **celestial sphere** pivoting on the **celestial poles** and rotating about a stationary Earth. We see only half of this sphere at one time, limited by the **horizon;** the point directly overhead is our **zenith.** The Sun's annual path on the celestial sphere is the **ecliptic,** a line that runs through the center of the **zodiac,** the 18°-wide strip of sky within which we always find the Moon and planets. The celestial sphere is organized into 88 **constellations,** or sectors.

1.2 Ancient Greeks such as Aristotle recognized that the Earth and Moon are spheres and understood the phases of the Moon, but because of their inability to detect stellar parallax, they rejected the idea that the Earth moves. Eratosthenes measured the size of the Earth with surprising precision. Hipparchus carried out many astronomical observations, making a star catalog, defining the system of stellar **magnitudes,** and discovering **precession** from the apparent shift in the position of the north celestial pole. Ptolemy of Alexandria summarized classical astronomy in his *Almagest;* he explained planetary motions, including **retrograde motion,** with remarkably good accuracy using a model centered on the Earth. This geocentric model, based on combinations of uniform circular motion using **epicycles,** was accepted as authority for more than a thousand years.

1.3 The ancient religion of **astrology,** whose main contribution to civilization was a heightened interest in the heavens, began in Mesopotamia. It reached its peak in the Greco-Roman world, especially as recorded in the *Tetrabiblos* of Ptolemy. Natal astrology is based on the assumption that the positions of the planets at the time of our birth, as described by a **horoscope,** determine our future. However, modern tests clearly show that there is no evidence for this, even in a broad statistical sense, and there is no verifiable theory to explain what might cause such an astrological influence.

1.4 Nicolaus Copernicus introduced the **heliocentric cosmology** to Renaissance Europe in his book *De Revolutionibus.* Although he retained the Aristotelian idea of uniform circular motion, Copernicus suggested that the Earth is a planet and that the planets all circle about the Sun, dethroning the Earth from its position at the center of the universe. Galileo Galilei was the father of both modern experimental physics and telescopic astronomy. He studied the **acceleration** of moving objects, and in 1610 began telescopic observations, discovering the nature of the Milky Way, the large-scale features of the Moon, the phases of Venus, and four satellites of Jupiter. Although he was accused of heresy for his support of the heliocentric cosmology, Galileo is credited with observations and brilliant writings that convinced most of his scientific contemporaries of the reality of the Copernican theory.

INTER-ACTIVITY

(This section in each chapter will be devoted to activities that small groups of students can collaborate on, either during class, in a discussion section, or as independent projects. Your instructor may assign some of these or you can use them in a study group to extend your understanding of astronomy.)

A With your group, consider the question with which we began this chapter: How many ways can you think of to prove to a member of the Flat Earth Society that our planet is indeed round?

 B Make a list of ways in which a belief in astrology (the notion that your life path or personality is controlled by the position of the Sun, Moon, and planets at the time of your birth) might be harmful to an individual or to society at large.

 C Members of the group should compare their experiences with the night sky. Have you seen the Milky Way? Can you identify any constellations? Make a list of reasons you think so many fewer people know the night sky today than at the time of the ancient Greeks. What reasons can you think of that a person today may want to be acquainted with the night sky?

 D Constellations commemorate great heroes, dangers, or events in the legends of the people who name them. Suppose we had to start from scratch today, naming the patterns of stars in the sky. What would you choose to commemorate by naming a constellation after it/him/her and why (begin with people from history; then, if you have time, include living people as well)? Can the members of your group agree on any choices?

E Although astronomical mythology no longer holds a powerful sway over the modern imagination, we still find proof of the power of astronomical images in the number of products in the marketplace that have astronomical names. How many can your group come up with? (Think of things like "Milky Way" candy bars, "Saturn" cars, and "Comet" cleanser.)

REVIEW QUESTIONS

1. From where on Earth could you observe all of the stars during the course of a year? What fraction of the sky can be seen from the North Pole?
2. Describe a practical way to determine in which constellation the Sun is found at any time of the year.
3. What is a constellation as astronomers define it today? What does it mean when an astronomer says, "I saw a comet in Orion last night"?
4. Give four ways to demonstrate that the Earth is round.
5. Explain why we see retrograde motion of the planets, according to both geocentric and heliocentric cosmologies.

6. Draw a picture that explains why Venus goes through phases the way the Moon does, according to the heliocentric cosmology. Does Jupiter also go through phases as seen from the Earth? Why?
7. In what ways did the work of Copernicus and Galileo differ from the traditional views of the ancient Greeks and of the Catholic Church?
8. What were five of Galileo's discoveries that were important to science?

THOUGHT QUESTIONS

9. Show with a simple diagram how the lower parts of a ship disappear first as it sails away from you on a spherical Earth. Use the same diagram to show why lookouts on old sailing ships could see farther from the masthead than from the deck. Would there be any advantage to posting lookouts on the mast if the Earth were flat? (Note that these nautical arguments for a spherical Earth were quite familiar to Columbus and other mariners of his time.)

10. Parallaxes of stars were not observed by ancient astronomers. How can this fact be reconciled with the heliocentric hypothesis?

11. Design an experiment to test whether or not the planets and their motions influence human behavior.

12. Why do you think so many people believe in astrology? What psychological needs does such a belief system satisfy?

13. Consider three cosmological perspectives: the geocentric perspective, the heliocentric perspective, and the modern perspective in which the Sun is a minor star on the outskirts of one galaxy among billions. Discuss some of the cultural and philosophical implications of each point of view.

In this section of the book, we will be offering (or reviewing) some basic formulas for doing astronomical calculations and then giving you some examples to figure out on your own.

A circle consists of 360 degrees (°). When we measure the angle in the sky that something moves, we can use the formula: speed = distance covered/time taken. This is true whether the motion is measured in kilometers per hour or degrees per hour; we just need to use consistent units.

14. The Moon moves relative to the background stars. Go outside at night and note the position of the Moon relative to nearby stars. Repeat the observation a few hours later. How far has the Moon moved? (For reference, the diameter of the Moon is about 1/2°.) Based on your estimate of its motion, how long will it take for the Moon to return to the position relative to the stars in which you first observed it?

15. The north celestial pole appears at an altitude above the horizon that is equal to the observer's latitude. Identify Polaris, the North Star, which lies very close to the north celestial pole. Measure its altitude. (This can be done with a protractor. Alternatively, your fist, extended at arm's length, spans a distance approximately equal to 10°.) Compare this estimate with your latitude. (Note that this experiment cannot be easily performed in the Southern Hemisphere because Polaris itself is not visible in the south and no bright star is located near the south celestial pole.)

16. Suppose Eratosthenes had found that in Alexandria at noon on the first day of summer, the line to the Sun makes an angle of 30° with the vertical. What then would he have found for the Earth's circumference?

17. Suppose Eratosthenes' results for the Earth's circumference were quite accurate. If the diameter of the Earth is 12,740 km, what is the length of his stadium in kilometers?

18. A small asteroid makes a "near miss" of the Earth. At its closest, it is seen to move across the sky at a speed (rate) of 1° per minute. Approximately how long will it take to cross the sky, from horizon to horizon?

19. Suppose you are on a strange planet and observe, at night, that the stars do not rise and set but circle parallel to the horizon. Now you walk in a constant direction for 8000 miles, and at your new location on the planet you find that all stars rise straight up in the east and set straight down in the west, perpendicular to the horizon.

a. How could you determine the circumference of the planet without any further observations?

b. What evidence is there that the Greeks could have done what you suggest?

c. What is the circumference, in miles, of that planet?

SUGGESTIONS FOR FURTHER READING

Culver, B. and Ianna, P. *Astrology: True or False.* 1988, Prometheus Books. The best skeptical book about astrology.

Ferris, T. *Coming of Age in the Milky Way.* 1988, Morrow. A general history of our understanding of the organization of the universe.

Fraknoi, A. "Your Astrology Defense Kit" in *Sky & Telescope*, Aug. 1989, p. 146. A review of the tenets and tests of astrology, along with some skeptical questions.

Gingerich, O. "From Aristarchus to Copernicus" in *Sky & Telescope*, Nov. 1983, p. 410.

Gingerich, O. "How Galileo Changed the Rules of Science" in *Sky & Telescope,* Mar. 1993, p. 32.

Gurshtein, A. "In Search of the First Constellations" in *Sky & Telescope,* June 1997, p. 47. This and an earlier article in the Oct. 1995 issue of *Sky & Telescope* present intriguing ideas about the origins of the ancient star groupings.

Krupp, E. *Beyond the Blue Horizon: Myths and Legends of the Sun, Moon, Stars, and Planets.* 1991, HarperCollins. Superb introductory book on the sky stories of ancient cultures.

Krupp, E. *Skywatchers, Shamans, and Kings.* 1997, Wiley. An excellent primer on ancient monuments and astronomical systems, and the terrestrial purposes they served.

Kuhn, T. *The Copernican Revolution.* 1957, Harvard U. Press. Classic study of the changes wrought by Copernicus' work.

Reston, J. *Galileo: A Life.* 1994, HarperCollins. A well-written, popular-level biography by a journalist.

Sagan, C. *Cosmos.* 1980, Random House. The chapter titled "Backbone of Night" focuses on the ancient Greek astronomers.

In February 1984, astronaut Bruce McCandless, having let go of the tether to the Space Shuttle, became the first human satellite of planet Earth. Here he is shown riding the Manned Maneuvering Unit, which had nitrogen gas jets that enabled him to move around in orbit and ultimately return to the shuttle. As he circled our planet, he was subject to the same basic laws of motion that govern all other satellites.

Orbits and Gravity

One had to be a Newton to notice that the Moon is falling, when everyone sees that it doesn't fall.

Paul Valery in "Analects" (Collected Works, Vol. 14, 1966)

THINKING AHEAD

How would you find a new planet that is too dim to be seen with the unaided eye and is so far away that it moves very slowly among the stars? This was the problem confronting astronomers during the 19th century as they tried to pin down a full inventory of our solar system.

If we could look down on the solar system from somewhere out in space, far above the plane of the planets' orbits, interpreting planetary motions would be much simpler. But the fact is, we must observe the positions of all the other planets from our own moving planet. Scientists of the Renaissance did not know the details of the Earth's orbit any better than the orbits of the other planets. Their problem, as we saw in Chapter 1, was that they had to deduce the nature of all planetary motion using only their observations of the other planets' positions in the sky as seen from the Earth. To solve this complex problem more fully, better observations and models of the planetary system were needed.

2.1 THE LAWS OF PLANETARY MOTION

At about the time that Galileo was beginning his experiments with falling bodies, the efforts of two other scientists dramatically advanced our understanding of the motions of the planets. These two astronomers were the observer Tycho Brahe and the mathematician Johannes Kepler. Together, they placed the speculations of Copernicus on a sound mathematical basis and paved the way for the work of Isaac Newton in the next century.

2.1.1 Tycho Brahe's Observatory

Three years after the publication of Copernicus' *De Revolutionibus*, Tycho Brahe (1546–1601) was born to a family of Danish nobility. He developed an early interest in astronomy and as a young man made significant astronomical observations. Among these was a careful study of what we now know was an exploding star that flared up to great brilliance in the night sky. His growing reputation gained him the patronage of the Danish King Frederick II, and at the age of 30 Tycho was able to establish a fine astronomical observatory on the North Sea island of Hven (Figure 2.1). Tycho was the last and greatest of the pre-telescopic observers in Europe.

At Hven, Tycho made a continuous record of the positions of the Sun, Moon, and planets for almost 20 years. His extensive and precise observations enabled him to note that the positions of the planets varied from those given in published tables, which were based on the work of Ptolemy. But Tycho was an extravagant and cantankerous fellow, and he accumulated enemies among government officials. When his patron, Frederick II, died in 1597, Tycho lost his political base and decided to leave Denmark. He took up residence in Prague, where he became court astronomer to the Emperor Rudolf of Bohemia. There, in the year before his death, Tycho found a most able young mathematician, Johannes Kepler, to assist him in analyzing his extensive planetary data.

2.1.2 Kepler

Kepler (1571–1630) (Figure 2.2) was born into a poor family in the German province of Württemberg and lived much of his life amid the turmoil of the Thirty Years' War. He attended university at Tübingen and studied for a theological career. There he learned the principles of the Copernican system and became converted to the heliocentric hy-

Granger Collection, New York

■ **FIGURE 2.1**

Tycho at Hven   A stylized engraving showing Tycho Brahe using his instruments to measure the altitude of a celestial object above the horizon. Note that the scene includes hints of the grandeur of Tycho's observatory.

AIP/Niels Bohr Library

■ **FIGURE 2.2**

Kepler   Johannes Kepler (1571–1630) was a German mathematician and astronomer. His discovery of the basic laws that describe planetary motion placed the heliocentric cosmology of Copernicus on a firm mathematical basis.

pothesis. Eventually, Kepler went to Prague to serve as an assistant to Tycho, who set him to work trying to find a satisfactory theory of planetary motion—one that was compatible with the long series of observations made at Hven.

Tycho was reluctant to provide Kepler with much material at any one time, for fear that Kepler would discover the secrets of the universal motions by himself, thereby robbing Tycho of some of the glory. Only after Tycho's death in 1601 did Kepler get full possession of the priceless records. Their study occupied most of Kepler's time for more than 20 years.

Kepler's most detailed study was of Mars, for which the observational data were the most extensive. In 1609 he published the first results in *The New Astronomy*, where we find his first two laws of planetary motion. Their discovery was a profound step in the development of modern science.

2.1.3 The Orbit of Mars

Kepler began his research under the assumption that the orbits of planets were circles, but the observations contradicted this idea. Working with data for Mars, he eventually discovered that the orbit of that planet had the shape of a somewhat flattened circle, or **ellipse.** Next to the circle, the ellipse is the simplest kind of closed curve, belonging to a family of curves known as *conic sections* (Figure 2.3).

You might recall from math classes that in a circle, the distance from a special point called the center to anywhere on the circle is exactly the same. In an ellipse, the *sum* of the

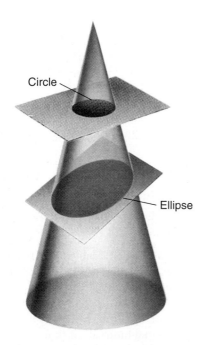

FIGURE 2.3

Conic Sections The circle and the ellipse are both formed by the intersection of a plane with a cone. This is why both curves are called conic sections.

FIGURE 2.4

Drawing an Ellipse We can construct an ellipse with two tacks and a string, where the tacks are the foci of the ellipse. Note that the length of the string remains the same, so that the sum of the distances from any point on the ellipse to the foci is always constant.

distances from two special points inside the ellipse to any point on the ellipse is always the same. These two points inside the ellipse are called its **foci** (singular: **focus**), a word invented for this purpose by Kepler.

This property suggests a simple way to draw an ellipse (Figure 2.4). We wrap the ends of a loop of string around two tacks pushed through a sheet of paper into a drawing board, so that the string is slack. If we push a pencil against the string, making the string taut, and then slide the pencil against the string all around the tacks, the curve that results is an ellipse. At any point where the pencil may be, the sum of the distances from the pencil to the two tacks is a constant length—the length of the string. The tacks are at the two foci of the ellipse.

The widest diameter of the ellipse is called its **major axis.** Half this distance—that is, the distance from the center of the ellipse to one end—is the **semimajor axis,** which is usually used to specify the size of the ellipse. For example, the semimajor axis of the orbit of Mars, which also turns out to be the planet's average distance from the Sun, is 228 million kilometers (km).

The shape (roundness) of an ellipse depends on how close together the two foci are, compared with the major axis. The ratio of the distance between the foci to the length of the major axis is called the **eccentricity** of the ellipse.

If the foci (or tacks) are in the same place, the eccentricity is zero and the ellipse is just a circle; thus, a circle is an ellipse of zero eccentricity. In a circle, the semimajor axis would be the radius. We can make ellipses of various shapes by varying the spacing of the tacks (as long as they are not farther apart than the length of the string). The greater the eccentricity, the more elongated is the ellipse, up to a maximum eccentricity of 1.0.

The size and shape of an ellipse are completely specified by its semimajor axis and its eccentricity. Kepler found that Mars has an elliptical orbit, with the Sun at one focus (the other focus is empty). The eccentricity of the orbit of Mars is only about 0.1; its orbit, drawn to scale, would be practically indistinguishable from a circle. Yet the difference is critical for understanding planetary motions.

Kepler generalized this result in his first law and said that *the orbits of all the planets are ellipses.* Here was a decisive moment in the history of human thought: It was not necessary to have only circles in order to have an acceptable cosmos. The universe could be a bit more complex than the Greek philosophers had wanted it to be.

Kepler's second law deals with the speed with which each planet moves along the ellipse. Working with Tycho's observations of Mars, Kepler discovered that the planet speeds up as it comes closer to the Sun and slows down as it pulls away from the Sun. He expressed the precise form of this relationship by imagining that the Sun and Mars are connected by a straight, elastic line (Figure 2.5).When Mars is closer to the Sun (positions 1 and 2 in Figure 2.5), the elastic line is not stretched as much, and the planet moves rapidly. Farther from the Sun, as in positions 3 and 4, the line is stretched a lot, and the planet does not move so fast. As Mars travels in its elliptical orbit around the Sun, the elastic line sweeps out areas of the ellipse (the colored regions in our figure). Kepler found that in equal intervals of time, the *areas* swept out in space by this imaginary line are always equal; that is, the area of the region from 1 to 2 is the same as that from 3 to 4.

This is also a general property of the orbits of all planets. A planet in a circular orbit always moves at a constant speed, but in an elliptical orbit the planet's speed varies considerably, following the rule Kepler found.

2.1.4 Laws of Planetary Motion

Kepler's first two laws of planetary motion describe the shape of a planet's orbit and allow us to calculate the speed of its motion at any point in the orbit. Kepler was pleased to have discovered such fundamental rules, but they did not satisfy his quest to understand planetary motions. He wanted to know why the orbits of the planets were spaced as they are, and to find a mathematical pattern in their orbital periods—a "harmony of the spheres," as he called it. For many years he worked to discover mathematical relationships governing planetary spacings and the time each planet took to go around the Sun.

In 1619, Kepler succeeded in finding a basic relationship that links the semimajor axes of the planets' orbits and their periods of revolution. The relationship is now known as Kepler's third law. It applies to all of the planets, including the Earth, and it provides a means for calculating their relative distances from the Sun.

For the solar system, Kepler's third law takes its simplest form when the period is expressed in years (the revo-

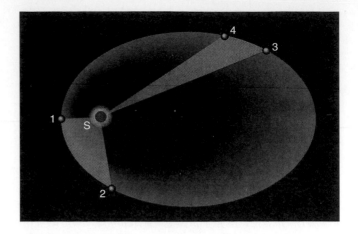

■ **FIGURE 2.5**

Kepler's Second Law: The Law of Equal Areas
A planet moves most rapidly on its elliptical orbit when at position 1, nearest the Sun, which is at one focus of the ellipse. The orbital speed of the planet varies in such a way that in equal intervals of time a line between the Sun and the planet sweeps out equal areas. Thus, the area swept out from 1 to 2 is the same as that from 3 to 4. Note that the eccentricities of the planets' orbits are substantially less than shown here.

lution period of the Earth) and the semimajor axis of the orbit is expressed in terms of the Earth's average distance from the Sun, called the **astronomical unit (AU).** One astronomical unit is equal to 1.5×10^8 km. With these units, Kepler's third law is

$$(\text{distance})^3 = (\text{period})^2$$

For example, this relationship tells us how to calculate Mars' average distance from the Sun (the semimajor axis of its orbit) from its period of 1.88 years. The period squared is $1.88 \times 1.88 = 3.53$, and this equals the cube of the semimajor axis. So what number must be cubed to give 3.53? The answer is 1.52 ($1.52 \times 1.52 \times 1.52 = 3.53$). Thus, the planet's semimajor axis in astronomical units must be 1.52. In other words, to go around the Sun in a little less than 2 years, Mars must be about 50 percent farther from the Sun than Earth is.

Kepler's three laws of planetary motion can be summarized as follows:

Kepler's First Law Each planet moves about the Sun in an orbit that is an ellipse, with the Sun at one focus of the ellipse.

Kepler's Second Law The straight line joining a planet and the Sun sweeps out equal areas in space in equal intervals of time.

Kepler's Third Law The squares of the planets' periods of revolution are in direct proportion to the cubes of the semimajor axes of their orbits.

These three laws provided a precise geometric description of planetary motion within the framework of the

Copernican system. With these tools, it was possible to calculate planetary positions with undreamed-of precision. But Kepler's laws are purely descriptive; they do not help us understand what forces of nature constrain the planets to follow this particular set of rules. That step was left to Newton.

NEWTON'S GREAT SYNTHESIS

It was the genius of Isaac Newton (1643–1727) that found a conceptual framework that completely explained the observations and rules assembled by Galileo, Tycho, Kepler, and others. Newton was born in Lincolnshire, England, in the year after Galileo's death (Figure 2.6). Against the advice of his mother, who wanted him to stay home and help with the family farm, he entered Trinity College at Cambridge in 1661 and eight years later was appointed Professor of Mathematics there. Among Newton's contemporaries in England were architect Christopher Wren, authors Samuel Pepys and Daniel Defoe, and composer G. F. Handel.

2.2.1 Newton's Laws of Motion

As a young man in college, Newton became interested in natural philosophy, as science was then called. He worked out some of his first ideas on mechanics and optics during the plague years of 1665 and 1666, when students were sent home from college. Newton, a moody and often difficult man, continued to work on his ideas in private, even inventing new mathematical tools to help him deal with the complexities involved. Eventually, his friend Edmund Halley (profiled in Chapter 12) prevailed on him to collect and publish the results of his remarkable investigations on motion and gravity. The result was a volume that set out the underlying system of the physical world, *Philosophiae Naturalis Principia Mathematica.* The *Principia,* as the book is generally known, was published at Halley's expense in 1687.

At the very beginning of the *Principia,* Newton proposes three laws that he presumes to govern the motions of all objects:

Newton's First Law Every body continues doing what it is already doing—being in a state of rest, or moving uniformly in a straight line—unless it is compelled to change by an outside force.

Newton's Second Law The change of motion of a body is proportional to the force acting on it, and is made in the direction in which that force is acting.

Newton's Third Law To every action there is an equal and opposite reaction (*or* the mutual actions of two bodies upon each other are always equal and act in opposite directions).

In the original Latin, the three laws contain only 59 words, but those few words set the stage for modern science. Let us examine them more carefully.

■ FIGURE 2.6

Newton Isaac Newton (1643–1727), whose work on the laws of motion, gravity, optics, and mathematics laid the foundations of much of physical science.

2.2.2 Interpretation of Newton's Laws

Newton's first law is a restatement of one of Galileo's discoveries, called the **conservation of momentum—** where momentum is a measure of a body's motion. The law states that in the absence of any outside influence, a body's momentum remains unchanged. We use the word *momentum* in everyday expressions as well, as in "This bill in Congress has a lot of momentum: It seems unstoppable." In other words, a stationary object stays put, and a moving object keeps moving unless some force intervenes. Momentum depends on three factors. The first is *speed*—how fast a body moves (zero if it is stationary). The second is the direction in which the body is moving. Scientists use the term *velocity* to describe both speed and direction. For example, 20 kilometers per hour (km/h) due south is velocity; 20 km/h is speed. The third factor in momentum is what Newton called *mass.* Mass is a measure of the amount of matter in a body, as we will discuss later. Momentum then is mass × velocity.

As mentioned in Chapter 1, it is not so easy to see this rule in action in the everyday world. On Earth, objects in motion do not remain in motion because outside forces are always acting. One important force is friction, which slows things down. If you roll a ball down the sidewalk, it eventually comes to a stop because the rubbing of the ball against the sidewalk exerts a force. But out in space between the stars, where friction is negligible, objects could in fact continue to move (coast) indefinitely. Newton's first law is sometimes called the *law of inertia,* inertia being the tendency of objects (and legislatures) to keep doing what they are already doing.

The momentum of a body can change only under the action of an outside influence. Newton's second law defines *force* in terms of its ability to *change momentum*. A force (a push or a pull) has both size and direction. When force is applied to a body, the momentum changes in the direction of the applied force. This means that a force is required to change either the speed or the direction of a body, or both—that is, to start it moving, to speed it up, to slow it down, to stop it, or to change its direction.

The rate of change in an object's velocity is called **acceleration.** Newton showed that the acceleration of a body was proportional to the force being applied to it. Suppose, after a hard exam, you push your astronomy textbook away from you in disgust on a long, smooth table. The harder you keep pushing the book, the more it will speed up. How much a force will accelerate an object is also determined by the object's mass. If you kept pushing a pen with the same force with which you pushed the textbook, the pen—having less mass—would be accelerated to a greater speed. (Note that we said above that the table was smooth because we wanted to minimize the effect of friction. You'd have even better luck accelerating the textbook on ice or out in space.)

Newton's third law is the most profound. Basically, it is a generalization of the first law, but it also gives us a way to define mass. If we consider a system of two or more objects isolated from outside influences, Newton's first law says that the total momentum of the system of objects should remain constant. Therefore, any change of momentum within the system must be balanced by another change that is equal and opposite to it, so that the momentum of the entire system is not changed.

This means that forces in nature do not occur alone: We find that in each situation there is always a pair of forces that are equal to and opposite each other. If a force is exerted on an object, it must be exerted by something else, and the object will exert an equal and opposite force back on that something.

Suppose that during that same difficult exam, another student runs screaming from the room and jumps out a (not very high) window. The force pulling him down after he jumps (as we will see in the next section) is the gravitational force between him and the Earth. Both he and the Earth must experience the same total change of momentum because of the influence of this mutual force. So, both the student and the Earth are accelerated by each other's pull. However, the student does much more of the moving! Because the Earth has enormously greater mass, it can experience the same change of momentum by accelerating only a very small amount. Things fall toward the Earth all the time, but the acceleration of our planet as a result is far too small to be noticed or measured.

A more obvious example of the mutual nature of forces between objects is familiar to all who have batted a baseball. The recoil you feel as you swing your bat shows that the ball exerts a force on it during the impact, just as the bat does on the ball. Similarly, when a rifle is discharged, the force pushing the bullet out of the muzzle is equal to the

■ **FIGURE 2.7**

Demonstrating Newton's Third Law The U.S. Space Shuttle at launch, powered by three liquid fuel engines burning liquid oxygen and liquid hydrogen, with two solid fuel boosters.

force pushing backward upon the gun and the shoulder of the person shooting it.

This is the principle behind jet engines and rockets: The force that discharges the exhaust gases from the rear of the rocket is accompanied by force that shoves the rocket forward. The exhaust gases need not push against air or the Earth; a rocket actually operates best in a vacuum (Figure 2.7).

2.2.3 Mass, Volume, and Density

Before we go on to discuss Newton's other work, we want to take a brief look at some terms that will be important to sort out clearly. We begin with *mass*, which is a measure of the amount of material in an object.

Volume is a measure of the physical space occupied by a body—say, in cubic centimeters or liters. In short, the volume is the "size" of an object; it has nothing to do with its mass. A penny and an inflated balloon may both have the same mass, but they have very different volumes.

The penny and the balloon are also very different in **density,** which is a measure of how much mass we have per unit volume. Specifically, density is the ratio of mass to volume. Note that often in everyday language we use "heavy"

TABLE 2.1 *Densities of Materials*	
Material	**Density (g/cm^3)**
Gold	19.3
Lead	11.4
Iron	7.9
Earth (bulk)	5.6
Rock (typical)	2.5
Water	1.0
Wood (typical)	0.8
Insulating foam	0.1
Silica gel	0.02

■ **FIGURE 2.8**

The Conservation of Angular Momentum When a spinning figure skater brings her arms in, their distance from the spin center is small, so her speed increases. When her arms are out, their distance from the spin center is greater, so she slows down.

and "light" as indications of density (rather than weight), as, for instance, when we say that iron is heavy or that puff pastry is light.

The units of density that will be used in this book are grams per cubic centimeter (g/cm^3).[1] If a block of some material has a mass of 300 g and a volume of 100 cm^3, its density is 3 g/cm^3. Familiar materials span a considerable range in density, from artificial materials such as plastic insulating foam (less than 0.1 g/cm^3) to gold (19.3 g/cm^3) (Table 2.1). In the astronomical universe, much more remarkable densities can be found, all the way from a comet's tail (10^{-16} g/cm^3) to a neutron star (10^{15} g/cm^3).

To sum up, then, mass is "how much," volume is "how big," and density is "how tightly packed."

2.2.4 Angular Momentum

The concept of **angular momentum** is a bit more complex, but it is important for understanding many astronomical objects. Angular momentum is a measure of the momentum of an object as it rotates or revolves about some fixed point. Whenever we deal with the revolution of spinning objects, from planets to galaxies, we have to consider angular momentum. The angular momentum of an object is defined as the product of three quantities: its mass, its velocity, and its distance from the fixed point around which it turns.

If these three quantities remain constant—that is, if the motion of a particular object takes place at a constant speed and at a fixed distance from the spin center—then the angular momentum is also a constant. So, ignoring air resistance, if you tie a string to your astronomy textbook and twirl it around your head at constant speed, you will have a system with constant angular momentum.

More generally, angular momentum is constant, or conserved, in any rotating system in which no external forces act, or in which the only forces are directed toward or away from the spin center. An example of such a system is a planet orbiting the Sun. Kepler's second law is an example of the conservation of angular momentum. When a planet approaches the Sun on its elliptical orbit, the distance to the spin center decreases; the planet speeds up to keep the angular momentum the same. Similarly, when the planet is farther from the Sun, it revolves more slowly.

The concept is illustrated by figure skaters, who bring their arms and legs in to spin more rapidly, and extend their arms and legs to slow down (Figure 2.8). (You can duplicate this yourself on a well-oiled piano stool, by starting yourself spinning slowly with your arms extended and then pulling your arms in.) And just as a planet speeds up when it approaches the Sun, a shrinking cloud of dust or a star collapsing on itself (both situations you will encounter as you read on) increases its spin rate as it contracts. There is less distance to the spin center, so the speed increases to keep the angular momentum the same.

[1] Generally we use the standard metric (or SI) units in this book. The proper metric unit of density is kg/m^3. But to most people g/cm^3 provides a more meaningful unit because the density of water is exactly 1 g/cm^3. Density expressed in g/cm^3 is sometimes called specific density or specific weight.

 UNIVERSAL GRAVITY

2.3.1 The Law of Gravity

Newton's laws of motion show that, left to themselves, objects at rest stay at rest, and those in motion continue moving uniformly in a straight line. Thus, it is the *straight line,* not the *circle,* that defines the most natural state of motion. But the planets move in ellipses, not straight lines. Some force must be bending their paths. That force, Newton proposed, is **gravity.**

In Newton's time gravity was something associated with the Earth alone. Everyday experience shows us that the Earth exerts a gravitational force upon objects at its surface. If you drop something off a leaning tower in Pisa, it falls toward the Earth, accelerating as it falls. Newton's insight was that the Earth's gravity might extend as far as the Moon and produce the acceleration required to curve the Moon's path from a straight line and keep it in its orbit. He further hypothesized that gravity is not limited to the Earth, but that there is a general force of attraction between all material bodies. If so, the attractive force between the Sun and each of the planets could keep each in its orbit.

Once Newton boldly hypothesized that there is a *universal attraction* among all bodies everywhere in space, he had to determine the exact nature of the attraction. The precise mathematical description of that gravitational force had to dictate that the planets move exactly as Kepler had observed them to (as expressed in Kepler's three laws). Also, the law of gravity had to predict the correct behavior of falling bodies on the Earth, as observed by Galileo. How must gravitational force depend on distance in order for these conditions to be met?

The answer to this question required mathematical tools that had not yet been developed. But this did not deter Isaac Newton, who invented what we today call *calculus* to deal with this problem. Eventually he was able to conclude that the force of gravity must drop off with increasing distance between the Sun and a planet (or between any two objects) in proportion to the *inverse square* of their separation. In other words, if a planet were twice as far from the Sun, the force would be $1/2^2$ or 1/4 as large. Put the planet three times farther away, and the force is $1/3^2$ or 1/9 as strong.

Newton also concluded that the gravitational attraction between two bodies must be proportional to their masses. The more mass an object has, the stronger the pull of its gravity. The gravitational attraction between any two objects is given by one of the most famous formulas in all of science:

$$\text{force} = GM_1M_2/R^2$$

where M_1 and M_2 are the masses of the two objects and R is their separation. The number represented by G is called the constant of gravitation. With such a force and the laws of motion, Newton was able to show mathematically that the only orbits permitted were exactly those described by Kepler's laws.

Newton's law of gravity works for the planets, but is it really universal? The gravitational theory should also predict the observed acceleration of the Moon toward the Earth, falling around the Earth at a distance of about 60 times the radius of the Earth, as well as of an object (say, an apple) dropped near the Earth's surface. The falling of an apple is something we can measure quite easily; can we use it to predict the motions of the Moon?

Newton's theory says that the force on (and therefore the acceleration of) an object toward the Earth should be inversely proportional to the square of its distance from the center of the Earth. Objects like apples at the surface of the Earth (R = 1 Earth-radius, the distance from its center) are observed to accelerate downward at 9.8 meters per second per second (9.8 m/s^2).

Therefore, if the law holds, the Moon, 60 Earth-radii from its center, should experience an acceleration toward the Earth that is $1/60^2$, or 3600 times less—that is, about 0.00272 m/s^2. This is precisely the observed acceleration of the Moon in its orbit. What a triumph! Imagine the thrill Newton must have felt to realize he had discovered and verified a law that holds for the Earth, apples, the Moon, and, as far as we know, everything in the universe!

Gravity is a "built-in" property of mass. Wherever masses occur, they will interact via the force of gravity. The more mass there is, the greater the force it can exert. Here on Earth, the largest concentration of mass is, of course, the planet we stand on, and its pull dominates the gravitational interactions we experience. But everything with mass attracts everything else with mass anywhere in the universe.

For example, you and this textbook (each having mass) attract each other via gravity. (We certainly hope it's not the *only* way you find our book attractive!) If you let the book fall, however, it is pulled much more strongly by the Earth, so it falls to the Earth, not toward you. But if you and the textbook somehow found yourselves out in space, far from any stars or planets, you would discover that gravitational attraction would start to pull you together.

Newton's law also suggests that gravity never becomes zero. It quickly gets weaker with distance, but it continues to act to some degree no matter how far away you get. The pull of the Sun is stronger at Mercury than at Pluto, but it can be felt far beyond Pluto, where (as we shall see) we have good evidence that it continues to make enormous numbers of smaller bodies move around huge orbits. And its pull joins with the pull of billions of other stars to create the gravitational pull of the Milky Way Galaxy in which we live. That force, in turn, can make other smaller galaxies orbit around the Milky Way, and so forth.

Why is it then, you may ask, that the astronauts aboard the Space Shuttle appear "weightless" and we see images on television of people and objects floating in the spacecraft? After all, the astronauts in the shuttle are only a few hundred kilometers above the surface of the Earth, not a significant distance compared to the size of the Earth. Gravity is certainly not a great deal weaker that much farther away. The astronauts feel weightless for the same reason that passengers in an elevator whose cable has broken or in an airplane whose engines no longer work feel weightless[2]

---

[2] In the film *Apollo 13,* the scenes where the astronauts were "weightless" were actually filmed in a falling airplane. As you might imagine, the plane fell for only short periods before the engines cut in again.

■ FIGURE 2.9

Astronaut in Free Fall Aboard the Space Shuttle *Columbia* in July 1997, astronaut Susan Still floats before the Spacelab control center. Both she and the shuttle are falling freely around the Earth, and so she experiences no force relative to the shuttle (just as people in a freely falling elevator would feel weightless for the brief time they had left to live).

(Figure 2.9). They are *falling,* and in free fall, they accelerate at the same rate as everything around them, including their spacecraft or a camera with which they are taking photographs of Earth. Thus, *relative* to the shuttle, the astronauts experience no additional force and so feel "weightless." Unlike the falling elevator passengers, however, the astronauts are falling *around* the Earth, not *to* the Earth (and will continue falling).

2.3.2 Orbital Motion and Mass

Kepler's laws are descriptions of the orbits of objects moving according to Newton's laws of motion and the law of gravity. Knowing that gravity is the force that attracts planets toward the Sun, however, allowed Newton to rethink Kepler's third law. Recall that Kepler had found a relationship between the period of a planet's revolution and its distance from the Sun. Newton's law of gravity can be used to show mathematically that this relationship is actually

$$D^3 = (M_1 + M_2) \times P^2$$

As explained in Section 2.1.4, we express distances in the solar system in astronomical units and periods in years. But Newton's formulation introduces the additional factor of the masses of the Sun (M_1) and planet (M_2), both expressed in units of the Sun's mass.

How did Kepler miss this factor? In units of the Sun's mass, the mass of the Sun is 1; in units of the Sun's mass, the mass of a typical planet is a negligibly small fraction, and so ($M_1 + M_2$) is very, very close to 1. This makes Newton's formula look almost the same as Kepler's. The tiny mass of the planets is the reason that Kepler did not realize that both masses had to be included in the calculation. There are many cases in astronomy, however, in which we do need to include the two mass terms—for example, when two stars or two galaxies orbit each other.

Including the mass term allows us to use this formula in a new way. If we can measure the motions (distances and periods) of objects acting under their mutual gravity, then the formula will permit us to deduce their masses. For example, we can calculate the mass of the Sun by using the distances and periods of the planets, or the mass of Jupiter by noting the motions of its moons. Indeed, Newton's reformulation of Kepler's third law is one of the most powerful concepts in astronomy. Our ability to deduce the masses of objects from their motions is key to understanding the nature and evolution of many astronomical bodies. We will use this law repeatedly throughout this text in calculations such as the orbits of comets.

2.4 ORBITS IN THE SOLAR SYSTEM

2.4.1 Description of an Orbit

The path of an object through space is called its **orbit,** whether that object is a spacecraft, planet, star, or galaxy. An orbit, once determined, allows the future positions of the object to be calculated.

Two points in any orbit have been given special names. The place where the planet is closest to the Sun (*helios* in Greek) is called the **perihelion** of its orbit, and the place where it is farthest away and moves the most slowly is the **aphelion.** For a satellite orbiting the Earth (*geos* in Greek), the corresponding terms are **perigee** and **apogee.**

2.4.2 Orbits of the Planets

Today, Newton's work enables us to calculate and predict the orbits of the planets with marvelous precision. We

Astronomy and the Poets

When Copernicus, Kepler, Galileo, and Newton formulated the fundamental rules that underlie everything in the physical world, they changed much more than the face of science. For some, they gave humanity the courage to let go of old superstitions and see the world as rational and manageable; for others, they upset comforting, ordered ways that had served humanity for centuries, leaving only a dry, mechanical clockwork universe in their wake.

Poets of the time reacted to such changes in their own work and debated whether the new world picture was an appealing or frightening one. John Donne (1573–1631), in a poem called *Anatomy of the World,* laments the passing of the old certainties:

> The new philosophy [science] calls all in doubt,
> The element of fire is quite put out;
> The Sun is lost, and th' earth, and no man's wit
> Can well direct him where to look for it.

(Here the "element of fire" refers also to the sphere of fire, which medieval thought placed between the Earth and the Moon.)

By the next century, poets like Alexander Pope were celebrating Newton and the Newtonian world view. Pope's famous couplet, written upon Newton's death, goes:

> Nature, and nature's laws lay hid in night.
> God said, Let Newton be! and all was light.

In his 1733 poem, *An Essay on Man,* Pope revels in the complexity of the new views of the world, incomplete though they are:

> Of man, what see we, but his station here,
> From which to reason, to which refer? . . .
> He, who thro' vast immensity can pierce,
> See worlds on worlds compose one universe,
> Observe how system into system runs,
> What other planets circle other suns,
> What vary'd being peoples every star,
> May tell why Heav'n has made us as we are . . .
> All nature is but art, unknown to thee;
> All chance, direction, which thou canst not see;
> All discord, harmony not understood;
> All partial evil, universal good:
> And, in spite of pride, in erring reason's spite,
> One truth is clear, whatever is, is right.

Poets and philosophers continued to debate whether humanity was exalted or debased by the new views of science. The 19th-century poet Arthur Hugh Clough (1819–1861) cries out in his poem *The New Sinai:*

> And as of old from Sinai's top God said that God is one,
> By science strict so speaks He now to tell us, there is None!
> Earth goes by chemic forces; Heaven's a Mécanique Celeste!
> And heart and mind of humankind a watchwork as the rest!

(A "mécanique celeste" is a clockwork model to demonstrate celestial motions.)

The 20th-century poet Robinson Jeffers (whose brother was an astronomer) saw it differently in a poem called *Star Swirls:*

> There is nothing like astronomy to pull the stuff out of man.
> His stupid dreams and red-rooster importance:
> Let him count the star-swirls.

---

know nine planets, beginning with Mercury closest to the Sun and extending outward to Pluto. The average orbital data for the planets are summarized in Table 2.2. (Ceres is the largest of the *asteroids;* see the next section.)

According to Kepler's laws, Mercury must have the shortest period of revolution (88 Earth-days); thus it has the highest orbital speed (averaging 48 km/s). At the opposite extreme, Pluto has a period of 249 years and an average orbital speed of just 5 km/s.

All of the planets have orbits of rather low eccentricity. The most eccentric orbits are those of Mercury (0.21) and Pluto (0.25); all of the rest have eccentricities less than 0.1. It is fortunate for the development of science that Mars has an eccentricity greater than many of the other planets. Otherwise the pre-telescopic observations of Tycho would not have been sufficient for Kepler to deduce that its orbit had the shape of an ellipse rather than a circle.

TABLE 2.2 *Orbital Data for the Planets*

Planet	Semimajor Axis (AU)	Period (yr)	Eccentricity
Mercury	0.39	0.24	0.21
Venus	0.72	0.62	0.01
Earth	1.00	1.00	0.02
Mars	1.52	1.88	0.09
(Ceres)	2.77	4.60	0.08
Jupiter	5.20	11.86	0.05
Saturn	9.54	29.46	0.06
Uranus	19.19	84.07	0.05
Neptune	30.06	164.80	0.01
Pluto	39.60	248.60	0.25

The planetary orbits are also confined close to a common plane, which is near the plane of the Earth's orbit (the ecliptic). The strange orbit of Pluto is inclined about 17° to the average, but all of the other planets lie within 10° of the common plane of the solar system.

2.4.3 Orbits of Asteroids and Comets

In addition to the nine planets, there are many smaller objects in the solar system. Some of these are natural satellites that orbit all of the planets except Mercury and Venus. In addition, there are two classes of smaller objects in heliocentric orbits: the *asteroids* and the *comets.* Both asteroids and comets are believed to be small chunks of material left over from the formation process of the solar system.

In general, the asteroids have orbits with smaller semimajor axes than do the comets (Figure 2.10). The great majority of them lie between 2.2 and 3.3 AU, in the region known as the **asteroid belt.** As you can see in Table 2.2, the asteroid belt (represented by its largest member, Ceres) is in the middle of a gap between the orbits of Mars and Jupiter. It is because these two planets are so far apart that stable orbits of small bodies can exist in the region between them.

Comets generally have orbits of larger size and greater eccentricity than those of the asteroids. Typically, the eccentricities of their orbits are 0.8 or higher. According to Kepler's second law, therefore, they spend most of their

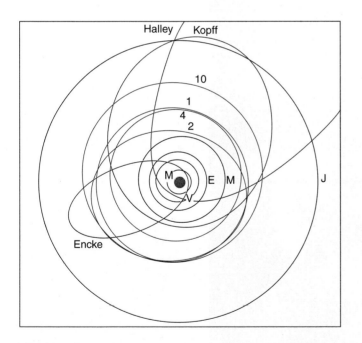

■ **FIGURE 2.10**

Solar System Orbits We see the orbits of typical comets and asteroids compared with those of the planets Mercury, Venus, Earth, Mars, and Jupiter (black circles). Shown in red are three comets: Halley, Kopff, and Encke. In blue are the four largest asteroids: 1 Ceres, 2 Pallas, 4 Vesta, and 10 Hygeia.

time far from the Sun, moving very slowly. As they approach perihelion, the comets speed up and whip through the inner parts of their orbits more rapidly.

2.5 MOTIONS OF SATELLITES AND SPACECRAFT

2.5.1 Space Flight and Satellite Orbits

The law of gravity and Kepler's laws describe the motions of Earth satellites and interplanetary spacecraft as well as the planets. Sputnik, the first artificial Earth satellite, was launched by what was then called the Soviet Union on October 4, 1957. Since that time, thousands of satellites have been placed into orbit around the Earth, and spacecraft have also orbited the Moon, Venus, Mars, Jupiter, and the asteroid Eros.

Once an artificial satellite is in orbit, its behavior is no different from that of a natural satellite, such as our Moon. If the satellite is high enough to be free of atmospheric friction, it will remain in orbit forever, following Kepler's laws in a perfectly respectable way. However, although there is no difficulty in maintaining a satellite once it is in orbit, a great deal of energy is required to lift the spacecraft off the Earth and accelerate it to orbital speed.

To illustrate how a satellite is launched, imagine a gun firing a bullet horizontally from the top of a high mountain (Figure 2.11a—adapted from a similar diagram by Newton, shown in Figure 2.11b). Imagine, further, that the friction of the air could be removed and that nothing can get in the bullet's way. Then the only force that acts on the bullet after it leaves the muzzle is the gravitational force between the bullet and Earth.

If the bullet is fired with some velocity we can call v_a, it continues to have that forward speed, but meanwhile the gravitational force acting upon it pulls it downward toward the Earth, where it strikes the ground at point *a*. However, if it is given a higher muzzle velocity, v_b, its higher forward speed carries it farther before it hits the ground. This is because, regardless of its forward speed, the downward gravitational force is the same. Thus, this faster-moving bullet strikes the ground at point *b*.

If our bullet is given a high enough muzzle velocity, v_c, the curved surface of the Earth causes the ground to tip out from under it so that it remains the same distance above the ground and falls around the Earth in a complete circle. The speed needed to do this—called the **circular satellite velocity**—is about 8 km/s, or about 17,500 miles per hour (mph) in more familiar units.

Each year more than 50 new satellites are launched into orbit by such nations as Russia, the United States, China, Japan, India, and Israel, as well as by the European Space Agency (ESA), a consortium of European nations (Figure 2.12). Most satellites are launched into low Earth orbit,

(a)

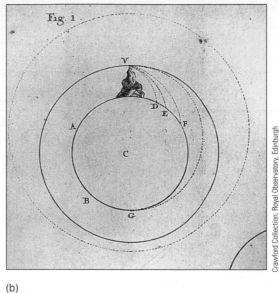

(b)

Crawford Collection, Royal Observatory, Edinburgh

■ **FIGURE 2.11**
Firing a Bullet into Orbit
(a) For paths *a* and *b,* the velocity is not enough to prevent gravity from pulling the bullet back to Earth; in case *c,* the velocity allows the bullet to fall completely around the Earth. (b) A diagram by Newton in his *De Mundi Systematic,* 1731 edition, illustrating the same concept shown in panel (a).

since this requires the minimum launch energy. At the orbital speed of 8 km/s, they circle the planet in about 90 minutes. Low Earth orbits are not stable indefinitely because the drag generated by friction with the thin upper atmosphere eventually leads to a loss of energy and "decay" of the orbit.

2.5.2 Interplanetary Spacecraft

The exploration of the solar system has been carried out largely by robot spacecraft sent to the other planets. To escape Earth, these craft must achieve **escape velocity,** the

U.S. Space Command, NORAD

■ **FIGURE 2.12**
Satellites in Earth Orbit
A plot of all known unclassified satellites and satellite debris in Earth orbit larger than about the size of a softball.

speed needed to move away from the Earth forever, which is about 11 km/s (about 25,000 mph). After this they coast to their targets, subject only to minor trajectory adjustments provided by small thruster rockets on board. In interplanetary flight, these spacecraft follow Keplerian orbits around the Sun, modified only when they pass near one of the planets.

As it comes close to its target, a spacecraft is deflected by the planet's gravitational force into a modified orbit, either gaining or losing energy in the process. By carefully choosing the aim point in a planetary encounter, controllers have actually been able to use a planet's gravity to redirect a flyby spacecraft to a second target. Voyager 2 used a series of gravity-assisted encounters to yield successive flybys of Jupiter (1979), Saturn (1980), Uranus (1986), and Neptune (1989). The Galileo spacecraft, launched in 1989, flew past Venus once and Earth twice to gain the energy required to reach its ultimate goal of orbiting Jupiter.

If we wish to orbit a planet, we must slow the spacecraft with a rocket when the spacecraft is near its destination, allowing it to be captured into an elliptical orbit. Additional rocket thrust is required to bring a vehicle down from orbit for a landing on the surface. Finally, if a return trip to Earth is planned, the landed payload must include enough propulsive power to repeat the entire process in reverse.

■ FIGURE 2.13

Modern Computing Power   Supercomputers at NASA Ames Research Center are capable of tracking the motions of more than a million objects under their mutual gravitation.

GRAVITY WITH MORE THAN TWO BODIES

Until now we have considered the Sun and a planet (or a planet and one of its satellites) as nothing more than a pair of bodies revolving around each other. In fact, all the planets exert gravitational forces upon one another as well. These interplanetary attractions cause slight variations from the orbits that would be expected if the gravitational forces between planets were neglected. Unfortunately, the problem of treating the motion of a body that is under the gravitational influence of two or more other bodies is very complicated and can be handled properly only with large computers. Fortunately, astronomers have such computers at their disposal in universities and government research institutes.

2.6.1 The Interactions of Many Bodies

As an example, suppose you have a cluster of a thousand stars all orbiting a common center (such clusters are quite common). If we know the exact position of each star at any given instant, we can calculate the combined gravitational force of the entire group on any one member of the cluster. Knowing the force on the star in question, we can therefore find how it will accelerate. If we know how it was moving to begin with, we can then calculate how it will move in the next instant of time, thus tracking its motion.

However, the problem is complicated by the fact that the other stars are also moving and thus changing the effect they will have on our star. Therefore, we must simultaneously calculate the acceleration of each star produced by the combination of the gravitational attractions of all the others in order to track the motions of all of them, and hence of any one. Such complex calculations have been carried out with modern computers to track the evolution of hypothetical clusters of stars with up to a million members (Figure 2.13).

Within the solar system, the problem of computing the orbits of planets and spacecraft is somewhat simpler. We have seen that Kepler's laws, which do not take into account the gravitational effects of the other planets on an orbit, really work quite well. This is because these additional influences are very small in comparison with the dominant gravitational attraction of the Sun. Under such circumstances, it is possible to treat the effects of other bodies as small *perturbations* (or disturbances). During the 18th and 19th centuries, mathematicians developed many elegant techniques for calculating perturbations, permitting them to predict very precisely the positions of the planets. Such calculations eventually led to the prediction and discovery of a new planet in 1846.

2.6.2 The Discovery of Neptune

The discovery of the eighth planet, Neptune, was one of the high points in the development of gravitational theory. In 1781, William Herschel, a musician and unpaid astronomer,

(a)

a, Yerkes Observatory; b, Corbis/Bettmann

(b)

■ **FIGURE 2.14**
Mathematicians Who Discovered a Planet (a) John Couch Adams (1819–1892) and (b) Urbain J. J. Leverrier (1811–1877) share the credit for discovering the planet Neptune.

accidentally discovered the seventh planet, Uranus. It happens that Uranus had been observed a century before, but in none of those earlier sightings was it recognized as a planet; rather, it was simply recorded as a star. Herschel's discovery showed that there could be planets in the solar system too dim to be visible to the unaided eye, but ready to be discovered with a telescope if we just knew where to look.

By 1790, an orbit had been calculated for Uranus using observations of its motion in the decade following its discovery. Even after allowance was made for the perturbing effects of Jupiter and Saturn, however, it was found that Uranus did not move on an orbit that exactly fit the earlier observations of it made since 1690. By 1840, the discrepancy between the positions observed for Uranus and those predicted from its computed orbit amounted to about 0.03°—an angle barely discernible to the unaided eye but still larger than the probable errors in the orbital calculations. In other words, Uranus just did not seem to move on the orbit predicted from Newtonian theory.

In 1843, John Couch Adams, a young Englishman who had just completed his studies at Cambridge, began a detailed mathematical analysis of the irregularities in the motion of Uranus to see whether they might be produced by the pull of an unknown planet. His calculations indicated the existence of a planet more distant than Uranus from the Sun. In October 1845, Adams delivered his results to George Airy, the British Astronomer Royal, informing him where in the sky to find the new planet. We now know that Adams' predicted position for the new body was correct to within 2°, but for a variety of reasons, Airy did not follow up right away.

Meanwhile, French mathematician Urbain Jean Joseph Leverrier, unaware of Adams or his work, attacked the same problem and published its solution in June 1846. Airy, noting that Leverrier's predicted position for the unknown planet agreed to within 1° with that of Adams, suggested to James Challis, Director of the Cambridge Observatory,

that he begin a search for the new object. The Cambridge astronomer, having no up-to-date star charts of the Aquarius region of the sky where the planet was predicted to be, proceeded by recording the positions of all the faint stars he could observe with his telescope in that location. It was Challis' plan to repeat such plots at intervals of several days, in the hope that the planet would distinguish itself from a star by its motion. Unfortunately, he was negligent in examining his observations; although he had actually seen the planet, he did not recognize it.

About a month later, Leverrier suggested to Johann Galle, an astronomer at the Berlin Observatory, that he look for the planet. Galle received Leverrier's letter on September 23, 1846, and, possessing new charts of the Aquarius region, found and identified the planet that very night. It was less than a degree from the position Leverrier predicted. The discovery of the eighth planet, now known as Neptune (the Latin name for the god of the sea), was a major triumph for gravitational theory, for it dramatically confirmed the generality of Newton's laws. The honor for the discovery is properly shared by the two mathematicians, Adams and Leverrier (Figure 2.14).

We should note that the discovery of Neptune was not a complete surprise to astronomers, who had long suspected the existence of the planet based on the "disobedient" motion of Uranus. On September 10, 1846, two weeks before Neptune was actually found, John Herschel, son of the discoverer of Uranus, remarked in a speech before the British Association, "We see [the new planet] as Columbus saw America from the shores of Spain. Its movements have been felt trembling along the far-reaching line of our analysis with a certainty hardly inferior to ocular demonstration."

This discovery was a major step forward in combining Newtonian theory with painstaking observations. Such work continues in our own times with the discovery of planets around other stars—work that we will discuss in chapters to come.

Images of Tycho Brahe:
www.mhs.ox.ac.uk/tycho/index.htm
Virtual exhibit from the Museum of the History of Science in Oxford, England, with images of and information about Tycho and his contemporaries.

Mathematical Discovery of Planets Site:
www-groups.dcs.st-and.ac.uk/~history/HistTopics/Neptune_and_Pluto.html
Has the fuller story of how Neptune was predicted and discovered, and who did what to whom. A spirited defense of George B. Airy (who comes across as the "villain" in many versions of the story), written by historian Alan Chapman, can be found at: www.u-net.com/ph/lassell/adams-airy.htm

Isaac Newton Sites:

- *Biography on the Math Tutor Site:*
  www-groups.dcs.st-and.ac.uk/~history//Mathematicians/Newton.html

- *Luminarium Newton Pages:*
  www.luminarium.org/sevenlit/newton/index.html

Sites for Observing Earth Satellites:

- *Visual Satellite Observer's Home Page:*
  www.satellite.eu.org/satintro.html
  Has lots of instructions and links.

- *The Satellite Observing Resources Page:*
  www.znark.com/sat/sattrack.html
  Also has links and tutorials.

- *Orbitessera:* www.mindspring.com/~n2wwd/
  A more advanced site.

- *Satellite Passes over North American Cities:*
  www.bester.com/satpasses.html

SUMMARY

2.1 Tycho Brahe was the most skillful of the pre-telescopic astronomical observers. His accurate observations of planetary positions provided the data used by Johannes Kepler to derive the three fundamental laws of planetary motion that bear his name: (1) Planetary orbits are **ellipses** (a figure described by its **semimajor axis** and **eccentricity**) with the Sun at one **focus;** (2) in equal intervals, a planet's orbit sweeps out equal areas; and (3) if times are expressed in years and distances in **astronomical units,** the relationship between the period (P) and semimajor axis (D) of an orbit is given by $D^3 = P^2$.

2.2 In his *Principia,* Isaac Newton established the three laws that govern the motion of objects: (1) Bodies continue at rest or in uniform motion unless acted upon by an outside force; (2) an outside force causes an acceleration (and changes the **momentum**) of an object; and (3) for each action there is an equal and opposite reaction. Momentum is a measure of the motion of an object and depends on both its mass and its velocity. **Angular momentum** is a measure of the motion of a spinning or revolving object. The **density** of an object is its mass divided by its volume.

2.3 Gravity, the attraction of all mass for all other mass, is the force that keeps the planets in orbit. Newton's law of gravity relates gravitational force to mass and distance ($F = GM_1M_2/R^2$). Newton was able to show the equivalence of gravitational force (weight) on Earth to the gravitational force between objects in space. When Kepler's laws are re-examined in the light of gravitational theory, it becomes clear that the masses of both Sun and planet are important for the third law, which becomes $D^3 = (M_1 + M_2) \times P^2$. Mutual gravitational effects permit us to calculate the masses of astronomical objects, from comets to galaxies.

2.4 The lowest point in a satellite **orbit** around the Earth is its **perigee,** and the highest point is its **apogee** (corresponding to **perihelion** and **aphelion** for an orbit about the Sun). The planets follow orbits about the Sun that are nearly circular and in the same plane. Most asteroids are found between Mars and Jupiter in the **asteroid belt,** whereas comets generally follow orbits of high eccentricity.

2.5 The orbit of an artificial satellite depends on the circumstances of its launch. The **circular satellite velocity**

at the Earth's surface is 8 km/s, and the **escape velocity** is 11 km/s. There are many possible interplanetary trajectories, including those that use gravity-assisted flybys of one object to redirect the spacecraft toward its next target.

2.6 Gravitational problems that involve more than two interacting bodies are much more difficult to deal with than two-body problems. They require large computers for accurate solutions. If one object dominates gravitationally, it is possible to calculate the effects of a second object in terms of small perturbations. This approach was used by Adams and Leverrier to predict the position of Neptune from its perturbations of the orbit of Uranus and thus discover a new planet mathematically.

INTER-ACTIVITY

A An eccentric, but very rich, alumnus of your college makes a bet with the dean that if you drop a marble and a bowling ball from the top of a tall water tower, the bowling ball would hit the ground first. Have your group discuss whether you would make a side bet that the alum is right. How would you decide who is right?

B Suppose a member of your group felt unhappy about his or her weight. Where could a person go to weigh one-fourth as much as he or she does now? Where would the person weigh even less? Would changing the unhappy person's weight have any effect on his or her mass?

C When the Apollo astronauts landed on the Moon, some commentators complained that it ruined the mystery and "poetry" of the Moon forever (and that lovers could never gaze at the full moon in the same way again). Others felt that knowing more about the Moon could only enhance its interest to us as we see it from Earth. How does your group feel? Why?

D Figure 2.12 shows an impressive swarm of satellites in orbit around the Earth. What do you think all these satellites do? How many categories of functions for Earth satellites can your group come up with?

REVIEW QUESTIONS

1. State Kepler's three laws in your own words.
2. Why did Kepler need Tycho Brahe's data to formulate his laws?
3. Which has more mass: an armful of feathers or an armful of lead? Which has more volume: a kilogram of feathers or a kilogram of lead? Which has higher density: a kilogram of feathers or a kilogram of lead?
4. Explain how Kepler was able to find a relationship (his third law) between the periods and distances of the planets that did not depend on the masses of the planets or the Sun.

5. Write out Newton's three laws of motion in terms of what happens with the momentum of objects.
6. What planet has the largest:
   a. semimajor axis
   b. speed of revolution around the Sun
   c. period of revolution around the Sun
   d. eccentricity
7. Why do we say that Neptune was the first planet to be discovered through the use of mathematics?

THOUGHT QUESTIONS

8. Is it possible to escape the force of gravity by going into orbit about the Earth? How does the force of gravity in the International Space Station (orbiting 500 km above the Earth's surface) compare with that on the ground? (*Hint:* The Earth's gravity acts as if all the mass were concentrated at the center of the Earth. Is the station significantly farther from the Earth's center than the Earth's surface is?)

9. What is the momentum of a body whose velocity is zero? How does Newton's first law of motion include the case of a body at rest?

10. Evil space aliens drop you and your astronomy instructor 1 km apart out in space, very far from any star or planet. Discuss the effects of gravity on each of you.

11. A body moves in a perfectly circular path at constant speed. Are there forces acting in such a system? How do you know?

12. As air friction causes a satellite to spiral inward closer to the Earth, its orbital speed increases. Why?

13. Use a history book or an encyclopedia to find out what else was happening in England during Newton's lifetime, and discuss what trends of the time might have contributed to his accomplishments and the rapid acceptance of his work.

FIGURING FOR YOURSELF

We have given several formulas in this chapter, and in this section we will ask you to apply them. Go back to the appropriate sections of the chapter if you need a reminder of such formulas as Kepler's third law.

14. What is the semimajor axis of a circle of diameter 24 cm? What is its eccentricity?

15. If 24 g of material fills a cube 2 cm on a side, what is the density of the material?

16. Draw an ellipse by the procedure described in the text, using a string and two tacks. Arrange the tacks so that they are separated by 1/10 the length of the string. Comment on the appearance of your ellipse. This (if you have been careful in your construction) is approximately the shape of the orbit of Mars.

17. The Earth's distance from the Sun varies from 147 million to 152 million km. What is the eccentricity of its orbit? (*Hint:* The distance between the foci of the ellipse is twice the distance between the Sun and the center of the ellipse. To find this, you need to first find the center of the ellipse by dividing the major axis in two. It helps to draw a diagram for yourself.)

18. Look up the revolution periods and distances from the Sun for Venus, Earth, Mars, and Jupiter in Table 2.2. Calculate D^3 and P^2 (with D in units of AU and P in units of years), and verify that they obey Kepler's third law.

19. What would be the period of a planet whose orbit has a semimajor axis of 4 AU? Of an asteroid with a semimajor axis of 10 AU?

20. What is the distance from the Sun (in astronomical units) of an asteroid with a period of revolution of 8 years? What is the distance of a planet with a period of 45.66 days?

21. In 1996 astronomers discovered an ice dwarf beyond Pluto that now bears the unromantic name of 1996 TL 66. It has a semimajor axis of 84 AU. What is its period according to Kepler's third law?

22. Newton showed that the periods and distances in Kepler's third law depend on the masses of the objects. What would be the period of revolution of the Earth (at 1 AU from the Sun) if the Sun had twice its present mass?

23. By what factor would a person's weight at the surface of the Earth be reduced if the Earth had its present mass but eight times its present volume? What if it had its present size but only one-third its present mass?

SUGGESTIONS FOR FURTHER READING

Christianson, G. "Newton's *Principia*: A Retrospective" in *Sky & Telescope*, July 1987, p. 18.

Christianson, G. "The Celestial Palace of Tycho Brahe" in *Scientific American*, Feb. 1961, p. 118.

Cohen, I. "Newton's Discovery of Gravity" in *Scientific American*, Mar. 1981, p. 166.

Gingerich, O. *The Eye of Heaven: Ptolemy, Copernicus and Kepler.* 1993, American Institute of Physics Press.

King-Hele, D. and Eberst, R. "Observing Artificial Satellites" in *Sky & Telescope*, May 1986, p. 457.

Koestler, A. *The Sleepwalkers: A History of Man's Changing Vision of the Universe.* 1959, Macmillan. A journalist's recreation of Renaissance astronomical developments.

Standage, T. *The Neptune File: Planet Detectives and the Discovery of Worlds Unseen.* 2000, Walker.

Thoren, V. *The Lord of Uraniborg.* 1990, Cambridge U. Press. Definitive modern study of Brahe's life and work.

Wilson, C. "How Did Kepler Discover His First Two Laws?" in *Scientific American*, Mar. 1972.

Southern Summer As captured with a fish-eye lens aboard the Space Shuttle on December 9, 1993, the Earth hangs above the Hubble Space Telescope as it is repaired. The reddish continent is Australia, its size and shape distorted by the special lens. Because the seasons in the Southern Hemisphere are opposite ours, it is summer in Australia on this December day.

3 Earth, Moon, and Sky

THINKING AHEAD

If the Earth's orbit is nearly a perfect circle (as we saw in earlier chapters), why is it hotter in summer and colder in winter in many places around the globe? And why are the seasons in Australia or Peru the opposite of those in the United States or Europe?

The story is told that Galileo, as he left the Hall of the Inquisition following his retraction of the doctrine that the Earth rotates and revolves about the Sun, said under his breath, "But nevertheless it moves." Historians are not sure whether the story is true, but certainly Galileo knew that the Earth was in motion, whatever church authorities said.

It is the motions of the Earth that produce the seasons and give us our measures of time and date. The Moon's motions around us provide the concept of the month and the cycle of lunar phases. In this chapter we examine some of the basic phenomena of our everyday world in their astronomical context.

Virtual Laboratories

 Tides and Tidal Forces in Astronomy

How to Observe Solar Eclipses

It is extremely dangerous to look at the Sun: Even a brief exposure can damage your eyes. Normally, few sane people are tempted to do this because it is painful (and something your mother told you never to do!). But during a solar eclipse, the temptation to take a look is strong. Think before you give in. The fact that the Moon is covering part of the Sun doesn't make the uncovered part any less dangerous to look at. Still, there are perfectly safe ways to follow the course of a solar eclipse, if you are lucky enough to be in the path of the shadow.

The easiest technique is to make a pinhole projector. Take a piece of cardboard with a small (1-mm) hole punched in it and hold it several feet above a light surface, such as a concrete sidewalk or a white sheet of paper, so that the hole is "aimed" at the Sun. The hole produces a fuzzy but adequate image of the eclipsed Sun (see diagram). Alternatively, you can let the tiny spaces between leaves form multiple pinhole images against a wall or sidewalk. Watching hundreds of little crescent Suns dancing in the breeze can be captivating.

■ ■ ■ ■ ■ ■ ■ ■ ■ ■ ■ ■ ■ ■ ■ ■ ■ ■ ■

It is safe to look at the Sun directly when it is totally eclipsed, even through binoculars or telescopes.

■ ■ ■ ■ ■ ■ ■ ■ ■ ■ ■ ■ ■ ■ ■ ■ ■ ■ ■

Although there are safe filters for looking at the Sun directly, people have suffered eye damage by looking through improper filters (or no filter at all). For example, neutral-density photographic filters are not safe because they transmit infrared radiation that can cause severe damage to the retina. Also unsafe are smoked glass, completely exposed color film, and many other homemade filters.

It is safe to look at the Sun directly when it is totally eclipsed, even through binoculars or telescopes. Unfortunately the total phase, as we discussed, is all too brief. But if you know when it is coming (and going), be sure you look, for it's an unforgettably beautiful sight. And, despite the ancient folklore that presents them as dangerous times to be outdoors, the partial phases of eclipses—as long as you are not looking directly at the Sun—are certainly not any more dangerous than being out in sunlight.

During past eclipses, unnecessary panic has been created by uninformed public officials acting with the best intentions. There were two marvelous total eclipses in Australia in the 20th century during which townspeople held newspapers over their heads for protection and schoolchildren cowered indoors with their heads under their desks. What a pity that all those people missed what would have been one of the most memorable experiences of their lifetimes!

How to watch a solar eclipse safely during its partial phases using a pinhole projector.

SURFING THE WEB

 Calendar Zone: www.calendarzone.com/
Everything you've ever wanted to ask or know about calendars and timekeeping, with links from around the world.

 Calendars and Their History:
astro.nmsu.edu/~lhuber/leaphist.html
A guide to different calendar systems and how they are based on the motions of celestial objects. Includes sections on the Hebrew, Islamic, Indian, and Chinese calendars.

 The Reasons for the Seasons:
astrosociety.org/education/publications/tnl/29/29.html
An issue of a newsletter for astronomy teachers that goes into more depth on what causes the seasons and how the different effects combine.

 Eclipse Home Page:
sunearth.gsfc.nasa.gov/eclipse/eclipse.html
This site, by NASA's eclipse expert Fred Espenak, contains a wealth of information on lunar and solar eclipses, past and future, as well as observing and photography hints.

SUMMARY

3.1 The terrestrial system of latitude and longitude makes use of the **great circles** called **meridians.** An analogous celestial coordinate system is called **right ascension** (RA) and **declination,** with the vernal equinox serving as reference point (like the prime meridian at Greenwich on the Earth). These coordinate systems help us locate any object on the celestial sphere. The Foucault pendulum is a way to demonstrate that the Earth rather than the sky is turning.

3.2 The familiar cycle of the seasons results from the 23° tilt of the Earth's axis of rotation. At the *summer solstice,* the Sun is higher in the sky and its rays strike the Earth more directly. The Sun is in the sky for more than half the day and can heat the Earth longer. At the *winter solstice,* the Sun is low in the sky and up for fewer than 12 hours. At the *vernal* and *autumnal equinoxes,* the Sun is on the celestial equator and we get 12 hours of day and night. The seasons are different at different latitudes.

3.3 The basic unit of astronomical time is the day (either the **solar day** or the **sidereal day**). **Apparent solar time** is based on the position of the Sun in the sky, and **mean solar time** is based on the average value of a solar day during the year. By international agreement, we define 24 time zones around the world, each with its own standard time. The convention of the **international date line** is necessary to reconcile times in different parts of the Earth.

3.4 The fundamental problem of the calendar is to reconcile the incommensurable lengths of the day, the month, and the year. Most modern calendars, beginning with the Roman (Julian) calendar of the 1st century B.C.E., neglect the problem of the month and concentrate on achieving the correct number of days in a year by using such conventions as the leap year. Today, most of the world has adopted the Gregorian calendar established in 1582.

3.5 The Moon's monthly cycle of **phases** results from the changing angle of its illumination by the Sun. The full moon is visible in the sky only during the night; other phases are visible during the day as well. Because its period of revolution is the same as its period of rotation, the Moon always keeps the same face toward the Earth.

3.6 The twice-daily ocean *tides* are primarily the result of the Moon's differential gravitational force on the material of the Earth's crust and ocean. These tidal forces cause ocean water to flow into two tidal bulges on opposite sides of the Earth; each day the Earth rotates through these bulges. Actual ocean tides are complicated by the additional effects of the Sun, and by the shape of the coasts and ocean basins.

3.7 The Sun and Moon have nearly the same angular size (about 1/2°). A **solar eclipse** occurs when the Moon moves between the Sun and the Earth, casting its shadow on a part of the Earth's surface. If the eclipse is total, the observer is in the Moon's *umbra,* the light from the bright disk of the Sun is completely blocked, and the solar atmosphere (the **corona**) comes into view. Solar eclipses take place rarely in any one location, but they are among the most spectacular sights in nature. A **lunar eclipse** takes place when the Moon moves into the Earth's shadow; it is visible (weather permitting) from the entire night hemisphere of the Earth.

A Have your group brainstorm about other ways (besides the Foucault pendulum) you could prove that it is our Earth that is turning once a day, and not the sky turning around us. (*Hint:* How does the spinning of the Earth affect the oceans and the atmosphere?)

 B What would the seasons on Earth be like if the Earth's axis were not tilted? How many things about life on Earth can you think of that would be different in this case?

 C After college and graduate training, members of your group are asked to set up a school in New Zealand. Describe some ways your school schedule in the Southern Hemisphere would have to differ from what we are used to in the North.

 D During the traditional U.S. Christmas vacation weeks, you are sent to the vicinity of the South Pole on a research expedition (depending on how well you did on your astronomy midterm, either as a research assistant or as a short-order cook!). Have your group discuss how the days and nights will be different there and how these differences might affect you.

 E Discuss with your group all the stories you have heard about the Moon and crazy behavior. Why do you think people associate crazy behavior with the full Moon? What other legends besides vampire stories are connected with the phases of the Moon?

F Your college town becomes the founding site for a strange new cult that worships the Moon. These true believers gather regularly around sunset and do a dance in which they must extend their arms in the direction of the Moon. Have your group discuss which way their arms will be pointing *at sunset* when the Moon is new, first quarter, full, and third quarter.

REVIEW QUESTIONS

1. Discuss how latitude and longitude on Earth are similar to declination and right ascension in the sky.
2. What is the latitude of the North Pole? The South Pole? Why does longitude have no meaning at the North and South Poles?
3. Make a table showing each main phase of the Moon and roughly when the Moon rises and sets for each phase. During which phase can you see the Moon in the middle of the morning? In the middle of the afternoon?
4. What are the advantages and disadvantages of apparent solar time? How is the situation improved by introducing mean solar time and standard time?

5. What are the two ways that the tilt of the Earth's axis causes the summers in the United States to be warmer than the winters?
6. Why is it difficult to construct a practical calendar based on the Moon's cycle of phases?
7. Explain why there are two high tides and two low tides every day. Strictly speaking, should the period during which there are two high tides be 24 hours? If not, what should the interval be?
8. What is the phase of the Moon during a total solar eclipse? During a total lunar eclipse?

THOUGHT QUESTIONS

9. Where are you on the Earth if you experience the following? (Refer back to Chapter 1 as well as this chapter.)
a. The stars rise and set perpendicular to the horizon.
b. The stars circle the sky parallel to the horizon.
c. The celestial equator passes through the zenith.
d. In the course of a year, all stars are visible.
e. The Sun rises on September 23 and does not set until March 21 (ideally).

10. In countries at far northern latitudes, the winter months tend to be so cloudy that astronomical observations are nearly impossible. Why can't good observations of the stars be made at those places during the summer months?

11. What is the phase of the Moon if it
a. rises at 3:00 P.M.?
b. is highest in the sky at 7:00 A.M.?
c. sets at 10:00 A.M.?

12. A car accident occurs around midnight on the night of a full moon. The driver at fault claims he was blinded momentarily by the Moon rising on the eastern horizon. Should the police believe him?

13. The secret recipe to the ever-popular veggie burgers in the college cafeteria is hidden in a drawer in the director's office. Two students decide to break in and get their hands on it, but they want to do It a few hours before dawn

on a night when there is no Moon, so they are less likely to be caught. What phases of the Moon would suit their plans?

14. Your great uncle, who often exaggerates events in his own life, tells you about a terrific adventure he had on February 29, 1900. Why would this story make you suspicious?

15. One year, when money is no object, you enjoy your birthday so much that you want to have another one right away. You get into your supersonic jet. Where should you and the people celebrating with you travel? From what direction should you approach? Explain.

16. Suppose you lived in the crater Copernicus on the side of the Moon facing the Earth.
a. How often would the Sun rise?
b. How often would the Earth set?

c. During what fraction of the time would you be able to see the stars?

17. In a lunar eclipse, does the Moon enter the shadow of the Earth from the east or west side? Explain why.

18. Describe what an observer at the crater Copernicus would see while the Moon is eclipsed. What would the same observer see during what would be a total solar eclipse as viewed from the Earth?

19. The day on Mars is 1.026 Earth-days long. The martian year lasts 686.98 Earth-days. The two moons of Mars take 0.32 Earth-day (for Phobos) and 1.26 Earth-days (for Deimos) to circle the planet. You are given the task of coming up with a martian calendar for a new Mars colony. What might you do?

FIGURING FOR YOURSELF

Now that you understand a bit more about why our systems of time and calendar work the way they do, you can ask what would happen if things on Earth had turned out differently.

20. Suppose the tilt of the Earth's axis were only 16°. What, then, would be the difference in latitude between the Arctic Circle and the Tropic of Cancer? What would be the effect on the seasons compared with that produced by the actual tilt of 23°?

21. Suppose the Earth took exactly 300.0 days to go around the Sun, and everything else (the day, the month) was the same? What kind of calendar would we have? How would this affect the seasons?

22. Consider a calendar based entirely on the day and the month (the Moon's period from full phase to full phase). How many days are there in a month? Can you figure out a scheme analogous to leap year to make this calendar work? Can you also incorporate the idea of a week into your lunar calendar?

23. a. If a star rises at 8:30 P.M. tonight, approximately what time will it rise two months from now?
b. What is the altitude of the Sun at noon on December 22, as seen from a place on the Tropic of Cancer?

24. Show that the Gregorian calendar will be in error by 1 day in about 3300 years.

SUGGESTIONS FOR FURTHER READING

Aveni, A. *Empires of Time: Calendars, Clocks, and Cultures.* 1989, Basic Books.

Bartky, I. and Harrison, E. "Standard and Daylight Saving Time" in *Scientific American,* May 1979.

Brunier, S. and Luminet, J. *Glorious Eclipses: Their Present, Past, and Future.* 2000, Cambridge U. Press.

Coco, M. "Not Just Another Pretty Phase" in *Astronomy,* July 1994, p. 76. Moon phases explained.

Gingerich, O. "Notes on the Gregorian Calendar Reform" in *Sky & Telescope,* Dec. 1982, p. 530.

Harris, J. and Talcott, R. *Chasing the Shadow: An Observer's Guide to Eclipses.* 1994, Kalmbach.

Kluepfel, C. "How Accurate Is the Gregorian Calendar?" in *Sky & Telescope,* Nov. 1982, p. 417.

Krupp, E. "Calendar Worlds" in *Sky & Telescope,* Jan. 2001, p. 103. On how the days of the week got their names.

Krupp, E. "Behind the Curve" in *Sky & Telescope,* Sept. 2002, p. 68. On the reform of the calendar by Pope Gregory XIII.

Pasachoff, J. "Solar Eclipse Science: Still Going Strong" in *Sky & Telescope,* Feb. 2001, p. 40. On what we have learned and are still learning from eclipses.

Pasachoff, J. and Ressmeyer, R. "The Great Eclipse" in *National Geographic,* May 1992, p. 30. About the July 1991 solar eclipse, with spectacular photographs.

Rey, H. *The Stars: A New Way to See Them.* 1976, Houghton Mifflin. Good introduction to time, the seasons, and celestial coordinates by the author of the "Curious George" children's stories.

Sobel, D. *Longitude.* 1995, Walker. Story of John Harrison, who solved the problem of how to measure longitude at sea.

Steel, D. *Marking Time: The Epic Quest to Invent the Perfect Calendar.* 2001, Wiley.

The Very Large Array of Radio Telescopes in New Mexico Seen with a Rainbow Just as water droplets in our atmosphere can disperse sunlight into a rainbow of color, so instruments called spectrometers allow astronomers to break the light from planets, stars, or galaxies into their component colors. The details of these colors can help us decipher what conditions are like in the objects from which they originate.

Photo by Martha Haynes and Riccardo Giovanelli, Cornell University

4 Radiation and Spectra

In the last hundred years the power of interpreting the messages brought by light has increased greatly. We are in touch with the stars.

Sir William Bragg in *The Universe of Light* (1933, G. Bell and Sons)

THINKING AHEAD

The nearest star is so far away that the fastest spacecraft humans have built would take almost 100,000 years to get there. Yet we very much want to know what material this neighbor star is composed of and how it differs from our own Sun. How can we learn about the chemical makeup of stars that we cannot hope to visit or sample?

In astronomy, many of the objects that we study are completely beyond our reach. The temperature of the Sun is so high that a spacecraft would be fried long before it reached the solar surface, and the stars are much too far away to visit in our lifetimes with the technology now available. Even light, which travels at a speed of 300,000 kilometers/second (km/s), takes more than four years to reach us from the *nearest* star. If we want to learn about the Sun and stars, or even about many members of our solar system, we must rely on techniques that allow us to analyze them from a distance.

Here we are in luck. Coded into the light and other kinds of radiation that reach us from objects in the universe is a wide range of information about what those objects are like and how they work. If we can decipher this "cosmic code" and read the messages it contains, we can learn an enormous amount about the cosmos without ever having to leave Earth or its immediate environment.

Virtual Laboratories

 Properties of Light and Its Interaction with Matter

The Doppler Effect

The light and other radiation we receive from the stars and planets is generated by processes at the atomic level—by changes in the way the parts of the atom interact and move. Thus, to appreciate how light is generated, we must first explore how atoms work. There is a bit of irony in the fact that in order to understand some of the largest structures in the universe, we must become acquainted with some of the smallest.

Notice that we have twice used the phrase *light and other radiation.* One of the key ideas explored in this chapter is that visible light is not unique; it is merely the most familiar example of a much larger family of radiation that can carry information to us.

The word *radiation* will be used frequently in this book, so it is important to understand what it means. In everyday language, *radiation* is often used to describe certain kinds of energetic subatomic particles released by radioactive materials in our environment. (An example is the kind of radiation used to treat cancer.) But this is not what we mean when we use the word *radiation* in an astronomy text. *Radiation,* as used in this book, is a general term for the waves (including light) that provide our primary link with the universe beyond our own solar system.

■ FIGURE 4.1

James Clerk Maxwell Maxwell (1831–1879) unified the rules governing electricity and magnetism into a coherent theory.

THE NATURE OF LIGHT

As we saw in earlier chapters, Newton's theory of gravity accounts for the motions of the planets as well as the objects on Earth. Application of this theory to a variety of problems dominated the work of scientists for nearly two centuries. In the 19th century, many physicists turned to the study of electricity and magnetism, which—as we shall see in a moment—are intimately connected with the production of light.

The scientist who played a role in this field comparable to Newton's role in the study of gravity was physicist James Clerk Maxwell (1831–1879), born and educated in Scotland (Figure 4.1). Inspired by a number of ingenious experiments that showed an intimate relationship between electricity and magnetism, Maxwell developed a theory that describes both with only a small number of elegant equations. It is this theory that allows us to understand the nature and behavior of light.

4.1.1 Maxwell's Theory of Electromagnetism

We will look at the structure of the atom in more detail in Section 4.4, but we begin by noting that the typical atom consists of several types of particles, a number of which have not only mass but an additional property called electric charge. In the nucleus (central part) of every atom are *protons,* which are positively charged; outside the nucleus are *electrons,* which have a negative charge.

Maxwell's theory deals with these electric charges and their effects, especially when they are moving. In the vicin-

ity of an electric charge, another charge feels a force of attraction or repulsion: Opposite charges attract; like charges repel. When charges are not in motion, we observe only this electric attraction or repulsion. If charges are in motion, however (as they are inside every atom and in a wire carrying a current), then we measure another force called *magnetism.*

Magnetism was well known for much of recorded human history, but its cause was not understood until the 19th century. Experiments with electric charges demonstrated that magnetism was the result of moving charged particles. Sometimes the motion is clear, as in the coils of heavy wire that make an industrial electromagnet. Other times it is more subtle, as in the kind of magnet you buy in a hardware store, in which many of the electrons inside the atoms are spinning in roughly the same direction; it is the alignment of their motion that causes the material to become magnetic.

Physicists use the word *field* to describe the action of forces that one object exerts on other distant objects. For example, we say the Sun produces a *gravitational field* that controls Earth's orbit, even though the Sun and Earth do not come directly into contact. Using this terminology, we can say that stationary electric charges produce *electric fields,* and moving electric charges also produce *magnetic fields.*

Actually, the relationship between electric and magnetic phenomena is even more profound. Experiments showed that changing magnetic fields could produce electric currents (and thus changing electric fields), and changing electric currents could (in turn) produce changing magnetic fields. So once begun, electric and magnetic field changes could continue to trigger each other.

Maxwell analyzed what would happen if electric charges were *oscillating* (moving constantly back and forth) and found that the resulting pattern of electric and

■ FIGURE 4.2

Making Waves An oscillation in a pool of water creates an expanding disturbance called a wave.

© 1993, Comstock, Inc.

magnetic fields would spread out and travel rapidly through space. Something similar happens when your finger or a nervous frog moves up and down in a pool of water. The disturbance moves outward and creates a pattern we call a *wave* in the water (Figure 4.2). You might at first think that there must be very few situations in nature where electric charges oscillate, but this is not at all the case. As we shall see, atoms and molecules (which consist of charged particles) oscillate back and forth all the time. The resulting electromagnetic disturbances are among the most common phenomena in the universe.

Maxwell was able to calculate the speed at which an electromagnetic disturbance moves through space; he found that it is equal to the speed of light, which had been measured experimentally. On that basis, he speculated that light was one form of a family of possible electric and magnetic disturbances called **electromagnetic radiation,** a conclusion that was again confirmed in laboratory experiments. When light (reflected from the pages of an astronomy textbook, for example) enters a human eye, its changing electric and magnetic fields stimulate nerve endings, which then transmit the information contained in these changing fields to the brain. The science of astronomy is primarily about analyzing radiation from distant objects to understand what they are and how they work.

4.1.2 The Wave-Like Characteristics of Light

The changing electric and magnetic fields in radiation are similar (as hinted earlier) to the waves that can be set up in a quiet pool of water. In both cases the disturbance travels rapidly outward from the point of origin and can use its energy to disturb other things farther away. (For example, in

water, the expanding ripples moving away from our twitching frog could disturb the peace of a grasshopper sleeping on a leaf in the same pool.) In the case of electromagnetic waves, the radiation generated by a transmitting antenna full of charged particles at your local radio station can, sometime later, disturb a group of electrons in your home radio antenna and bring you the news and weather while you are getting ready for class or work in the morning.

The waves generated by charged particles differ from water waves in some profound ways, however. Water waves require water to travel in. Sound waves, to give another example, are pressure disturbances that require air to travel through. But electromagnetic waves do not require water or air: The fields generate each other and so can move through a vacuum (such as outer space). This was such a disturbing idea to 19th-century scientists that they actually made up a substance to fill all of space—one for which there was not a single shred of evidence—just so light waves could have something to travel through: They called it the *aether*. Today we know that there is no aether and that electromagnetic waves have no trouble at all moving through empty space (as all the starlight visible on a clear night must surely be doing).

The other difference is that *all* electromagnetic waves move at the same speed in space (the speed of light), which turns out to be the fastest possible speed in the universe. No matter where electromagnetic waves are generated and no matter what other properties they have, when they are moving (and not interacting with matter), they move at the speed of light. Yet you know from everyday experience that waves like light are not all the same. For example, we perceive that light waves differ from one another in a property we call color. Let's see how we can denote the differences among the whole broad family of electromagnetic waves.

FIGURE 4.3

Characterizing Waves Electromagnetic radiation has wave-like characteristics. The wavelength (λ) is the distance between crests, the frequency (f) is the number of cycles per second, and the speed (c) is the distance the wave covers over time.

The nice thing about a wave is that it is a repeating phenomenon. Whether it is the up-and-down motion of a water wave or the changing electric and magnetic fields in a wave of light, the pattern of disturbance repeats in a cyclical way. Thus, any wave motion can be characterized by a series of crests and troughs (Figure 4.3). Moving up a full crest and down a full trough completes one cycle. The horizontal length covered by one cycle—the distance between one crest and the next, for example—is called the **wavelength.** Ocean waves provide an analogy: The wavelength is the distance that separates successive wave crests.

For visible light, our eyes (and minds) perceive different wavelengths as different colors: Red, for example, is the longest visible wavelength and violet is the shortest. The main colors of visible light from longest to shortest wavelength can be remembered using the mnemonic ROY G. BIV—for red, orange, yellow, green, blue, indigo, violet. Other invisible forms of electromagnetic radiation have different wavelengths, as we will see in the next section.

We can also characterize different waves by their **frequency,** the number of wave cycles that pass per second. If you count ten crests moving by each second, for example, then the frequency is 10 cycles per second (cps). In honor of Heinrich Hertz, the physicist who—inspired by Maxwell's work—discovered radio waves, a cps is also called a *hertz (hz).* Take a look at the dial of your radio, for example, and you will see the channel assigned to each radio station on the dial characterized by its frequency, usually in units of Khz (kilohertz, or thousands of hertz) or Mhz (megahertz, or millions of hertz).

Wavelength and frequency are related because all electromagnetic waves travel at the same speed. To see how this works, imagine a parade in which everyone is forced by prevailing traffic conditions to move at exactly the same speed. You stand on a corner and watch the waves of marchers come by. First you see row after row of very skinny fashion models. Because they are not very wide, a good number of the models can fit by you each minute; we can say they have a high frequency. Next, however, come several rows of circus elephants. The elephants are large and marching at the

same speed as the models, so far fewer of them can march past you per minute: Because they have a wider spacing, they must be content with a lower frequency.

Mathematically, we can express this relationship as

$$c = \lambda f$$

where the Greek letter for "l"—lambda, λ—is used to denote wavelength and c is the scientific symbol for the speed of light. The formula can be expressed as follows: For any wave motion, the speed at which a wave moves equals the frequency times the wavelength. Waves with longer wavelengths have lower frequencies.

4.1.3 Light as a Photon

The electromagnetic wave model of light (as formulated by Maxwell) was one of the great triumphs of 19th-century science. When Heinrich Hertz actually made invisible electromagnetic waves (what today are called radio waves) on one side of a room and detected them on the other in 1887, it ushered in a new era that led to the modern age of telecommunications. However, by the beginning of the 20th century, more sophisticated experiments had revealed that light behaves in certain ways that cannot be explained by the wave model. Reluctantly, physicists had to accept that sometimes light (and all other electromagnetic radiation) behaves more like a "particle"—or at least a self-contained packet of energy—than a wave. We call such packets of electromagnetic energy **photons.**

The fact that light behaves like a wave in certain experiments and like a particle in others was a very surprising and unlikely idea; at first, it was as confusing to physicists as it seems to students. After all, our common sense says that waves and particles are opposite concepts. On one hand, a wave is a repeating disturbance that by its very nature is not in only one place, but spreads out. A particle, on the other hand, is something that can be in only one place at any given time. Strange as it sounds, though, countless experiments now confirm that electromagnetic radiation can sometimes behave like a wave and at other times like a particle.

Then, again, perhaps we shouldn't be surprised that something that always travels at the "speed limit" of the universe and doesn't need a substance to travel through might not obey our everyday commonsense ideas. The confusion that this wave–particle duality of light caused in physics was eventually resolved by the introduction of a more complicated theory of waves and particles, now called *quantum mechanics.* (This is one of the most interesting fields of modern science, but it is mostly beyond the scope of our book. If you get interested in it, see some of the references at the end of this chapter.)

In any case, you should now be prepared when scientists (or the authors of this book) sometimes discuss electromagnetic radiation as if it consisted of waves and at other times refer to it as a stream of photons. A photon (being a packet of energy) carries a specific amount of energy. We can use the idea of energy to connect the photon and wave models. How much energy a photon has depends on its fre-

■ **FIGURE 4.4**

The Inverse-Square Law for Light
As light energy radiates away from its source, it spreads out in such a way that the energy decreases as the square of the distance from its source.

quency when you think about it as a wave. A low-energy radio wave has a low frequency as a wave, while a high-energy x ray at your dentist's office is a high-frequency wave. Among the colors of visible light, violet-light photons have the highest energy and red has the lowest.

Test whether the connection between photons and waves is clear to you. In the above example, which photon would have the longer wavelength as a wave: the radio wave or the x ray? If you answered the radio wave, you are correct. Radio waves have lower frequency, so the wave cycles are longer (they are elephants, not fashion models).

4.1.4 Propagation of Light

Let's think for a moment about how light from a light bulb moves through space. The expansion of light waves is extremely democratic: They travel away from the bulb, not just toward *your* eyes but in all directions. They must therefore cover an ever-widening space. Yet the total amount of light available can't change once the light has left the bulb. This means that, as the same expanding shell of light covers a larger and larger area, there must be less and less of it in any given place. Light (and all other electromagnetic radiation) gets weaker and weaker as it gets farther from its source.

The increase in the area that the radiation must cover is proportional to the square of the distance that the radiation has traveled (Figure 4.4). If we stand twice as far from the source, our eyes will intercept 2-squared (2×2), or 4 times less light. If we stand 10 times farther from the source, we get 10-squared, or 100 times less light. You can see how this weakening means trouble for sources of light at astronomical distances! One of the nearest stars, called Alpha Centauri, emits about the same total energy as does the Sun. But it is about 270,000 times farther away, and so it appears about 73 billion times fainter. No wonder that the stars, which close-up would look more or less like the Sun, look like faint pinpoints of light from far away.

This idea—that the apparent brightness of a source (how bright it looks to us) gets weaker with distance in the way we have described—is known as the **inverse-square law** of light propagation. In this respect, the propagation of radiation is similar to the effects of gravity. Remember that the force of gravity between two attracting masses is also inversely proportional to the square of their separation.

4.2 THE ELECTROMAGNETIC SPECTRUM

Objects in the universe send us an enormous range of electromagnetic radiation. Scientists call this range the **electromagnetic spectrum,** which they have divided into a number of categories. The spectrum is shown in Figure 4.5, with some information about the waves in each part or band.

4.2.1 Types of Electromagnetic Radiation

Electromagnetic radiation with the shortest wavelengths, no longer than 0.01 nanometer (nm), is called **gamma rays** (1 nm = 10^{-9} m; see Appendix 5). The name *gamma* comes from the third letter of the Greek alphabet; gamma rays were the third kind of radiation discovered coming from radioactive atoms when physicists first investigated their behavior. Because gamma rays carry a lot of energy, they can be dangerous for living tissues. Gamma radiation is generated deep in the interior of stars. They are also generated by some of the most violent phenomena in the universe, such as the deaths of stars and the merging of stellar corpses. Gamma rays coming to Earth are absorbed by our atmosphere before they reach the ground (which is a good thing for our health); thus, they can only be studied using instruments in space.

Electromagnetic radiation with wavelengths between 0.01 nm and 20 nm is referred to as **x rays.** These are now familiar to all of us from visits to the doctor and dentist; being more energetic than light, x rays are able to penetrate soft tissues and so allow us to make images of the shadows of the bones within them. While x rays can penetrate a short length of human flesh, they are stopped by the large numbers of atoms in the Earth's atmosphere with which they interact. Thus, x-ray astronomy (like gamma-ray astronomy) could not develop until we invented ways of sending instruments above our atmosphere (Figure 4.6).

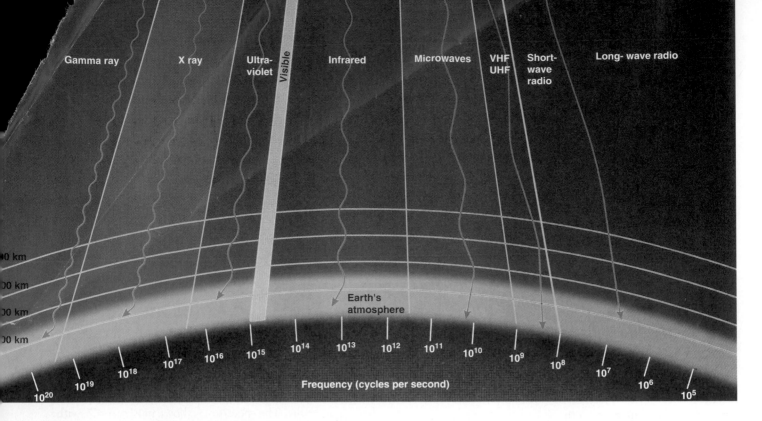

Gamma ray   X ray   Ultra-violet   Visible   Infrared   Microwaves   VHF UHF   Short-wave radio   Long- wave radio

Earth's atmosphere

0 km
0 km
0 km
0 km

10^{20} 10^{19} 10^{18} 10^{17} 10^{16} 10^{15} 10^{14} 10^{13} 10^{12} 10^{11} 10^{10} 10^{9} 10^{8} 10^{7} 10^{6} 10^{5}

Frequency (cycles per second)

■ **FIGURE 4.5**

Radiation and the Earth's Atmosphere   The bands of the electromagnetic spectrum and how well Earth's atmosphere transmits them. Note that high-frequency waves from space do not make it to the surface and must therefore be observed from space. Some infrared and microwaves are absorbed by water and thus are best observed from high altitudes. Low-frequency radio waves are blocked by Earth's ionosphere.

■ **FIGURE 4.6**

The X-Ray Sky   What the sky would look like if we had eyes tuned to certain types of x rays (and were standing above Earth's atmosphere). This map of the sky (with the disk of our Milky Way Galaxy running across its center) was constructed from data gathered by the European ROSAT satellite. Each color (red, yellow, and blue) shows x rays of different frequencies or energies. The red outlines the glow from a hot local bubble of gas all around us, blown by one or more exploding stars in our cosmic vicinity. The yellow and blue show more distant sources of x rays, such as the remnants of other exploded stars or the center of our Galaxy (in the middle of the picture), as well as a diffuse background of x rays that result from the blending of x rays from many more distant sources in the universe.

■ FIGURE 4.7

The Action of a Prism
When we pass a beam of white sunlight through a prism, we see a rainbow-colored band of light that we call a continuous spectrum.

Radiation intermediate between x rays and visible light is **ultraviolet** (meaning higher energy than violet). Outside the world of science, ultraviolet light is sometimes called "black light" because our eyes cannot see it. Ultraviolet radiation is mostly blocked by a section of Earth's atmosphere called the *ozone* layer, but a small fraction of the ultraviolet rays from our Sun do penetrate to cause sunburn or, in extreme cases, skin cancer in human beings. Ultraviolet astronomy is also best done from space.

Electromagnetic radiation with wavelengths between roughly 400 and 700 nm is called **visible light** because these are the waves that human vision can perceive. As we shall see, this is also the band of the electromagnetic spectrum where the Sun gives off the greatest amount of radiation. These two observations are not coincidental: Human eyes evolved to see the kinds of waves that the Sun produced most effectively. The more light there was to see by, the more efficiently we could evade predators and make babies before we got stepped on or eaten (which, in a crude way, is what life is all about from the point of view of your genetic material). Visible light penetrates the Earth's atmosphere effectively, except when it is temporarily blocked by passing clouds.

In 1672, in the first paper that he submitted to the Royal Society, Newton described an experiment in which he permitted sunlight to pass through a small hole and then through a prism. Newton found that sunlight, which gives the impression of being white, is actually made up of a mixture of all the colors of the rainbow (Figure 4.7). Although all the colors of light may impinge on objects in our world, not all of them are reflected. When white light from the Sun or from a light fixture hits a pair of blue jeans, the dye in the jeans absorbs all the colors except blue. Some of the blue colors are reflected back into our eyes, and our brain interprets that particular set of wavelengths as blue color.

Between visible light and radio waves are the wavelengths of **infrared** or heat radiation. Astronomer William Herschel first discovered infrared in 1800 while trying to measure the temperatures of different colors of sunlight spread out into a spectrum. He noticed that when he accidentally pushed his thermometer beyond the reddest color, it still registered heating due to some invisible energy coming from the Sun. This was the first hint about the existence of the other (invisible) bands of the electromag-

netic spectrum, although it would take many decades for our full understanding to develop.

A "heat lamp" radiates mostly infrared radiation, and the nerve endings in our skin are sensitive to this band of the electromagnetic spectrum. Infrared waves are absorbed by water and carbon dioxide molecules in Earth's atmosphere, and for this reason, infrared astronomy is best done from high mountaintops, high-flying airplanes, and spacecraft.

All electromagnetic waves longer than infrared are called **radio waves,** but this is so broad a category that we generally divide it into several subsections. Among the most familiar of these are *microwaves,* used in short-wave communication and microwave ovens; radar waves, used at airports and by the military; FM and television waves, which carry the news and entertainment without which it is now hard to imagine modern culture; and AM radio waves, which were the first to be developed for broadcasting. The wavelengths of these different categories range from a few millimeters to hundreds of meters. Other radio radiation can have wavelengths as long as several kilometers.

With such a wide range of wavelengths, you will not be surprised to learn that not all radio waves interact with Earth's atmosphere the same way. Microwaves are absorbed by water vapor (which makes them effective in heating foods with high water contents in microwave ovens). FM and TV waves are not absorbed and can travel easily through our atmosphere. AM radio waves are absorbed or reflected by a layer in Earth's atmosphere called the *ionosphere.* Given some of the truly dumb things people are saying these days on AM radio, maybe it's just as well that this type of radiation can not get out of our planet's shielding atmosphere!

We hope this brief survey has left you with one strong impression. Although visible light is what most people associate with astronomy, the light that our eyes can see is only a tiny fraction of the broad range of waves that the universe sends us. Today we understand that judging some astronomical phenomenon by using only the light we can see is like hiding under the table at a big dinner party and judging all the guests by nothing but their shoes. There's a lot more to each person than meets our eye under the table. It is very important for those who study astronomy today to avoid being *visible-light chauvinists*—to respect only the information seen by their eyes while ignoring the

TABLE 4.1 *Types of Electromagnetic Radiation*

Type of Radiation	Wavelength Range (nm)	Radiated by Objects at This Temperature	Typical Sources
Gamma rays	Less than 0.01	More than 10^8 K	Produced in nuclear reactions; requires very-high-energy processes
X rays	0.01–20	106–10^8 K	Gas in clusters of galaxies; solar corona
Ultraviolet	20–400	10^4–10^6 K	Very hot stars
Visible	400–700	10^3–10^4 K	Stars
Infrared	10^3–10^6	10–10^3 K	Cool clouds of dust and gas; planets; satellites
Radio	More than 10^6	Less than 10 K	No astronomical objects this cold; radio emission produced by electrons moving in magnetic fields (synchrotron radiation)

information gathered by instruments sensitive to other bands of the electromagnetic spectrum.

Table 4.1 summarizes the bands of the electromagnetic spectrum and indicates the temperatures and typical astronomical objects that emit each kind of electromagnetic radiation. Many of the objects in the table will not be part of your course in planetary astronomy, but we mention them so you can get the scope of astronomical observation.

4.2.2 Radiation and Temperature

Some astronomical objects emit mostly infrared radiation, others mostly visible light, and still others mostly ultraviolet radiation. What determines the type of electromagnetic radiation emitted by the Sun, stars, and other astronomical objects? The answer often turns out to be their *temperature*.

At the microscopic level, everything in nature is in motion. A *solid* is composed of molecules and atoms that are in continuous vibration: They move back and forth in place, but their motion is much too small for our eyes to make out. A *gas* consists of molecules that are flying about freely at high speed, continually bumping into one another and bombarding the surrounding matter. The energy of these random molecular and atomic motions is called *heat*. The hotter the solid or gas, the more rapid the motion of its molecules or atoms. The temperature of something is thus a measure of the average motion energy of the particles that make it up.

This motion at the microscopic level is responsible for much of the electromagnetic radiation on Earth and in the universe. As atoms and molecules move about and collide, or vibrate in place, their charged particles generate electromagnetic radiation (as we saw in Section 4.1). The characteristics of this radiation are determined by the temperature of those atoms and molecules. In a hot solid or gas, for example, the individual particles vibrate in place or move rapidly from collision to collision, so the emitted waves are, on average, more energetic. In very cool material, the particles have low-energy atomic and molecular motions and thus generate lower-energy waves.

4.2.3 Radiation Laws

To understand in more quantitative detail the relationship between temperature and electromagnetic radiation, we may consider an idealized object that (unlike your sweater or your astronomy instructor's head) does not reflect or scatter any radiation but absorbs all the electromagnetic energy that falls on it. The energy that is absorbed causes the atoms and molecules in it to vibrate or move around at increasing speeds. As it gets hotter, this object will radiate electromagnetic waves until absorption and radiation are in balance. We want to discuss such an idealized object because, as you will see, stars behave in very nearly the same way.[1]

The radiation from our ideal object has several characteristics, as illustrated in Figure 4.8. The graphs show the power emitted at each wavelength by objects of different temperatures. In science, the word *power* means the energy coming off per second (and it is typically measured in *watts*, which you are probably familiar with from buying light bulbs.)

First of all, notice that the white curves show that at each temperature our object emits a *continuous spectrum* of waves; that is, it emits radiation at all wavelengths. This is because in any solid or denser gas, some molecules or atoms vibrate or move between collisions slower than average, and some move faster than average. So when we look at the electromagnetic waves emitted, we find a broad range or spectrum of energies and wavelengths. More waves are emitted at the *average* vibration or motion rate (the tall part of each curve), but if we have a large number of atoms or molecules, some waves will be detected at each wavelength.

Second, note that an object at a higher temperature emits *more* power at all wavelengths than does a cooler one. In a hot gas (the taller curves in Figure 4.8), for example, the atoms have more collisions and give off more waves at each possible energy or wavelength. In the real

---

[1] Such an object is called a *blackbody* in more advanced courses, but since blackbodies in nature—like stars—do not look black, we prefer not to use the term.

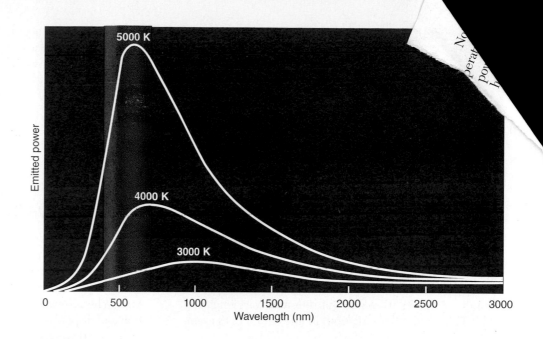

FIGURE 4.8

Radiation Laws Illustrated
These graphs show a count of the number of photons per second (or the total amount of power) given off at different wavelengths for objects at three different temperatures (three white curves). Note that at hotter temperatures, more energy is emitted at all wavelengths. The higher the temperature, the shorter the wavelength at which the peak amount of energy is radiated (this is called Wien's law).

world of stars, this means that hotter stars give off more energy at every wavelength than do cooler stars.

Third, the graph shows us that the higher the temperature, the shorter the wavelength at which the maximum power is emitted. Remember that shorter wavelength means higher frequency and energy. It makes sense, then, that hot objects give off their average waves at shorter wavelengths (higher energies) than do cool objects. You may have observed examples of this rule in everyday life. When a burner on an electric stove is turned on low, it emits only heat, which is infrared radiation, but does not glow with visible light. If the burner is set to a higher temperature, it starts to glow a dull red. At a still higher setting, it glows a brighter orange-red (shorter wavelength). At even higher temperatures, which cannot be reached with ordinary stoves, metal can appear brilliant yellow or even blue-white.

We can use these ideas to come up with a rough sort of "thermometer" for measuring the temperatures of stars. Because many stars give off most of their waves in visible light, the color of light that dominates a star's appearance is a rough indicator of its temperature. If one star looks red and another looks blue, which one has the higher temperature? Because blue is the shorter-wavelength color, it is the sign of a hotter star. (Note that the temperatures we associate with different colors in science are not the same as the ones artists use. In art, red is often called a "hot" color and blue a "cool" color, but in nature, it's the other way around.)

We can develop a more precise star thermometer by measuring how much energy a star gives off at each wavelength and by constructing diagrams like Figure 4.8. The location of the peak (or maximum) in the power curve of each star can tell us its temperature. The temperature at the surface of the Sun, which is where the radiation that we see is emitted, turns out to be 5800 K. (Throughout this text we use the Kelvin or absolute temperature scale. On

this scale, water freezes at 273 K and boils at 373 K. All molecular motion ceases at 0 K. The various temperature scales are described in Appendix 5.) There are stars cooler than the Sun and stars hotter than the Sun; there are even some stars so hot that most of their energy is given off at ultraviolet wavelengths.

The wavelength at which maximum power is emitted can be calculated according to the equation

$$\lambda_{max} = \frac{3 \times 10^6}{T}$$

where the wavelength is in nanometers (one billionth of a meter) and the temperature is in Kelvins. This relationship is called **Wien's law.** For the Sun, the wavelength at which the maximum energy is emitted is 520 nm, which is near the middle of that portion of the electromagnetic spectrum called visible light. Characteristic temperatures of other astronomical objects, and the wavelengths at which they emit most of their power, are listed in Table 4.1.

We can also describe our observation that hotter objects radiate more power at all wavelengths in a mathematical form. If we sum up the contributions from all parts of the electromagnetic spectrum, we obtain the total energy emitted by a blackbody. What we usually measure from a large object like a star is the **energy flux,** the power emitted per square meter. The word *flux* means "flow" here: We are interested in the flow of power into an area (like the area of a telescope mirror). It turns out that the energy flux from a blackbody at temperature T is proportional to the fourth power of its absolute temperature. This relationship is known as the **Stefan-Boltzmann law** and can be written in the form of an equation as

$$F = \sigma T^4$$

where F stands for the energy flux and σ (Greek letter sigma) is a constant number.

...ice how impressive this result is. Increasing the temperature of a star would have a tremendous effect on the power it radiates. If the Sun, for example, were twice as hot—that is, if it had a temperature of 11,600 K—it would radiate 2^4, or 16, times more power than it does now. Tripling the temperature would raise the power output 81 times! Hot stars really shine away a tremendous amount of energy.

4.3 SPECTROSCOPY IN ASTRONOMY

As we hinted at the beginning of the chapter, electromagnetic radiation carries a lot of information about the nature of stars and other astronomical objects. To extract this information, however, astronomers must be able to study the amounts of energy we receive at different wavelengths of visible light (and other radiation) in fine detail. Let's examine how we can do this and what we can learn.

4.3.1 Optical Properties of Light

Visible light and other forms of electromagnetic energy exhibit certain behaviors that are important to the design of telescopes and other instruments. For example, light can be *reflected* from a surface. If the surface is smooth and shiny, as in a mirror, the direction of the reflected light beam can be calculated accurately from a knowledge of the shape of the reflecting surface. Light is also bent, or *refracted*, when it passes from one kind of transparent material into another—say, from the air into a glass lens.

Reflection and refraction of light are the basic properties that make possible all optical instruments—from eyeglasses to giant astronomical telescopes. Such instruments are generally combinations of glass lenses, which bend light according to the principles of refraction, and curved mirrors, which depend on the properties of reflection. Small optical devices, such as eyeglasses or binoculars, generally use lenses, while large telescopes depend almost entirely on mirrors for their main optical elements. In Chapter 5 we will discuss astronomical instruments and their uses. For now, we turn to another behavior of light, one that is essential for the decoding of light.

When light passes from one transparent material to another, an interesting effect occurs in addition to simple refraction. Because the bending of the beam depends on the wavelength of the light as well as the properties of the material, different wavelengths or colors of light are bent by different amounts and are therefore separated. This phenomenon is called **dispersion.**

Figure 4.7 shows how light is separated into different colors with a *prism*—a piece of glass in the shape of a triangle. Upon entering one face of the prism, light is refracted once, the violet light more than the red, and upon leaving the opposite face, the light is bent again and further dispersed. If the light leaving the prism is focused on a screen, the different wavelengths or colors that make up white light are lined up side by side just like a rainbow (Figure 4.9). (In fact, a rainbow is formed by the dispersion of light through raindrops; see Making Connections: *The Rainbow*.) Because this array of colors is a spectrum of light, the instrument used to disperse the light and form the spectrum is called a **spectrometer.**

4.3.2 The Value of Stellar Spectra

If the spectrum of the white light from the Sun and stars were simply a continuous rainbow of colors, astronomers would have little interest in the detailed study of a star's spectrum once they had learned its approximate temperature. When Newton, who first described the laws of refraction and dispersion in optics, observed the solar spectrum, all he could see was a continuous band of colors.

In 1802, however, William Wollaston built an improved spectrometer that included a lens to focus the Sun's spectrum on a screen. With this device Wollaston saw that the colors were not spread out uniformly, but instead some ranges of color were missing, appearing as dark bands in the solar spectrum. He mistakenly attributed these lines to natural boundaries between the colors. In 1815 the German physicist Joseph Fraunhofer, upon a more careful examination of the solar spectrum, found about 600 such dark lines (missing colors), which pretty quickly ruled out the boundary hypothesis (Figure 4.10).

These physicists called the missing colors *dark lines* because the rainbow-colored (continuous) spectrum as they viewed it was interrupted at various wavelengths, and in their spectrometers such interruptions looked like dark lines. Later, researchers found that similar dark lines could be produced in the spectra (*spectra* is the plural of *spectrum*) of artificial light sources. They did this by passing their light through various apparently transparent substances—usually containers with just a bit of thin gas in them.

These gases turned out not to be transparent at *all* colors: They were quite opaque at a few sharply defined wavelengths. Something in each gas had to be absorbing just a

400   450   500   550   600   650   700   750
Nanometers

■ FIGURE 4.9

A Continuous Spectrum   When white light passes through a prism, its dispersion forms a continuous spectrum of all the colors. Although it is hard to see in this printed version, in a well-dispersed spectrum, many subtle gradations in color are visible as your eye scans from one end (the violet) to the other (the red).

The Rainbow

Rainbows are an excellent illustration of the refraction of sunlight. You have a good chance of seeing a rainbow anytime you are between the Sun and a rain shower; this situation is illustrated in Figure 4A. The raindrops act like little prisms and break white light into the spectrum of colors. Suppose a ray of sunlight encounters a raindrop and passes into it. The light changes direction—is refracted (Figure 4B)—when it passes from air to water; the blue and violet light are refracted more than the red. Some of the light is then reflected at the backside of the drop and reemerges from the front, where it is again refracted. As a result, the white light is spread out into a rainbow of colors.

Note that in Figure 4B violet light lies above the red light after it emerges from the raindrop. When you look at a rainbow, however, the red light is higher in the sky. Why? Look again at Figure 4A. If the observer looks at a raindrop that is high in the sky, the violet light passes over her head and the red light enters her eye. Similarly, if the observer looks at a raindrop that is low in the sky, the violet light reaches her eye and the drop appears violet, whereas the red light from that same drop strikes the ground and is not seen. Colors of intermediate wavelengths are refracted to the eye by drops that are intermediate in altitude between the drops that appear violet and the ones that appear red. Thus, a single rainbow always has red on the outside and violet on the inside.

For an even simpler example of refraction, put a pencil at a slanted angle in a glass of water. What do you see? Can you offer an explanation?

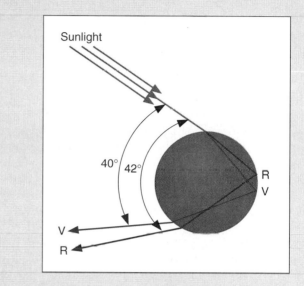

■ FIGURE 4B

A diagram showing the path of light passing through a raindrop. Refraction separates white light into its component colors.

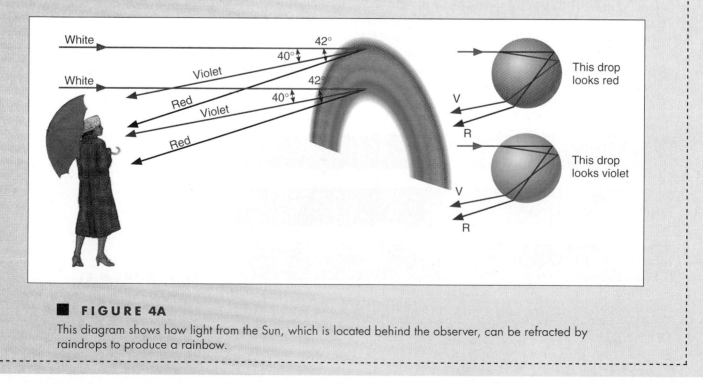

■ FIGURE 4A

This diagram shows how light from the Sun, which is located behind the observer, can be refracted by raindrops to produce a rainbow.

National Solar Observatory/National Optical Astronomy Observatories

■ **FIGURE 4.10**

The Visible Spectrum of the Sun Our star's spectrum is crossed by dark lines produced by atoms in the solar atmosphere that absorb light at certain wavelengths.

few colors of light and no others. All gases did this, but each different element absorbed a different set of colors and thus showed different dark lines. If the gas in a container consisted of two elements, then light passing through it was missing the colors (showing dark lines) that go with both of the elements. So it became clear that certain lines in the spectrum "go with" certain elements. This discovery, as we shall see, was one of the most important steps forward in the history of astronomy.

What would happen if there were no continuous spectrum for our gases to remove light from? What if, instead, we heated the same thin gases until they were hot enough to glow with their *own light?* When the gases were heated, a spectrometer revealed no continuous spectrum, but several separate *bright lines.* That is, these hot gases *emitted* light only at certain specific wavelengths or colors. The colors at which they emitted when they were heated were the very same colors at which they had absorbed when a continuous source of light was behind them.

When the gas was pure hydrogen, it would absorb or emit one pattern of colors; when it was sodium, it would absorb or emit a different pattern. A mixture of hydrogen and sodium emitted both sets of spectral lines. Thus, scientists began to see that different substances showed distinctive

spectral signatures by which their presence could be detected (Figure 4.11). Just as your signature allows the bank to identify you, the unique pattern of colors for each type of atom can help us identify which element or elements are in the gas.

4.3.3 Types of Spectra

In such experiments, then, we can distinguish among the three different types of spectra. A **continuous spectrum** (formed when a solid or very dense gas gives off radiation) is an array of all wavelengths or colors of the rainbow. A continuous spectrum can serve as a backdrop from which the atoms of a much less dense gas can absorb light. A dark line, or **absorption line spectrum,** consists of a series or pattern of dark lines—missing colors—superimposed upon the continuous spectrum of a source. A bright line, or **emission line spectrum,** appears as a pattern or series of bright lines; it consists of light in which only certain discrete wavelengths are present. (Figure 4.18—a little further ahead in this chapter—is a picture of what each of these looks like.)

When it's a gas, each particular chemical element or compound produces its own characteristic pattern of spectral lines—its spectral signature. No two patterns are alike.

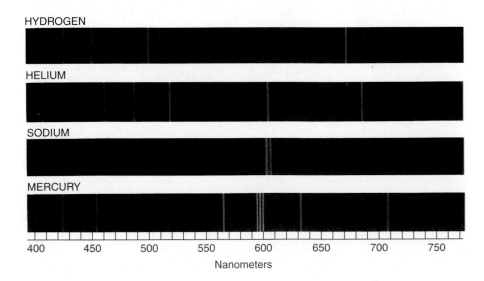

HYDROGEN

HELIUM

SODIUM

MERCURY

| 400 | 450 | 500 | 550 | 600 | 650 | 700 | 750 |

Nanometers

■ **FIGURE 4.11**

Line Spectra from Different Elements Each type of hot glowing gas (each element) produces its own unique pattern of lines, so the composition of a gas can be identified by its spectrum.

In other words, each particular gas can absorb or emit only certain wavelengths of light peculiar to that gas. The temperature and other conditions determine whether the lines are bright or dark, but the wavelengths for any element are the same in either case. It is the precise pattern of wavelengths that makes the signature of each element unique. (Liquids and solids can also generate spectral lines or bands, but they are broader and less well defined and hence more difficult to interpret.)

The dark lines in the solar spectrum thus give evidence of certain chemical elements between us and the Sun, absorbing those wavelengths of sunlight. Because the space between us and the Sun is pretty empty, astronomers quickly realized that the atoms doing the absorbing must be in a thin atmosphere of cooler gas around the Sun. Since the Sun is made entirely of gas, this outer atmosphere is not all that different from the rest of the Sun, just thinner and cooler. Thus, we can use what we learn about its composition as an indicator of what the whole Sun is made of. Similarly, we can use the presence of absorption and emission lines to analyze the composition of other stars and clouds of gas in space.

Such analysis of spectra is the key to modern astronomy. Only in this way can we "sample" the stars, which are too far away for us to visit. Even if we could visit them, they are so hot that it would be very difficult to find graduate students willing to go take a sample. But there is no need to ask for such volunteers: Encoded in the electromagnetic radiation from celestial objects is clear information about the chemical makeup of these objects. Only by understanding what the stars were made of could astronomers begin to form theories about what made them shine and how they evolved.

In 1860 the German physicist Gustav Kirchhoff became the first person to use spectroscopy to identify an element in the Sun when he found the spectral signature of sodium gas. In the years that followed, astronomers found many other chemical elements in the Sun and stars. In fact, the element helium was found first in the Sun from its spectrum and only later identified on Earth. (The word *helium* comes from *helios*, the Greek name for the Sun.)

And why are there specific lines for each element? The answer to that question was not found until the 20th century; it required the development of a model for the atom. We therefore turn next to a closer examination of the atoms that make up all matter.

4.4 THE STRUCTURE OF THE ATOM

The idea that matter is composed of tiny particles called atoms is at least 25 centuries old. It took until the 20th century, however, for scientists to invent instruments that permitted them to probe inside an atom and find that it is not, as had been thought, hard and indivisible. Instead, the atom is a complex structure composed of still smaller particles.

4.4.1 Probing the Atom

The first of these smaller particles was discovered by British physicist James Thomson in 1897. Named the *electron*, this particle is negatively charged. (It is the flow of these particles that produces currents of electricity, whether in lightning bolts or in the wires leading to your hair dryer.) Because an atom in its normal state is electrically neutral, each electron in an atom must be balanced by the same amount of positive charge.

The next problem was to determine where in the atom the positive and negative charges are located. In 1911 British physicist Ernest Rutherford devised an experiment that provided part of the answer to this question. He bombarded an extremely thin piece of gold foil, only about 400 atoms thick, with a beam of *alpha particles* emitted from a radioactive material (Figure 4.12). We now know that alpha particles are helium atoms that have lost their electrons and thus are positively charged. Most of the alpha particles passed through the gold foil just as if it and the atoms in it were nearly empty space. About 1 in 8000 of the alpha particles, however, completely reversed direction and bounced backward from the foil. Rutherford wrote, "It was quite the most incredible event that has ever happened to me in my life. It was almost as incredible as if you fired a 15-inch shell at a piece of tissue paper and it came back and hit you."

The only way to account for the alpha particles that reversed direction when they hit the gold foil was to assume that nearly all of the mass as well as all of the positive charge in each individual gold atom was concentrated in a tiny center or **nucleus.** When an alpha particle strikes a nucleus, it reverses direction, much as a cue ball reverses direction when it hits another billiard ball. Rutherford's model placed the other type of charge—the negative electrons—in orbit around this nucleus.

Rutherford's model required that the electrons be in motion. Positive and negative charges attract each other, so stationary electrons would fall into the nucleus. Also, because both the electrons and the nucleus are extremely small, most of the atom is empty, which is why nearly all of Rutherford's alpha particles were able to pass right through the gold foil without colliding with anything. Rutherford's model was a very successful explanation of the experiments he conducted, although eventually scientists would discover that the nucleus has structure as well.

4.4.2 The Atomic Nucleus

The simplest possible atom (and the most common one in the Sun and stars) is hydrogen: The nucleus of ordinary hydrogen contains a single positively charged particle called a *proton.* Moving around this proton is a single *electron.* The mass of an electron is nearly 2000 times smaller than the mass of a proton; the electron carries an amount of charge exactly equal to that of the proton but opposite in sign (Figure 4.13). Opposite charges attract each other, so it is electromagnetic force that holds the proton and electron

Gold foil

Source of
alpha particles

Orbits followed
by negative
charge

(b)

■ FIGURE 4.12

Rutherford's Experiment (a) When Rutherford allowed alpha particles from a radioactive source to strike a target of gold foil, he found that, although most of them went straight through, some rebounded back in the direction from which they came. (b) From this experiment he concluded that the atom must be constructed like a miniature solar system, with the positive charge concentrated in the nucleus and the negative charge orbiting in the large volume around the nucleus.

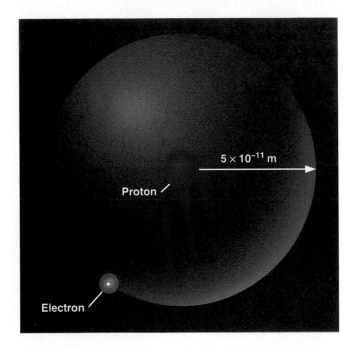

5×10^{-11} m

Proton

Electron

Electron

Electron

2 Protons

2 Neutrons

10^{-10} m

■ FIGURE 4.13

The Hydrogen Atom This is a schematic diagram of a hydrogen atom in its lowest energy state, also called the *ground state*. The proton and electron have equal but opposite charges, which exert an electromagnetic force that binds the hydrogen atom together. In this sketch the size of the particles is exaggerated for educational purposes; they are not to scale.

■ FIGURE 4.14

The Helium Atom Here we see a schematic diagram of a helium atom in its lowest energy state. Two protons are present in the nucleus of all helium atoms. In the most common variety of helium, the nucleus also contains two neutrons, which have nearly the same mass as the proton but carry no charge. Two electrons can be seen orbiting the nucleus.

together, just as gravity is the force that keeps the planets in orbit around the Sun.

There are, of course, other types of atoms. Helium, for example, is the second most abundant element in the Sun. Helium has two protons in its nucleus, instead of the single proton that characterizes hydrogen. In addition, the he-

lium nucleus contains two *neutrons*, particles with a mass comparable to that of the proton but with no electric charge. Moving around this nucleus are two electrons, so the total net charge of the helium atom is also zero (Figure 4.14).

From this description of hydrogen and helium, perhaps you have guessed the pattern for building up all the

■ FIGURE 4.15

Isotopes of Hydrogen A single proton in the nucleus defines the atom to be hydrogen, but there may be zero, one, or two neutrons. The most common isotope is the one with only a single proton. A hydrogen nucleus with one neutron is called deuterium; one with two neutrons is called tritium. The symbol for hydrogen is H; the superscript to the left indicates the total number of particles (protons plus neutrons) in the nucleus.

elements (different types of atoms) that we find in the universe. The type of element is determined by the *number of protons* in the nucleus of the atom. For example, any atom with 6 protons is the element carbon, with 8 protons is oxygen, with 26 is iron, and with 92 is uranium. On Earth, a typical atom has the same number of electrons as protons, and these electrons follow complex orbital patterns around the nucleus. Deep inside stars, however, it is so hot that the electrons get loose from the nucleus and (as we shall see) live separate, productive lives.

The number of neutrons in the nucleus is usually approximately equal to the number of protons, but the number of neutrons is not necessarily the same for all atoms of a given element. For example, most hydrogen atoms contain no neutrons at all. There are, however, hydrogen atoms that contain one proton and one neutron, and others that contain one proton and two neutrons. The various types of hydrogen nuclei with different numbers of neutrons are called **isotopes** of hydrogen (Figure 4.15), and other elements have isotopes as well. You can think of isotopes as siblings in the same element "family"—closely related but with different characteristics and behaviors.

4.4.3 The Bohr Atom

Rutherford's model for atoms has one serious problem. Maxwell's theory of electromagnetic radiation says that when electrons change either their speed or the direction of motion, they must emit energy. Orbiting electrons constantly change their direction of motion, so they should emit a constant stream of energy. According to Maxwell's theory, all electrons should spiral into the nucleus of the atom as they lose energy, and this collapse should happen very quickly—in about 10^{-16} seconds.

It was the Danish physicist Niels Bohr (1885–1962) who solved the mystery of how electrons remain in orbit. He was trying to develop a model of the atom that would also explain certain regularities observed in the spectrum

of hydrogen (see Section 4.5). He suggested that the spectrum of hydrogen can be understood if we assume that orbits of only certain sizes are possible for the electron. Bohr further assumed that as long as the electron moves in only one of these allowed orbits, it radiates no energy. Its energy would change only if it moved from one orbit to another.

This suggestion, in the words of science historian Abraham Pais, was "one of the most audacious hypotheses ever introduced in physics." If something equivalent were at work in the everyday world, you might find that, as you went for a walk after astronomy class, nature permitted you to walk 2 steps per minute, 5 steps per minute, and 12 steps per minute, but no speeds in between. No matter how you tried to move your legs, only *certain* walking speeds would be permitted. To make things more bizarre, it would take no effort to walk at one of the allowed speeds, but it would be difficult to change from one speed to another. Luckily, no such rules apply at the level of human behavior, but at the microscopic level of the atom, experiment after experiment has confirmed the validity of Bohr's strange idea. Bohr's suggestions became one of the foundations of the new (and much more sophisticated) model of the subatomic world called *quantum mechanics.*

In Bohr's model, if the electron moves from one orbit to another closer to the atomic nucleus, it must give up some energy in the form of electromagnetic radiation. If the electron goes from an inner orbit to one farther from the nucleus, however, it requires some additional energy. One way to obtain the necessary energy is to absorb electromagnetic radiation that may be streaming past the atom from an outside source.

A key feature of Bohr's model is that each of the permitted electron orbits around a given atom has a certain energy value. To move from one orbit to another (which will have *its* own specific energy value) requires a change in the electron's energy—a change determined by the difference between the two energy values. If the electron goes to a lower level, the energy difference will be given off; if the electron goes to a higher level, the energy difference must be obtained from somewhere else. Each jump (or transition) to a different level has a fixed and definite energy change associated with it.

A crude analogy for this situation might be life in a tower of luxury apartments where the rent is determined by the quality of the view. Such a building has certain definite numbered levels or floors on which apartments are located. No one can live on floor 5.37 or 22.5. In addition, the rent gets higher as you go up to higher floors. If you want to exchange an apartment on the 20th floor for one on the 2nd floor, you will not owe as much rent and you will get a refund from the landlord. But if you want to move from the 3rd floor to the 25th floor, you'd better be prepared to find some extra resources because your rent will increase. In the atom, too, the "cheapest" place for an electron to live is the lowest possible level, and energy is required to move to a higher level.

Here we have one of the situations where it is easier to think of electromagnetic radiation as photons than as waves (see p. 88). As electrons move from one level to another, they give off or absorb little packets of energy. When an electron moves to a higher level, it absorbs a photon of just the right energy (provided one is available). When it moves to a lower level, it emits a photon with the exact amount of energy it no longer needs in its "lower-cost living situation."

The photon and wave perspectives must be equivalent: Light is light, no matter how we look at it. Thus each photon carries a certain amount of energy that is proportional to the frequency of the wave it represents. The value of its energy is given by the formula

$$E = h \times f$$

where the constant of proportionality, h, is called Planck's constant. It is named for Max Planck, the German physicist who was one of the originators of the quantum theory. If metric units are used (that is, if energy is measured in joules and frequency in hertz), then Planck's constant has the value $h = 6.626 \times 10^{-34}$ joules. Higher-energy photons correspond to higher-frequency waves (which have a shorter wavelength); lower-energy photons are waves of lower frequency.

To take a specific example, consider a calcium atom inside the Sun's atmosphere in which an electron jumps from a lower level to a higher level. To do this, it needs about 5×10^{-19} joules of energy, which it can conveniently obtain by absorbing a passing photon of that energy coming from deeper inside the Sun. This photon is equivalent to a wave of light whose frequency is about 7.5×10^{14} hz and whose wavelength is about 3.9×10^{-7} m (393 nm), in the deep violet part of the visible light spectrum. Although it may seem strange at first to switch from picturing light as a photon (or energy packet) to picturing it as a wave, such switching has become second nature to astronomers and can be a handy tool for doing calculations about spectra.

4.5 FORMATION OF SPECTRAL LINES

We can use Bohr's model of the atom to understand how spectral lines are formed. The concept of energy levels for the electron orbits in an atom leads naturally to an explanation of why atoms absorb or emit only specific energies or wavelengths of light.

4.5.1 The Hydrogen Spectrum

Let's look at the hydrogen atom from the perspective of the Bohr model. Suppose a beam of white light (which consists of photons of all wavelengths) shines through a gas of atomic hydrogen. A photon of wavelength 656 nm has just the right energy to raise an electron in a hydrogen atom from the second to the third orbit. Thus, as all the photons of different energies (or wavelengths or colors) stream by the hydrogen atoms, photons with *this* particular wave-

length can be absorbed by those atoms whose electrons are orbiting on the second level. When they are absorbed, the electrons on the second level will move to the third level, and a number of the photons of this wavelength and energy will be missing from the general stream of white light.

Other photons will have the right energies to raise electrons from the second to the fourth orbit, or from the first to the fifth orbit, and so on. Only photons with exactly these correct energies can be absorbed. All of the other photons will stream past the atoms untouched. Thus, the hydrogen atoms absorb light at only certain wavelengths and produce dark lines at those wavelengths in the spectrum we see.

Suppose we have a container of hydrogen gas through which a whole series of photons is passing, allowing many electrons to move up to higher levels. When we turn off the light source, these electrons "fall" back down from larger to smaller orbits and emit photons of light—but, again, only light of those energies or wavelengths that correspond to the energy difference between permissible orbits. The orbital changes of hydrogen electrons that give rise to spectral lines are shown in Figure 4.16. (Note that in a real hydrogen atom the electron orbits are not as evenly spaced as those shown in this diagram.)

Similar pictures can be drawn for atoms other than hydrogen. However, because these other atoms ordinarily have more than one electron each, the orbits of their electrons are much more complicated, and the spectra are more complex as well. For our purposes, the key conclusion is this: *Each type of atom has its own unique pattern of electron orbits, and no two sets of orbits are exactly alike.* This means that each type of atom shows its own unique set of spectral lines, produced by electrons moving between its unique set of orbits.

Astronomers and physicists have worked hard to learn the lines that go with each element by studying the way atoms absorb and emit light in laboratories here on Earth. Then they can use this knowledge to identify the elements in celestial bodies. In this way, we now know the chemical makeup of the atmospheres of the planets and even galaxies of stars so distant that their light started on its way to us long before Earth had even formed.

4.5.2 Energy Levels and Excitation

Bohr's model of the hydrogen atom was a great step forward in our understanding of the atom. However, we know today that atoms cannot be represented by quite so simple a picture; even the concept of sharply defined electron orbits is not really correct. We can still retain the concept that only certain discrete energies are allowable for an atom. These energies, called **energy levels,** can be thought of as representing certain average distances of the electron's possible orbits from the atomic nucleus.

Ordinarily, an atom is in the state of lowest possible energy, its **ground state.** In the Bohr model, the ground state corresponds to the electron being in the innermost orbit.

FIGURE 4.16

The Bohr Model for Hydrogen   Here we follow the emission or absorption of photons by a hydrogen atom according to the Bohr model. Several different series of spectral lines are shown, corresponding to transitions of electrons from or to certain allowed orbits. Each series of lines that terminates on a specific inner orbit is named for the physicist who studied it. At the top, for example, you see the Balmer series; arrows show electrons jumping from the second orbit ($n = 2$) to the third, fourth, fifth, and sixth orbits. Each time a "poor" electron from a lower level wants to rise to a higher position in life, it must absorb energy to do so. It can absorb the energy it needs from passing waves (or photons) of light. The next set of arrows (Lyman series) shows electrons falling down to the first orbit from different (higher) levels. Each time an electron goes downward toward the nucleus, it can afford to give off (emit) some energy it no longer needs.

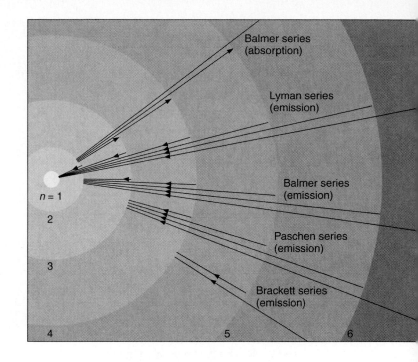

An atom can absorb energy, which raises it to a higher energy level (corresponding, in the Bohr picture, to an electron's movement to a larger orbit). The atom is then said to be in an **excited state.** Generally, an atom remains excited for only a very brief time. After a short interval, typically a hundred-millionth of a second or so, it drops back spontaneously to its ground state, with the simultaneous emission of light. The atom may return to its lowest state in one jump, or it may make the transition in steps of two or more jumps, stopping at intermediate levels on the way down. With each jump it emits a photon of the wavelength that corresponds to the energy difference between the levels at the beginning and end of that jump.

An energy-level diagram for a hydrogen atom and several possible atomic transitions are shown in Figure 4.17; it shows the same series of emission lines as the diagram of the Bohr model in Figure 4.16. When we measure the energies involved as the atom jumps between levels, we find that the transitions to or from the ground state, called the *Lyman series* of lines, result in the emission or absorption of ultraviolet photons. But the transitions to or from the first excited state (labeled $n = 2$ in Figure 4.16), called the *Balmer series,* produce emission or absorption in visible light. In fact, it was to explain this Balmer series that Bohr first suggested his model of the atom.

We mentioned that atoms that have absorbed specific photons from a passing beam of white light and have thus become excited generally de-excite themselves and emit that light again in a very short time. You might wonder, then, why *dark* spectral lines are ever produced. In other words, why doesn't this reemitted light quickly "fill in" the darker absorption lines?

Imagine a beam of white light coming toward you through some cooler gas. Some of the reemitted light *is* ac-

tually returned to the beam of white light you see, but this fills in the absorption lines only to a slight extent. The reason is that the atoms in the gas reemit light *in all directions* and only a small fraction of the reemitted light is in the direction of the original beam (toward you). In a star, much of the reemitted light actually goes in directions leading

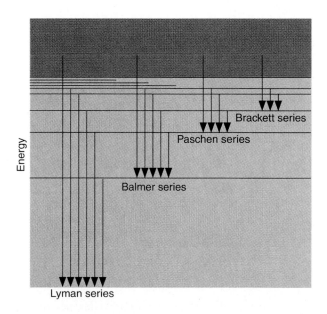

FIGURE 4.17

Energy-Level Diagram for Hydrogen   At higher and higher energy levels, the levels become more and more crowded together, approaching a limit. The shaded region represents energies at which the atom is ionized (the electron is no longer attached to the atom). Each series of arrows represents electrons falling from higher levels to lower ones, releasing photons or waves of energy in the process.

■ FIGURE 4.18

Three Kinds of Spectra When we see a light bulb or other source of continuous radiation (a), all the colors are present. When the continuous spectrum is seen through a thinner gas cloud, the cloud's atoms produce absorption lines in the continuous spectrum (b). When the excited cloud is seen without the continuous source behind it, its atoms produce emission lines (c). We can learn which types of atoms are in the cloud from the pattern of the absorption or emission lines.

back into the star, which does observers outside the star no good whatsoever.

Figure 4.18 summarizes the different kinds of spectra we have discussed. A light bulb produces a continuous spectrum (a in the figure). When that continuous spectrum is viewed through a thinner cloud of gas, an absorption line spectrum can be seen superimposed on the continuous spectrum (b). If we look only at a cloud of excited gas atoms (with no continuous source seen behind it), we see that the excited atoms give off an emission line spectrum (c).

Atoms in a hot gas are moving at high speeds and continually colliding with one another and with any loose electrons. They can be excited and de-excited by these collisions as well as by absorbing and emitting light. The speed of atoms in a gas depends on the temperature. When the temperature is higher, so are the speed and energy of the collisions. The hotter the gas, therefore, the more likely that electrons will occupy the outermost orbits, which correspond to the highest energy levels. This means that the level where electrons *start* their upward jumps in a gas can serve as an indicator of how hot that gas is. In this way, the absorption lines in a spectrum give astronomers information about the temperature of the regions where the lines originate.

4.5.3 Ionization

We have described how certain discrete amounts of energy can be absorbed by an atom, raising it to an excited state and moving one of its electrons farther from its nucleus. If enough energy is absorbed, the electron can be completely removed from the atom. The atom is then said to be **ionized.** The *minimum amount* of energy required to remove one electron from an atom in its ground state is called its *ionization energy.*

Still greater amounts of energy must be absorbed by the now-ionized atom (called an **ion**) to remove an additional electron deeper in the structure of the atom. Successively greater energies are needed to remove the third, fourth, fifth, and so on, electrons from the atom. If enough energy is available, an atom can become completely ionized, losing all of its electrons. A hydrogen atom, having only one electron to lose, can be ionized only once; a helium atom can be ionized twice; and an oxygen atom up to eight times. When we examine regions of the cosmos where there is a great deal of energetic radiation, such as the neighborhoods where hot young stars have recently formed, we see a lot of ionization going on.

An atom that has become ionized has lost a negative charge—the missing electron—and thus is left with a net positive charge. It therefore exerts a strong attraction on any free electron. Eventually, one or more electrons will be captured and the atom will become neutral (or ionized to one less degree) again. During the electron capture process, the atom emits one or more photons. Which photons are emitted depends on whether the electron is captured at once to the lowest energy level of the atom or stops at one or more intermediate levels on its way to the lowest available level.

Just as the excitation of an atom can result from a collision with another atom, ion, or electron (collisions with electrons are usually most important), so can ionization. The rate at which such collisional ionizations occur depends on the speeds of the atoms and hence on the temperature of the gas.

The rate at which ions and electrons recombine also depends on their relative speeds—that is, on the temperature. In addition, it depends on the density of the gas: The higher the density, the greater the chance for recapture because the different kinds of particles are crowded more closely together. From a knowledge of the temperature and density of a gas, it is possible to calculate the fraction of atoms that have been ionized once, twice, and so on. In the Sun, for example, we find that most of the hydrogen and helium atoms in its atmosphere are neutral, whereas most of the calcium atoms, as well as many other heavier atoms, are ionized once.

The energy levels of an ionized atom are entirely different from those of the same atom when it is neutral. Each time an electron is removed from the atom, the energy levels of the ion, and thus the wavelengths of the spectral lines it can produce, change. This helps astronomers differentiate the ions of a given element. Ionized hydrogen, having no electron, can produce no absorption lines.

4.6 THE DOPPLER EFFECT

Although the last two sections contain many new concepts, we hope you have seen one major idea emerge. Astronomers can learn about the elements in planetary atmospheres, stars, and galaxies by decoding the information in their spectral lines. There is a complicating factor in learning how to decode the message of starlight, however. If a planet or star is moving toward or away from us, its lines will be in a slightly different place in the spectrum from where they would be at rest. And most objects in the universe do have some motion relative to the Sun.

4.6.1 Motion Affects Waves

In 1842 Christian Doppler pointed out that if a light source is approaching or receding from the observer, the light waves will be, respectively, crowded more closely together or spread out. (This is actually true for all kinds of waves, including sound: Doppler first measured the effect of motion on waves by hiring a group of musicians to play on an open railroad car as it was moving along the track.) The general principle, now known as the **Doppler effect,** is illustrated in Figure 4.19.

In part (a) of the figure, the light source (S) is at rest with respect to the observer. The source gives off a series of waves with crests labeled 1, 2, 3, and 4. The light waves spread out evenly in all directions, like the ripples from a splash in a pond. The crests are separated by a distance, λ, where λ is the wavelength. The observer, who happens to be located in the direction of the bottom of the page, sees the light waves coming nice and evenly, one wavelength apart. Observers located anywhere else would see the same thing.

On the other hand, if the source is moving with respect to the observer, as in part (b), the situation is more complicated. Between the time one crest is emitted and the next is ready to come out, the source has moved a bit—in our case, toward the bottom of the page and toward our original observer. By moving, the source has decreased the

■ **FIGURE 4.19**

The Doppler Effect (a) A source S makes waves whose numbered crests (1, 2, 3, 4) wash over a stationery observer. (b) The source S now moves toward observer *A* and away from observer *C*. Wave crest 1 was omitted when the source was at position S1, crest 2 at position S2, and so forth. Observer *A* sees the waves compressed by this motion and sees a blueshift (if the waves are light). Observer *C* sees the waves stretched out by the motion and sees a redshift. Observer *B*, whose line of sight is perpendicular to the source's motion, sees no change in the waves (and feels left out).

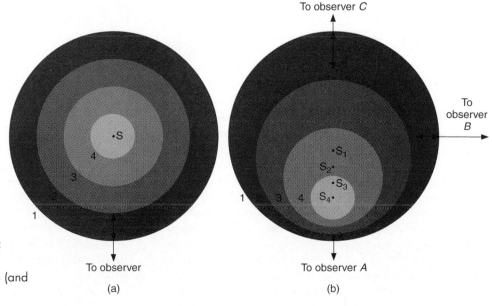

(a)

(b)

distance between crests—squeezed the crests together a bit, you might say.

In part (b), the sources are in four positions, S_1, S_2, S_3, and S_4, each corresponding to the emission of one wave crest. To our original observer, labeled observer *A,* the waves seem to follow one another more closely, at a decreased wavelength and thus increased frequency. (Remember, all light waves travel at the speed of light, no matter what. This means that motion cannot affect the speed, but only the wavelength and the frequency. As the wavelength decreases, the frequency must increase. If the waves are shorter, more will be able to fit by during each second.)

The situation is not the same for other observers. Let's look at the situation from the point of view of observer *C,* located opposite observer *A* in the figure. For her, the source is moving away from her location. As a result, the waves are not squeezed together but instead are spread out by the motion of the source. The crests arrive with an increased wavelength and decreased frequency. To observer *B,* in a direction at right angles to the motion of the source, no effect is observed. The wavelength and frequency remain the same as they were in part (a) of the figure.

We can see from this illustration that the Doppler effect is produced only by a motion toward or away from the observer, a motion called **radial velocity.** Observers between *A* and *B* and between *B* and *C* would observe, respectively, some shortening or lengthening of the light waves for that part of the motion of the source that is along their line of sight.

You have probably heard the Doppler effect with sound waves. When a train whistle or police siren approaches you and then moves away, you will notice a decrease in the pitch (or frequency) of the sound waves. They have changed from slightly more frequent than at rest when coming toward you, to slightly less frequent than at rest when moving away from you. (A nice example of this change in the sound of a train whistle can be heard at the end of the Beach Boys song "Caroline, No" on their album *Pet Sounds.*)

4.6.2 Color Shifts

When the source of waves moves toward you, the wavelengths decrease a bit. If the waves involved are visible light, then the colors of the light change slightly. As wavelengths decrease, they shift toward the blue end of the spectrum: Astronomers call this a *blueshift.* (Since the end of the spectrum is really violet, the term should probably be *violetshift,* but blue is a more common color.) When the source moves away from you and the wavelengths get longer, we call the change in colors a *redshift.* Because the Doppler effect was first used with visible light in astronomy, the terms *blueshift* and *redshift* became well established. Today astronomers use these words to describe changes in radio wavelengths or x-ray wavelengths as comfortably as they use them to describe changes in visible light.

The greater the motion toward or away from us, the greater the Doppler shift. If the relative motion is entirely along the line of sight, the formula for the Doppler shift of light is

$$\frac{\Delta\lambda}{\lambda} = \frac{v}{c}$$

where λ is the wavelength emitted by the source, $\Delta\lambda$ is the difference between λ and the wavelength measured by the observer, c is the speed of light, and v is the relative speed of the observer and the source in the line of sight. The variable v is counted as positive if the velocity is one of recession, and negative if it is one of approach. Solving this equation for the velocity, we find

$$v = c \times \frac{\Delta\lambda}{\lambda}$$

If a star approaches or recedes from us, the wavelengths of light in its continuous spectrum appear shortened or lengthened, respectively, as do those of the dark lines. However, unless its speed is tens of thousands of kilometers per second, the star does not appear noticeably bluer or redder than normal. The Doppler shift is thus not easily detected in a continuous spectrum and cannot be measured accurately in such a spectrum. The wavelengths of the absorption lines can be measured accurately, however, and their Doppler shift is relatively simple to detect.

This may sound like a horrible note on which to end the chapter. If all the stars are moving and motion changes the wavelength of each spectral line, won't this be a disaster for astronomers trying to figure out what elements are present in the stars? After all, it is the precise wavelength (or color) that tells astronomers which lines belong to which element. And we first measure these wavelengths in containers of gas in our laboratories, which are not moving. If every line in a star's spectrum is now shifted by its motion to a different wavelength (color), how can we be sure which lines and which elements we are looking at in a star whose speed we do not know?

Take heart: This situation sounds worse than it really is. Astronomers rarely judge the presence of an element in an astronomical object by a single line. It is the *pattern* of lines unique to hydrogen or calcium that enables us to determine that those elements are part of the star or galaxy we are observing. The Doppler effect does not change the pattern of lines from a given element—it only shifts the whole pattern slightly toward redder or bluer wavelengths. The shifted pattern is still quite easy to recognize. Best of all, when we do recognize a familiar element's pattern, we get a bonus: The amount the pattern is shifted now tells us the speed of the object in our line of sight.

We hope you can see why the training of new astronomers includes so much work on learning to decode light (and other electromagnetic radiation). We have seen that a skillful "decoder" can learn the temperature of a star, what elements are in it, and even its speed in a direction toward us or away from us. That's really an impressive bit of decoding for stars that are light years away.

SkyView: skyview.gsfc.nasa.gov/
Multi-wavelength "virtual observatory" that allows you to dial up any part of the sky and ask to see what it looks like in various bands of the electromagnetic spectrum.

Tutorial on Temperature:
www.unidata.ucar.edu/staff/blynds/tmp.html
An introduction to the idea and measurement of temperature, with useful historical information and applications in the realms of astronomy.

Radiation Demonstration:
www-astro.phast.umass.edu/courseware/vrml/bb/

Lets you see blackbody curves for different temperatures and explore some of the radiation laws we mention briefly in this chapter; can get technical in places.

Doppler Effect Activity:
www.explorescience.com/news/june2001.cfm
Gives background information on the Doppler effect and lets you play with wave sources moving at different speeds.

Introduction to Bohr Theory:
www.bclp.net/~kdrews/bohr/bohr.html
Gives good information about many aspects of Bohr's work and its application to spectra.

SUMMARY

4.1 James Clerk Maxwell showed that whenever charged particles change their motion, as they do in every atom and molecule, they give off waves of energy. Light is one form of this **electromagnetic radiation.** The **wavelength** of light determines the color of visible radiation. Wavelength (λ) is related to **frequency** (f) and the speed of light (c) by the equation $c = \lambda f$. Electromagnetic radiation sometimes behaves like waves, but at other times it behaves as if it were a little packet of energy, called a **photon.** The apparent brightness of a source of electromagnetic energy decreases with increasing distance from that source in proportion to the square of the distance, a relationship known as the **inverse-square law.**

4.2 The **electromagnetic spectrum** consists of **gamma rays, x rays, ultraviolet radiation** (all forms of electromagnetic radiation with wavelengths shorter than that of visible light), **visible light, infrared,** and **radio radiation** (the last two with wavelengths longer than that of visible light). Many of these wavelengths cannot penetrate the layers of the Earth's atmosphere and must be observed from space. The emission of electromagnetic radiation is intimately connected to the temperature of the source. The higher the temperature of an idealized emitter of electromagnetic radiation, the shorter is the wavelength at which the maximum amount of radiation is emitted. The mathematical equation describing this relationship $(\lambda_{max} = 3 \times 10^6/T)$ is known as **Wien's law.** The total power emitted per square meter increases with increasing temperature. The relationship between emitted **energy flux** and temperature $(F = \sigma T^4)$ is known as the **Stefan–Boltzmann law.**

4.3 A **spectrometer** is a device that forms a spectrum, often utilizing the optical phenomenon of **dispersion.** The light from an astronomical source can consist of a **contin-** uous spectrum, a bright line or **emission line spectrum,** or a dark line or **absorption line spectrum.** Because each element leaves its spectral "signature" in the pattern of lines we observe, spectral analyses reveal the composition of the Sun and stars.

4.4 Atoms consist of a **nucleus** containing one or more positively charged protons. All atoms except hydrogen also contain one or more neutrons in the nucleus. Negatively charged electrons orbit the nucleus. The number of protons defines an element (hydrogen, helium, and so on) of the atom. Nuclei with the same number of protons but different numbers of neutrons are different **isotopes** of the same element.

4.5 When an electron moves from one **energy level** to another, a photon is emitted and a spectral emission line is formed. Absorption lines are formed when an electron moves to a higher energy level. Since each atom has its own characteristic set of energy levels, each is associated with a unique pattern of spectral lines. An atom in its lowest energy level is in the **ground state.** If an electron is in an orbit other than the least energetic one possible, the atom is said to be **excited.** If an atom has lost one or more electrons, it is called an **ion** and is said to be **ionized.** The spectra of different ions look different and can tell astronomers about the temperatures of the source they are observing.

4.6 If an atom is moving toward us when an electron changes orbits and produces a spectral line, we see that line shifted slightly toward the blue of its normal wavelength in a spectrum. If the atom is moving away, we see the line shifted toward the red. This shift is known as the **Doppler effect** and can be used to measure the **radial velocities** of distant objects by the formula $v = c(\Delta\lambda/\lambda)$.

A Have your group make a list of all the electromagnetic wave technology you use during a typical day.

 B How many applications of the Doppler effect can your group think of in everyday life? For example, why would the highway patrol find it useful?

C Have members of your group go home and "read" the face of your radio set and then compare notes. What do all the words and symbols mean? What frequencies can your radio tune in on? What is the frequency of your favorite radio station? What is its wavelength?

D If your instructor gave you a spectrometer, what kind of spectra does your group think you would see from each of the following: (1) a household lightbulb, (2) the Sun, (3) the "neon lights of Broadway," (4) an ordinary household flashlight, and (5) a streetlight on a busy shopping street?

E Suppose astronomers wanted to send a message to an alien civilization that is living on a planet with an atmosphere very similar to that of Earth's. This message must travel through space, make it through the other planet's atmosphere, and be noticeable to the residents of that planet. Have your group discuss what band of the electromagnetic spectrum might be best for this message and why. (Some people, including an earlier Congress, have warned scientists not to send such messages and reveal the presence of our civilization to a possible hostile cosmos. Do you agree with this concern?)

REVIEW QUESTIONS

1. What distinguishes one type of electromagnetic radiation from another? What are the main categories (or bands) of the electromagnetic spectrum?
2. What is a wave? Use the terms *wavelength* and *frequency* in your definition.
3. Is your textbook the kind of idealized object (described in Section 4.2.3) that absorbs all the radiation falling on it? Explain. How about the black sweater worn by one of your classmates?
4. Where in an atom would you expect to find electrons? Protons? Neutrons?
5. Explain how emission lines and absorption lines are formed. In what sorts of cosmic objects would you expect to see each?
6. Explain how the Doppler effect works for sound waves, and give some familiar examples.
7. What kind of motion for a star does not produce a Doppler effect? Explain why.
8. Describe how Bohr's model used the work of Rutherford and Maxwell. Why was Bohr's model considered a radical notion?

THOUGHT QUESTIONS

9. Make a list of some of the many practical consequences of Maxwell's theory of electromagnetic waves (television is one example).

10. Explain how you would calculate the total amount of power radiated into space by the Sun. What information about the Sun would you need to make this calculation?

11. What type of electromagnetic radiation is best suited to observing a star with:
a. a temperature of 5800 K
b. a gas heated to a temperature of 1 million K
c. a person on a dark night

12. Why is it dangerous to be exposed to x rays but not (or at least much less) dangerous to be exposed to radio waves?

13. Go outside on a clear night and look carefully at the brightest stars. Some should look slightly red and others slightly blue. The primary factor that determines the color of a star is its temperature. Which is hotter: a blue star or a red one? Explain.

14. Water faucets are often labeled with a red dot for hot water and a blue dot for cold. Given Wien's law, does this labeling make sense?

15. The planet Jupiter appears yellow and Mars is red. Does this mean that Mars is cooler than Jupiter? Explain your answer.

16. Suppose you are standing at the exact center of a park surrounded by a circular road. An ambulance drives completely around this road, with siren blaring. How does the pitch of the siren change as it circles around you?

17. How could you measure Earth's orbital speed by photographing the spectrum of a star at various times throughout the year? (*Hint:* Suppose the star lies in the plane of Earth's orbit.)

FIGURING FOR YOURSELF

The formula on page 88 for the relationship between the speed and other characteristics of a wave can be derived from our basic understanding of motion. The speed of anything that is moving is

$$\text{speed} = \frac{\text{distance traveled}}{\text{time taken}}$$

(So, for example, a car on the highway traveling at a speed of 100 km/h covers 100 km during the time of 1 h.) For an electromagnetic wave to travel the distance of one of its wavelengths, λ, at the speed of light, c, we have $c = \lambda/t$. The frequency of a wave is the number of cycles per second. If a wave has a frequency of a million cycles per second, then the time for each cycle to go by is a millionth of a second. So $t = 1/f$. Substituting into our wave equation, we get $c = \lambda \times f$.

18. What is the wavelength of the carrier wave of a campus radio station, broadcasting at a frequency of 97.2 Mhz (million cps)?

19. What is the frequency of a red laser beam, with a wavelength of 670 nm, that your astronomy instructor might use to point to slides during a lecture on galaxies?

20. You go to a dance club to forget how hard your astronomy midterm was. What is the frequency of a wave of ultraviolet light coming from a blacklight in the club, if its wavelength is 150 nm?

21. What is the energy of the photon with the frequency you calculated in problem 20?

22. "Tidal waves," or tsunamis, are waves caused by earthquakes that travel rapidly through the ocean. If tsunamis travel at the speed of 600 km/h and approach a shore at the rate of one wave crest every 15 min, what would be the distance between those wave crests at sea?

23. How many times brighter or fainter would a star appear if it were moved to:
a. twice its present distance
b. ten times its present distance
c. half its present distance

24. Two stars with identical diameters are the same distance away. One has a temperature of 5800 K; the other has a temperature of 2900 K. Which is brighter? How much brighter is it?

25. If the emitted infrared radiation from Pluto has a wavelength of maximum intensity at 50,000 nm (50 μm), what is the temperature of Pluto (assuming it follows Wien's law)?

26. What is the temperature of a star whose maximum light is emitted at a wavelength of 290 nm?

27. Suppose a spectral line of some element, normally at 500 nm, is observed in the spectrum of a star to be at 500.1 nm. How fast is the star moving toward or away from the Earth?

SUGGESTIONS FOR FURTHER READING

Augensen, H. and Woodbury, J. "The Electromagnetic Spectrum" in *Astronomy*, June 1982, p. 6.

Bova, B. *The Beauty of Light.* 1988, Wiley. A readable introduction to all aspects of the production and decoding of light by a science writer.

Connes, P. "How Light Is Analyzed" in *Scientific American*, Sept. 1968.

Darling, D. "Spectral Visions: The Long Wavelengths" in *Astronomy*, Aug. 1984, p. 16; "The Short Wavelengths" in *Astronomy*, Sept. 1984, p. 14.

Gingerich, O. "Unlocking the Chemical Secrets of the Cosmos" in *Sky & Telescope*, July 1981, p. 13.

Gribbin, J. *In Search of Schroedinger's Cat.* 1984, Bantam; *Schroedinger's Kittens and the Search for Reality.* 1995, Little, Brown. Clear, basic introductions to the fundamental ideas of quantum mechanics by a British physicist/science writer.

Hearnshaw, J. *The Analysis of Starlight.* 1986, Cambridge U. Press. A history of spectroscopy.

Sobel, M. *Light.* 1987, U. of Chicago Press. An excellent nontechnical introduction to all aspects of light.

Stencil, R. et al. "Astronomical Spectroscopy" in *Astronomy*, June 1978, p. 6.

The Hubble Space Telescope (HST) Astronauts Story Musgrave and Jeffrey Hoffman are seen refurbishing the orbiting telescope in space. Australia's west coast can be seen in the background.

NASA

5 Astronomical Instruments

Beyond the [faintest stars the eye can see] you will behold through the telescope a host of other stars, which escape the unassisted sight, so numerous as to be almost beyond belief . . .

Galileo Galilei in *Siderius Nuncius*, 1610, reporting on his first observations of the night sky with a telescope

THINKING AHEAD

When you look at the night sky far from city lights on a camping trip, there seem to be an overwhelming number of stars up there. In reality, only about 6000 stars are visible to the unaided eye. The light from most stars is so weak by the time it reaches Earth that it cannot register on human vision. How can we learn about the vast majority of objects in the universe that our unaided eyes simply cannot see?

In this chapter we describe the tools astronomers use to extend their vision. We have learned almost everything we know about the universe from studying electromagnetic radiation (see Chapter 4). In the 20th century, access to space made it possible to detect electromagnetic radiation at all wavelengths from gamma rays to radio waves. The different wavelengths carry different kinds of information, and the appearance of any given object often depends on the wavelength at which the observations are made (Figure 5.1).

There are three basic components in a modern system for measuring radiation from astronomical sources. First, there is a **telescope,** which serves as a "bucket" for collecting visible light (or radiation at other wavelengths). Just as you can catch more rain with a garbage can than with a coffee cup, large telescopes gather much more light than your eye can. Second, there is an instrument attached to the telescope that sorts the incoming radiation according to wavelength. Sometimes the sorting is fairly crude. For example, we might simply want to measure blue light separately from red light so that we

VISIBLE LIGHT

(a)

Infrared Processing and Analysis Center/JPL

INFRARED

(c)

Infrared Processing and Analysis Center/JPL

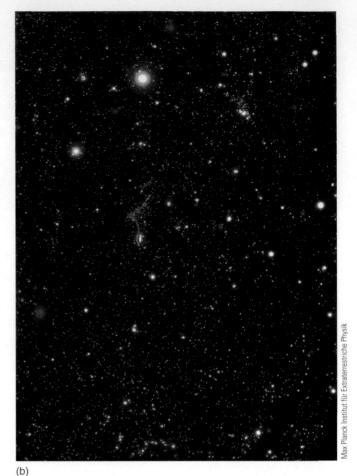

(b)

Max Planck Institut für Extraterrestriche Physik

FIGURE 5.1

The Orion Region of the Sky at Different Wavelengths   The same part of the sky looks different when observed with instruments sensitive to different bands of the spectrum. (a) Visible light. This shows part of the Orion region as the human eye sees it. Note the stars that form the constellation of Orion, the hunter. (b) X rays. Here the view emphasizes the pointlike x-ray sources in the vicinity. The colors change from yellow to white to blue with increasing energy of the x rays. The bright, hot stars in Orion are still seen in this image, but so are many other objects located at very different distances, including other stars, white dwarfs, galaxies, and quasars at the edge of the observable universe. (The belt and sword of Orion are shown for reference.) *(ROSAT)* (c) Infrared radiation. Here we see the glowing dust in this region.

can determine the color of a star and estimate its temperature. But at other times we want to look at very narrow wavelength regions so that we can see individual spectral lines (see Chapter 4) and determine what an object is made of. Third, there is some kind of **detector,** a device that can sense the radiation in the wavelength regions we have chosen and make a permanent record of the observations.

The history of the development of astronomical telescopes is really just the story of how new technologies have been applied to improve the efficiency of each of these three basic components: the telescopes, the wavelength-sorting devices, and the detectors. Let's look at each in turn.

5.1   TELESCOPES

Many ancient cultures constructed special sites for observing the sky (Figure 5.2). At these ancient *observatories* they could measure the positions of celestial objects, mostly to keep track of time and date. Many of these ancient observatories had religious and ritual functions as well. The eye

(a)

(b)

■ **FIGURE 5.2**

Two Pretelescopic Observatories   (a) The Jantar Mantar, built in 1724 by Maharaja Jai Singh in Delhi, India. (b) Seventeenth-century bronze instruments from the old Chinese imperial observatory, Beijing.

was the only device available to gather light, all of the colors in the light were observed at once, and the only permanent record of the observations was made by human beings writing down or sketching what they saw.

In 1610 Galileo first used a simple tube with lenses that he could hold in his hand to observe the sky and gather more light than his eyes alone could. Even this small telescope revolutionized ideas about the nature of the planets and the position of the Earth, and it got Galileo into a lot of trouble with church authorities (as we explained in Chapter 1).

5.1.1   How Telescopes Work

Telescopes have come a long way since Galileo's time. Now they tend to be huge devices; the most expensive cost hundreds of millions of dollars. The reason astronomers keep building bigger and bigger telescopes is that celestial objects, such as planets, stars, and galaxies, send much more light to Earth than any human eye (with its tiny opening) can catch. There is plenty of starlight to go around: If you have ever watched the stars with a group of friends, you know that each of you can see each of the stars. If a thousand more people were watching, each of them would also catch a bit of each star's light. Yet, as far as you are concerned, the light not shining into your eye is wasted. It would be great if some of this "wasted" light could also be captured and brought to your eye. This is precisely what a telescope does.

The most important functions of a telescope are (1) to *collect* the faint light from an astronomical source and (2) to *focus* all the light gathered up into an image. Most objects of interest to astronomers are extremely faint; the

more light we can collect, the better we can study such objects. (Although we use the word *light,* some types of telescopes collect not visible light, but each of the various forms of electromagnetic radiation. We want no light-chauvinists among our readers!)

Telescopes that collect visible radiation use a lens or mirror (which we will discuss in a moment) to gather up the light. Other types of telescopes may use collecting devices that look very different from the lenses and mirrors that we are familiar with, but they serve the same function. In all types of telescopes, the light-gathering ability is determined by the area of the device acting as the light-gathering bucket. Since most telescopes have round mirrors or lenses, we can compare their light-gathering power by comparing their **apertures** (or diameters).

The amount of light a telescope can collect increases with the square of the aperture. A telescope with a mirror that is 4 meters (m) in diameter can collect 16 times as much light as a telescope that is 1 m in diameter. (The diameter is squared because the area of a circle equals $\pi d^2/4$, where d is the diameter of the circle.)

After the telescope forms an image, we need to have some way to detect and record it so that we can measure, reproduce, and analyze the image in a host of ways. Before the 19th century, astronomers simply viewed images with their eyes and wrote descriptions of what they saw. This was very inefficient and did not lead to a very reliable long-term record; eyewitness testimony is notoriously inaccurate!

In the 19th century, the use of photography became widespread. In those days, photographs were a chemical record of an image on a specially treated glass plate. Today the image is generally detected with sensors similar to those in television cameras, recorded electronically, and stored in computers. This permanent record can then be used for detailed studies. Professional astronomers rarely look through the large telescopes that they use for their research.

5.1.2 Formation of an Image by a Lens or a Mirror

Whether or not you wear glasses, you see the world through lenses; they are key elements of your eyes. A lens is a transparent piece of material that bends the rays of light passing through it. If the light rays are parallel, the lens brings them together in one place to form an image (Figure 5.3). If the curvatures of the lens surfaces are just right, all parallel rays of light (say, from a star) are bent, or *refracted*, in such a way that they converge toward a point, called the **focus** of the lens. At the focus, an image of the light source appears. In the case of parallel light rays, the distance of the focus, or image, behind the lens is called the *focal length* of the lens.

As you look at Figure 5.3, you may ask why two rays of light from the same star would be parallel to each other. After all, if you draw a picture of a star shining in all directions, the rays of light coming from the star don't look parallel at all. But remember that the stars (and other astronomical objects) are all outrageously far away. By the time the rays of light headed our way actually arrive at Earth, they are—for all practical purposes—parallel to each other. Put another way, any rays that were *not* more or less parallel to the ones pointed at Earth are now heading in some very different direction in the universe.

To view the image formed by the primary lens in a telescope, we can install another lens called an *eyepiece* (Fig-

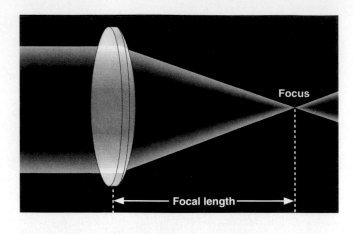

■ **FIGURE 5.3**
Formation of an Image by a Simple Lens Parallel rays from a distant source are bent by the convex lens so that they all come together in a single place (the focus) to form an image.

ure 5.4). This lens can magnify the image. Stars are points of light and magnifying them makes little difference, but the image of a planet or galaxy, which has structure, can often benefit from magnification, just as a tiny image in a magazine can sometimes look clearer through a magnifying glass. Using different eyepieces, we can change the magnification of the image.

A telescope can also be built by using a concave *mirror*—one curved like the inner surface of a sphere—to form an image (see Figure 5.4). Telescope mirrors are coated with a shiny metal, usually silver or aluminum or oc-

(a) Refractor (b) Reflector

■ **FIGURE 5.4**
Refracting and Reflecting Telescopes (a) Light enters a *refracting* telescope through a lens at the upper end, which focuses the light near the bottom of the telescope. An eyepiece then magnifies the image so that it can be viewed by the eye, or a detector like a photographic plate can be placed at the focus. (b) The upper end of a *reflecting* telescope is open, and the light passes through to the primary mirror, located at the bottom of the telescope. The mirror then focuses the light at the top end, where it can be detected. Alternatively, as in this sketch, a second mirror may reflect the light to a position outside the telescope structure, where an observer can have easier access to it.

casionally gold, to make them highly reflecting. If the mirror has the correct shape, all parallel rays are reflected back to the same point, the focus of the mirror. Thus, images are produced by a mirror exactly as they are by a lens.

Many people, when thinking of a telescope, picture a long tube with a large glass lens at one end. This design is called a **refracting telescope.** Galileo's telescopes were refractors, as are binoculars and opera glasses today (Figure 5.5). There is, however, a limit to the size of refracting telescopes. The largest one ever built is the 40-in. refractor at Yerkes Observatory in Wisconsin. The main problem is that the light must pass *through* the lens of a refractor. That means the glass must be perfect all the way through, but it has proven to be impossible to make large pieces of glass without flaws and bubbles in them.

Also, since the light must pass through the lens, it can be supported only around the edges. The force of gravity will cause a large lens to sag and distort the path of the light rays as they pass through it. Finally, again because the light passes through it, both sides of the lens must be manufac-

tured to precisely the right shape in order to produce sharp images.

Telescopes designed with mirrors avoid all of these problems. Because the light is reflected from the front surface only, flaws and bubbles within the glass do not affect the path of the light. Only the front surface has to be manufactured to a precise shape, and the mirror can be supported from the back. For these reasons, most astronomical telescopes (both amateur and professional) use a mirror rather than a lens as their primary optical part; these are called **reflecting telescopes** (see Figure 5.4).

The first successful reflecting telescope was built by Newton in 1668. The concave mirror is placed at the bottom of a tube or open framework. The mirror reflects the light back up the tube to form an image near the front end at a location called the *prime focus*. The image can be observed at the prime focus, or additional mirrors can intercept the light and redirect it to a position where the observer can get at it easily (Figure 5.6). Since an astronomer at the prime focus can block quite a bit of the light coming to the main mirror, the use of a small *secondary mirror* allows more light to get through the system.

Eyepiece

Prism

Objective

Light path in binoculars

■ **FIGURE 5.5**

Binoculars Binoculars are common examples of refracting telescopes. The main light-collecting element in each side is the initial (objective) lens. After following a path through several prisms (used to bend the light and shorten the length of the instrument), the light is viewed through a magnifying eyepiece. Binoculars make the image both brighter and larger, helping you see what is happening on stage or on the field—even if you can't afford really good tickets.

(a) Prime focus   (b) Newtonian focus   (c) Cassegrain focus

■ **FIGURE 5.6**

Focus Arrangements for Reflecting Telescopes
Standard reflecting telescopes have three options for where the light is brought to a focus. (a) Prime focus: Light is detected where it comes to a focus after reflecting from the primary mirror. (b) Newtonian focus: Light is reflected by a small secondary mirror off to one side, where it can be detected (see also Figure 5.4b). (c) Cassegrain focus: Light is reflected by a small secondary mirror back down through a hole in the primary mirror and detected at a station below.

Since the time of Newton, when the sizes of the mirrors in telescopes were measured in inches, reflecting telescopes have grown ever larger. In 1948, U.S. astronomers built a telescope with a 5-m (200-in.) diameter mirror on Palomar Mountain in southern California. It remained the largest visible-light telescope in the world for several decades. The giants of today, however, have primary mirrors (i.e., the largest mirrors in the telescope) that are 8 to 10 m in diameter (Figure 5.7).

5.2.1   Modern Visible-Light and Infrared Telescopes

The decade starting in 1990 saw telescope building around the globe at an unprecedented rate (see Table 5.1, which also includes Web sites for each telescope in case you want to visit or learn more about them). Breakthroughs in technology made it possible at long last to build telescopes significantly larger than the 5-m telescope at Palomar for a reasonable cost. All of these new telescopes have been designed to work well in the infrared as well as visible wavelengths.

The differences between the Palomar telescope and the modern Gemini North telescope are easily seen in Figure 5.8. The Palomar telescope is a massive steel structure designed to hold the 14.5-ton primary mirror. Glass tends to sag under its own weight, and a sagging telescope mirror will distort the image. The Palomar telescope was designed to minimize the sagging with a mirror so thick that it is stiff and rigid. Thick also means heavy, however, and hence a huge steel structure is needed to hold the mirror. A mirror 8 m in diameter built using the same technology would have to weigh at least eight times as much and would require an enormous steel structure to support it.

The 8-m Gemini North telescope looks like a featherweight by contrast, and indeed it is. The mirror is only about 8 in. thick and weighs 24.5 tons, less than twice as much as the Palomar mirror. This mirror does sag, but with modern computers it is possible to measure that sag many times each second and apply forces at 120 different locations on the back of the mirror to correct the sag. This *active control* of the mirror shape makes it possible to build huge lightweight telescopes; 17 telescopes with mirrors 6.5 m in diameter and larger have been constructed since 1990.

The twin 10-m Keck telescopes on Mauna Kea, which were the first of these new-technology instruments, use precision control in an entirely novel way. Instead of a single primary mirror 10 m in diameter, each Keck telescope achieves its large aperture by combining the light from 36 separate hexagonal mirrors, each 1.8 m wide (Figure 5.9). Computer-controlled actuators constantly adjust these 36 mirrors so that the overall reflecting surface acts like a single mirror with just the right shape to collect and focus the light into a sharp image.

In addition to holding the mirror, the steel structure of a telescope is designed so that the entire telescope can be pointed quickly toward any object in the sky. Since the Earth is rotating, the telescope must have a motorized drive system that moves it backward very smoothly at

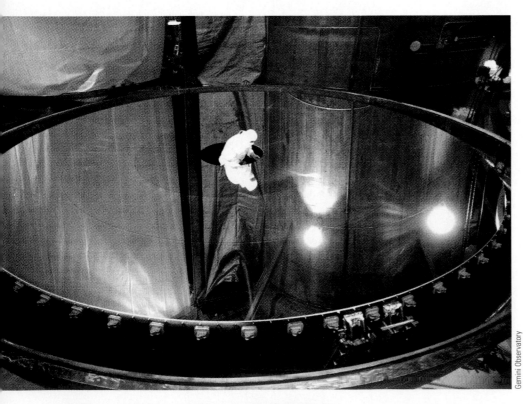

Gemini Observatory

■ FIGURE 5.7

A Large Telescope Mirror This image shows the primary mirror of the Gemini North telescope just after it was coated with aluminum. The mirror is 8 m in diameter; note the person In the center hole in the mirror.

TABLE 5.1 *Large Optical Telescopes Being Built or in Operation*

Aperture (m)	Telescope Name	Location	Status	Web Address
16.4	Very Large Telescope (four 8.2-m telescopes)	Cerro Paranal, Chile°	Four telescopes complete	www.eso.org/vlt/
11.8	Large Binocular Telescope (two 8.4-m telescopes)	Mount Graham, Arizona	First light 2004	medusa.as.arizona.edu/lbtwww/lbt.html
10.0	Keck I and II (two 10-m telescopes)	Mauna Kea, Hawaii	Completed 1993–96	www2.keck.hawaii.edu
10.4	Gran Telescopio Canarias (GTC)	La Palma, Canary Islands	2004	www.gtc.iac.es/home.html
9.1	Hobby–Eberly (HET)	Mount Locke, Texas	Completed 1997	www.as.utexas.edu/mcdonald/het/het.html
9.1	Southern African Large Telescope	Sutherland, South Africa	2003	www.salt.ac.za
8.3	Subaru (Pleiades)	Mauna Kea, Hawaii	First light 1998	www.naoj.org/
8.0	Gemini (North)	Mauna Kea, Hawaii[†]	First light 1999	www.gemini.edu
8.0	Gemini (South)	Cerro Pachon, Chile[†]	First light 2000	www.gemini.edu
6.5	Multi-Mirror (MMT)	Mount Hopkins, Arizona	First light 1998	sculptor.as.arizona.edu/foltz/www/mmt.html
6.5	Magellan (twin telescopes)	Las Campanas, Chile	First light 1997 and 2002	www.ociw.edu/magellan_lco/
6.0	Bolshoi Teleskop	Mount Pastukhov, Russia	Completed 1976	—
5.0	Hale Telescope	Mount Palomar, California	Completed 1948	www.astro.caltech.edu/palomar/
4.2	Herschel Telescope	Canary Islands, Spain	Completed 1987	www.ing.iac.es/Astronomy/telescopes/wht/
4.2	SOAR	Cerro Pachon, Chile	First light 2003	www.soartelescope.org
4.0	Blanco Telescope	Cerro Tololo, Chile[†]	Completed 1974	www.ctio.noao.edu/telescopes/4m/base4m.html
4.0	Visible and Infrared Survey Telescope (Vista)	Cerro Paranal, Chile	2006	www.vista.ac.uk/
3.9	Anglo-Australian (AAT)	Siding Spring, Australia	Completed 1975	www.aao.gov.au/index.html
3.8	Mayall Telescope	Kitt Peak, Arizona[†]	Completed 1973	www.noao.edu/kpno
3.8	United Kingdom Infrared	Mauna Kea, Hawaii	Completed 1979	www.jach.hawaii.edu/JACpublic/UKIRT/home.html
3.7	Advanced Electro-Optical System Telescope	Haleakala, Hawaii	Completed 2000	—
3.6	Telescopio Nazionale Galileo	La Palma, Canaries	Completed 1998	www.tng.iac.es
3.6	Canada–France–Hawaii	Mauna Kea, Hawaii	Completed 1979	www.cfht.hawaii.edu/
3.6	ESO Telescope	Cerro La Silla, Chile°	Completed 1976	www.ls.eso.org/
3.6	ESO New Technology	Cerro La Silla, Chile°	Completed 1989	www.ls.eso.org/
3.5	Max Planck Institut	Calar Alto, Spain	Completed 1983	www.mpia-hd.mpg.de/Public/CAHA/index.html
3.5	WIYN	Kitt Peak, Arizona[†]	Completed 1993	www.noao.edu/wiyn/wiyn.html
3.5	Astrophysical Research Corporation	Apache Point, New Mexico	Completed 1993	www.apo.nmsu.edu/
3.5	Starfire Optical Range	Kirtland AFB, New Mexico	Completed 1994	www.de.afrl.af.mil/SOR/

° Part of the European Southern Observatory (ESO).
[†] Part of the U.S. National Optical Astronomy Observatories (NAOA).

(a)

(b)

■ FIGURE 5.8

Modern Reflecting Telescopes (a) The Palomar 5-m reflector. The Hale telescope on Palomar Mountain has a complex mounting structure that enables the telescope (in the open "tube" pointing upward in this photo) to swing easily into any position. The people in the bottom foreground give you a sense of scale. (b) The Gemini North 8-m telescope. The Gemini North telescope has a mirror twice the diameter of the Palomar mirror, but note how much less massive the whole instrument seems. The Gemini North telescope was completed about 50 years after the Palomar telescope. Engineers took advantage of new technologies to build a telescope that is much lighter in weight relative to the size of the primary mirror.

■ FIGURE 5.9

Thirty-Six Eyes Are Better Than One A close-up of the 10-m Keck telescope's mirror as it was being assembled. The finished mirror is composed of 36 hexagonal sections. In this view, 18 sections have been installed. To get a sense of the mirror's size, note the person working at the center of the mirror assembly.

exactly the same rate the Earth is moving forward, so it can continue to point at the object being observed. All this machinery must be housed in a dome to protect it from the elements. The dome has an opening in it that can be positioned in front of the telescope and moved along with it so that the light from the objects being observed is not blocked.

5.2.2 Picking the Best Observing Sites

A telescope like the Gemini or Keck telescope costs about $100 million to build. That kind of investment simply demands that the telescope be placed in the best possible site. Since the end of the 19th century, astronomers have realized that the best observatory sites are on mountains far from the lights and pollution of cities. Although a number of urban observatories remain, especially in the large cities of Europe, they have become administrative centers or museums. The real action takes place far away, often on desert mountains or isolated peaks in the Atlantic and Pacific Oceans, where we find the living quarters, computers, electronic and machine shops, and of course the telescopes themselves. A large observatory today requires a supporting staff of 20 to 100 people in addition to the astronomers.

The performance of a telescope is determined not only by the size of its mirror but also by its location. The Earth's atmosphere, so vital to life, is a big headache for the observational astronomer. In at least four ways our air imposes limitations on the usefulness of telescopes:

1. The most obvious limitation is weather—clouds, wind, rain, and the like. At the best sites the weather is clear as much as 75 percent of the time.
2. Even on a clear night, the atmosphere filters out a certain amount of starlight, especially in the infrared, where the absorption is due primarily to water vapor. Astronomers therefore prefer dry sites, generally found at the highest altitudes.
3. The sky should also be dark. Near cities the air scatters the glare from lights, producing an illumination that hides the faintest stars and limits the distances that can be probed by telescopes. (Astronomers call this effect *light pollution.*) Observatories are best located at least a hundred miles from the nearest large city.
4. Finally, the air is often unsteady; light passing through this turbulent air is disturbed, resulting in blurred star images. Astronomers call these effects "bad **seeing.**" When seeing is bad, images of celestial objects are distorted by the constant twisting and bending of light rays by turbulent air.

The best observatory sites are therefore high, dark, and dry. The world's largest telescopes are found in such remote mountain locations as the Andes Mountains of Chile (Figure 5.10), the desert peaks of Arizona, the Canary Islands in the Atlantic Ocean, and Mauna Kea in Hawaii, a dormant volcano 13,700 ft (4200 m) high.

5.2.3 The Resolution of a Telescope

In addition to gathering as much light as they can, astronomers also want to have the sharpest images possible. **Resolution** refers to the fineness of detail present in the image. As you can imagine, astronomers are always eager to make out more detail in the images they study, whether they are following the weather on Jupiter or trying to peer into the violent heart of a cannibal galaxy that recently ate its neighbor for lunch.

One factor that determines how good the resolution will be is the size of the telescope. Larger apertures produce sharper images. Until very recently, however, optical and infrared telescopes on the Earth's surface never produced images as sharp as optics theory said they should. The problem—as we saw above—is the atmosphere. The light from an astronomical source has to move through many blobs or cells of gas, ranging from inches to several feet in size, each having a slightly different temperature from its neighbor. Each cell acts like a lens, bending (refracting) the path of the light by a small amount. This slightly changes the position where each light ray finally reaches the detector in a telescope. The cells of air are in motion, constantly being blown through the light path of the telescope by winds,

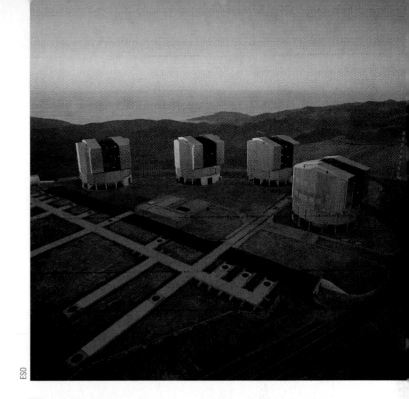

■ **FIGURE 5.10**

A High and Dry Site   Cerro Paranal, a mountain summit 2.7 km above sea level in Chile's Atacama Desert, is the site of the European Southern Observatory's Very Large Telescope. This photograph shows the four 8-m telescope buildings on the site and vividly illustrates that astronomers prefer high and dry sites for their instruments.

often in different directions at different altitudes. As a result, the path followed by the light is constantly changing.

For an analogy, think about watching a ticker tape parade from a window high up in a skyscraper. You decide to throw some confetti down toward the marchers. Even if you drop a handful all at the same time and in the same direction, air currents will toss the pieces around and they will reach the ground in different places. As we described earlier, we can think of the light from the star as a series of parallel beams, each making its way through the atmosphere. Each path will be slightly different, and each will reach the detector of the telescope in a slightly different place. The result is a blurred image, and because the cells are being blown by the wind, the nature of the blur will change many times each second. (You have probably noticed this effect as the "twinkling" of stars seen from Earth. The light beams are bent enough that part of the time they reach your eye, and part of the time some of them miss, thereby making the star seem to vary in brightness. In space, the light of the stars is steady.)

Astronomers scour the world for locations where this atmospheric blurring is as small as possible. It turns out the best sites are in coastal mountain ranges and on isolated volcanic peaks in the middle of an ocean. Air that has flowed long distances over water before it encounters land is especially stable.

The resolution of an image is measured in units of angle on the sky, typically in units of *arcseconds*. One arcsecond is 1/3600 degree, and there are 360 degrees in a full circle. So we are talking about tiny angles on the sky. To give you a sense of just how tiny, we might note that 1 arcsecond is how big a quarter would look when seen from a distance of 5 *kilometers*. The best images obtained from the ground with traditional techniques reveal details as small as several tenths of an arcsecond across. This image size is remarkably good. Astronomers are never satisfied, however, and one of the major reasons for launching the Hubble Space Telescope was to escape the Earth's atmosphere and obtain even sharper images.

Recently, thanks to fast computers, astronomers have devised a technique called **adaptive optics** that can beat the atmosphere at its own game of blurring. This technique (which is most effective in the infrared region of the spectrum with our current technology) makes use of a small flexible mirror placed in the beam of a telescope. A sensor measures how much the atmosphere has distorted the image, and as often as 500 times each second sends instructions to the flexible mirror on how it should change shape in order to compensate exactly for the distortions produced by the atmosphere. The light is thus brought back to a nearly perfectly sharp focus at the detector. Figure 5.11

shows just how effective this new technique is. With adaptive optics, ground-based telescopes can achieve resolutions of 0.1 arcsecond or a little better in the infrared. This impressive figure is the equivalent of the resolution that the Hubble Space Telescope achieves in the visible-light region of the spectrum.

ASTRONOMY BASICS
How Astronomers Really Use Telescopes

In the popular view, an astronomer spends most nights in a cold observatory peering through a telescope, but this picture, which may have had some validity in previous centuries, is not very accurate today. Most astronomers do not live at observatories, but near the universities or laboratories where they work. A typical astronomer might spend only a week or so each year observing at the telescope and the rest of the time measuring or analyzing the data acquired during that week. Many astronomers use radio telescopes or space experiments, which work just as well during the day. Still others work at purely theoretical problems (often using high-speed computers) and never observe at a telescope of any kind.

Even when astronomers are observing with a large telescope, they seldom peer through it. Electronic detectors record the data permanently for detailed analysis after the observations are completed. At some observatories, it is now possible to conduct observations remotely, with the astronomer sitting at an office computer terminal, which can be thousands of miles away from the telescope.

Time on the major telescopes is at a premium, and an observatory director will typically receive many more requests for telescope time than can be accommodated during the year. Astronomers must therefore write a convincing proposal explaining how they would like to use the telescope and why their observations will be important to the progress of astronomy. A committee of astronomers is then asked to judge and rank the proposals, and time is assigned to those with the greatest merit. Even if your proposal is among the high-rated ones, you may have to wait many months for your turn. If the skies are cloudy on the nights you have been assigned, it may be more than a year before you get another chance.

Some older astronomers still remember long, cold nights spent alone in an observatory dome, with only music from a tape recorder or all-night radio station for company. The sight of the stars shining brilliantly hour after hour through the open slit in the observatory dome was unforgettable. So was the relief as the first pale light of dawn announced the end of a 12-hour observing session. Astronomy is much easier today, with teams of observers working together, perhaps without ever leaving a computer workstation in a warm room. But some of the romance has gone from the field, too.

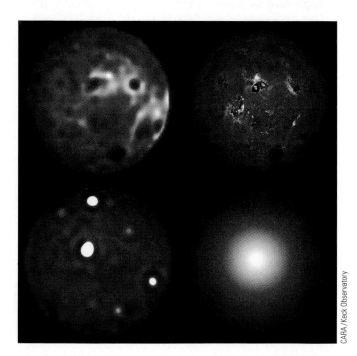

CARA/Keck Observatory

■ FIGURE 5.11

The Power of Adaptive Optics The lower right image shows Jupiter's satellite Io as it appears through the 10-m Keck telescope without adaptive optics; the image is blurred by the Earth's atmosphere. For comparison, the upper right picture was taken by the Galileo spacecraft when it was passing close to the surface of Io. The two left-hand images were taken in the infrared with the Keck telescope plus adaptive optics, which makes corrections for the blurring by the Earth's atmosphere. The top one looks very much like the image taken by Galileo. The lower was taken at a wavelength that is very sensitive to heat, and we see the hot lava erupted by Io's volcanoes.

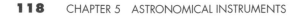

5.3 VISIBLE-LIGHT DETECTORS AND INSTRUMENTS

After a telescope captures its share of the radiation from an astronomical source, the radiation must be *detected* and measured. The first detector used for astronomical observations was the human eye, but it suffers from being connected to an imperfect recording and retrieving device, the human brain. Photography and modern electronic detectors have eliminated the foibles and quirks of human memory by making a permanent record of the information from the cosmos.

The eye also suffers from having a very short *integration time;* it takes only a fraction of a second to add light energy together before sending the image to the brain. One important advantage of modern detectors is that the light from astronomical objects can be collected by the detector over longer periods of time; this technique is called "taking a long exposure." Exposures of several hours are required to detect very faint objects in the cosmos.

Before the light reaches the detector, astronomers normally use some instrument to sort the light according to wavelength. The instrument may be as simple as colored filters, which transmit light within a specified range of wavelengths. Red cellophane is an everyday example of a filter that transmits only red light. After the light passes through the filter, it forms an image, which astronomers can then use to measure the *brightness* and *color* of objects. We will show you many examples of such images in the later chapters of this book, and we will describe what can be learned from them.

Alternatively, the instrument may be a complicated set of optics that spread the light out into its rainbow of colors so that the astronomer can measure individual lines in the spectrum. Such an instrument is called a *spectrometer*, since it allows astronomers to measure (meter) the spectrum of a source of radiation. Both types of wavelength-sorting instruments still have to use detectors to record and measure the properties of light.

5.3.1 Photographic and Electronic Detectors

Throughout most of the 20th century, photographic film or *plates* served as the prime astronomical detectors, whether for direct imaging or for photographing spectra. In a plate, a light-sensitive chemical coating is applied to a piece of glass that, when developed, provides a lasting record of the image. At observatories around the world, vast collections of photographs preserve what the sky has looked like during the past 100 years. Photography represents a huge improvement over the human eye, but it still has serious limitations. Photographic films are inefficient: Only about 1 percent of the light that actually falls on the film contributes to the image; the rest is wasted.

Astronomers today are delighted to have much more efficient electronic detectors to record astronomical images. Most often, these are **charge-coupled devices (CCDs),** which are similar to the detectors used in video camcorders or in digital cameras. In a CCD, photons of radiation generate a stream of charged particles (electrons) that are stored on the detector and counted at the end of the exposure. Because CCDs record as much as 60 to 70 percent of all the photons that strike them, we can detect much fainter objects. Among these are a host of small moons around the outer planets, icy miniplanets beyond Pluto, and dwarf galaxies of stars. CCDs also provide more accurate measurements of the brightness of astronomical objects than photography, and their output can go directly into a computer for analysis.

5.3.2 Infrared Observations

Observing the universe in the infrared band of the spectrum presents some additional challenges. The infrared extends from wavelengths near 1 micrometer (μm), which is about the longwave sensitivity limit of both CCDs and photography, out to 100 μm or longer. Recall from Chapter 4 that infrared radiation is heat radiation. The main challenge to infrared astronomers is to distinguish the tiny amount of heat that reaches the Earth from stars and galaxies from the much greater heat radiated by the telescope itself and our planet's atmosphere.

Typical temperatures on the Earth's surface are near 300 K, and the atmosphere through which observations are made is only a little cooler. According to Wien's law (see Section 4.2), the telescope, the observatory, and even the sky are radiating infrared energy with a peak wavelength of about 10 μm. To infrared eyes, everything on Earth is brightly aglow. The challenge is to detect faint cosmic sources against this sea of light. The infrared astronomer must always contend with the situation that a visible-light observer would face if working in broad daylight with a telescope and optics lined with bright fluorescent lights.

To solve this problem, astronomers must protect the infrared detector from nearby radiation, just as you would shield photographic film from bright daylight. Since anything warm radiates infrared energy, the detector must be isolated in very cold surroundings; often it is held near absolute zero (1 to 3 K) by immersing it in liquid helium. The second step is to reduce the radiation emitted by the telescope structure and optics, and to block this heat from reaching the infrared detector.

Like the infrared, each band of the electromagnetic spectrum presents its own challenges to the astronomer. It is a challenge, for example, to design "mirrors" to reflect such penetrating radiation as x rays and gamma rays, which normally pass straight through matter. However, although the technical details of design are complicated, the three basic components of an observing system are the same at all wavelengths: a telescope to gather up the radiation, filters or an instrument to sort the radiation according to wavelength, and some method for detecting and making a permanent record of the observations.

5.3.3 Spectroscopy

Spectroscopy is one of the astronomer's most powerful tools, providing information about the composition, temperature, motion, and other characteristics of celestial objects. More than half the time spent on most large telescopes is used for spectroscopy.

The many different wavelengths present in light can be separated by passing it through a spectrometer to form a spectrum. The design of a simple spectrometer is illustrated in Figure 5.12. Light from the source (actually, the image of a source produced by the telescope) enters the instrument through a small hole or narrow slit and is collimated (made into a beam of parallel rays) by a lens. The light then passes through a prism, producing a spectrum: Different wavelengths leave the prism in different directions because each wavelength is bent by a different amount when it enters and leaves the prism. A second lens placed behind the prism focuses the many different images of the slit or entrance hole onto a CCD or other detecting device. This collection of images (spread out by color) is the spectrum that astronomers can then analyze at leisure.

In practice, astronomers today are more likely to use a different device, called a *grating*, to disperse the spectrum. A grating is a piece of material with thousands of grooves in its surface. Like a prism, a grating also spreads light out into a spectrum.

5.4 RADIO TELESCOPES

In addition to visible and infrared radiation, radio waves from astronomical objects can also be detected from the surface of the Earth (see Chapter 4). In the early 1930s Karl G. Jansky, an engineer at Bell Telephone Laboratories, was experimenting with antennas for long-range radio communication when he encountered some mysterious static—radio radiation coming from an unknown source (Figure 5.13). He discovered that this radiation came in strongest about 4 min earlier on each successive day and correctly concluded that since the Earth's sidereal rotation period is 4 min shorter than a solar day (see Section 3.3), the radiation must be originating from some region fixed on the celestial sphere. Subsequent investigation showed that the source of this radiation was part of the Milky Way; Jansky had discovered the first source of cosmic radio waves.

In 1936 Grote Reber, who was an amateur astronomer interested in radio communications, built from galvanized iron and wood the first antenna specifically designed to receive cosmic radio waves. Over the years, Reber built several such antennas and used them to carry out pioneering surveys of the sky for celestial radio sources; he remained active in radio astronomy for more than 30 years. During the first decade, he worked practically alone because

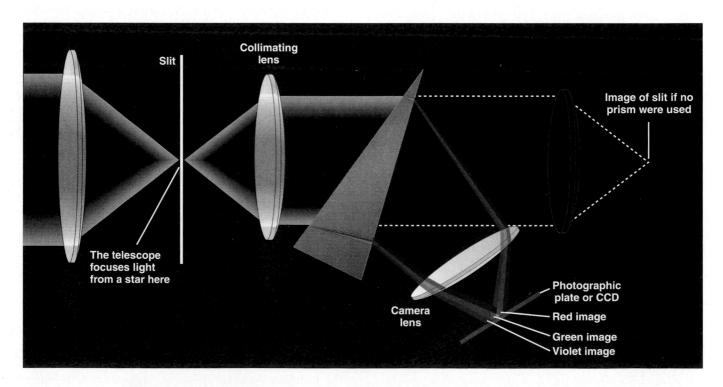

■ **FIGURE 5.12**

A Prism Spectrometer The light from the telescope is focused on a slit, so the light of one object at a time enters the spectrometer. A prism (or grating) disperses the light into a spectrum, which is then photographed or recorded electronically.

George Ellery Hale: Master Telescope Builder

All of the major research telescopes in the world today were begun in the 20th century. The giant among early telescope builders was George Ellery Hale (1868–1938). Not once but four times he initiated projects that led to the construction of what was the world's largest telescope at the time. And he was a master at finding and winning over wealthy benefactors to underwrite the construction of these new instruments.

Hale's training and early research were in solar physics. In 1892, at age 24, he was named Associate Professor of Astral Physics and director of the astronomical observatory at the University of Chicago. At that time, the largest telescope in the world was the 36-in. refractor at the Lick Observatory near San Jose, California. Taking advantage of an existing glass blank for a 40-in. telescope, Hale set out to raise the money for a larger telescope than the one at Lick. One prospective donor was Charles T. Yerkes, who, among other things, ran the trolley system in Chicago.

Hale wrote to Yerkes, encouraging him to support construction of the giant telescope by saying that "the donor could have no more enduring monument. It is certain that Mr. Lick's name would not have been nearly so widely known today were it not for the famous observatory established as a result of his munificence." Yerkes agreed, and the new telescope was completed in May 1897; it remains the largest refractor in the world.

Even before the completion of the Yerkes refractor, Hale was not only dreaming of building a still larger telescope but was also taking concrete steps to achieve that goal. In the 1890s there was a major controversy about the relative quality of refracting and

George Ellery Hale

reflecting telescopes. Hale realized that 40 in. was close to the maximum feasible aperture for refracting telescopes. If telescopes with significantly larger apertures were to be built, they would have to be reflecting telescopes.

Using funds borrowed from his own family, Hale set out to construct a 60-in. reflector. For a site, he left the Midwest for the much better conditions on Mount Wilson, at that time a wilderness peak above the small city of Los Angeles. In 1904, at the age of 36, Hale received funds from the Carnegie Foundation to establish the Mount Wilson Observatory. The 60-in. mirror was placed in its mount in December 1908.

Two years earlier, in 1906, Hale had already approached John D. Hooker, who had made his fortune in hardware and steel pipe, with a proposal to build a 100-in. telescope. The technological risks were substantial. The 60-in. telescope was not yet complete, and the usefulness of large reflectors for astronomy had yet to be demonstrated. George Ellery Hale's brother called him "the greatest gam-

bler in the world." Once again, Hale was successful in obtaining funds, and the 100-in. telescope was completed in November 1917. (It was with this telescope that Edwin Hubble was able to establish that the spiral nebulae were separate islands of stars—or galaxies—quite removed from our own Milky Way.)

Hale was not through dreaming. In 1926, he wrote an article in *Harper's Magazine* about the scientific value of a still larger telescope. This article came to the attention of the Rockefeller Foundation, which granted $6 million (a very large sum in those days) for the construction of a 200-in. telescope. Hale died in 1938, but the 200-in. (5-m) telescope on Palomar Mountain was dedicated ten years later and is now named in Hale's honor.

The Yerkes 40-in. (1-m) refracting telescope.

Choosing Your Own Telescope

If the astronomy course you are taking whets your appetite for exploring the sky further, you may be thinking about buying your own telescope. The good news is that many excellent amateur telescopes are now on the market, and prices are much more reasonable than they were 20 years ago. The bad news is that a good telescope, like a good camera or video recorder, is still not inexpensive, and some research is required to find the model that is best for you. The best sources of information about personal telescopes are the two popular-level magazines published for amateur astronomers: *Sky & Telescope* and *Astronomy*. Both carry regular articles of advice, reviews, and ads from reputable telescope dealers around the country. (See also some Web sites listed at the end of the chapter.)

In some ways, choosing a telescope is like choosing a car: Personal preference can play a major role in the decision. A certain shape or brand name may be important to some car buyers but no big deal to others. In the same way, some of the factors that determine which telescope is right for you depend on your preferences:

- Will you be setting up the telescope in one place and leaving it there, or do you want an instrument that is portable and can come with you on camping trips? Does it have to be carried some distance by a person, or can it get to its destination in a car?
- Do you want to observe the sky with your eyes only, or do you want to take photographs? (Long-exposure photography, for example, requires a good clock-drive to turn your telescope to compensate for the Earth's rotation.)
- What sorts of objects will you be observing? Are you interested primarily in comets, planets, star clusters, or galaxies? Or do you want to observe all kinds of celestial sights?

You may not know the answers to some of these questions yet. For this reason, our number-one recommendation is that you "test-drive" some telescopes first. Most communities have amateur astronomy clubs that sponsor star parties open to the public. The members of these clubs often know a lot about telescopes and can share their ideas with you. Your instructor may know where the nearest amateur astronomy club meets; many of the clubs in the country are also listed on the Web sites for *Sky & Telescope* (skyandtelescope.com) and *Astronomy* magazine (www.astronomy.com).

Furthermore, you may already have an instrument like a telescope at home (or have access to one through a relative or friend). Many amateur astronomers recommend starting your survey of the sky with a good pair of binoculars. These are easily carried around and can show you many objects not visible (or clear) to the unaided eye.

When you are ready to purchase a telescope, you might find the following thoughts useful:

- The key characteristic of a telescope is the aperture of the main mirror or lens; when someone says they have a 6-in. or 8-in. telescope, they mean the diameter of the collecting surface. The larger the aperture, the more light you can gather and the fainter the objects you can see or photograph.
- Telescopes of a given aperture that use lenses (refractors) are typically more expensive than those using mirrors (reflectors) because both sides of a lens must be polished to great accuracy. And, because the light passes through it, the lens must be made of high-quality glass throughout. In contrast, only the front surface of a mirror must be accurately polished.
- Magnification is not one of the criteria on which to base your choice of a telescope. As we discussed in the main text, the magnification of the image is done by a smaller eyepiece, so the magnification can be changed by changing eyepieces. However, a telescope will magnify not only the astronomical object you are viewing but also the turbulence of the Earth's atmosphere. If the magnification is too high, your image will shimmer and shake and be difficult to view. A good telescope will come with a variety of eyepieces that stay within the range of useful magnifications.
- The mount of a telescope is one of its most critical elements. Because a telescope shows a tiny field of view, which is magnified significantly, even the smallest vibration or jarring of the telescope can move the object you are viewing around or out of your field of view. A sturdy and stable mount is essential for serious viewing or photography (although it clearly affects how portable your telescope can be).
- A telescope requires some practice to set up and use effectively. Don't expect everything to go perfectly on your first try. Take some time to read the instructions. If a local amateur astronomy club is nearby, use it as a resource.

---

and Britain. IRAS was equipped with a 0.6-m telescope cooled to a temperature of less than 10 K. For the first time the infrared sky could be seen as if it were night, rather than through a bright foreground of atmospheric and telescope emissions. IRAS carried out a rapid but comprehensive survey of the entire infrared sky over a ten-month period, cataloging more than 250,000 sources of infrared radiation. Since then, several other infrared telescopes have operated in Earth orbit with much better sensitivity and resolution due to improvements in infrared detectors. The most powerful of these infrared telescopes is SIRTF (Space Infrared Telescope Facility), which was launched in 2003. Watch for the results of this new mission—they should be coming out just as you are reading this text!

5.5.2 Hubble Space Telescope

In April 1990, a great leap forward (or should we say upward) in astronomy was made with the launch of the Hubble Space Telescope (HST). This telescope has an aperture of 2.4 m, the largest put into space so far. (Its aperture was limited by the size of the payload bay in the Space Shuttle, which was its launch vehicle.) It was named for Edwin Hubble, the astronomer who discovered the expansion of the universe in the 1920s.

HST is operated jointly by the NASA Goddard Space Flight Center and the Space Telescope Science Institute in Baltimore. It is the first orbiting observatory designed to be serviced by shuttle astronauts, and they have made several visits to improve or replace its initial instruments and to repair some of the systems that operate the spacecraft (see the opening figure for this chapter).

HST's mirror was ground and polished to a remarkable degree of accuracy. If we were to scale up its mirror to the size of the entire continental United States, there would be no hill or valley larger than about 6 cm in its smooth surface. Unfortunately, after it was launched, scientists discovered that the primary mirror had a slight error in its shape equal to roughly 1/50 the width of a human hair. Small as that sounds, it was enough to ensure that much of the light entering the telescope did not come to a clear focus and that all the images were blurry. In a misplaced effort to save money, a complete test of the optical system had not been carried out before launch, so the error was not discovered until HST was in orbit.

The solution was to do something very similar to what we do for astronomy students with blurry vision: put corrective optics in front of their eyes. In December 1993, in one of the most exciting and difficult space missions ever flown, astronauts captured the orbiting telescope and brought it back into the shuttle payload bay. There they installed a package containing compensating optics as well as a new improved camera before releasing HST back into orbit. The telescope now works as it was designed to. We have sprinkled some new images and discoveries from HST throughout this book and invite you to enjoy them for their beauty as well as for their information content.

5.5.3 High-Energy Observatories

Ultraviolet, x-ray, and gamma-ray observations can be made only from space. Such observations first became possible in 1946, with V2 rockets captured from the Germans. The U.S. Naval Research Laboratory put instruments on these rockets for a series of pioneering flights used initially to detect ultraviolet radiation from the Sun. Since then, many other rockets have been launched to make x-ray and ultraviolet observations of the Sun, and later of other celestial objects as well.

Beginning in the 1960s, a steady stream of high-energy observatories has been launched into orbit to reveal and explore the universe at short wavelengths. The most recent of these is the Chandra x-ray telescope, which was launched in 1999 (Figure 5.18). It is producing x-ray images with

■ **FIGURE 5.18**

The Chandra X-Ray Satellite This painting shows NASA's Chandra X-ray Observatory in orbit. Chandra, the world's most powerful x-ray telescope, was deployed in July 1999.

NASA

TABLE 5.3 *Some Recent Observatories in Space*

Observatory	Dates of Operation	Bands of the Spectrum	Notes	Web Site
Einstein (HEAO-2)	1978–1981	x rays	first to take x-ray images	heasarc.gsfc.nasa.gov/docs/ einstein/heao2.html
International UV Explorer (IUE)	1978–1996	UV	UV spectroscopy	www.vilspa.esa.es/iue/iue.html
Infrared Astron. Satellite (IRAS)	1983–1984	IR	mapped 250,000 sources	irsa.ipac.caltech.edu/IRASdocs/
Hipparcos	1989–1993	visible light	measured over 100,000 precise positions	astro.esa.int/Hipparcos/ hipparcos.html
Cosmic Background Explorer (COBE)	1989–1993	IR, mm	observed the 3-degree background radiation	space.gsfc.nasa.gov/astro/cobe/
Compton Gamma-Ray Observatory	1990–2000	gamma rays	gamma-ray sources and spectra	cossc.gsfc.nasa.gov/
Hubble Space Telescope (HST)	1990–1993	visible, UV, IR	2.4-m mirror; images and spectra	hubblesite.org
Roentgen Satellite (ROSAT)	1990–1998	x rays	x-ray images and spectra	wave.xray.mpe.mpg.de/rosat
Infrared Space Observatory (ISO)	1995–1998	IR	infrared images and spectra	isowww.estec.esa.nl/
Rossi X-Ray Timing Explorer	1995–	x rays	variability of x-ray sources	heasarc.gsfc.nasa.gov/docs/ xte/XTE.html
BeppoSAX	1996–2002	x rays	observes over wide spectral range	bepposax/gsfc.nasa.gov/ bepposax
HALCA	1997–2000	radio	8-m radio dish, made array with dishes on the ground	www.vsop.isas.ac.jp/
Far UV Spectroscopic Explorer (FUSE)	1998–	UV	far-UV spectroscopy	fuse.pha.jhu.edu/
Chandra X-Ray Astronomy Facility	1999–	x rays	x-ray images and spectra	xrtpub.harvard.edu/
XMM Newton	1999–	x rays	x-ray spectroscopy	xmm.vilspa.esa.es
Microwave Anisotropy Probe (MAP)	2001–	microwaves	examine the background radiation from the Big Bang	map.gsfc.nasa.gov
Internat'l Gamma Ray Astrophys. Laboratory (INTEGRAL)	2002–	x and gamma rays	higher resolution gamma-ray images	sci.esa.int/home/integral
Space Infrared Telescope Facility (SIRTF)	2003–	IR	0.85-m telescope in space	sirtf.caltech.edu
Stratospheric Observatory for IR Astron. (SOFIA)	2004–	IR	airborne 2.5-m IR telescope	sofia.arc.nasa.gov/

unprecedented resolution and sensitivity. The 2002 Nobel Prize in physics was awarded to Riccardo Giacconi, who was a pioneer in the field of building and launching sophisticated x-ray instruments. Table 5.3 lists some of the most important space observatories.

THE FUTURE OF LARGE TELESCOPES

Each new generation of large telescopes, either in space or on the ground, simply whets the appetite of both scientists and the public. When we go on a hike, we are eager to see what lies just around the next bend in the path. Researchers are no different, and astronomers and engineers are now working on the technologies that would allow us to explore even more distant parts of the universe and to see them more clearly.

The premier space facility planned for the next decade is the James Webb Space Telescope, which (in a departure from tradition) is named after one of the early administrators of NASA instead of a scientist. This telescope will have a mirror 6 m in diameter and made up, like the Keck telescopes, of 36 small hexagons. The telescope is scheduled for launch in 2010 and should have the sensitivity needed to detect the very first generation of stars, formed when the universe was only a few hundred million years old.

■ FIGURE 5.19

Artist's Conception of the Overwhelmingly Large Telescope
The mirror in this telescope, which is currently being designed in Europe, is 100 m across—slightly wider than a football field is long. The telescope is being designed to operate in the open air without a protective enclosure. Protective mirror covers deploy from four surrounding buildings when the telescope is not in use.

On the ground, radio astronomers are planning the Atacama Large Millimeter Array (ALMA), which will consist of 64 radio telescopes, each 12 m in diameter. ALMA will be an interferometer and will be able to achieve spatial resolution ten times better than the Hubble Space Telescope but at millimeter wavelengths. The array will be located at an altitude of 16,400 ft on a plateau in the Atacama Desert of northern Chile, one of the driest spots on Earth. ALMA will be a particularly good tool for studying molecules, which are found in cool regions of gas where star formation is taking place.

And finally, several groups of infrared and optical astronomers around the globe are exploring the feasibility of building ground-based telescopes with mirrors 30–100 m across. Stop and think what this means: 100 m is longer than a football field! It is technically impossible to build and transport a single mirror that is 30 m or larger in diameter. These giant telescopes will follow the Keck design and be made up of a thousand or more smaller hexagonal mirrors, all held precisely in position so that they form a continuous surface.

The most ambitious of these projects is the OWL project (Figure 5.19), which is being developed in Europe. OWL stands for Overwhelmingly Large Telescope, and the design calls for a 100-m primary mirror. The cost is likely to exceed a billion dollars. These giant telescopes will combine light-gathering power with high-resolution imaging and will be able to attack many important astronomical problems. For example, they should be able to tell us when, where, and how often planets form around other stars. They should even be able to provide us images and spectra of planets, and thus perhaps give us the first real evidence that life exists elsewhere (see Chapter 15).

SURFING THE WEB

Web sites for the major telescopes and observatories on the ground and in space are given in Tables 5.1, 5.2, and 5.3.

 Multiwavelength Astronomy:
www.ipac.caltech.edu/Outreach/Multiwave/multiwave.html
Good images of the same celestial object taken with telescopes collecting different wavelength bands. You can see radio, visible-light, and x-ray images side by side.

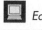 *Early Radio Astronomy:*
www.nrao.edu/intro/ham.connection.html
A brief summary of the history of radio astronomy, with emphasis on the connection with amateur (or "ham") radio.

Whispers from the Cosmos:
www.ncsa.uiuc.edu/Cyberia/Bima/BimaHome.html
Online "exhibit" about radio astronomy in general and the
Berkeley–Illinois–Maryland Association array of radio telescopes in particular.

 Orbiting Astronomical Observatories:
www.seds.org/~spider/oaos/oaos.html
A good annotated list, with fine links, of the telescopes in Earth
orbit.

 History of High-Energy Astrophysics:
heasarc.gsfc.nasa.gov/docs/heasarc/headates/heahistory.html
A good chronology with links detailing all major experiments and
space missions relevant to the study of gamma rays, x rays, or ultraviolet waves from space.

 Sites with Information on Buying Your Own Telescope:

- *Astronomy magazine:*
  www.astronomy.com/content/static/beginners/
  scopebuyingguide/sbg_1.asp

- *Sky & Telescope magazine:*
  http://skyandtelescope.com/howto/scopes/article_
  241_1.asp

- *The Telescope Review Site:*
  www.weatherman.com/BEGINNER.HTM

- *Weber State Planetarium Buying Your First Telescope:*
  physics.weber.edu/planet/telescope.html
  What we like about this site is that it recommends NOT
  buying a telescope at first.

- *The Heretic's Guide to Buying Your First Telescope:*
  www.findascope.com

SUMMARY

5.1 A **telescope** collects the faint light from astronomical sources and brings it to a **focus,** where an instrument (filters or a *spectrometer*) sorts the light according to wavelength. Light is then directed to a **detector,** where a permanent record is made. The light-gathering power of a telescope is determined by its **aperture**—that is, by the area of its largest or primary lens or mirror. The aperture is a convex lens in a **refracting telescope** or a concave mirror in a **reflector** that brings the light to a focus. Most large telescopes are reflectors; it is easier to manufacture and support large mirrors because the light does not have to pass through the glass.

5.2 New technologies for supporting lightweight mirrors have led to the construction of a number of new large telescopes since 1990. The site for an astronomical observatory must be carefully chosen for clear weather, dark skies, low water vapor, and excellent atmospheric **seeing** (low atmospheric turbulence). The **resolution** (the fineness of detail that can be seen) of a visible-light or infrared telescope is degraded by turbulence in the Earth's atmosphere. The new technique of **adaptive optics,** however, can make corrections for this turbulence in real time and produce exquisite images.

5.3 Visible-light detectors include the eye, photographic film, and the **charge-coupled device (CCD).** Detectors sensitive to infrared radiation must be cooled to very low temperatures. A spectrometer disperses the light into a spectrum to be recorded for detailed analysis.

5.4 In the 1930s, radio astronomy was pioneered by Jansky and Reber. A radio telescope is basically a radio antenna (often a large curved dish) connected to a receiver. Significantly enhanced resolution can be obtained with **interferometers,** including **interferometer arrays** like the 27-element VLA. Expanding to very long baseline interferometers, radio astronomers can achieve resolutions as good as 0.0001 arcsec. **Radar** astronomy involves transmitting as well as receiving. The largest radar telescope is the 305-m bowl at Arecibo.

5.5 Infrared observations are made with telescopes aboard aircraft and in space, as well as from ground-based facilities on dry mountain peaks. Ultraviolet, x-ray, and gamma-ray observations must be made from above the atmosphere. Many orbiting observatories have been flown to observe in these bands of the spectrum in the last few decades. The largest-aperture instrument in space is the Hubble Space Telescope (HST). The premier x-ray satellite is called Chandra. The Space Infrared Telescope Facility, launched in 2003, will be delivering its results while you are reading this text.

5.6 New and even larger telescopes are on the drawing boards. The James Webb Space Telescope, a 6-m successor to Hubble, is currently scheduled for launch in 2010. Radio astronomers are planning to build an interferometric array for millimeter astronomy called ALMA in the Atacama Desert of northern Chile. And several groups around the world are developing designs for ground-based optical and infrared telescopes with primary mirrors 30–100 m in diameter.

INTER-ACTIVITY

A Most large telescopes get many more proposals for observing projects than there is night observing time available in a year. Suppose your group is the telescope time allocation committee reporting to an observatory director. What criteria would you use in deciding how to give out time on the telescope? What steps could you take to make sure all your colleagues thought the process was fair and people would still talk to you at future astronomy meetings?

B Your group is a committee of nervous astronomers about to make a proposal to the government of a small European country to chip in to build the world's largest telescope in the high dry desert of the Chilean Andes Mountains. You expect the government ministers to be very skeptical about supporting this project. What arguments would you make to convince them to participate?

C The same government ministers we met in activity B ask you to draw up a list of the pros and cons of having the world's largest telescope in the mountains of Chile (instead of a mountain in Europe). What would your group list in each column?

D Make a list of all the ways in which an observing session at a large optical telescope and a large radio telescope might differ. (*One hint:* Bear in mind that because the Sun is not especially bright at many radio wavelengths, observations with radio telescopes can often be done during the day.)

E Another "environmental threat" to astronomy (besides light pollution) comes from the spilling of terrestrial communications into the "channels"—wavelengths and frequencies—previously reserved for radio astronomy. For example, the demand for cellular phones means that more and more radio channels will be used for this purpose. The faint signals from cosmic radio sources could be drowned in a sea of earthly conversation (translated and sent as radio waves). Assume your group is a congressional committee being lobbied by both radio astronomers who want to save some clear channels for doing astronomy and the companies that stand to make a lot of money from expanding cellular phone use. What arguments would sway you to each side? [For a real-world example of where this issue is being debated, see *Science* magazine, Nov. 28, 1997, p. 1569.]

REVIEW QUESTIONS

1. Name the two spectral windows through which electromagnetic radiation easily reaches the surface of the Earth, and describe the largest-aperture telescope currently in use for each window.
2. List the six bands into which we commonly divide the electromagnetic spectrum, and list the largest-aperture telescope currently in use in each band.
3. When astronomers discuss the apertures of their telescopes, they say bigger is better. Explain why.
4. Why are the largest visible-light telescopes in the world made with mirrors rather than lenses?
5. Compare the eye, photographic film, and CCDs as detectors for light. What are the advantages and disadvantages of each?
6. Radio and radar observations are often made with the same antenna, but otherwise they are very different techniques. Compare and contrast radio and radar astronomy in terms of the equipment needed, the methods used, and the kind of results obtained.
7. Why do astronomers place telescopes in Earth orbit? What are the advantages for the different regions of the spectrum?
8. What was the problem with the Hubble Space Telescope and how was it solved?
9. Describe the techniques radio astronomers use to obtain a resolution comparable to what astronomers working with visible light can achieve.
10. What kind of visible-light and infrared telescopes on the ground are astronomers planning for the future? Why are they building them on the ground and not in space?
11. What kind of visible-light and infrared telescopes are astronomers planning to launch into space in the future?

12. What happens to the image produced by a lens if the lens is "stopped down" with an iris diaphragm—a device that covers its periphery?

13. What would be the properties of an *ideal* astronomical detector? How closely do the actual properties of a CCD approach this ideal?

14. Fifty years ago, the astronomers on the staff of Mount Wilson and Palomar Observatories each received about 60 nights per year for their observing programs. Today an astronomer feels fortunate to get 10 nights per year on a large telescope. Can you suggest some reasons for this change?

15. The largest observatory complex in the world is on Mauna Kea in Hawaii, at an altitude of 4.2 km. This is by no means the tallest mountain on Earth. What are some factors astronomers consider when selecting an observatory site? Don't forget practical ones. Should astronomers, for example, consider building an observatory on Mount McKinley (Denali) or Mount Everest?

16. Another site recently developed for astronomy is the Antarctic plateau. Discuss its advantages and disadvantages.

17. Suppose you are looking for sites for an optical observatory, an infrared observatory, and an x-ray observatory. What are the main criteria of excellence for each? What sites are actually considered the best today?

18. Radio astronomy involves wavelengths much longer than those of visible light, and many orbiting observatories have probed the universe for radiation of very short wavelengths. What sorts of objects and physical conditions would you expect to be associated with radiation emissions of very long and very short wavelengths?

19. The dean of a university located near the ocean (who was not a science major in college) proposes building an infrared telescope right on campus and operating it in a nice heated dome so that astronomers will be comfortable on cold winter nights. Criticize this proposal, giving your reasoning.

FIGURING FOR YOURSELF

The primary reason for building large telescopes is to gather light from a larger area and hence to see fainter objects. The human eye actually works on the same principle. Your pupil expands under dark conditions to capture more light. It takes a few seconds for the pupil to expand in response to the change in illumination as you go from a fully lighted room to the dark outdoors at night. Far more important for dark adaptation is the gradual release of a chemical called rhodopsin, which makes the cones in your eye more sensitive to low light levels. Reaching maximum dark adaption takes about 30 minutes. Have you noticed how your eyes gradually get better at seeing in the dark as the minutes pass—in a movie theater, for example?

The larger the area over which light is gathered, the more light can be captured and brought to a focus. Remember that the area A of a circle is given by

$$A = \frac{\pi d^2}{4}$$

where d is the diameter of the circle.

20. In broad daylight, the size of your pupil is typically 3 mm. In dark situations, it expands to about 7 mm. How much more light can it gather?

21. How much more light can be gathered by a telescope that is 8 m in diameter than by your fully dark-adapted eye (neglect the increased sensitivity because of rhodopsin)?

22. How much more light can the Keck telescope (with its 10-m-diameter mirror) gather than an amateur telescope whose mirror is 25 cm (0.25 m) across?

23. People are often bothered when they discover that reflecting telescopes have a second mirror in the middle to bring the light out to an accessible focus where big instruments can be mounted. "Don't you lose light?" people ask. Well, yes, you do, but there is no better alternative. You can estimate how much light is lost by such an arrangement. The primary mirror (the one at the bottom in Figure 5.6) of the Gemini North telescope is 8 m in diameter. The secondary mirror at the top is about 1 m in diameter. Use the formula for the area of a circle to estimate what fraction of the light is blocked by the secondary mirror.

24. Telescopes can now be operated remotely from a warm room, but until about 25 years ago, astronomers worked at the telescope to guide it so that it remained pointed in exactly the right place. In a large telescope, like the Palomar 200-in. telescope, astronomers sat in a cage at the top of the telescope, where the secondary mirror is located in Figure 5.6. Assume for the purpose of your calculation that the diameter of this cage was 40 in. What fraction of the light is blocked? Now you can see why these cages were made very cramped and uncomfortable—but what a ride!

25. The HST cost about $1.7 billion for construction and $300 million for its shuttle launch, and it costs $250 million per year to operate. If the telescope lasts for 20 years, what is the total cost per year? Per day? If the telescope can be used just 30 percent of the time for actual observations, what is the cost per hour and per minute for the as-tronomer's observing time on this instrument? What is the cost per person in the United States? Was your investment in Hubble worth it?

SUGGESTIONS FOR FURTHER READING

Bartusiak, M. "The Next Generation Space Telescope: Predicting the Past" in *Astronomy,* Feb. 2001, p. 40. On what the instrument might detect.

Chaisson, E. *The Hubble Wars.* 1994, HarperCollins. Controversial story of the building of the HST.

Dyer, A. "What's the Best Telescope for You?" in *Sky & Telescope,* Dec. 1997, p. 28.

Florence, R. *The Perfect Machine: Building the Palomar Telescope.* 1994, Harper Perennial. Highly readable account of how the 200-in. telescope was built.

Gehrels, N. et al. "The Compton Gamma-Ray Observatory" in *Scientific American,* Dec. 1993, p. 68.

Graham, R. "Astronomy's Archangel" in *Astronomy,* Nov. 1998, p. 56. On master mirror-maker Roger Angel of the University of Arizona.

Graham, R. "High Tech Twin Towers" in *Astronomy,* Oct. 2000, p. 36. On the Gemini telescopes.

Grunsfeld, J. "Remaking the Hubble" in *Sky & Telescope,* Mar. 2002, p. 30. On the spring 2002 upgrade mission, by one of the astronauts.

Harrington, P. "Back to Basics with Binoculars" in *Astronomy,* Aug. 2002, p. 64. Brief introduction to how binoculars work and what to look for.

Irion, R. "Prime Time" in *Astronomy,* Feb. 2001, p. 46. On how time is allotted on the major research telescopes.

Janesick, J. and Blouke, M. "Sky on a Chip: The Fabulous CCD" in *Sky & Telescope,* Sept. 1987, p. 238.

Jayawardhana, R. "The Age of the Behemoths" in *Sky & Telescope,* Feb. 2002, p. 30. On current and planned large telescopes.

Junor, B. et al. "Seeing the Details of the Stars with Next Generation Telescopes" in *Mercury,* Sept./Oct. 1998, p. 26. On adaptive optics.

Krisciunas, K. *Astronomical Centers of the World.* 1988, Cambridge U. Press. History of and guide to major observatories.

Krisciunas, K. "Science with the Keck Telescope" in *Sky & Telescope,* Sept. 1994, p. 20. What it's like to use the telescope, and what work is being done with it.

Pilachowski, C. and Trueblood, M. "Telescopes of the 21st Century" in *Mercury,* Sept./Oct. 1998, p. 10. On the telescopes now being built or planned on the ground.

Schilling, G. "Adaptive Optics Comes of Age" in *Sky & Telescope,* Oct. 2001, p. 30. Excellent 11-page review.

Shore, L. "VLA: The Telescope That Never Sleeps" in *Astronomy,* Aug. 1987, p. 15.

Tarenghi, M. "Eyewitness View: First Sight for a Glass Giant" in *Sky & Telescope,* Nov. 1998, p. 47. On the first observations with the Very Large Telescope.

Whitt, K. "Destination: La Palma" in *Astronomy,* Jan. 2001, p. 62. Photo album and brief introduction to the Newton, Kapteyn, Herschel, and other telescopes there.

Exploring the Surface of Mars The view from the Mars Pathfinder lander, part of which is visible at the bottom of the picture. Some distance away is the little six-wheeled Sojourner rover, about the size of a microwave oven. During the summer of 1996, the public avidly followed the adventures of this little robot on the martian surface.

6

Other Worlds: An Introduction to the Solar System

In all of the history of mankind, there will be only one generation that will be the first to explore the Solar System, one generation for which, in childhood, the planets are distant and indistinct discs moving through the night sky, and for which, in old age, the planets are places, diverse new worlds in the course of exploration.

Carl Sagan in *The Cosmic Connection* (1973, Doubleday)

THINKING AHEAD

Surrounding the Sun is a family of worlds, large and small: nine planets, dozens of moons, and smaller pieces too numerous to count. It is a complex system with a wide range of conditions on the various worlds. Yet the system has some remarkable regularities to it. Most objects circle the Sun in the same direction and in the same plane, for example. What caused the orbits of all these worlds to share a common motion?

During the past 40 years we have learned more about the solar system than anyone imagined before the space age. In addition to gathering information with powerful new telescopes, we have sent spacecraft directly to many members of the planetary system. (Planetary astronomy is the only branch of our science in which we can—at least vicariously—travel to the objects we want to study.) With evocative names like Voyager, Pioneer, and Pathfinder, our robot explorers have flown past or around every planet except Pluto, returning images and data that have dazzled both astronomers and the public. In the process, we have also investigated dozens of fascinating moons, four ring systems, four asteroids, and two comets.

Probes have penetrated the atmosphere of Jupiter and landed on the surfaces of Venus, Mars, the Moon, and the asteroid Eros. Humans have set foot on the Moon and returned samples of its surface soil for analysis in the laboratory (Figure 6.1). We have even discovered other places in our solar system that might be able to support life.

Virtual Laboratories

 Plantetary Geology

Planetary Atmospheres and Their Retention

NASA

■ **FIGURE 6.1**

Astronauts on the Moon The lunar lander and surface rover from the *Apollo 15* mission are seen in this view of the one place beyond Earth that has been explored directly by humans.

The chapters about planets in today's textbooks are nothing like the ones earlier generations of students were reading. We can now envision the members of the solar system as other worlds like our own, each with its own chemical and geological history, and each with unique sights that interplanetary tourists may someday want to visit. Some have called these last few decades the golden age of planetary exploration, comparable to another golden age of exploration—the period starting in the 15th century, when great sailing ships plied the Earth's oceans and humanity began to get familiar with our own planet's surface.

In this chapter we discuss our planetary system and introduce the idea of comparative planetology—studying how the planets work by comparing them with one another. We want to get to know the planets not only for their own sakes but also to see what they can tell us about the origin and evolution of the entire solar system. In the next seven chapters, we will describe the better-known members of the solar system in some detail, returning to the theme of our origins in Chapter 13.

6.1 **OVERVIEW OF OUR PLANETARY SYSTEM**

The solar system[1] consists of the Sun and many smaller objects: the planets, their satellites and rings, and such "debris" as asteroids, comets, and dust. Decades of observation and

[1] The generic term for a group of planets and other bodies circling a star is *planetary system*. Ours is called the *solar system* because our Sun is sometimes called *Sol*. Strictly speaking, then, there is only one solar system; planets circling other stars are planetary systems.

spacecraft exploration have revealed that most of these objects formed together with the Sun about 4.5 billion years ago. They represent clumps of material that condensed from an enormous cloud of gas and dust. The central part of this cloud became the Sun, and a small fraction of the material in the outer parts eventually formed the other objects.

6.1.1 An Inventory

The Sun, a rather ordinary star, is by far the most massive member of the solar system, as shown in Table 6.1. It is an enormous ball of incandescent gas, 1.4 million kilometers (km) in diameter, with an interior temperature of millions of degrees. The Sun will be discussed later in this book, as our first (and best-studied) example of a star.

Table 6.1 also shows that most of the material of the planets is actually concentrated in the largest one, Jupiter, which is more massive than all the rest of the planets com-

TABLE 6.1 *Mass of Members of the Solar System*	
Object	**Percentage of Total Mass**
Sun	99.80
Jupiter	0.10
Comets	0.05
All other planets	0.04
Satellites and rings	0.00005
Asteroids	0.000002
Cosmic dust	0.0000001

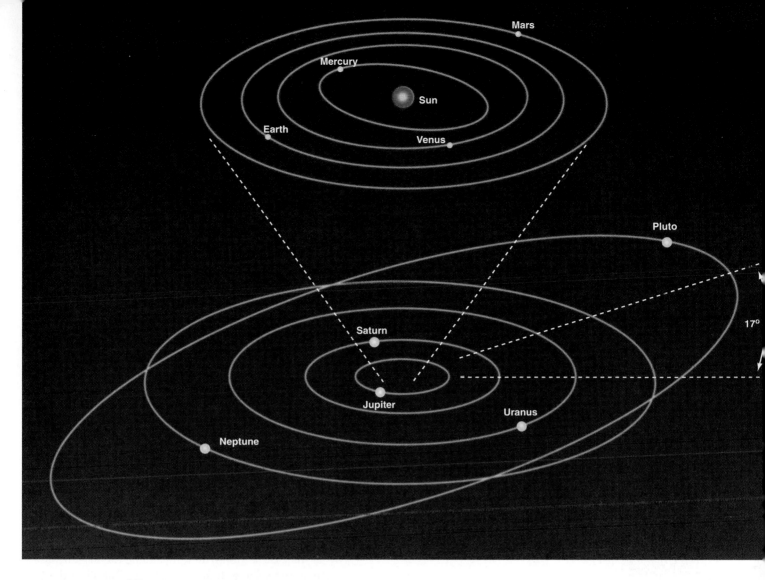

■ FIGURE 6.2

The Orbits of the Planets All the planets except Pluto orbit the Sun in roughly the same plane.

bined. Astronomers were able to measure the masses of the planets centuries ago, using Newton's laws of gravity, by measuring the planets' gravitational effects on one another or on moons that orbit about them. Today we make even more precise measurements of their mass by tracking their gravitational effects on the motion of spacecraft that pass near them.

Besides the Earth, five other planets (Mercury, Venus, Mars, Jupiter, and Saturn) were known to the ancients, and three (Uranus, Neptune, and Pluto) were discovered since the invention of the telescope. The nine planets all revolve in the same direction around the Sun (Figure 6.2). They orbit in approximately the same plane, like cars traveling on concentric tracks on a giant, flat racecourse. Each planet stays in its own traffic lane, following a nearly circular orbit about the Sun and obeying the "traffic laws" discovered by Galileo, Kepler, and Newton.

Each of the planets also rotates (spins) about an axis running through it, and in most cases the direction of rota-

tion is the same as the direction of revolution about the Sun. The exception is Venus, which rotates very slowly backward (that is, in a retrograde direction). Uranus and Pluto also have strange rotations, each spinning about an axis tipped nearly on its side.

The four planets closest to the Sun (Mercury through Mars) are called the inner or **terrestrial planets;** often the Moon is also discussed as a part of this group, bringing the total of terrestrial objects to five. (We generally call the Earth's satellite the *Moon,* with a capital *M,* and the other satellites *moons,* with lowercase *m*'s.) The terrestrial planets are relatively small worlds, composed primarily of rock and metal. All of them have solid surfaces that bear the records of their geological history in the forms of craters, mountains, and volcanoes (Figure 6.3).

The next four planets (Jupiter through Neptune) are much larger and are composed primarily of lighter ices, liquids, and gases. We call these four the jovian (after "Jove," another name for Jupiter in mythology) or **giant planets,**

NASA

◼ FIGURE 6.3

The Surface of Mercury The pockmarked face of a terrestrial world, more typical of the inner planets than the watery surface of Earth. This black-and-white image, taken with the Mariner 10 spacecraft, shows a region more than 400 km wide.

a name they richly deserve (Figure 6.4). More than 1400 Earths could fit inside Jupiter, for example. These planets do not have solid surfaces on which future explorers might land. They are more like vast, spherical oceans with much smaller, dense cores.

At the outer edge of the system lies little Pluto, which is neither a terrestrial nor a jovian planet; it is similar to the satellites of the outer planets and a number of icy objects that orbit the Sun beyond Pluto (see Chapter 12). It may have a common origin with them. Pluto was the last planet to have its mass measured—something astronomers could not do until Pluto's satellite was discovered in 1978. Table 6.2 summarizes some of the main facts about these nine planets.

6.1.2 Smaller Members of the Solar System

Most of the planets are accompanied by one or more satellites or moons; only Mercury and Venus move through space alone. There are more than 100 known satellites (see Appendix 8 for a listing of the larger ones), and undoubtedly many other small ones remain undiscovered. The largest of the satellites are as big as small planets and just as interesting; these include, in addition to our Moon, the four largest moons of Jupiter (called the Galilean satellites after their discoverer) and the largest moons of Saturn and Neptune (confusingly named Titan and Triton).

Each of the jovian planets also has rings made up of countless small bodies ranging in size from mountains down to mere grains of dust, all in orbit about the equator of the planet. The bright rings of Saturn are the widest and by far the easiest to see; they are among the most beautiful sights in the solar system (Figure 6.5). But all of the four

⊕ EARTH

NASA

◼ FIGURE 6.4

The Four Giant Planets A montage showing the four giant planets (as photographed by the Voyager cameras) and Earth to scale.

TABLE 6.2 *The Planets*

Name	Distance from Sun (AU)°	Revolution Period (years)	Diameter (km)	Mass (10^{23} kg)	Density (g/cm^3)†
Mercury	0.39	0.24	4,878	3.3	5.4
Venus	0.72	0.62	12,102	48.7	5.3
Earth	1.00	1.00	12,756	59.8	5.5
Mars	1.52	1.88	6,787	6.4	3.9
Jupiter	5.20	11.86	142,984	18,991	1.3
Saturn	9.54	29.46	120,536	5,686	0.7
Uranus	19.18	84.07	51,118	866	1.3
Neptune	30.06	164.82	49,660	1,030	1.6
Pluto	39.44	248.60	2,200	0.01	2.1

° An AU (or astronomical unit) is the distance from Earth to the Sun.
† We give densities in units where the density of water is 1. To get densities in units of kg/m^3, multiply the value we give by 1000.

ring systems are interesting to scientists because of their complicated forms, influenced by the pull of the satellites that also orbit these giant planets.

The solar system has many other less conspicuous members. The **asteroids** are rocky bodies that orbit the Sun like miniature planets, mostly in the space between Mars and Jupiter (Figure 6.6). Most asteroids are remnants of the initial population of the solar system that existed before the planets themselves formed. Some of the smallest satellites of the planets, such as the moons of Mars, are very likely captured asteroids.

Another class of small bodies is composed in part of ice, made of frozen gases such as water, carbon dioxide, carbon monoxide, and methyl alcohol; these objects are called **comets.** Comets also are remnants from the formation of the solar system, but they were formed and continue (with rare exceptions) to orbit the Sun in distant, cooler regions—stored in a sort of cosmic deep freeze.

Finally, there are countless grains of broken rock, which we call cosmic dust, scattered throughout the solar system. When these particles collide with Earth's atmosphere (as millions do each day), they burn up, producing brief flashes of light that we see as shooting stars or **meteors.** Occasionally, some larger chunk of material from space survives the passage through the atmosphere to become a **meteorite,** a piece of the solar system that has landed on Earth. (You can see meteorites on display in many natural history museums and can sometimes even purchase pieces of them from gem and mineral dealers.)

6.1.3 A Scale Model of the Solar System

Astronomy often deals with dimensions and distances that far exceed our ordinary experience. What does 1.4 billion kilometers—the distance from the Sun to Saturn—really mean to anyone? Sometimes it helps to visualize such large systems in terms of a scale model.

■ **FIGURE 6.5**

Saturn and Its Rings Voyager 1 image of a crescent Saturn and its complex system of rings, taken from a distance of about 1.5 million km. Notice that you can see the edge of the planet through the rings at left.

NASA/JPL

■ **FIGURE 6.6**

Asteroid Eros   Close-up view of this small Earth-approaching asteroid, taken by the NEAR-Shoemaker spacecraft from an altitude of about 100 km. This view of the heavily cratered surface is about 10 km wide. The spacecraft orbited Eros for a year before gently landing on its surface.

In our imagination, let us build a scale model of the solar system, adopting a scale factor of 1 billion (10^9)—that is, reducing the actual solar system by dividing every dimension by a factor of 10^9. Earth then has a diameter of 1.3 cm, about the size of a grape. The Moon is a pea orbiting this grape at a distance of 40 cm, or a little more than a foot away. The Earth-Moon system fits into a bookbag.

In this model the Sun is nearly 1.5 m in diameter, about the average height of an adult, and our Earth is at a distance of 150 m—about one city block—from the Sun. Jupiter is five blocks away from the Sun, and its diameter is 15 cm, about the size of a very large grapefruit. Saturn is 10 blocks from the Sun, Uranus 20 blocks, and Neptune 30 blocks. Pluto, whose distance varies quite a bit during its 249-year orbit, is currently just beyond 30 blocks and getting farther away with time. Most of the satellites of the outer solar system are the sizes of various kinds of seeds orbiting the grapefruit, oranges, and lemons that represent the outer planets.

In our scale model, a human is reduced to the dimensions of a single atom, and cars and spacecraft to the size of molecules. Sending the Voyager spacecraft to Neptune involves navigating a single molecule from the Earth-grape toward a lemon 5 km away with an accuracy equivalent to the width of a thread in a spider's web.

If that model represents the solar system, where would the nearest stars be? If we keep the same scale, the closest stars would be tens of thousands of kilometers away. If you built our scale model in the city where you live, you would have to place the representations of these stars on the other side of the Earth or beyond.

By the way, model solar systems like the one we just presented are being built in cities throughout the world. In Sweden, for example, Stockholm's huge Globe Arena has become a model for the Sun, and Pluto is represented by a 12-cm sculpture in the small town of Delsbo 300 km away. Another model solar system has been built on the Mall between the White House and Congress in Washington, D.C. (perhaps proving they are worlds apart?).

6.2   COMPOSITION AND STRUCTURE OF PLANETS

The fact that there are two distinct kinds of planets—the rocky terrestrial planets and the gas-rich jovian planets—leads us to believe that they formed under different conditions. Certainly their compositions are dominated by different elements. Let us look at each type in more detail.

6.2.1   The Giant Planets

The two largest planets, Jupiter and Saturn, have nearly the same chemical makeup as the Sun itself; they are composed primarily of the two elements hydrogen and helium, with 75 percent of the mass being hydrogen and 25 percent helium. On Earth, both hydrogen and helium are gases, so Jupiter and Saturn are sometimes called gas planets. But this name is misleading. Jupiter and Saturn are so large that the gas is compressed in their interior until the hydrogen becomes a liquid. Because the bulk of both planets consists of compressed, liquefied hydrogen, we should really call them liquid planets (Figure 6.7).

Under the force of gravity, the heavier elements sink toward the inner parts of a liquid or gaseous planet. Both Jupiter and Saturn thus have cores composed of heavier rock, metal, and ice, but we cannot see these regions directly. In fact, when we look down from above, all we see is the atmosphere with its swirling clouds. We must infer the existence of the denser core inside these planets from studies of each planet's gravity.

Uranus and Neptune are much smaller than Jupiter and Saturn, but each also has a core of metal, rock, and ice. In one way, however, Uranus and Neptune are very similar to Jupiter and Saturn: All four of these giant planets have cores of about the same mass—roughly ten times the entire mass of the Earth. Uranus and Neptune were less efficient at attracting hydrogen and helium gas, so they have much smaller atmospheres in proportion to their cores.

MAKING CONNECTIONS

Names in the Solar System

We humans just don't feel comfortable until something has a name. Types of butterflies, new elements, and mountains of Venus all need names for us to feel we are acquainted with them. How do we give names to objects and features in the solar system?

Planets and satellites are named after gods and heroes in Greek and Roman mythology (with a few exceptions among the moons of Uranus that are drawn from English literature). When William Herschel, a German immigrant to England, first discovered the planet we now call Uranus, he wanted to name it Georgium Sidus (George's star) after King George III of his adopted country. This caused such an outcry among astronomers in other countries, however, that the classical tradition was resumed and has been maintained ever since. Luckily, there were a lot of minor gods in the ancient pantheon, so plenty of names are left for the many small moons we are discovering around the giant planets. (Appendix 8 lists the larger satellites; how many names do you recognize?)

Comets are named after their discoverers (giving an extra incentive to comet hunters); asteroids are named by their discoverers after just about anyone or anything they want. Recently, asteroid names have been used to recognize people who have made significant contributions to astronomy, including two of the authors of this book.

That was pretty much all the naming that was needed while our study of the solar system was confined to Earth. But now our robot probes have surveyed and photographed many worlds in great detail, and each world has a host of features that suddenly need names. To make sure that naming things in space remains multinational, rational, and somewhat dignified, astronomers have given the responsibility of approving names to a special committee of the International Astronomical Union, the body that includes scientists from every country that does astronomy.

This committee has developed a set of rules for naming features on other worlds. For example, craters on Venus are named for women who have made significant contributions to human knowledge and welfare. Volcanic features on Jupiter's moon Io, which is in a constant state of volcanic activity, are named after gods of fire and thunder from the mythologies of many cultures. Craters on Mercury commemorate famous novelists, playwrights, artists, and composers. On Saturn's satellite Tethys, all the features are named after characters in Homer's great epic poem, *The Odyssey*. As we explore further, it may well turn out that more places in the solar system need names than Earth history can provide. Perhaps by then, explorers and settlers on these worlds will be ready to develop their own names for the places that (if only for a while) they will call home.

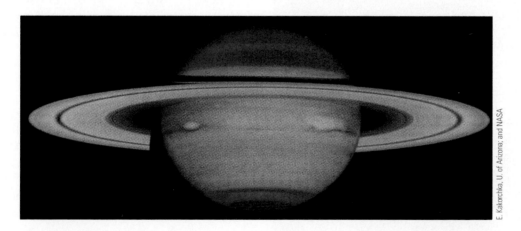

E. Kakorchka, U. of Arizona; and NASA

■ FIGURE 6.7

Saturn in False Color   This image of Saturn and its ring system were taken by the Hubble Space Telescope on January 4, 1998, celebrating the instrument's eighth anniversary. The colors are not real because the image was taken in infrared light and then processed by computer to make a picture our eyes can see. Like the other giant planets, Saturn is made mostly of liquid and gas. Two huge white storm systems are visible near Saturn's equator. The moon Dione is visible at the lower left, and the moon Tethys is seen at the upper right, having just crossed the face of Saturn.

Chemically, each of the giant planets is dominated by hydrogen and its many compounds. Nearly all of the oxygen present is chemically combined with hydrogen to form H_2O (water). Chemists call such a hydrogen-dominated composition *reduced*. Throughout the outer solar system, we find abundant water (mostly in the form of ice) and reducing chemistry.

6.2.2   The Terrestrial Planets

The terrestrial planets are quite different from the giants. In addition to being much smaller, they are composed primarily of rocks and metals. These, in turn, are made of elements that are less common in the universe as a whole. The most abundant rocks, called silicates, are made of silicon and oxygen, while the most common metal is iron. We can tell from their densities (see Table 6.2) that Mercury has the highest proportion of metals (which are denser) and the Moon has the lowest. (To refresh your memory about the definition of density, see Section 2.2.)

Earth, Venus, and Mars all have roughly similar bulk compositions: About one third of their mass consists of iron–nickel or iron–sulfur combinations; two thirds is made of silicates. Because these planets are largely composed of oxygen compounds (such as the silicate minerals of their crusts), their chemistry is said to be *oxidized*.

When we look at the internal structure of each of the terrestrial planets, we find that the densest metals are in a central core, with the lighter silicates near the surface. If these planets were liquid, like the giant planets, we could understand this effect as the result of the sinking of heavier materials due to the pull of gravity. Thus we believe that, although the terrestrial planets are solid today, they must once have been hot enough to melt.

Differentiation is the name given to the process by which gravity helps separate a planet's interior into layers of different compositions and densities. The heavier metals sink to form a core, while the lightest minerals float to the surface to form a crust. Later, when the planet cools, this layered structure is preserved. In order for a rocky planet to differentiate, it must be heated to the melting point of rocks, typically above 1300 K. (Temperature was defined in Chapter 4; see Appendix 5 for a discussion of the different temperature scales.)

6.2.3   Moons, Asteroids, and Comets

Chemically and structurally, Earth's Moon is like the terrestrial planets, but it is an exception among satellites. Most of the moons are in the outer solar system, and they have compositions similar to those of the cores of the giant planets around which they orbit. The three largest satellites, Ganymede and Callisto in the jovian system and Titan in the saturnian system, are composed half of frozen water, half of rocks and metals. Most of these moons differentiated during formation, and today they have cores of rock and metal, with upper layers and crusts of very cold and thus very hard ice (Figure 6.8).

The asteroids and comets, as well as the smallest moons, were probably never heated to the melting point. They therefore retain their original composition and structure. This is one of their main attractions for astronomers: They represent relatively unmodified material dating back to the time of the formation of the solar system. In a sense, they are ancient chemical "fossils," helping us to learn about a time whose traces have long been erased on larger worlds.

6.2.4   Temperatures: Going to Extremes

Generally speaking, the farther a planet or satellite is from the Sun, the cooler it is. The planets are heated by the radiant energy of the Sun, which gets weaker as the square of the distance. You know how rapidly the heating effect of a fireplace or an outdoor "radiant heater" diminishes as you walk away from it—the same effect applies to the Sun. Mercury, the closest planet to the Sun, has a blistering surface

JPL/NASA

■   **FIGURE 6.8**

Ganymede   A view of Jupiter's satellite Ganymede from the Galileo spacecraft, taken in June 1996. The brownish-gray color of the surface indicates a dusty mixture of rocky material and ice. The bright spots are places where recent impacts have uncovered fresh ice from underneath.

temperature of more than 500 K (230°C), as hot as the cleaning cycle on an electric oven, whereas the surface temperature on Pluto is only about 50 K, colder than liquid air.

Mathematically, the temperatures (expressed on the Kelvin scale) decrease approximately in proportion to the square root of the distance from the Sun. Pluto is 30 AU from the Sun, or 100 times the distance of Mercury. Thus, Pluto's temperature is less than that of Mercury by the square root of 100, or a factor of 10—from 500 K to 50 K.

Because of this strong dependence of temperature on distance from the Sun, Earth is the only planet where surface temperatures generally lie between the freezing and boiling points of water. Our planet is the only one with liquid water on its surface and therefore (as far as we know) the only one to support life.

ASTRONOMY BASICS
There's No Place Like Home

In the classic film *The Wizard of Oz,* Dorothy, our heroine, concludes after her many adventures in "alien" environments that "there's no place like home." The same can be said of the other worlds in our solar system. There are many fascinating places, large and small, that we might like to visit, but on none of them could human beings survive without a great deal of artificial assistance.

A thick atmosphere keeps the surface temperature on our neighbor Venus at a sizzling 700 K (near 900°F), hot enough to melt several metals. And the pressure of its mostly carbon dioxide atmosphere equals that found almost a kilometer underwater on Earth. Mars, on the other hand, has temperatures generally below freezing, with air (also mostly CO_2) so thin, it's like that found at an altitude of 30 km (100,000 ft) in Earth's atmosphere. And the red planet is so dry that it has not had any rain for billions of years.

In the outer layers of the jovian planets, it is neither warm enough nor solid enough for human habitation. Any bases we build in the systems of the giant planets may well have to be in space or on one of their satellites—none of which is particularly hospitable to a luxury hotel with swimming pool and palm trees. Perhaps we will find warmer havens deep inside the clouds of Jupiter or in the ocean under the frozen ice of its moon Europa.

All of this suggests that we had better take good care of the Earth because it is the only site where large numbers of our species can survive unaided. Recent human activity may be reducing the habitability of our planet by adding pollutants to the atmosphere or reducing the fraction of the Earth's surface where vegetation can thrive and continue to supply us with oxygen. In a solar system that seems unready to receive us, making the Earth less hospitable to life may be a grave mistake.

■ ■ ■ ■ ■ ■ ■ ■ ■ ■ ■ ■ ■ ■ ■

6.2.5 Geological Activity

The crusts of all of the terrestrial planets, as well as of the larger satellites, have been modified over their histories by both internal and external forces. Externally, each has been battered by a slow rain of projectiles from space, leaving surfaces pockmarked by impact craters of all sizes (see Figure 6.3). We have good evidence that this bombardment was far greater in the early history of the solar system, but it certainly continues to this day. (The collision of more than 20 large pieces of Comet Shoemaker-Levy 9 with Jupiter in the summer of 1994 is one recent example of this process. Figure 6.9 shows the aftermath of these collisions, when debris clouds larger than the Earth could be seen in Jupiter's atmosphere.)

At the same time, internal forces on the terrestrial planets have buckled and twisted their crusts, built up mountain ranges, erupted as volcanoes, and generally reshaped the surfaces in what we call geological activity. (The prefix *geo* means "Earth," so this is a bit of an "Earth-chauvinist" term, but it is so widely used that your authors bow to tradition.) Among the terrestrial planets, Earth and Venus have experienced the most geological activity over their histories,

STScI/NASA

■ FIGURE 6.9

Jupiter with Huge Dust Clouds The Hubble Space Telescope took this sequence of images of Jupiter in summer 1994, when the fragments of Comet Shoemaker-Levy 9 collided with the giant planet. Here we see the site hit by fragment G, from 5 minutes to five days after impact. Several of the dust clouds generated by the collisions became larger than the Earth.

although some of the satellites in the outer solar system are also surprisingly active. In contrast, our own Moon is a dead world where geological activity ceased billions of years ago.

Geological activity on a planet is the result of a hot interior. The forces of volcanism and mountain building are driven by heat escaping from the cores of the planets. As we will see, each of the planets was heated at the time of its birth, and this primordial heat initially powered extensive volcanic activity, even on our Moon. But small objects like the Moon soon cooled off. The larger the planet or satellite, the more likely it is to retain its internal heat, and therefore the more we expect to see surface evidence of continuing geological activity. The effect is similar to our own experience with a baked potato: The larger the potato, the more slowly it cools. If we want a potato to cool quickly, we cut it into small pieces.

For the most part, the history of volcanic activity on the terrestrial planets conforms to the predictions of this simple theory. The Moon, smallest of these objects, is a geologically dead world. Although we know much less about Mercury, it seems likely that this planet, too, ceased most volcanic activity about the same time the Moon did. Mars represents an intermediate case: It has been much more active than the Moon, but less so than the Earth. Earth and Venus, the largest terrestrial planets, still have molten interiors and high levels of geological activity even today, some 4.5 billion years after their birth.

NASA

■ FIGURE 6.10

Our Cratered Moon This image of the Moon's surface was taken during the *Apollo 16* mission and shows craters of many different sizes.

6.3 DATING PLANETARY SURFACES

How do we know the age of the surfaces we see on planets and satellites? If a world has a surface (as opposed to being mostly gas and liquid), we have developed some techniques for estimating how long ago that surface solidified. Note that the age of these surfaces is not necessarily the age of the planet as a whole. On geologically active objects (including the Earth), vast outpourings of molten rock or the erosive effects water and ice have erased evidence of earlier epochs and present us with only a relatively young surface for investigation.

6.3.1 Counting the Craters

One way to estimate the age of a surface is by counting the number of impact craters. This technique works because the rate at which impacts have occurred in the solar system has been roughly constant for several billion years. Thus, in the absence of forces to eliminate craters, the number of craters is simply proportional to the length of time the surface has been exposed. This technique has been successfully applied to planets, satellites, and even asteroids (Figure 6.10).

Bear in mind that crater counts can tell us only the time since the surface experienced a major change. Esti-

mating ages from crater counts is a little like walking along a sidewalk in a snowstorm after the snow has been falling steadily for a day or more. You may notice that in front of one house the snow is deep, while next door the sidewalk may be almost clear. Do you conclude that less snow has fallen in front of Ms. Jones's house than Mr. Smith's? More likely, you conclude that Jones has recently swept the walk clean and Smith has not. Similarly, the numbers of craters indicate how long it has been since a planetary surface was last "swept clean" by lava flows or by molten materials ejected when a large impact happened nearby.

Astronomers can thus use the numbers of craters on different parts of the same world to provide important clues about how regions on that world evolved. On a given planet or moon, the more heavily cratered terrain will generally be older (i.e., more time will have elapsed there since something swept the region clean).

6.3.2 Radioactive Rocks

Another way to trace the history of a solid world is to measure the ages of individual rocks. After samples were brought back from the Moon by the Apollo astronauts, the techniques that had been developed to date rocks on Earth could also be applied to establish a chronology for the Moon. In addition, a few samples of material from the Moon, Mars, and the large asteroid Vesta have fallen to Earth as meteorites and can also be examined directly (see Chapter 13).

Scientists measure the ages of rocks using the properties of natural **radioactivity.** Around the beginning of the

20th century, physicists began to understand that some atomic nuclei are not stable but can spontaneously split apart (decay) into smaller nuclei. The process of radioactive decay involves the emission of particles such as electrons, or of radiation in the form of gamma rays (see Chapter 4).

For any one radioactive nucleus, it is not possible to predict when the decay process will happen. Such decay is random in nature, just as the throw of dice is: As gamblers have all too often found out, it is impossible to say just when dice will come up 7 or 11. But for a very large number of dice tosses, we can calculate the odds that 7 or 11 will come up. Similarly, if you have a very large number of radioactive atoms of one type (say, uranium), there is a specific time period, called its **half-life,** during which the chances are fifty–fifty that decay will occur for any of the nuclei.

A particular nucleus may last a shorter or longer time than its half-life, but in a large sample almost exactly half will have decayed after a time equal to one half-life. Half of those remaining will have decayed in two half-lives, leaving only one half of a half—or one quarter—of the original sample (Figure 6.11).

If you had 1 gram of pure radioactive nuclei whose half-life was 100 years, then after 100 years you would have 1/2 gram, after 200 years 1/4 gram, after 300 years only 1/8 gram, and so forth. However, the material does not disappear. Instead, the radioactive atoms are replaced with their decay products. Sometimes the radioactive atoms are called *parents*, and the decay products are called *daughter* elements.

In this way, radioactive elements whose half-lives we have determined can provide accurate nuclear clocks. By comparing how much of a radioactive parent element is left in a rock with how much of its daughter products have accumulated, we can learn how long the decay process has been going on and hence how long ago the rock formed. Table 6.3 summarizes the decay reactions used most often to date lunar and terrestrial rocks.

When astronauts first flew to the Moon, one of their most important tasks was to bring back lunar rocks for ra-

dioactive age dating. Until then, astronomers and geologists had no real idea of the age of the lunar surface. Counting craters had given relative ages (for example, the heavily cratered lunar highlands were older than the dark lava plains), but scientists could not measure the actual age in years. Some thought that the ages were as young as those of Earth's surface, implying an active geology on the Moon. Only in 1969, when the first Apollo samples were dated, did we learn that the Moon is an ancient, geologically dead world. Using such dating techniques, we can determine the ages of both the Earth and the Moon: Each was formed about 4.5 billion years ago. (We show this date in the "Cosmic Calendar" at the end of the Prologue.)

By the way, note that the decay of radioactive nuclei generally releases energy in the form of heat. Although the energy from a single nucleus is not very large (in human terms), the enormous numbers of radioactive nuclei in a planet or satellite (especially early on) can be a significant source of internal energy for that world.

TABLE 6.3 *Radioactive Decay Reactions Used to Date Rocks*

Parent	Daughter	Half-Life (billion yr)
Samarium-147°	Neodymium-143	106
Rubidium-87	Strontium-87	48.8
Thorium-232	Lead-208	14.0
Uranium-238	Lead-206	4.47
Potassium-40	Argon-40	1.31

°The number after each element is its atomic weight, equal to the number of protons plus neutrons in its nucleus. This specifies the *isotope* of the element; different isotopes of the same element differ in the number of neutrons.

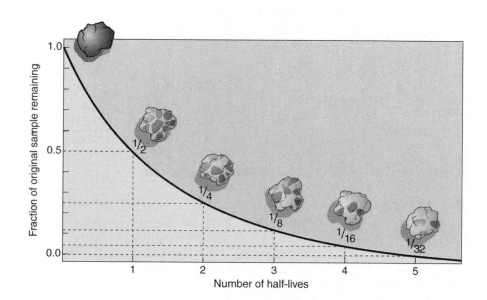

■ FIGURE 6.11

Radioactive Decay This graph shows (in red) the amount of a radioactive sample that remains after several half-lives have passed. After one half-life, half the sample is left; after two half-lives, one half of the remainder (or 1/4) is left; and after three half-lives, one half of that (or 1/8) is left. Note that in reality, the decay of radioactive materials in a rock sample would not cause any visible change in the appearance of the rock; the splashes of color are shown here for educational purposes only.

6.4 ORIGIN OF THE SOLAR SYSTEM

Much of astronomy is motivated by a desire to understand the origin of things—to find at least partial answers to the age-old questions of where the universe, the Sun, the Earth, and we ourselves came from. Each planet and satellite is a fascinating place that may stimulate our imaginations as we try to picture what it would be like to visit there. But taken together, the members of the solar system also preserve patterns that can tell us about the formation of the entire system. As we begin our exploration of the planets, we want to introduce our modern picture of how the solar system formed. We will return to this question in more detail in Chapter 13.

6.4.1 Looking for Patterns

One way to approach the question of origins is to look for regularities among the planets. We found, for example, that all the planets revolve in the same direction around the Sun and that they lie in nearly the same flat plane. The Sun also spins in the same direction about its own axis. Astronomers interpret this pattern as evidence that the Sun and planets formed together from a spinning system of gas and dust that we call the **solar nebula** (Figure 6.12).

The composition of the planets gives another clue. Spectroscopic analysis (see Chapter 4) allows us to determine what elements are present in the Sun and the planets.

Art by planetary astronomer William Hartmann

■ FIGURE 6.12

In the Solar Nebula An expert artist's conception of the solar nebula, the flattened cloud of gas and dust from which our planetary system formed. Icy and rocky planetesimals can be seen in the foreground. The bright center is where the Sun is forming.

The Sun has the same hydrogen-dominated composition as Jupiter and Saturn and therefore appears to have been formed from the same reservoir of material. In comparison, each of the terrestrial planets and satellites is relatively deficient in the light gases and the various ices that form out of the common elements oxygen, carbon, and nitrogen. Instead, on Earth and its neighbors we see mostly the rarer heavy elements like iron and silicon. This pattern suggests that the processes that led to planet formation in the inner solar system must somehow have rejected much of the lighter materials. These must have escaped, leaving just a residue of heavy stuff.

The reason for this is not hard to guess, bearing in mind the heat of the Sun. The inner planets (which are closer to the Sun) are now made of rock and metal, which can survive heat, but they contain very little ice or gas, which evaporate when the temperatures are high. (To see what we mean, just compare how long a rock and an ice cube survive when they are placed in the sunlight.) In the outer solar system, where it has always been cooler, the planets and their satellites are mostly composed of ice and gas.

6.4.2 The Evidence from Far Away

A second approach to understanding the origin of the solar system is to look outward for evidence that other, similar systems of planets are forming elsewhere. We cannot look back in time to the formation of our own system, but many stars in space are much younger than the Sun. In these systems, the process of planet formation might still be accessible to direct observation. One of the important astronomical discoveries of recent years is the presence of just such planetary nurseries. What we actually observe are other "solar nebulas" or *circumstellar disks*—flattened, spinning clouds of gas and dust surrounding young stars. These disks resemble our own solar system's initial stages of formation billions of years ago (Figure 6.13).

In addition to discovering these stellar nurseries, astronomers have recently developed tools to let them detect planets in orbit around other stars. So far we can find only giant planets, so we know nothing about the presence or absence of terrestrial planets in other planetary systems. But already we know that other planetary systems are not necessarily configured like our own. For example, many of the planets found around distant stars are in highly elliptical orbits, unlike the planets in the solar system, which follow nearly circular orbits. In other cases, giant planets are much closer to their stars than Mercury is to the Sun. An exciting new opportunity is opening as astronomers try to understand this bewildering variety of planetary systems, instead of focusing exclusively on our own solar system, which may not be typical at all.

6.4.3 Building Planets

Circumstellar disks are a common occurrence around very young stars, suggesting that disks and stars form together. Astronomers can use theoretical calculations to see how

STScI/NASA

■ FIGURE 6.13

Planetary Nurseries Hubble
Space Telescope photos of the Orion
nebula, a region of current star formation, and of an imbedded circumstellar
disk orbiting a very young star. This disk
is a contemporary analogue of our own
solar nebula—a location where planets
are probably being formed today.

solid bodies might form from the gas and dust in these disks as they cool. These models show that material begins to coalesce by first forming smaller objects, precursors of the planets, which we call **planetesimals.**

Fast computers simulate the way millions of planetesimals, each no larger than 100 km in diameter, might have gathered together under their mutual gravity to form the planets we see today. We are beginning to understand that this process was a violent one, with planetesimals crashing into each other and sometimes even disrupting the growing planets themselves. As a consequence of these violent impacts (as well as the heat from radioactive materials in them), all of the planets were heated and differentiated—which helps explain their present internal structures.

The process of impacts and collisions in the early solar system was complex and often apparently random. The so-

lar nebula model can explain many of the regularities we find in the solar system, but the random collisions of massive collections of planetesimals could be the reason for some exceptions to the "rules" of solar system behavior. For example, why do the planets Uranus and Pluto spin on their sides? Why does Venus spin slowly and in the opposite direction from all the other planets? Why does the composition of the Moon resemble the Earth in many ways and yet exhibit substantial differences? Although we cannot know for sure, the answers to such questions probably lie in enormous collisions that took place in the solar system long before life on Earth began.

Today, some 4.5 billion years after its origin, the solar system is—thank goodness—a much less violent place. As we will see, however, some planetesimals have continued to interact and collide, and their fragments move about the

Carl Sagan: Solar System Advocate

The best-known astronomer in the world during the 1970s and 1980s, Carl Sagan devoted most of his professional career to studying the planets, and much of his considerable energy to raising public awareness of what we can learn from exploring the solar system. Born in Brooklyn, New York, in 1934, Sagan became interested in astronomy as a youngster; he also credits science fiction stories for sustaining his fascination with what's out there. After earning both his bachelor's and PhD degrees from the University of Chicago, he worked at the University of California and Harvard before becoming professor of astronomy at Cornell University in 1968.

In the early 1960s, when many scientists still thought that Venus might turn out to be a hospitable place, Sagan demonstrated that the atmosphere of our neighborhood planet could act like a giant greenhouse, keeping the heat in and raising the temperature enormously. He showed that the seasonal changes astronomers had seen on Mars were caused not by vegetation but by wind-blown dust. He was a member of the scientific teams for many of the robotic missions that explored the solar system and was instrumental in getting NASA to put a message-bearing plaque aboard the Pioneer spacecraft, as well as audio-video records on the Voyager probes (as explained in Chapter 15).

To encourage public interest and public support of planetary exploration, Sagan helped found The Planetary Society, now the largest space-interest organization in the world. He was a tireless and eloquent exponent of the need to study the solar system close-up, and of the value of learning about other worlds in order to take better care of our own.

Carl Sagan

Sagan did experimental work simulating conditions on the early Earth and demonstrated how some of life's fundamental building blocks might have formed from the primordial "soup" of natural compounds on our

■ ■ ■ ■ ■ ■ ■ ■ ■ ■ ■ ■ ■

. . . He was a tireless and eloquent exponent of the need to study the solar system close-up . . .

planet. He and his students built "Jupiter boxes" to demonstrate how conditions in the giant planet's atmosphere may in some ways resemble what it was like on Earth before life began. In addition, he and his colleagues developed computer models showing that the consequences of nuclear war for the Earth would be even more devastating than anyone had thought (this is now called the *nuclear winter* hypothesis) and demonstrating some of the serious consequences of continued pollution of our atmosphere.

Sagan was perhaps best known, however, as a brilliant popularizer of astronomy and the author of many books on science, including the best-selling *Cosmos*, the Pulitzer-Prize-winning *Dragons of Eden*, and several evocative tributes to solar system exploration such as *The Cosmic Connection* and *Pale Blue Dot*. His book *The Demon Haunted World*, finished just before his death in 1996, is perhaps the best antidote to fuzzy thinking about pseudo-science and irrationality in print today. His intriguing science fiction novel, *Contact*, also a successful film, is recommended by many science instructors as a scenario for making contact with life elsewhere that is much more reasonable than most science fiction.

An appearance on *The Tonight Show* in 1973 (where his lucid explanations caught the attention of then-host Johnny Carson, an amateur astronomer) began a long series of television appearances that culminated in Sagan's 13-part public television series, *Cosmos*. It has been seen by an estimated 500 million people in 60 countries and has become one of the most-watched series in the history of public broadcasting. Sagan, by the way, was quick to point out that not once in the episodes of *Cosmos* did he say "billions and billions," a phrase now indelibly associated with him (although he did use it in other media appearances over the years). A few astronomers scoffed at a scientist who spent so much time in the public eye, but it is probably fair to say that Sagan's enthusiasm and skill as an explainer won more friends for the science of astronomy than anyone or anything else in the second half of the 20th century.

solar system as roving "transients" that can make trouble for the established members of the Sun's family, such as our own Earth. In Chapter 7 we will examine some of the devastation caused by impacts from space in the past along with some of the dangers that may face us in the future.

 The Nine Planets: A Multimedia Tour of the Solar System: seds.lpl.arizona.edu/nineplanets/nineplanets/nineplanets.html
Impressive amounts of information about each planet and moon in the solar system, with sections on space exploration, images, and links.

Views of the Solar System: www.solarviews.com/eng/homepage.htm
Another good site of information and images about the worlds in our solar system. The site is similar to *The Nine Planets*, but more focused on graphics. There is useful background information, a data browser, a glossary, and more.

Planetary Sciences Site: nssdc.gsfc.nasa.gov/planetary/planetary_home.html
NASA's National Space Science Data Center has a somewhat more technical, but still quite accessible, site with information and images on each world our spacecraft have explored, with concise descriptions of the various missions to the planets.

Gazetteer of Planetary Nomenclature: planetarynames.wr.usgs.gov
A helpful resource if you want to find out how objects and features in the solar system get their names.

Planetary Photojournal: photojournal.jpl.nasa.gov/
This NASA site features thousands of the best images from planetary exploration, with detailed captions and excellent indexing. You can dial up images by world, feature name, or date, and download them in a number of formats.

The Thousand-Yard Model: Earth as a Peppercorn: www.noao.edu/education/peppercorn/pcmain.html
Another scale model of the solar system, with careful instructions for doing a "walking tour" and other activities.

SUMMARY

6.1 Our solar system consists of the Sun, nine planets, more than 100 known satellites, and a host of smaller objects. The planets can be divided into two groups: the inner **terrestrial planets** and the outer **giant planets.** Pluto does not fit into either category. The giant planets are composed mostly of liquids and gases. Smaller members of the solar system include **asteroids,** rocky and metallic objects found mostly between Mars and Jupiter; **comets,** made mostly of frozen gases, which generally orbit far from the Sun; and countless smaller grains of cosmic dust. When a **meteor** survives its passage through our atmosphere and falls to Earth, we call it a **meteorite.**

6.2 The giant planets have dense cores roughly ten times the mass of Earth, surrounded by layers of hydrogen and helium. The terrestrial planets consist mostly of rocks and metals. They were once molten, which allowed their structures to **differentiate** (i.e., their denser materials sank to the center). The Moon resembles the terrestrial planets in composition, but most of the other satellites—which orbit the giant planets—have a large quantity of frozen ices within them. Generally, worlds closer to the Sun have higher temperatures. The surfaces of terrestrial planets have been

modified by impacts from space and by varying degrees of geological activity.

6.3 The ages of the surfaces of objects in the solar system can be estimated by counting craters: On a given world, a more heavily cratered region will always be older. We can also use samples of rocks with **radioactive** elements in them to obtain the time since the layer in which the rock formed last solidified. The **half-life** of a radioactive element is the time it takes for half the sample to decay; we determine how many half-lives have passed by how much of a sample remains the radioactive element and how much has become the decay product.

6.4 Regularities among the planets lead astronomers to hypothesize that the Sun and the planets formed together in a giant, spinning cloud of gas and dust called the **solar nebula.** Recent observations show tantalizingly similar circumstellar disks around other stars. We have also found giant planets orbiting a number of nearby stars. Within the solar nebula, material first coalesced into **planetesimals;** many of these gathered together to make the planets and satellites. The remainder can still be seen as comets and asteroids.

A Your group should discuss and make a list of the reasons humanity might have to explore other worlds. Does the group think such missions of exploration are worth the investment? Why?

B Your instructor will assign each group a world. Your task is to think about what it would be like to be there. Feel free to look ahead in the book to the relevant chapters. Discuss where on or around your world we would establish a foothold and what we would need to survive there.

 C In Astronomy Basics: *There's No Place Like Home* in this chapter, your authors discuss briefly how human activity is transforming our planet's overall environment. Can you think of other ways that this is happening?

 D Some scientists criticized Carl Sagan for "wasting his research time" popularizing astronomy. Do you think scientists should spend their time interpreting their field for the public? Why or why not? Are there ways that scientists who are not as eloquent or charismatic as Carl Sagan can still contribute to the public understanding of science?

E Your group has been named a special committee by the International Astronomical Union to suggest names of features (like craters, trenches, etc.) on a newly explored asteroid. Given the restrictions that any people after whom features are named must no longer be alive, what names or types of names would you suggest? (You are not restricted to names of people, by the way.)

F A member of your group has been kidnapped by a little-known religious cult that worships the planets. They will release him only if your group can tell which of the planets are currently visible in the sky during the evening and morning. You are forbidden from getting your instructor involved. How else could you find out the information you need? (Be as specific as you can; if your instructor says it's OK, feel free to continue answering this question in a library.)

REVIEW QUESTIONS

1. List some reasons that the study of the planets has progressed more in the last few decades than has any other branch of astronomy.
2. Imagine you are a travel agent in the next century. An eccentric billionaire asks you to arrange a "Guinness Book of Solar System Records" kind of tour. Where would you direct him to find the following (use this chapter and Appendices 7 and 8):
   a. the least-dense planet
   b. the densest planet
   c. the largest moon in the solar system
   d. the planet at whose surface you would weigh the most (*Hint:* Weight is directly proportional to surface gravity.)
   e. the smallest planet
   f. the planet that takes the longest time to rotate
   g. the planet that takes the shortest time to rotate
   h. the planet whose diameter is the closest to Earth's
   i. the moon that takes the shortest time to orbit its planet
   j. the densest moon
   k. the moon that orbits the farthest from its planet
   l. the most massive moon
   m. the most spectacular ring system
3. How do terrestrial and giant planets differ? List as many ways as you can think of.
4. Why are there so many craters on the Moon and so few on Earth?
5. How do asteroids and comets differ?
6. How and why is the Earth's Moon different from the larger satellites of the giant planets?
7. Where would you look for some "original" planetesimals left over from the formation of our solar system?
8. Describe how we use radioactive elements and their decay products to find the age of a rock sample. Is this necessarily the age of the entire world from which the sample comes? Explain.
9. What was it like in the solar nebula? Why did the Sun form at its center?
10. What can we learn about the formation of our solar system by studying other stars? Explain.

THOUGHT QUESTIONS

11. Using some of the astronomical resources in your college library or the World Wide Web, find five names of features on each of three other worlds that are named after people. In a sentence or two, describe who each of these people was.

12. Explain why the planet Venus is differentiated, but asteroid Fraknoi, a very boring and small member of the asteroid belt, is not.

13. Would you expect as many impact craters per unit area on the surface of Venus as on the surface of Mars? Why or why not?

14. Interview a sample of 20 people who are not taking an astronomy class and ask them if they can name a living astronomer. What percentage of those interviewed were able to name one? Typically, the one living astronomer the public knows these days is Stephen Hawking. Why is he better known than most astronomers? How would your results have differed if you had asked the same people to name a movie star or a professional football player?

FIGURING FOR YOURSELF

Let's compare the densities of several members of the solar system. The density of an object equals its mass divided by its volume. The volume of a sphere (like a planet) is calculated with the formula:

$$V = \frac{4}{3}\pi R^3$$

where π (Greek pi) has a value of approximately 3.14. Although planets are not perfect spheres, this formula works well enough. The masses and diameters of the planets are given in Table 6.2. For the data on the satellites, look in Appendix 8. We will select Saturn's satellite Mimas as our example, with a mass of 4×10^{19} kg and a diameter of approximately 400 km (radius 200 km = 2×10^5 m). The volume of Mimas is $4/3 \times 3.14 \times (2 \times 10^5)^3 = 3.3 \times 10^{16}$ m^3. Density is mass divided by volume: $4 \times 10^{19} / 3.3 \times 10^{16} = 1.2 \times 10^3$ kg/m^3. Note that the density of water in these units is 1000 kg/m^3, so Mimas must be mainly made of ice, not rock. (Note that the density of Mimas given in Appendix 8 is 1.2, but the units used there are different. In that table, we give density in units of g/cm^3, for which the density of water equals 1. Can you show, by converting units, that 1 g/cm^3 is the same as 1000 kg/m^3?)

15. Calculate the average density of our own planet, Earth. Show your work. (See Table 6.2 for data.)

16. Calculate the density of Jupiter. Show your work. Is it more or less dense than Earth? Why?

17. Calculate the density of Saturn. Show your work. How does it compare with the density of water? Explain how this can be?

18. What is the density of Jupiter's moon Europa (see Appendix 8 for data on moons)? Show your work.

19. Look at Appendices 7 and 8 and find the satellite whose diameter is the largest fraction of the diameter of the planet it orbits.

20. Barnard's Star, the second closest star to us, is about 56 trillion (5.6×10^{12}) km away. Calculate how far it would be using the scale model in Section 6.1.3.

21. A radioactive nucleus has a half-life of 5×10^8 years. Assuming that a sample of rock (say, in an asteroid) solidified right after the solar system formed, approximately what fraction of the radioactive element should be left in the rock today?

SUGGESTIONS FOR FURTHER READING

Barnes-Svarney, P. "The Chronology of Planetary Bombardments" in *Astronomy*, July 1988, p. 21.

Beatty, J. et al. *The New Solar System*, 4th ed. 1999, Sky Publishing. A compendium of articles on planetary astronomy by leading scientists in the field. Occasionally technical.

Burrows, W. E. *Exploring Space: Voyages in the Solar System and Beyond.* 1990, Random House. A science journalist looks at the broad political and social context of planetary exploration.

Greeley, R. and Batson, R. *The NASA Atlas of the Solar System.* 1997, Cambridge U. Press. Over-sized, coffee-table book with the best maps and images of the planets beautifully displayed.

Hartmann, W. "Piecing Together Earth's Early History" in *Astronomy*, June 1989, p. 24.

Jayawardhana, R. "Spying on Planetary Nurseries" in *Astronomy*, Nov. 1998, p. 62.

Kross, J. "What's In a Name?" in *Sky & Telescope*, May 1995, p. 28. Discusses the naming of features in the solar system.

Morrison, D. and Owen, T. *The Planetary System*, 3rd ed. 2003, Addison-Wesley. A thorough introduction to our modern knowledge.

O'Meara, S. "The World's Largest Solar System Scale Model" in *Sky & Telescope*, July 1993, p. 99. At the Lakeview Museum in Peoria, IL. (See an update in the March 1998 issue, p. 80, on a wide range of such models.)

Sagan, C. *Pale Blue Dot.* 1994, Random House. Eloquent introduction to planetary exploration and a sequel to Sagan's *Cosmos*.

Sheehan, W. *Worlds in the Sky.* 1992, U. of Arizona Press. History of how we came to understand the planets.

Stern, S. and Levinson, H. "Toward a Planet Paradigm" in *Sky & Telescope*, Aug. 2002, p. 42. On how to define a planet.

Molten lava fountaining from a volcanic vent in Hawaii—testament to the

7   Earth as a Planet

By coming face to face with alternative fates of worlds more or less like our own, we have begun to better understand the Earth. Every one of these worlds is lovely and instructive. But, so far as we know, they are also, every one of them, desolate and barren. Out there, there are no 'better places.' So far, at least.

Carl Sagan in Pale Blue Dot (1994, Random House)

THINKING AHEAD

Airless worlds in our solar system seem peppered with craters large and small. The Earth, on the other hand, has few craters, but a thick atmosphere and much surface activity. We believe that in the past the Earth, too, had many craters that have since been erased by forces in its crust and atmosphere. What can the persistent cratering on so many other worlds tell us about the history of our planet?

As our first step in exploring the solar system in more detail, we turn to the most familiar planet, our own Earth. The first humans to see the Earth as a blue sphere floating in the blackness of space were the *Apollo 8* astronauts who made the first voyage around the Moon in 1968. For many people, the historic images showing our world as a small, distant globe represent a pivotal moment in human history, when it became difficult for educated human beings to avoid viewing our world with a global perspective.

In earlier chapters we discussed our planet's size and orbit, and we described the seasons and tides, which are related to the motions of the Earth. In this chapter we examine the composition and structure of our planet with its envelope of ocean and atmosphere. We ask how our terrestrial environment came to be the way it is today. Because we live on it, we know the Earth much better than any other planet, and we can perform experiments on its surface that we cannot yet do elsewhere with any regularity. This direct knowledge helps us to use the Earth as a reference point for understanding other worlds, where we are mostly limited to remote observations.

Virtual Laboratories

Planetary Geology

Planetary Atmospheres and Their Retention

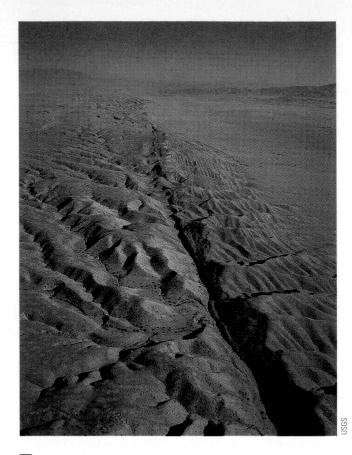

USGS

■ **FIGURE 7.8**

The San Andreas Fault   We see part of a very active region in California where one crustal plate is sliding sideways with respect to the other. The fault is marked by the valley running up the center of the photo. Major slippages along this fault can produce extremely destructive earthquakes.

One of the most famous faults is the San Andreas fault, lying on the boundary between the Pacific plate and the North American plate (Figure 7.8). This fault runs from the Gulf of California to the Pacific Ocean northwest of San Francisco. The Pacific plate, to the west, is moving northward, carrying Los Angeles, San Diego, and parts of the southern California coast with it. In several million years, Los Angeles will be an island off the coast of San Francisco.

Unfortunately for us, the motion along fault zones does not take place smoothly. The creeping motion of the plates against each other builds up stresses in the crust that are released in sudden, violent slippages to generate earthquakes. Because the average motion of the plates is constant, the longer the interval between earthquakes, the greater the stress and the more energy released when the surface finally moves.

For example, the part of the San Andreas fault near the central California town of Parkfield has slipped every 25 years or so during the past century, moving an average of about 1 m each time. In contrast, the average interval be-tween major earthquakes in the Los Angeles region is about 150 years, and the average motion is about 7 m. The last time the San Andreas fault slipped in this area was in 1857; tension has been building ever since, and sometime soon it is bound to be released. Sensitive instruments placed within the Los Angeles basin show that the basin is distorting and contracting in size as these tremendous pressures build up beneath the surface.

When two continental masses are moving on a collision course, they push against each other under great pressure. The Earth buckles and folds, dragging some rock deep below the surface and raising other folds to heights of many kilometers. This is the way most of the mountain ranges on Earth were formed; the Alps, for example, are a result of the African plate's bumping into Europe. As we will see, however, quite different processes produced the mountains on other planets.

Once a mountain range is formed by upthrusting of the crust, its rocks are subject to erosion by water and ice. The sharp peaks and serrated edges have little to do with the forces that initially make mountains. Instead, they result from the processes that tear mountains down. Ice is an especially effective sculptor of rock (Figure 7.9). In a planet without moving ice or running water (such as the Moon or Mars), mountains remain smooth and dull.

7.2.5   Volcanoes

Volcanoes mark locations where magma rises to the surface. One example is the mid-ocean ridges, which are long undersea mountain ranges formed by magma rising from the Earth's mantle at plate boundaries. A second major kind of volcanic activity is associated with subduction zones, and volcanoes sometimes also appear in regions where continental plates are colliding. In each case, the volcanic activity gives us a way to sample some of the material from deeper within our planet.

Other volcanic activity occurs above mantle "hot spots"—areas far from plate boundaries where heat is nevertheless rising from the interior of the Earth. One of the best-known such hot spots is under the island of Hawaii, where it currently supplies the magma to maintain three active volcanoes, two on land and one under the ocean. The Hawaiian hot spot has been active for at least 100 million years. As the Earth's plates have moved during that time, the hot spot has generated a 3500-km-long chain of volcanic islands. The tallest Hawaiian volcanoes are among the largest individual mountains on Earth, more than 100 km in diameter and rising 9 km above the ocean floor. As we saw in Chapter 5, one of these volcanic mountains, the now-dormant Mauna Kea, has become one of the world's great sites for doing astronomy.

Not all volcanic eruptions produce mountains. If the lava flows rapidly from long cracks, it can spread out to form lava plains. The largest known terrestrial eruptions, such as those that produced the Snake River basalts in the

NASA

■ **FIGURE 7.9**

Mountains on Earth   The Alps, a young region of the Earth's crust where sharp mountain peaks are being sculpted by glaciers. We owe the beauty of our young, steep mountains to the erosion of ice and water.

northwestern United States or the Deccan plains in India, are of this type. As we will see later, similar lava plains are found on the Moon and the other terrestrial planets.

THE EARTH'S ATMOSPHERE

We live at the bottom of the ocean of air that envelops our planet. The atmosphere, weighing down upon the Earth's surface under the force of gravity, exerts a pressure at sea level that scientists define as 1 **bar** (a term that comes from *barometer*, an instrument used to measure atmospheric pressure). A bar of pressure means that each square centimeter of the Earth's surface has a weight equivalent to 1.03 kg pressing down on it. Humans have evolved to live at this pressure; make the pressure a lot lower or higher, and we do not function well at all.

The total mass of our atmosphere is about 5×10^{18} kg; this sounds like a big number, but it is only about a millionth of the total mass of the Earth. The atmosphere represents a smaller fraction of the Earth than the fraction of your mass represented by the hair on your head.

7.3.1 Structure of the Atmosphere

The structure of the atmosphere is illustrated in Figure 7.10. Most of the atmosphere is concentrated near the surface of the Earth, within about the bottom 10 km—where clouds form and airplanes fly. Within this region, called the **troposphere,** warm air, heated by the surface, rises and is replaced by descending currents of cooler air—another example of convection. This circulation generates clouds and wind. Within the troposphere, temperature drops rapidly

with increasing elevation to values near 50°C below freezing at its upper boundary, where the **stratosphere** begins. Most of the stratosphere, which extends to about 50 km above the surface, is cold and free of clouds.

■ **FIGURE 7.10**

The Structure of the Earth's Atmosphere   Height increases up the left side of the diagram, and the names of the different layers are shown at the right. In the upper ionosphere, ultraviolet radiation from the Sun can strip electrons from their atoms, leaving the atmosphere ionized. (This layer can scatter or reflect AM radio waves, assuring that some of the dumb things said on AM radio never make it out of our planet's atmosphere.) The curving red line shows the temperature; see the scale for temperature at the bottom of the chart.

Near the top of the stratosphere is a layer of **ozone** (O_3), a heavy form of oxygen with three atoms per molecule instead of the usual two. Since ozone is a good absorber of ultraviolet light, it protects the surface from some of the Sun's dangerous ultraviolet radiation, making it possible for life to exist on land. Because ozone is essential to our survival, we react with justifiable concern to evidence that atmospheric ozone is being destroyed. Just such evidence has been accumulating; one especially important agent of this destruction is a series of industrial chemicals called CFCs (chlorofluorocarbons).

Each year a large ozone hole forms above the Antarctic continent, and by now the ozone loss has progressed into the temperate zones. The production of CFCs has been banned by international agreement, but these chemicals are destroyed so slowly that we can expect increasing loss of ozone well into this century.

At heights above 100 km, the atmosphere is so thin that orbiting satellites can pass right through it with very little friction. At these elevations, individual atoms can occasionally escape completely from the gravitational field of the Earth. There is a continuous, slow leaking of atmosphere—especially of lightweight atoms, which move faster than heavy ones. The Earth's atmosphere cannot, for example, hold on for long to hydrogen or helium, which escape into space.

7.3.2 Atmospheric Composition and Origin

At the Earth's surface, the atmosphere consists of 78 percent nitrogen (N_2), 21 percent oxygen (O_2), and 1 percent argon (A), with traces of water vapor (H_2O), carbon dioxide (CO_2), and other gases. Variable amounts of dust particles and water droplets are also found suspended in the air.

A complete census of the Earth's volatile materials, however, should look at more than the gas that is now present. *Volatile* materials are those that evaporate at a relatively low temperature. If the Earth were just a little bit warmer, some materials that are now liquid or solid might become part of the atmosphere. Suppose, for example, that our planet were heated to above the boiling point of water (100°C, or 373 K); that's a large change for humans, but a small change compared to the range of possible temperatures in the universe. At 100°C the oceans would boil and the water vapor would become a part of the atmosphere.

To estimate how much water vapor would be released, we note that there is enough water to cover the entire Earth to a depth of about 3000 m. Since the pressure exerted by 10 m of water is equal to about 1 bar, the average pressure at the ocean floor is about 300 bars. Since water weighs the same whether in liquid or in vapor form, if the oceans boiled away, the atmospheric pressure of the water would still be 300 bars. Water would therefore greatly dominate the Earth's atmosphere, with nitrogen and oxygen reduced to the status of trace constituents.

On a warmer Earth, another source of additional atmosphere would be found in the sedimentary carbonate rocks of the crust. These are minerals that contain abundant carbon dioxide. If all these rocks were heated, they would release about 70 bars of CO_2, far more than the current CO_2 pressure of only 0.0005 bar. Thus, the atmosphere of a warm Earth would be dominated by water vapor and carbon dioxide, with a surface pressure close to 400 bars.

Several lines of evidence convince scientists that the composition of the Earth's atmosphere has changed over our planet's history. Scientists can infer the amount of atmospheric oxygen, for example, by studying the chemistry of minerals that formed at various times. We examine this question in more detail in Section 7.4.

Today we see that CO_2, H_2O, sulfur dioxide (SO_2), and other gases are released from deeper within the Earth through the action of volcanoes. Much of this apparently new gas, however, is probably recycled material that has been subducted through plate tectonics. But where did our planet's original atmosphere come from?

Three possibilities exist for the original source of the Earth's atmosphere and oceans (since the early Earth was warmer, we can think of today's oceans as liquified early atmosphere, as just discussed): (1) The atmosphere could have been formed with the rest of the Earth, as it accumulated from the debris left over from the formation of the Sun; (2) it could have been released from the interior through volcanic activity, subsequent to the formation of the Earth; or (3) it may have been derived from impacts by comets or other icy materials from the outer parts of the solar system. Current evidence favors a combination of the interior and cometary sources.

7.3.3 Weather and Climate

All planets with atmospheres have *weather*, which is the name we give to the circulation of the atmosphere. The energy that powers the weather is derived primarily from the sunlight that heats the surface. Both the rotation of the planet and slower seasonal changes cause variations in the amount of sunlight striking different parts of the Earth. The atmosphere and oceans try to redistribute the heat from warmer to cooler areas. Weather on any planet represents the response of its atmosphere to changing inputs of energy from the Sun (see Figure 7.11 for a dramatic example).

Climate is a term used to refer to the effects of the atmosphere that last through decades and centuries. Changes in climate (as opposed to the random variations in weather from one year to the next) are often difficult to detect over short time periods, but as they accumulate, their effect can be devastating. Modern farming is especially sensitive to temperature and rainfall; for example, calculations indicate that a drop of only 2°C throughout the growing season would cut the wheat production by half in Canada and the United States.

■ FIGURE 7.11

A Storm from Space A large tropical storm marked by clouds swirling around a low-pressure region, photographed by astronauts orbiting the Earth. The combination of Earth's tilted axis of rotation, moderately rapid rotation, and oceans of liquid water can lead to violent weather on our planet.

The best-documented changes in the Earth's climate are the great ice ages, which have periodically lowered the temperature of the Northern Hemisphere over the past million years or so. The last ice age, which ended about 14,000 years ago, lasted some 20,000 years. At its height, the ice was almost 2 km thick over Boston and stretched as far south as New York City. Today we are in a relatively warm period, interpreted by many scientists as a fairly short-lived interglacial interval between major ice ages.

Scientists think that the ice ages are primarily the result of changes in the tilt of the Earth's rotational axis, produced by the gravitational effects of the other planets. They are less certain about recent evidence that on several occasions about a billion years ago the entire ocean froze over—a situation called "snowball Earth." As we will see in Chapter 9, there is also evidence of periodic climate changes on Mars, and modern calculations suggest that these also have their origin in slow changes of that planet's orbit and rotational axis.

7.4 LIFE AND CHEMICAL EVOLUTION

As far as we know, Earth seems to be the only planet in the solar system with life. The origin and development of life are an important part of the story of our planet. Life arose early in Earth's history, and living organisms have been interacting with their environment for billions of years. We recognize that life-forms have evolved to adapt themselves to the environment on Earth, and we are now beginning to realize that the Earth itself has been changed in important ways by the presence of living matter. The study of the coevolution of life and our planet is one of the subjects of the modern science of *astrobiology*.

7.4.1 The Origin of Life

The record of the birth of life on Earth has been lost in the restless motions of the crust. By the time the oldest surviving rocks were formed, about 3.9 billion years ago, life already may have existed, according to chemical evidence. At 3.5 billion years ago, life had achieved the sophistication to build large colonies called *stromatolites*, a form so successful that stromatolites still grow on Earth today (Figure 7.12). But little crust survives from these ancient times, and abundant fossils have been produced only during the past 600 million years—less than 15 percent of our planet's history.

There is little direct evidence about the actual origin of life. We know that the atmosphere of the early Earth, unlike today, contained abundant carbon dioxide but no oxygen gas. In the absence of oxygen, many complex chemical reactions are possible that lead to the production of amino acids, proteins, and other chemical building blocks of life. Therefore it seems likely that these chemical building

Meteor Crater, Northern Arizona

■ **FIGURE 7.17**

Meteor Crater in Arizona
Here we see a 50,000-year-old impact scar made by the collision of a 40-m lump of iron with our planet. Although impact craters are common on less-active bodies such as the Moon, this is one of the very few well-preserved craters on the Earth.

7.5.3 Extinction of the Dinosaurs

The impact that produced Meteor Crater would have been dramatic indeed to any humans who witnessed it, since the energy release was equivalent to a 15-megaton nuclear bomb. But such explosions are devastating only in their local areas; they have no *global* consequences. Much larger (and rarer) impacts, however, can disturb the ecological balance of the entire planet and thus influence the course of evolution.

The best-documented large impact took place 65 million years ago, at the end of what is now called the Cretaceous era of geological history. This point in the history of life on Earth was marked by a **mass extinction,** in which more than half of the species on our planet died out. There are a dozen or more mass extinctions in the geological record, but this particular event has always intrigued paleontologists because it marks the end of the dinosaur age. For tens of millions of years these great creatures had flourished. Then they suddenly disappeared (along with many other species), and thereafter the mammals began the development and diversification that ultimately led to the readers (and authors) of this book.

The object that collided with the Earth at the end of the Cretaceous era struck a shallow sea in what is now the Yucatan peninsula of Mexico. Its mass must have been more than a trillion tons (Figure 7.18) because scientists have found a worldwide layer of sediment deposited from the dust cloud that enveloped the planet after its impact. First identified in 1979, this sediment layer is rich in the

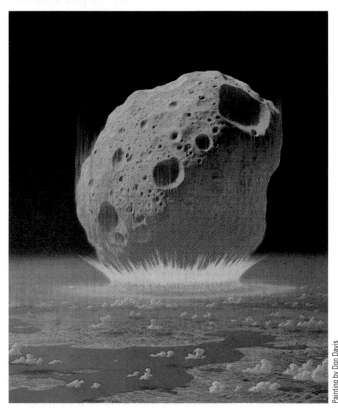

Painting by Don Davis

■ **FIGURE 7.18**

A 20-km Asteroid Hits the Earth   This painting shows an impact similar to the one that led to a mass extinction 65 million years ago.

rare metal iridium and other elements that are relatively abundant in asteroids and comets but exceedingly rare in the Earth's crust. Even though it was diluted by the material the explosion excavated from the surface of the Earth, this cosmic component is easily identified. In addition, this layer of sediment contains many minerals characteristic of the temperatures and pressures of a gigantic explosion.

The impact released energy equivalent to 5 billion Hiroshima-sized nuclear bombs and excavated a crater 200 km across and deep enough to penetrate through the Earth's crust. This large crater, named Chicxulub for a small town near its center, has subsequently been buried in sediment, but its outlines can still be identified (Figure 7.19). The explosion that created the Chicxulub crater lifted about *100 trillion tons* of dust into the atmosphere! We can determine this amount by measuring the thickness of the sediment layer that formed when this dust settled to the surface.

Such a quantity of airborne material would have blocked sunlight completely, plunging the Earth into a period of cold and darkness that lasted several months. Other worldwide effects included large-scale fires (started by the hot, flying debris from the explosion) that destroyed most of the planet's forests and grasslands, and a long period in which rainwater around the globe was acidic. It was these environmental effects, rather than the explosion itself, that were responsible for the mass extinction, including the death of the dinosaurs.

A demonstration of just such effects took place in July 1994, when pieces of a comet called Shoemaker–Levy 9 impacted the planet Jupiter. Some comet fragments were larger than a kilometer across. The comet fragments exploded as they penetrated the giant planet's atmosphere, and vast clouds of dark material were seen to expand over great distances (see Chapter 12).

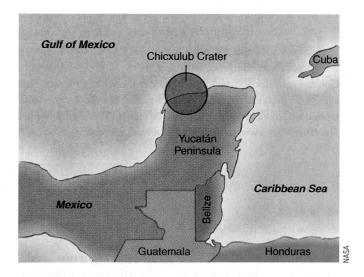

■ FIGURE 7.19

Site of the Chicxulub Crater Map showing the location of the buried crater from the impact 65 million years ago on Mexico's Yucatan peninsula. The crater is now buried under more than 500 m of sediment.

7.5.4 Impacts and the Evolution of Life

Several other mass extinctions in the geological record have been tentatively identified with large impacts, although none is so well documented as the blast that destroyed the dinosaurs. Even without such specific documentation, it is now clear that impacts of this size do occur on all planets and that their effects can be catastrophic for life. A catastrophe for one group of living things, however, may create opportunities for another group. Following each mass extinction, there is a sudden evolutionary burst as new species develop to fill the ecological niches opened by the event.

Impacts by comets and asteroids represent the only mechanisms we know of that could cause truly global catastrophes and seriously influence the evolution of life all over the planet. According to some estimates, the majority of all species extinctions may be due to such impacts. As paleontologist Stephen Jay Gould of Harvard noted, such a perspective fundamentally changes our view of biological evolution. The central issues for the survival of a species must now include more than just its success in competing with other species and adapting to slowly changing environments, as envisioned by Darwinian natural selection. Of at least equal importance is its ability to survive random global catastrophes due to impacts.

Still earlier in its history, the Earth was subject to even larger impacts from the leftover debris of planet formation. We know the Moon was repeatedly struck by objects larger than 100 km in diameter, a thousand times more massive than the object that wiped out most terrestrial life 65 million years ago (see Chapter 8). The Earth must have experienced similar large impacts during its first 700 million years of existence. Some of them were probably violent enough to strip the planet of most of its atmosphere and to boil away its oceans. Such events would sterilize the planet, destroying any life that had begun. Life may have formed and been wiped out several times before our own primitive ancestors took hold sometime about 4 billion years ago.

The fact that the oldest surviving microbes are thermophiles (adapted to very high temperatures) can also be explained by such large impacts. An impact that was just a bit too small to sterilize the planet would still have destroyed anything that lived in what we consider "normal" environments, and only the creatures that liked high temperatures would survive. Thus the oldest surviving terrestrial life-forms are probably the remnants of a sort of evolutionary bottleneck caused by repeated large impacts early in the planet's history.

7.5.5 Impacts in Our Future?

The impacts by asteroids and comets that have had such a major influence on life are not necessarily a thing of the past. In the full scope of planetary history, 65 million years ago was just yesterday. The Earth actually orbits the Sun within a sort of cosmic shooting gallery, and though major impacts are rare, they are by no means over. Humanity can

7. If you wanted to live where the chances of a destructive earthquake were small, would you pick a location near a fault zone, near a mid-ocean ridge, near a subduction zone, or on a volcanic island such as Hawaii? What are the relative risks of earthquakes at each of these locations?

8. If all life were destroyed on Earth by a large impact, would new life eventually form to take its place? Explain how conditions would have to change for life to start again on our planet.

9. Why will a decrease in the Earth's ozone be harmful to life?

10. Why are we concerned about the increases in CO_2 and other gases that cause the greenhouse effect in the Earth's atmosphere? What steps can we take in the future to reduce the levels of CO_2 in our atmosphere? What factors stand in the way of taking the steps you suggest (you may include technological, economic, and political factors in your answer).

11. Is there evidence of climate change in your area over the past century? How would you distinguish a true climate change from the random variations in weather that take place from one year to the next?

12. Do you think scientists should make plans to defend the Earth from future asteroid impacts? Is it right to intervene in the same evolutionary process that made the development of mammals (including us) possible after the big impact 65 million years ago?

FIGURING FOR YOURSELF

After scientists mapped the boundaries between tectonic plates in the Earth's crust and measured the annual rate at which the plates move (which is about 5 cm per year), we could estimate quite a lot about the rate at which the geology of the Earth is changing. As an example, let's suppose that the next slippage along the San Andreas fault in southern California takes place in the year 2007 and that it completely relieves the accumulated strain in this region. How much slippage is required for this to occur?

The speed of motion of the Pacific plate relative to the North American plate is 5 cm per year. That's 500 cm per century, or 5 m per century. The last southern California earthquake was in 1857. The time from 1857 to 2007 is 150 years, or 1.5 centuries, so the slippage to completely relieve the strain would be $5 \times 1.5 = 7.5$ m.

13. Europe and North America are also moving apart by about 5 m per century. As the continents separate, new ocean floor is created along the Mid-Atlantic Rift. If the rift is 5000 km long, what is the total area of new ocean floor created in the Atlantic each century? (Remember that 1 km equals 1000 m.)

14. Over the entire Earth there are 60,000 km of active rift zones, with average separation rates of 5 m per century. How much area of new ocean crust is created each year over the entire planet? (This area is approximately equal to the amount of ocean crust that is subducted, since the total area of the oceans remains about the same.)

15. With the information from problem 14, you can calculate the average age of the ocean floor. First find the total area of the ocean floor (equal to about 60 percent of the surface area of the Earth). Then compare this with the area created (or destroyed) each year. The average lifetime is the ratio of these numbers—the total area of ocean crust compared to the amount created (or destroyed) each year.

16. What is the volume of new oceanic basalt added to the Earth's crust each year? Assume that the thickness of the new crust is 5 km, that there are 60,000 km of active rifts, and that the average speed of plate motion is 5 cm per year. What fraction of the Earth's entire volume does this annual addition of new material represent?

17. Suppose a major impact that produces a mass extinction takes place on the Earth once every 5 million years. Further suppose that if such an event occurred today, you and most other humans would be killed (this would be true even if the human species as a whole survived). Such impact events are random, and one could take place at any time. Calculate the probability that such an impact will occur within the next 50 years (within your lifetime).

18. How do the risks of dying from the impact of an asteroid or comet compare with other risks we are concerned about, such as dying in a car accident or from heart disease or some other "natural" cause? (*Hint:* To find the annual risk, go to the library or the Web and look up the annual number of deaths from a particular cause in a particular country, and then divide by the population of that country.)

SUGGESTIONS FOR FURTHER READING

Allegre, C. and Schneider, S. "The Evolution of the Earth" in *Scientific American*, Oct. 1994, p. 66.

Becker, L. "Repeated Blows" in *Scientific American*, Mar. 2002, p. 60. On whether there have been repeated giant impacts and extinctions on Earth.

Chyba, C. "Death from the Sky: Tunguska" in *Astronomy*, Dec. 1993, p. 38. Excellent review article.

Frankel, C. *The End of the Dinosaurs: Chicxulub Crater and Mass Extinctions*. 1999, Cambridge U. Press. Approaches extinctions by focusing on the story of the crater.

Gallant, R. "Journey to Tunguska" in *Sky & Telescope*, June 1994, p. 38. A planetarium director tells of his trip to the site and fills in its history.

Hartmann, W. "Piecing Together Earth's Early History" in *Astronomy*, June 1989, p. 24.

Hartmann, W. and Miller, R. *The History of the Earth*. 1993, Workman. Lavishly illustrated chronicle of our planet's evolution.

Heppenheimer, T. "Journey to the Center of the Earth" in *Discover*, Nov. 1987, p. 86.

Jones, P. and Wigley, T. "Global Warming Trends" in *Scientific American*, Aug. 1990, p. 84. Three hundred years of data analyzed.

Lewis, J. *Rain of Iron and Ice: The Very Real Threat of Comet and Asteroid Bombardment*. 1995, Addison-Wesley. Good overview of impacts.

Lunine, J. *Earth: Evolution of a Habitable World*. 1998, Cambridge U. Press. History of Earth in a cosmic context.

Meissner, R. *The Little Book of Planet Earth*. 2002, Copernicus/Springer Verlag.

Morrison, D. and Chapman, C. "Target Earth: It Will Happen" in *Sky & Telescope*, Mar. 1990, p. 261. (See an updated version in *Astronomy*, Oct. 1995, p. 34.)

Murphy, J. and Nance, R. "Mountain Belts and the Supercontinent Cycle" in *Scientific American*, Apr. 1992, p. 84. There has been more than one supercontinent through geological time.

Orgel, L. "The Origin of Life on Earth" in *Scientific American*, Oct. 1994, p. 77.

Powell, J. *Night Comes to the Cretaceous*. 1998, Harcourt Brace. Fascinating look at the controversy surrounding the proposal that an impact killed off the dinosaurs, illustrating the reactions of scientists to a revolutionary new idea.

Steel, D. *Target Earth*. 2000, Readers Digest Books. On impacts and the people who study them.

York, D. "The Earliest History of the Earth" in *Scientific American*, Jan. 1993, p. 90.

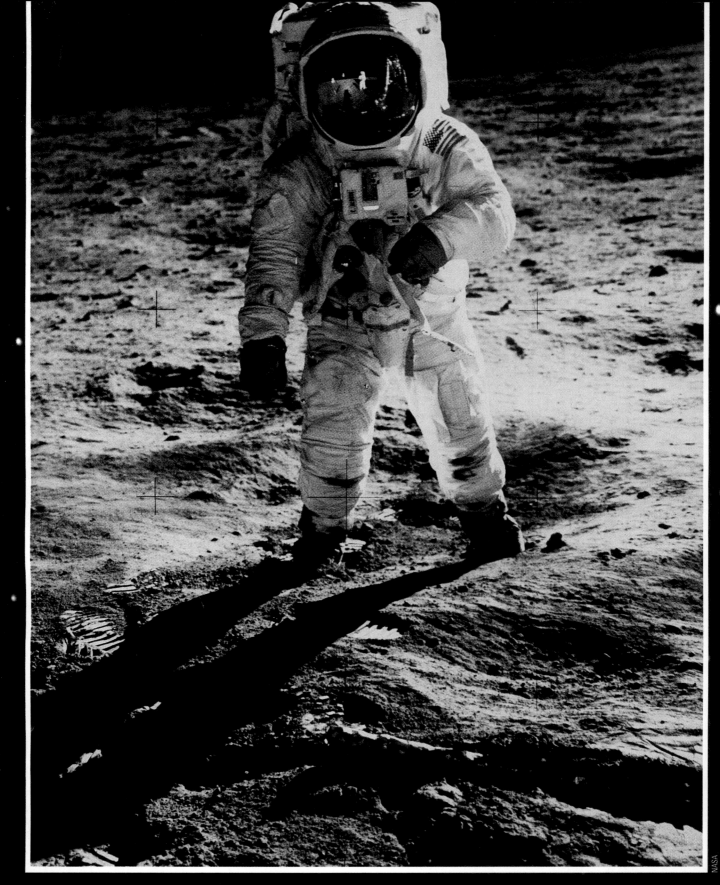

NASA

***Apollo 11* Astronaut Buzz Aldrin on the Surface of the Moon.**
Because there is no atmosphere, ocean, or geological activity on the Moon
today, the footprints you see in this image will likely be preserved in the lunar
soil for millions of years.

Cratered Worlds: The Moon and Mercury

This is one small step for a man, one giant leap for mankind.

What astronaut Neil Armstrong meant to say when he became the first human being to walk on another world. (In his nervousness, he left out the word "a," making the statement a bit less effective.)

THINKING AHEAD

The Moon is the only other world human beings have ever visited. What is it like to stand on the surface of our natural satellite? And what can we learn from going there and bringing home pieces of a different world?

We begin our discussion of the planets as worlds with two relatively simple objects: the Moon and Mercury. Unlike the Earth, the Moon is geologically dead, a place that has exhausted its internal energy sources. Because its airless surface preserves the events that happened long ago, the Moon provides us with a window on the earlier epochs of solar system history. The planet Mercury is in many ways similar to the Moon, which is why the two are discussed together. Both are relatively small, lacking in atmospheres, deficient in geological activity, and dominated by the effects of impact cratering. Still, the processes that have molded their surfaces are not unique to these two worlds; they have acted on many other members of the planetary system as well.

Virtual Laboratories

Planetary Geology

Planetary Atmospheres and Their Retention

8.1 GENERAL PROPERTIES OF THE MOON

The Moon has only 1/80 the mass of the Earth and about 1/6 the Earth's surface gravity—too low to retain an atmosphere (Figure 8.1). Moving molecules of a gas can escape from a planet just the way a rocket does, and the lower the gravity, the easier it is for the gas to leak away into space. If, early in its history, the Moon acquired an atmosphere from its hot interior or collected a temporary envelope of gases from impacting comets, such an atmosphere was lost before it could leave any recognizable evidence of its short existence. Indeed, the Moon is dramatically deficient in a wide range of *volatiles*, those elements and compounds that evaporate at relatively low temperatures. Some of the Moon's properties are summarized in Table 8.1, along with comparative values for Mercury.

TABLE 8.1 *Properties of the Moon and Mercury*

	Moon	Mercury
Mass (Earth = 1)	0.0123	0.055
Diameter (km)	3476	4878
Density (g/cm^3)	3.3	5.4
Surface gravity (Earth = 1)	0.17	0.38
Escape velocity (km/s)	2.4	4.3
Rotation period (days)	27.3	58.65
Surface area (Earth = 1)	0.27	0.38

8.1.1 Exploration of the Moon

Most of what we know about the Moon today derives from the U.S. Apollo program, which sent nine piloted spacecraft to our satellite between 1968 and 1972, landing 12 astronauts on its surface (see the photograph that opens this chapter). Before the era of spacecraft studies, astronomers had mapped the side of the Moon that faces the Earth with telescopic resolution of about 1 km, but lunar geology hardly existed as a scientific subject. All that changed beginning in the early 1960s.

Initially the Russians took the lead in lunar exploration with Luna 3, which returned the first photos of the lunar far side in 1959, and then with Luna 9, which landed on the surface in 1966 and transmitted pictures and other data to Earth. However, these efforts were overshadowed on July 20, 1969, when the first American astronaut set foot on the Moon.

Table 8.2 summarizes the nine Apollo flights—six that landed and three others that circled the Moon but did not land. The initial landings were on flat plains selected for safety reasons. But with increasing experience and confidence, NASA targeted the last three missions to more geologically interesting locales. The level of scientific exploration also increased with each mission, as the astronauts spent longer times on the Moon and carried more elaborate equipment. Finally, on the last Apollo landing, NASA included one scientist, geologist Jack Schmitt, among the astronauts (Figure 8.2).

In addition to landing on the lunar surface and studying it at close hand, the Apollo missions accomplished three objectives of major importance for lunar science. First, the astronauts collected nearly 400 kg of samples for detailed laboratory analysis on Earth (Figure 8.3). These samples, still being studied, have probably revealed more about the Moon and its history than all other lunar studies combined. Second, each Apollo landing after the first one deployed

FIGURE 8.1

The Moon from Earth The Moon as photographed by astronauts from a spacecraft near the Earth. The left side of the image shows part of the hemisphere that faces the Earth; several dark maria are visible. The right side shows part of the hemisphere that faces away from the Earth; it is dominated by highlands. The resolution of this image is several kilometers, similar to that of high-powered binoculars or a small telescope.

TABLE 8.2   *Apollo Flights to the Moon*

Flight	Date	Landing Site	Main Accomplishment
Apollo 8	Dec. 1968	—	First humans to fly around the Moon
Apollo 10	May 1969	—	First spacecraft rendezvous in lunar orbit
Apollo 11	July 1969	Mare Tranquillitatis	First human landing on the Moon; 22 kg of samples returned
Apollo 12	Nov. 1969	Oceanus Procellarum	First ALSEP; visit to Surveyor 3 lander
Apollo 13	Apr. 1970	—	Landing aborted due to explosion in command module
Apollo 14	Jan. 1971	Mare Nubium	First "rickshaw" on the Moon
Apollo 15	July 1971	Imbrium/Hadley	First "rover"; visit to Hadley Rille; astronauts traveled 24 km
Apollo 16	Apr. 1972	Descartes	First landing in highlands; 95 kg of samples returned
Apollo 17	Dec. 1972	Taurus Mountains	Geologist among the crew; 111 kg of samples returned

an Apollo Lunar Surface Experiment Package (ALSEP), which continued to operate for years after the astronauts departed. (The ALSEPs were turned off by NASA in 1978 as a cost-cutting measure.) Third, the orbiting Apollo Command Modules carried a wide range of instruments to photograph and analyze the lunar surface from above.

The last human left the Moon in December 1972, just a little more than three years after Neil Armstrong took his "giant leap for mankind." The program of lunar exploration was cut off in midstride due to political and economic pressures. It had cost just about $100 per American, spread over ten years—the equivalent of one large pizza per person per year. Yet for many people, the Moon landings were one of the central events in 20th-century history.

The giant Apollo rockets built to travel to the Moon were left to rust on the lawns of NASA centers in Florida, Texas, and Alabama, although recently some have at least been moved indoors to museums (Figure 8.4). Today no nation on Earth has the capability of returning people to the Moon; NASA estimates that it would require more than a decade to mount such an effort, and the financial and political will to do so is lacking. (In a bizarre piece of irony, a few people even question whether we went to the Moon at all, proposing instead that the Apollo program was a fake,

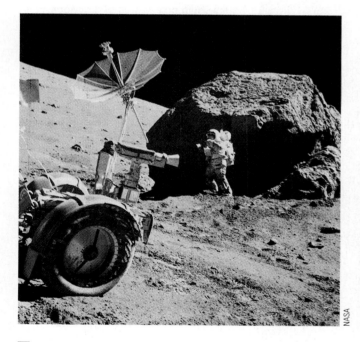

■ **FIGURE 8.2**

A Scientist on the Moon   Geologist (and later U.S. Senator) Jack Schmitt in front of a large boulder in the Litrow Valley at the edge of the lunar highlands. The "rover" vehicle the astronauts used for driving around on the Moon is seen at left. Note how black the sky is on the airless Moon. No stars are visible because the surface is brightly lit by the Sun, and the exposure therefore is not long enough to reveal stars.

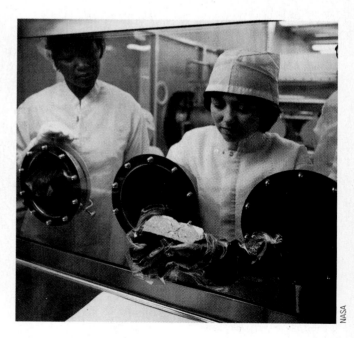

■ **FIGURE 8.3**

Handling Moon Rocks   Lunar samples collected in the Apollo project are analyzed and stored in NASA facilities at the Johnson Space Center in Houston, Texas. Here a technician is examining a rock sample using gloves in a sealed environment to avoid contaminating the sample.

filmed on a Hollywood sound stage! See the Moon hoax Web sites at the end of this chapter for some replies from scientists to such claims.) Having reached our nearest neighbor in space, humanity has for now retreated back to our own planet. It is hard to guess how long it will be before humans again venture out into the solar system.

The scientific legacy of Apollo remains, however, as discussed in the following sections of this chapter. The Apollo results have also been recently supplemented by three small scientific orbiters, one launched by NASA, one by Japan, and the third built by the U.S. military as a byproduct of the "star wars" space defense effort (Figure 8.5).

■ **FIGURE 8.4**

Moon Rocket on Display   One of the unused Saturn 5 rockets built to go to the Moon, now a tourist attraction at NASA's Johnson Space Center in Houston.

8.1.2   Composition and Structure of the Moon

The composition of the Moon is not the same as that of the Earth. With an average density of only 3.3 g/cm^3, the Moon must be made almost entirely of silicate rock. Compared to the Earth, it is depleted in iron and other metals. It is as if the Moon were composed of the same basic silicates as the Earth's mantle and crust, with the core metals

■ **FIGURE 8.5**

Lunar South Pole   This image, made of approximately 1500 individual pictures taken with the Clementine orbiter spacecraft, shows the south pole region of the Moon. Don't let the round border of the image fool you; you are looking at only a small portion of the Moon. For a sense of scale, note that the largest visible impact feature, the Schrodinger Basin at the lower right, is 320 km in diameter. Clementine gave astronomers their first close-up look at the Moon's polar regions. The south pole is of particular interest because ice was found in the permanent shadow zone (center).

and the volatiles selectively removed. These differences in composition between Earth and Moon provide important clues concerning the origin of the Moon, a topic we return to later in this chapter.

Probes of our satellite's interior carried out with seismometers taken to the Moon as part of the Apollo program confirm the absence of a large metal core. The Moon also lacks a global magnetic field like that of the Earth, a result consistent with the theory that such a field must be generated by motions in a liquid metal core. Not only is the metal lacking, but in addition the Moon is cold and has a solid interior, with very little seismic activity.

We also know from the study of lunar samples that water and other volatiles are largely absent from most of the crust. Very near the poles, however, NASA's Lunar Prospector mission measured the presence of hydrogen and interpreted this as an indication of ice mixed with the soil. The total quantity of water ice was estimated to be at least 6 billion tons. As liquid, this would only be enough water to fill a lake a few miles across, but compared with the rest of the dry lunar crust, so much water is remarkable. Presumably the water was carried to the Moon by comets and asteroids that hit our satellite. Some small fraction froze in a few extremely cold regions where the Sun never shines. One reason this discovery could be important is that it raises the possibility of future human habitations near the lunar poles, or even of a lunar base as a way station on the routes to Mars and the rest of the solar system.

8.2 THE LUNAR SURFACE

8.2.1 General Appearance

If you look at the Moon through a telescope, you can see that it is covered by impact craters of all sizes. The most conspicuous of the Moon's surface features—those that can be seen with the unaided eye and that make up the feature often called "the man in the Moon"—are vast splotches of darker lava flows.

Centuries ago, the early lunar observers thought that the Moon had continents and oceans and that it was a possible abode of life. Early names given to the lunar "seas"—Mare Nubium (Sea of Clouds), Mare Tranquillitatis (Tranquil Sea), and so on—are still in use today. In contrast, the "land" areas between the seas do not have names. Thousands of individual craters have been named, however, mostly for great scientists and philosophers. Among the most prominent are those named for Plato, Copernicus, Tycho, and Kepler. Galileo has only a very small crater, however, reflecting his low standing among the Roman Catholic scientists who made some of the first lunar maps.

We know today that the resemblance of lunar features to terrestrial ones is superficial. Even when they look somewhat similar, the origins of lunar features such as craters and mountains may be very different from their terrestrial counterparts. The Moon's relative lack of internal

activity, together with the absence of air and water, make most of its geological history unlike anything we know on Earth. And much of the lunar surface is older than the rocks of the Earth's crust, as we learned when the first lunar samples were analyzed in the laboratory.

8.2.2 Lunar History

To trace the detailed history of the Moon or of any planet, we must be able to estimate the ages of individual rocks. Once lunar samples were brought back by the Apollo astronauts, the radioactive dating techniques that had been developed for the Earth (see Section 6.3) were applied to them. The solidification ages of the samples ranged from about 3.3 to 4.4 billion years, substantially older than most of the rocks on the Earth. As we saw in Chapter 6, it now seems clear that both the Earth and the Moon were formed 4.5 billion years ago.

Most of the surface of the Moon (83 percent) consists of silicate rocks called *anorthosites;* these regions are known as the lunar **highlands.** They are made of relatively low-density rock that solidified on the cooling Moon like slag floating on the top of a smelter. Because they formed so early in lunar history (between 4.1 and 4.4 billion years ago), the highlands are also extremely heavily cratered, bearing the scars of billions of years of impacts by interplanetary debris (Figure 8.6).

The most prominent lunar features are the **maria** (Latin for "seas") that we mentioned earlier. *Maria* is the plural term, by the way; the singular is *mare* (pronounced "mah-ray"). These dark, round plains, which are much less cratered than the highlands, cover just 17 percent of the

■ **FIGURE 8.6**
Lunar Highlands The old, heavily cratered lunar highlands make up 83 percent of the Moon's surface. The width of this image is 250 km.

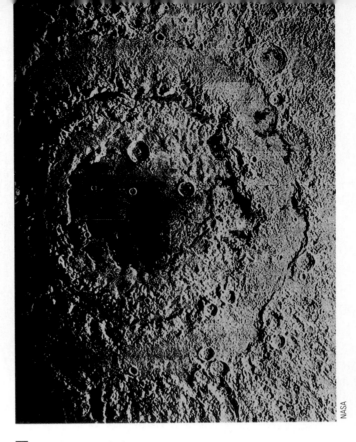

■ **FIGURE 8.7**

Lunar Maria About 17 percent of the Moon's surface consists of the maria—flat plains of basaltic lava. This view of Mare Imbrium also shows numerous secondary craters and evidence of material ejected from the large crater Copernicus on the upper horizon. Copernicus is an impact crater almost 100 km in diameter that was formed long after the lava in Imbrium had already been deposited.

■ **FIGURE 8.8**

Mare Orientale The youngest of the large lunar impact basins is Orientale, formed 3.8 billion years ago. Its outer ring is about 1000 km in diameter, roughly the distance between New York City and Detroit, Michigan. Unlike most of the other basins, Orientale has not been filled in with lava flows, so it retains its striking "bull's-eye" appearance. It is located on the edge of the Moon as seen from Earth.

lunar surface, mostly on the side facing the Earth (Figure 8.7). Today we know that the maria are volcanic plains, laid down in eruptions billions of years ago. They partly fill huge depressions called *impact basins*, which were produced by collisions of large chunks of material with the Moon relatively early in its history (Figure 8.8).

The lunar maria are all composed of basalt, very similar in composition to the crust under the oceans of the Earth or to the lavas erupted by many terrestrial volcanoes. A series of large eruptions on the Moon between 3.3 and 3.8 billion years ago (dated from laboratory measurements of returned samples) formed smooth flows typically a few meters thick but extending over distances of hundreds of kilometers. Eventually, these lava flows filled in the lowest parts of the giant impact basins to form the mare surfaces we see today (Figure 8.9).

Volcanic activity may have begun very early in the Moon's history, although most evidence of the first half-billion years is lost. What we do know is that the major mare volcanism, which involved the release of liquid magma from hundreds of kilometers below the surface, ended about 3.3 billion years ago. After that the Moon's interior cooled, and by 3 billion years ago all volcanic activity had ceased. Since then our satellite has been a geologically dead world, changing only slowly as the result of random impacts.

■ **FIGURE 8.9**

Rock from a Lunar Mare Sample of basalt from the mare surface. The gas bubbles are characteristic of rock formed from lava. All lunar rocks are chemically distinct from terrestrial rocks, a fact that has recently allowed scientists to identify a few lunar samples among the thousands of meteorites that reach Earth.

■ **FIGURE 8.10**

Lunar Mountain Mt. Hadley on the edge of Mare Imbrium, photographed by the *Apollo 15* astronauts. Note the smooth contours of the lunar mountains, which have not been sculpted by water or ice.

The major lunar mountains are all the result of the Moon's history of impacts. Long, arc-shaped ranges that border the maria are debris ejected from the impacts that formed these giant basins. These mountains have low, rounded profiles that resemble old, eroded mountains on Earth (Figure 8.10). But appearances can be deceiving. The mountains of the Moon have not been eroded, except for the effects of meteoritic impacts. They are rounded because that is the way they formed, and there has been no water or ice to carve them into cliffs and sharp peaks. (Compare Figure 8.10 with Figure 7.9.)

8.2.3 On the Lunar Surface

> The surface is fine and powdery. I can pick it up loosely with my toe. But I can see the footprints of my boots and the treads in the fine sandy particles.
> —Neil Armstrong, *Apollo 11* astronaut, immediately after stepping onto the Moon for the first time

The surface of the Moon is buried under a fine-grained soil of tiny, shattered rock fragments. The dark basaltic dust of the lunar maria was kicked up by every astronaut footstep, and this dust eventually worked its way into all of the astronauts' equipment. The upper layers of the surface are porous, consisting of loosely packed dust into which their boots sank several centimeters (Figure 8.11). This lunar dust, like so much else on the Moon, is the product of impacts. Each cratering event, large or small, breaks up the rock of the lunar surface and scatters the fragments. Ultimately, billions of years of impacts have reduced much of the surface layer to particles about the size of dust or sand.

In the absence of any air, the lunar surface experiences much greater temperature extremes than the surface of the Earth, even though the Earth is virtually the same distance from the Sun. Near local noon, when the Sun is highest in the sky, the temperature of the dark lunar soil rises to near the boiling point of water. During the long lunar night

■ **FIGURE 8.11**

Footprint in Moondust Apollo photo of an astronaut's bootprint in the lunar soil.

(which, like the lunar day, lasts two Earth weeks) it drops to about 100 K ($-173°C$). The extreme cooling is a result not only of the absence of air but also of the porous nature of the dusty soil, which cools more rapidly than solid rock would.

8.3 IMPACT CRATERS

The Moon provides an important benchmark for understanding the history of the planetary system. Most solid worlds show the effects of impacts, often extending back to the era when a great deal of debris from our system's formation process was still present. On the Earth, this long history has been erased by our active geology. On the Moon, in contrast, most of the impact history is preserved. If we can understand what has happened on the Moon, we may be able to apply this knowledge to the other planets and satellites.

8.3.1 Volcanic Versus Impact Origin of Craters

Until the middle of the 20th century, scientists did not generally recognize that lunar craters were the result of impacts. Since impact craters are extremely rare on Earth, geologists did not expect them to be the major feature of lunar geology. They reasoned (perhaps unconsciously) that since the craters we have on Earth are volcanic, the lunar craters must have a similar origin.

One of the first geologists to propose that lunar craters were the result of impacts was Grove K. Gilbert, a scientist with the U.S. Geological Survey in the 1890s. He pointed out that the large lunar craters, which are mountain-rimmed, circular features with floors generally below the level of the surrounding plains, are larger and have different shapes from known volcanic craters. Terrestrial volcanic craters are smaller and deeper and almost always occur at the tops of volcanic mountains (Figure 8.12). Gilbert therefore concluded that the lunar craters were not volcanic. The only alternative was an impact origin. His careful reasoning, although not accepted at the time, laid the foundations for the modern science of lunar geology.

Gilbert concluded that the lunar craters were produced by impacts, but he didn't understand why all of them were circular and not oval. The reason lies in the speed with which projectiles approach the Earth or Moon. Attracted by the gravity of the larger body, the incoming chunk strikes with at least escape velocity, which is 11 km/s for the Earth and 2.4 km/s (5400 mi/h) for the Moon. To this escape velocity is added whatever speed the projectile already had with respect to the Earth or Moon, typically 10 km/s or more.

At these speeds, the energy of impact produces a violent explosion that excavates a large volume of material in a symmetrical way. Photographs of bomb and shell craters on Earth confirm that explosion craters are always essen-

(a)   Terrestrial volcano

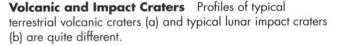

(b)   Lunar impact crater

■ FIGURE 8.12

Volcanic and Impact Craters   Profiles of typical terrestrial volcanic craters (a) and typical lunar impact craters (b) are quite different.

tially circular. Only following World War I did scientists recognize the similarity between impact craters and explosion craters, but Gilbert did not live to see his impact hypothesis widely accepted.

8.3.2 The Cratering Process

Let's consider how an impact at these high speeds produces a crater. When such a fast projectile strikes a planet, it penetrates two or three times its own diameter before stopping. During these few seconds its energy of motion is transferred into a shock wave, which spreads through the target body, and into heat, which vaporizes most of the projectile and some of the surrounding target. The shock wave fractures the rock of the target, while the hot silicate vapor generates an explosion similar to that of a nuclear bomb detonated at ground level (Figure 8.13). The size of the excavated crater depends primarily on the speed of impact, but generally it is 10 to 15 times the diameter of the projectile.

(a)   (b)   (c)   (d)

■ FIGURE 8.13

Stages in the Formation of an Impact Crater   (a) The impact occurs. (b) The projectile vaporizes and a shock wave spreads through the lunar rock. (c) Ejecta are thrown out of the crater. (d) Most of the ejected material falls back to fill the crater and form an ejecta blanket.

Observing the Moon

The Moon is one of the most beautiful sights in the sky, and it is the only object close enough to reveal its *topography* (mountains and valleys) to us without a visit from a spacecraft. A fairly small amateur telescope easily shows craters or mountains on the Moon as small as a few kilometers across.

Even as seen through a good pair of binoculars, the appearance of the Moon's surface changes dramatically with its phase (see Chapter 3 for an explanation of the phases). At full phase it shows almost no topographic detail, and you must look closely to see more than a few craters. This is because the sunlight illuminates the surface straight on, and in this flat lighting no shadows are cast. Much more revealing is the view near first or third quarter, when sunlight streams in from the side and topographic features cast sharp shadows. It is almost always more rewarding to study a planetary surface under such oblique lighting, when the maximum information about the surface relief can be obtained.

The flat lighting at full phase does, however, accentuate brightness contrasts on the Moon, such as those between the maria and highlands. Notice on the accompanying photo that several of the large mare craters seem to be surrounded by white material, and that the light streaks or rays that can stretch for hundreds of kilometers across the surface are clearly visible. These lighter features are ejecta, splashed out from the crater-forming impact.

By the way, there is no danger in looking at the Moon with binoculars or telescopes. The reflected sunlight is never bright enough to hurt you. In fact, the sunlit surface of the Moon has about the same brightness as a sunlit landscape on Earth. So explore away. Several guides to observing the Moon are listed in the recommendations for further reading at the end of this chapter.

One interesting thing about the Moon that you can see without binoculars or telescopes is popularly called "the new Moon in the old Moon's arms." Look at the Moon when it is a thin crescent and you can often make out the faint circle of the entire lunar disk, even though the sunlight shines on only the crescent. The rest of the disk is illuminated not by sunlight but by earthlight—sunlight reflected from the Earth. The light of the full Earth on the Moon is about 50 times brighter than that of the full Moon shining on the Earth.

Lick Observatory

The Appearance of the Moon at Different Phases Illumination from the side (*left*) brings craters and other topographic features into sharp relief. At full phase (*right*) there are no shadows and it is more difficult to see such features. However, the flat lighting at full phase brings out some surface features, such as the bright rays of ejecta that stretch out from a few large young craters.

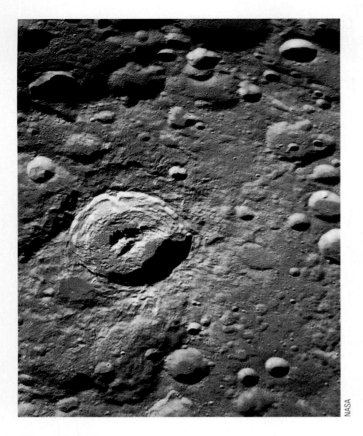

■ FIGURE 8.14

Typical Impact Crater King Crater on the far side of the Moon, a fairly recent lunar crater 75 km in diameter, shows most of the features associated with large impact structures.

An impact explosion of the sort described above leads to a characteristic kind of crater, as illustrated in Figure 8.14. The central cavity is initially bowl-shaped (the word *crater* comes from the Greek word for "bowl"), but the rebound of the crust partially fills it in, producing a flat floor and sometimes creating a central peak. Around the rim, landslides create a series of terraces.

The rim of the crater is turned up by the force of the explosion, so it rises above both the floor and the adjacent terrain. Surrounding the rim is an *ejecta blanket* consisting of material thrown out by the explosion. This debris falls back to create a rough, hilly region, typically about as wide as the crater diameter. Additional, higher-speed ejecta fall at greater distances from the crater, often digging small *secondary craters* where they strike the surface (see Figure 8.7).

Some of these streams of ejecta can extend from the crater for hundreds or even thousands of kilometers, creating on the Moon the bright *crater rays* that are prominent in photos taken near full phase (see Seeing for Yourself: *Observing the Moon*). The brightest lunar crater rays are associated with large young craters such as Kepler and Tycho.

8.3.3 Using Crater Counts

As discussed in Chapter 6, if a world has had little erosion or internal activity, like the Moon during the past 3 billion years, it is possible to use the number of impact craters on its surface to estimate the age of that surface. By "age" we mean the time since a major disturbance occurred on that surface (such as the volcanic eruptions that produced the lunar maria).

We can't directly measure the rate at which craters are being formed on the Earth and Moon, since (as we saw in Chapter 7) the average interval between large crater-forming impacts is longer than the span of human history. Our best-known example of such a large crater, Meteor Crater in Arizona, is 50,000 years old. However, the cratering rate can be estimated from the number of craters on the lunar maria or calculated from the number of potential "projectiles" (asteroids and comets) present in the solar system today, as will be discussed in Chapter 12. Both lines of reasoning lead to about the same answers.

For the Moon these calculations indicate that a crater 1 km in diameter should be produced about every 200,000 years, a 10-km crater every few million years, and one or two 100-km craters every billion years. If the cratering rate stayed the same over the millennia, we can figure out how long it must have taken to make all the craters we see in the lunar maria. Our calculations show that it would have taken several billion years. This result is similar to the age determined for the maria from radioactive dating—3.3 to 3.8 billion years old. The fact that these two calculations agree suggests that our assumption was right: Comets and asteroids in approximately their current numbers have been impacting planetary surfaces for at least 3.8 billion years. Calculations carried out for other planetary bodies (and their satellites) indicate that they also have been subject to about the same number of interplanetary impacts during this time.

Earlier than 3.8 billion years ago, however, we have good reason to believe that the impact rates must have been a great deal higher. This becomes immediately evident when we compare the numbers of craters on the lunar highlands with those on the maria. Typically, there are ten times more craters on the highlands than on a similar area of maria. Yet the radioactive dating of highland samples showed that they are only a little older than the maria, typically 4.2 billion years rather than 3.8 billion years. If the rate of impacts had been constant throughout the Moon's history, the highlands would have had to be at least ten times older. They would thus have had to form 38 billion years ago—long before the universe itself began.

In science, when an assumption leads to an unbelievable conclusion, we must go back and re-examine that assumption—in this case, the constant impact rate. The contradiction is resolved if the impact rate varied over time, with a much heavier bombardment earlier than 3.8 billion years ago (Figure 8.15). This "heavy bombardment" produced most of the craters we see today in the highlands.

■ FIGURE 8.15

Cratering Rates over Time The number of craters being made on the Moon's surface has varied with time over the past 4.3 billion years.

8.4 THE ORIGIN OF THE MOON

It is characteristic of modern science to ask how things originated. Understanding the origin of the Moon has proved to be challenging for planetary scientists, however. Part of the difficulty is simply that we know so much about our satellite (quite the opposite of our usual problem in astronomy). Put in simple terms, the Moon is both tantalizingly similar to the Earth and frustratingly different.

8.4.1 Ideas for the Origin of the Moon

Most of the hypotheses for the Moon's origin follow one of three general ideas:

1. The fission theory—the Moon was once part of the Earth but separated from it early in their history.
2. The sister theory—the Moon formed together with (but independently of) the Earth, as we believe many satellites of the outer planets formed.
3. The capture theory—the Moon formed elsewhere in the solar system and was captured by the Earth.

Unfortunately, there seem to be fundamental problems with each of these ideas. Perhaps the easiest hypothesis to reject is the capture theory. Its primary drawback is that no one knows of any way that the early Earth could have captured a large satellite from elsewhere. One body approach-

ing another cannot go into orbit around it without a substantial loss of energy; this is the reason that spacecraft destined to orbit other planets are equipped with retro-rockets. Furthermore, if such a capture did take place, the satellite would go into a very eccentric orbit rather than the nearly circular orbit the Moon occupies today. Finally, there are too many compositional similarities between the Earth and the Moon, particularly an identical fraction of the major isotopes of oxygen, to justify seeking a completely independent origin.

The fission hypothesis, which states that the Moon separated from the Earth, was suggested in the late 19th century by George Darwin, profiled in Chapter 3. More modern calculations have shown that the sort of spontaneous fission or splitting imagined by Darwin is impossible. Furthermore, it is difficult to understand how a Moon made out of terrestrial material in this way could have developed the many distinctive chemical differences now known to characterize our satellite.

Scientists have therefore been left with the sister hypothesis—that the Moon formed alongside the Earth—or with some modification of the fission hypothesis that finds a more acceptable way for the lunar material to have separated from the Earth. But the more we learn about our satellite, the less these old ideas seem to fit the bill.

8.4.2 The Giant Impact Hypothesis

In an effort to resolve these apparent contradictions, scientists developed a fourth hypothesis for the origin of the Moon, one that involves a giant impact early in the Earth's history. There is increasing evidence that large chunks of material—objects of essentially planetary mass—were orbiting in the inner solar system at the time that the terrestrial planets formed. The giant impact hypothesis envisions the Earth being struck obliquely by an object approximately one-tenth the Earth's mass—a "bullet" about the size of Mars. This is very nearly the largest impact the Earth could experience without being shattered.

Such an impact would disrupt much of the Earth by penetrating into the core, ejecting a vast amount of material into space, and releasing almost enough energy to break the planet apart (Figure 8.16). The simulations done on large computers indicate that material totaling several percent of the Earth's mass could be ejected in such an impact. Most of this material would be from the stony mantles of the Earth and the impacting body, not from their metal cores. This ejected rock vapor then cooled and formed a ring of material orbiting around the Earth. It was this ring that ultimately condensed into the Moon.

While we do not have any current way of showing that the giant impact hypothesis is the correct model of the Moon's origin, it does offer potential solutions to most of the major problems raised by the chemistry of the Moon. First, since the Moon's raw material is derived from the *mantles* of the Earth and the projectile, the absence of metals is easily understood. Second, most of the volatile elements would have been lost during the high-temperature

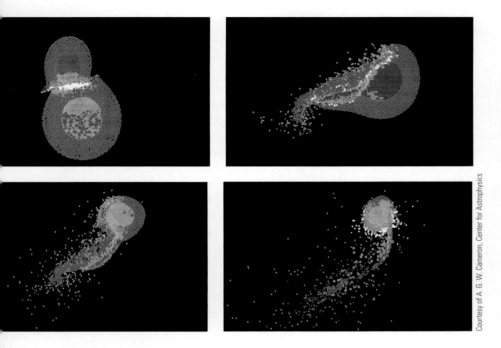

Courtesy of A. G. W. Cameron, Center for Astrophysics

■ **FIGURE 8.16**

How an Impact May Have Formed the Moon Four frames from a series of computer simulations showing a giant impact with the Earth. Iron is shown as spheres of blue and green. Rock is represented by red, pink, yellow, and white spheres, in order of increasing temperature. In this particular simulation, the impactor has 1/5 the mass in the system, and the Earth has 4/5. After the oblique impact, a plume of material is ejected into space. It is this plume that, under the right circumstances, can condense to form the Moon. Note that the plume eventually has much more pink (rock) than blue or green (iron), which explains why the Moon resembles the mantle of the Earth.

phase following the impact, explaining the lack of these materials on the Moon. Yet by making the Moon primarily of terrestrial mantle material, it is also possible to understand similarities such as identical abundances of various oxygen isotopes.

This idea that large impacts (especially during the early history of the solar system) have played a major role in shaping the worlds we see is not unique to our study of the Moon. As you read through the other chapters about the planets and their satellites, you will see further indications that a number of the present-day characteristics of our system may be due to our violent past.

8.5 MERCURY

The planet Mercury is similar to the Moon in many ways. Like the Moon, it has no atmosphere and its surface is heavily cratered. As described later in this chapter, it also shares with the Moon the likelihood of a violent birth history.

8.5.1 Mercury's Orbit

Mercury is the nearest to the Sun of the nine planets and, in accordance with Kepler's third law, it has the shortest period of revolution about the Sun (88 of our days) and the highest average orbital speed (48 km/s). It is appropriately named for the fleet-footed messenger god of the Romans. Because Mercury remains close to the Sun, it can be difficult to pick out in the sky. As you might imagine, it's best seen when its eccentric orbit takes it as far from the Sun as possible.

The semimajor axis of Mercury's orbit—that is, the planet's average distance from the Sun—is 58 million km, or 0.39 astronomical units (AU). However, because its

orbit has the high eccentricity of 0.206, Mercury's actual distance from the Sun varies from 46 million km at perihelion to 70 million km at aphelion. (If some of these terms are new to you, you may want to look at Chapter 2 or the glossary at the end of the book.) Pluto is the only planet with a more eccentric orbit than Mercury. Furthermore, the 7° inclination of Mercury's orbit to the plane of the ecliptic is also greater than that of any other planet except Pluto.

8.5.2 Composition and Structure

Table 8.1 compares some basic properties of Mercury with those of the Moon. Mercury's mass is 1/18 that of the Earth; again Pluto is the only planet with a smaller mass. Mercury is also the second smallest of the planets, having a diameter of 4878 km, less than half that of the Earth. Mercury's density is 5.4 g/cm^3, much greater than the density of the Moon, indicating that the composition of these two objects differs substantially.

Its composition is one of the most interesting things about Mercury; it is unique among the planets. Mercury's high density tells us that it must be composed largely of heavier materials such as metals. The most likely models for Mercury's interior suggest a metallic iron–nickel core amounting to 60 percent of the total mass, with the rest of the planet made up primarily of silicates. The core has a diameter of 3500 km and extends out to within 700 km of the surface. We could think of Mercury as a metal ball the size of the Moon surrounded by a rocky crust 700 km thick (Figure 8.17). Unlike the Moon, Mercury also has a weak magnetic field. The existence of this field is consistent with a large metal core, and it suggests that at least part of the core must be liquid in order to generate the observed field. (See Section 7.1 for the explanation of how magnetic fields and liquid cores are connected.)

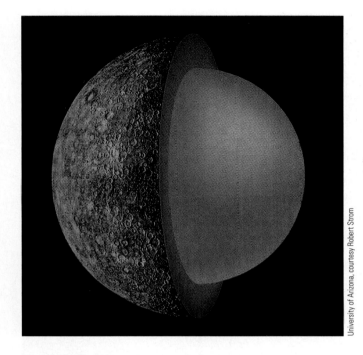

FIGURE 8.17

Mercury's Internal Structure The interior of Mercury is dominated by a metallic core about the same size as our Moon.

University of Arizona, courtesy Robert Strom

8.5.3 Mercury's Strange Rotation

Visual studies of Mercury's indistinct surface markings were once thought to indicate that the planet kept one face to the Sun (as the Moon does to the Earth). Thus, for many years it was widely believed that Mercury's rotation period was equal to its revolution period of 88 days, making one side perpetually hot while the other was always cold.

Radar observations of Mercury in the mid-1960s, however, showed conclusively that Mercury does not keep one side fixed to the Sun. Recall from Section 5.4 that the frequency of a transmitted radar pulse can be controlled precisely. Any motion of the target then introduces a measurable change in this frequency as a result of the Doppler effect (see Section 4.6). Such target motions can include rotation: If a planet is turning, one side seems to be approaching the Earth while the other is moving away from it. The result is to spread or broaden the precise transmitted frequency into a range of frequencies in the reflected signal (Figure 8.18). The degree of broadening provides an exact measurement of the rotation rate of the planet.

Mercury's sidereal period of rotation (how long it takes to turn with respect to the distant stars) was found to be about 59 days, which is just two thirds of the planet's period of revolution. Subsequently astronomers found that a situation where the spin and the orbit of a planet are in a 2-to-3 ratio is stable. That is, it can continue for a long time, like the situation where the time to spin equals the time to orbit (a relationship Pluto exhibits, for example).

Mercury is very hot on its daylight side, but because it has no appreciable atmosphere, it gets surprisingly cold during the long nights. The temperature on the surface climbs to 700 K at noontime. After sunset, however, the

FIGURE 8.18

Doppler Radar Measures Rotation When a radar beam is reflected from a rotating planet, the motion of one side of the planet's disk toward us and the other away from us causes Doppler shifts in the reflected signal. The effect is to cause both a "redshift" and a "blueshift," and thus to widen the spread of frequencies in the radio beam.

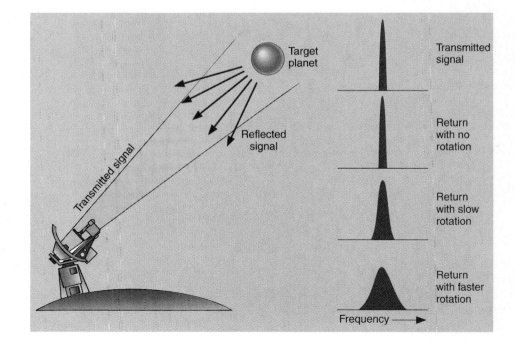

temperature drops, reaching 100 K just before dawn. The *range* in temperature on Mercury is thus 600 K (or 600°C), more than on any other planet.

8.5.4 The Surface of Mercury

The first close-up look at Mercury came in 1974, when the U.S. spacecraft Mariner 10 passed 9500 km from the surface of the planet and transmitted more than 2000 photographs to Earth, revealing details with a resolution down to 150 m. Mercury strongly resembles the Moon in appearance (Figures 8.19 and 8.20). It is covered with thousands of craters and larger basins up to 1300 km in diameter. Some of the brighter craters are rayed, like Tycho and Copernicus on the Moon, and many have central peaks. There are also *scarps* (cliffs) more than a kilometer high and hundreds of kilometers long, as well as ridges and plains.

Most of the mercurian features have been named in honor of artists, writers, composers, and other contributors to the arts and humanities, in contrast with the scientists commemorated on the Moon. Among the most prominent craters are Bach, Shakespeare, Tolstoy, Mozart, and Goethe.

There is no evidence of plate tectonics on Mercury. Distinctive long scarps, however, can sometimes be seen cutting across craters; this means the scarps must have formed later than the craters (Figure 8.21). These long cliffs appear to have their origin in the slight compression of Mercury's crust. Apparently, at some point in its history this planet shrank, wrinkling the crust, and it must have done so after most of the craters on its surface had already

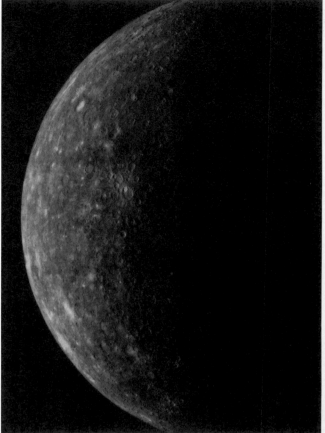

Mosaic courtesy of the Astrogeology Team, USGS, Flagstaff

■ **FIGURE 8.19**

Best Photo of Mercury The planet Mercury as photographed by Mariner 10 in 1974. The heavily cratered surface resembles that of the Moon.

NASA/JPL

■ **FIGURE 8.20**

Caloris Basin Part of the Caloris basin on Mercury, 1300 km in diameter, can be seen at the bottom. This partially flooded impact basin is the largest structural feature on Mercury seen by Mariner 10. Compare this photo with Figure 8.8, the Orientale basin on the Moon.

What a Difference a Day Makes

Mercury rotates three times for each two orbits around the Sun. It is the only planet that exhibits this relationship between its spin and its orbit, and there are some interesting consequences for any observers who might someday be stationed on the surface of Mercury.

Here on Earth, we take it for granted that days are much shorter than years. Therefore, the two ways of defining the local "day"—how long the planet takes to rotate, and how long the Sun takes to return to the same position in the sky (see Chapter 3)—are the same for most practical purposes. But this is not the case on Mercury. While Mercury rotates in 59 Earth days, the time for the Sun to return to the same place in Mercury's sky turns out to be two Mercury years, or 176 Earth days. (Note that this result is not intuitively obvious, so don't be upset with yourself if you didn't come up with it.) Thus, if one day at noon a Mercury explorer suggests to her companion that they should meet at noon the next day, this could mean a very long time apart!

To make things even more interesting, recall that Mercury has an eccentric orbit, meaning that its distance from the Sun varies significantly during each mercurian year. By Kepler's law, the planet moves fastest in its orbit when closest to the Sun. Let's examine how this affects the way we would see the Sun in the sky during one 176-Earth-day cycle. We'll look at the situation as if we were standing on the surface of Mercury in the center of a giant basin that astronomers call Caloris (see Figure 8.20).

At the location of Caloris, Mercury is most distant from the Sun at sunrise; this means the rising Sun looks smaller in the sky (although still more than twice the size it appears from the Earth). As the Sun rises higher and higher, it looks bigger and bigger; Mercury is now getting closer to the Sun in its eccentric orbit. At the same time, the apparent motion of the Sun slows down as Mercury's faster motion in orbit begins to catch up with its rotation.

At noon the Sun is now three times larger than it looks from Earth and hangs almost motionless in the sky. As the afternoon wears on, the Sun eventually looks smaller and smaller again, and moves faster and faster in the sky. At sunset, a full Mercury year (or 88 Earth days) after sunrise, the Sun is back to its smallest apparent size as it dips out of sight. Then it takes another Mercury year before the Sun rises again. (Sunrise and sunset are much more sudden on Mercury, since there is no atmosphere to bend or scatter the rays of sunlight.)

Astronomers call locations like the Caloris basin the "hot longitudes" on Mercury because the Sun is closest to the planet at noon, just when it is lingering overhead for many Earth days. This makes them the hottest places on the planet.

We bring all this up not because the exact details of this scenario are so important, but to illustrate how many of the things we take for granted on Earth are not the same on other worlds. As we've mentioned before, one of the best things about taking an astronomy class should be ridding you forever of any "Earth chauvinism" you might have. The way things are on our planet is just one of the many ways nature can arrange reality.

NASA/JPL

formed. If the standard cratering chronology applies to Mercury, this shrinkage must have taken place during the past 4 billion years. If we understood the interior structure and composition of the planet better, we could probably calculate how internal changes in temperature might have led to this global compression, which has no counterpart on the Moon or the other terrestrial planets.

Although it has become something of a forgotten planet in the two decades since the Mariner 10 flybys, two spacecraft missions are now planned to revisit the planet:

■ **FIGURE 8.21**

Discovery Scarp on Mercury This long cliff, nearly 1 km high and more than 100 km long, cuts across several craters. We conclude that the compression that made "wrinkles" like this in the planet's surface must have taken place after the craters were formed.

the Messenger probe from NASA and BepiColombo from ESA (the European Space Agency). Astronomical studies from the Earth are also still useful. For example, enhanced radar reflectivity from both poles of the planet was discovered in 1992, indicating the presence of ice just beneath the surface. Ice on a planet so close to the Sun may sound surprising, but the temperature very near the poles remains low enough for ice to survive. Still, the discovery of these icy polar caps was unexpected, since they seem to imply that the polar regions of Mercury have remained below freezing since the formation of the planet. Probably there are cold traps there similar to the deep craters that shelter ice at the poles of our Moon.

8.5.5 The Origin of Mercury

The problem with understanding how Mercury formed is the reverse of the problem posed by the composition of the Moon. We have seen that, unlike the Moon, Mercury is composed mostly of metal. However, astronomers think that Mercury should have formed with about the same ratio of metal to silicate as that found on the Earth or Venus. How did it lose so much of its rocky material?

The most probable explanation for Mercury's silicate loss may be similar to the explanation for the Moon's lack of a metal core. Mercury is likely to have experienced several giant impacts very early in its youth, and one or more of these may have torn away a fraction of its mantle and crust, leaving a body dominated by its iron core.

Today astronomers recognize that the early solar system was a chaotic place, with the final stages of planet formation characterized by impacts of great violence. Some objects of planetary mass may have been destroyed, while others could have fragmented and then reformed, perhaps more than once. Both the Moon and Mercury, with their strange compositions, bear testimony to the catastrophes that must have characterized the solar system during its youth.

SURFING THE WEB

The Moon—General Sites About the Moon, with Good Links:

- *The Nine Planets Site:*
  seds.lpl.arizona.edu/nineplanets/nineplanets/luna.html

- *Views of the Solar System Site:*
  www.solarviews.com/eng/moon.htm

- *Planetary Sciences Site:*
  nssdc.gsfc.nasa.gov/planetary/planets/moonpage.html

- *Exploring the Moon:*
  cass.jsc.nasa.gov/pub/expmoon/lunar_missions.html
  An extensive site from NASA's Johnson Space Flight Center, with information on U.S. and Russian Moon exploration and good links to many other Web resources.

- *Apollo Lunar Surface Journal:*
  www.hq.nasa.gov/office/pao/History/alsj/
  Information, interviews, maps, photos, video and audio clips, and much more on each of the Apollo landing missions.

Recent Missions to the Moon:

- *Clementine Mission Site:*
  www.nrl.navy.mil/clementine/

- *Lunar Prospector Home Page:* lunar.arc.nasa.gov/

Touring the Moon with Binoculars:
skyandtelescope.com/observing/objects/moon/
An introductory guide, with a map, for viewing the Moon with simple equipment.

Sites That Take a Skeptical Look at Claims That the Moon Landings Were a Hoax:

- *The Bad Astronomy Site:*
  www.badastronomy.com/bad/tv/foxapollo.html

- *Science@NASA Site:*
  science.nasa.gov/headlines/y2001/ast23Feb_2.htm

- *Comments by Astronomer Jim Scotti:*
  pirlwww.lpl.arizona.edu/~jscotti/NOT_faked/

 Mercury:

- *The Nine Planets Site:*
  seds.lpl.arizona.edu/nineplanets/nineplanets/mercury.html

- *Views of the Solar System Site:*
  www.solarviews.com/eng/mercury.htm

- *Planetary Sciences Site:*
  nssdc.gsfc.nasa.gov/planetary/planets/mercurypage.html

- *Mercury Unveiled:*
  www.psrd.hawaii.edu/PSRdiscoveries/Jan97/MercuryUnveiled.html
  Planetary scientist G. Jeffrey Taylor updates our current understanding of Mercury based on reanalyzed Mariner 10 images.

- *Messenger Mission:* messenger.jhuapl.edu
  Previews the NASA probe to Mercury.

SUMMARY

8.1 Most of what we know about the Moon derives from the Apollo program, including 400 kg of lunar samples still being intensively studied. The Moon has 1/80 the mass of the Earth and is severely depleted in both metals and volatile materials. It is made almost entirely of silicates like those in the Earth's mantle and crust. However, spacecraft have found evidence of a small amount of water near the lunar poles, most likely deposited by comet and asteroid impacts.

8.2 Lunar rocks can be dated by their radioactivity. Like the Earth, the Moon was formed about 4.5 billion years ago. The heavily cratered **highlands** are made of rocks more than 4 billion years old. The darker volcanic plains of the **maria** were erupted between 3.3 and 3.8 billion years ago. Generally, the surface is dominated by impacts, including continuing small impacts that produce its fine-grained soil.

8.3 A century ago Gilbert suggested that the lunar craters were of impact origin, but the cratering process was not well understood until more recently. High-speed impacts produce explosions and excavate craters with raised rims, ejecta blankets, and often central peaks. Cratering rates have been roughly constant for the past 3 billion years but earlier were much greater. Crater counts can be used to derive approximate ages for geological features on the Moon and other planets.

8.4 The three standard hypotheses for the origin of the Moon are the fission hypothesis, the sister hypothesis, and the capture hypothesis. All have problems, and recently they have been supplanted by the giant impact hypothesis, which ascribes the origin of the Moon to the impact of a Mars-sized projectile with the Earth 4.5 billion years ago.

8.5 Mercury is the nearest planet to the Sun and the fastest moving. Mercury is similar to the Moon in having a heavily cratered surface and no atmosphere, but it differs in having a very large metal core. Early in its evolution, it apparently lost part of its silicate mantle, probably due to one or more giant impacts. Long scarps on its surface testify to a global compression of Mercury's crust during the past 4 billion years.

INTER-ACTIVITY

 A We mentioned that no nation on Earth now has the capability to send a human being to the Moon, even though the United States once sent 12 astronauts to land there. How does your group feel about this? Should we continue the exploration of space with human beings? Should we put habitats on the Moon? Should we go to Mars? Does humanity have a "destiny in space"? Whatever your answer to these questions is, make a list of the arguments and facts that support your position.

B When they first hear about the giant impact hypothesis for the origin of the Moon, many students are intrigued and wonder why we can't cite more evidence for it. Could your group make a list of reasons we cannot find any traces on the Earth of the great impact that formed the Moon?

C We discussed that the ice (mixed into the soil) that is found on the Moon was most likely delivered by comets. Have your group make a list of all the reasons the Moon would not have any ice of its own left from its early days.

 D Can your group make a list of all the things that would be different if our Earth had no Moon? Don't restrict your answer to astronomy and geology. (You may want to review Chapter 3.)

E If, one day, humanity decides to establish a colony on the Moon, where should we put it? Make a list of the advantages and disadvantages of locating such a human habitat on the near side, the far side, or at the poles. What site would be best for visible-light and radio astronomy?

 F A member of the class (but luckily, not a member of your group) suggests that he has always dreamed of building a vacation home on the planet Mercury. Can your group make a list of all the reasons such a house would be hard to build and keep in good repair?

1. What is the composition of the Moon, and how does it compare to the composition of the Earth? Of Mercury?
2. Outline the main events in the Moon's geological history.
3. Explain how high-speed impacts form circular craters. How can this explanation account for the characteristic features of impact craters?
4. Explain the evidence for a period of heavy bombardment on the Moon about 4 billion years ago. What might have been the source of the impacting debris during this early era?
5. How did our exploration of the Moon differ from that of Mercury (and the other planets)?
6. Summarize the four main theories for the origin of the Moon. Give pros and cons for each.
7. What do current ideas about the origins of the Moon and Mercury have in common? How do they differ?

THOUGHT QUESTIONS

8. One of the primary scientific objectives of the Apollo program was the return of lunar material.
   a. Why was this so important?
   b. What can be learned from samples?
   c. Are they still of value now?

9. Apollo astronaut David Scott dropped a hammer and a feather together on the Moon, and both reached the ground at the same time. What are the two distinct advantages that this experiment on the Moon had over the same experiment as performed by Galileo on the Earth?

10. Galileo thought the lunar maria might be seas of water. If you had no better telescope than the one he had, could you prove that they are not composed of water?

11. Why did it take so long for geologists to recognize that the lunar craters had an impact origin rather than a volcanic one?

12. How might a crater made by the impact of a comet with the Moon differ from a crater made by the impact of an asteroid?

13. Why are the lunar mountains smoothly rounded rather than having sharp, pointed peaks (as they were almost al-ways depicted in science-fiction illustrations and films before the first lunar landings)?

14. The lunar highlands have about ten times more craters on a given area than do the maria. Does this mean that the highlands are ten times older? Explain your reasoning.

15. At the end of Section 8.2, your authors say that lunar night and day each last about 2 Earth weeks. After looking over the information in Chapter 3 and this chapter about the motions of the Moon, can you explain why? (It helps to draw a diagram for yourself.)

16. Give several reasons Mercury would be a particularly unpleasant place to build an astronomical observatory.

17. If, in the remote future, we establish a base on Mercury, keeping track of time will be a challenge. Discuss how to define a year on Mercury, and the two ways to define a day. Can you come up with ways that humans raised on Earth might deal with time cycles on Mercury?

18. The Moon has too little iron, Mercury too much. How can both of these anomalies be the result of giant impacts? Explain how the same process can yield such apparently contradictory results.

FIGURING FOR YOURSELF

The average density of a body equals its mass divided by its volume:

$$\text{density} = \frac{\text{mass}}{\frac{4}{3}\pi \times \text{radius}^3}$$

Astronomers can measure both mass and radius accurately when a spacecraft flies by a body.

19. Using the information in this chapter, calculate the average density of the Moon. Show your work. Does your calculation agree with the figure we give in this chapter?

20. Using the information in this chapter, calculate the average density of Mercury. Show your work. Does your calculation agree with the figure we give in this chapter?

21. The Moon was once closer to the Earth than it is now. When it was at half its present distance, how long was its period of revolution? (See Chapter 2 for the formula to use.)

22. Astronomers believe that the deposit of lava in the giant mare basins did not happen in one flow but in many different eruptions spanning some time. Indeed, in any one mare we find a variety of rock ages, typically spanning about 100 million years. The individual lava flows as seen in Hadley Rill by the *Apollo 15* astronauts were about 4 m thick. Estimate the average time interval between the beginnings of successive lava flows if the total depth of the lava in the mare is 2 km.

23. The Moon requires about one month (0.08 year) to orbit the Earth. Its distance from us is about 400,000 km (0.0027 AU). Use Kepler's third law, as modified by Newton, to calculate the mass of the Earth relative to the Sun.

SUGGESTIONS FOR FURTHER READING

On the Moon

Chaikin, A. *A Man on the Moon.* 1994, Viking Press. A well-reviewed history of manned lunar exploration.

Coco, M. "Staging a Moon Shot" in *Astronomy,* Aug. 1992, p. 62. Describes photographing the Moon.

Foust, J. "NASA's New Moon" in *Sky & Telescope,* Sept. 1998, p. 48. On results from the Lunar Prospector mission.

Jayawardhana, R. "Deconstructing the Moon" in *Astronomy,* Sept. 1998, p. 40. An update on the giant impact hypothesis for the origin of the Moon.

Kitt, M. *The Moon: An Observing Guide for Backyard Telescopes.* 1992, Kalmbach. Eighty-page illustrated primer for beginners.

Ryder, G. "Apollo's Gift: The Moon" in *Astronomy,* July 1994, p. 40. Good evolutionary history of the Moon.

Schmitt, H. "Exploring Taurus–Littrow: Apollo 17" in *National Geographic,* Sept. 1973. First-person account by the only scientist to walk on the Moon.

Spudis, P. "The Giant Holes of the Moon" in *Astronomy,* May 1996, p. 50. On the results of the Clementine mission. (See also reports in *Sky & Telescope,* July 1995, p. 32; Feb. 1997, p. 24.)

Spudis, P. *The Once and Future Moon.* 1998, Smithsonian Institution Press. A geologist discusses what our exploration of the Moon has taught us, and what we might do in the future to know and use the Moon better.

Taylor, G. "The Scientific Legacy of Apollo" in *Scientific American,* July 1994, p. 40.

On Mercury

Armstrong, J. "Heading Back to the Forgotten Planet" in *Astronomy,* Oct. 2002, p. 40. Preview of Messenger Mission.

Beatty, J. "Mercury's Cool Surprise" in *Sky & Telescope,* Jan. 1992, p. 35.

Chapman, C. "Mercury's Heart of Iron" in *Astronomy,* Nov. 1988, p. 22. Good introduction to our modern view of the planet.

Nelson, R. "Mercury: The Forgotten Planet" in *Scientific American,* Nov. 1997, p. 56.

Sheehan, W. and Dobbins, T. "Mesmerized by Mercury" in *Sky & Telescope,* June 2000, p. 109. History of Mercury observations and how amateur astronomers can contribute.

Strom, R. *Mercury: The Elusive Planet.* 1987, Smithsonian Institution Press. Best book on the inner planet.

Heavily Eroded Canyonlands on Mars This Viking spacecraft view, taken from Mars orbit, looks down on a small part of the Valles Marineris canyon complex and shows an area about 60 km across.

9 Earth-Like Planets: Venus and Mars

One deduction from this thin air [on Mars] we must be careful not to make— that because it is thin, it is incapable of supporting intelligent life. That beings constituted physically as we are would find it a most uncomfortable habitat is pretty certain. But lungs are not wedded to logic, as public speeches show, and there is nothing in the world or beyond it to prevent, so far as we know, a being with gills, for example, from being a most superior person.

Percival Lowell, in Mars (1895)

THINKING AHEAD

For decades science-fiction writers have conjured up visions of Venus and Mars as Earth-like worlds where humans would someday visit and (perhaps) come face to face with alien life. Now that our robot probes have explored both planets, we know that neither has conditions similar to our own. How did it happen that these three neighboring terrestrial planets have turned out to be so different from each other?

The Moon and Mercury are geologically dead. In contrast, the larger terrestrial planets—the Earth, Venus, and Mars—are more active and interesting worlds. We have already discussed the Earth, and we now turn to Venus and Mars. These are the nearest planets and the most accessible to spacecraft. Not surprisingly, the greatest effort in planetary exploration has been devoted to these fascinating worlds. In this chapter we discuss some of the results of four decades of scientific exploration of Mars and Venus.

Virtual Laboratories

 Planetary Geology

Planetary Atmospheres and Their Retention

THE NEAREST PLANETS: AN OVERVIEW

As you might expect from close neighbors, Mars and Venus are among the brightest objects in the night sky. The average distance of Mars from the Sun is 227 million km (1.52 AU), or about half again as far from the Sun as the Earth. However, Mars has a relatively large orbital eccentricity of 0.093, which means its distance varies by 42 million km as it goes around the Sun.

Venus' orbit, in contrast, is very nearly circular, at a distance of 108 million km (0.72 AU) from the Sun. Like Mercury, Venus sometimes appears as an "evening star" and sometimes as a "morning star." The closest Mars ever gets to Earth is about 56 million km. Venus approaches the Earth more closely than does any other planet: At its nearest it is only 40 million km away from us.

9.1.1 Appearance

Venus looks very bright, and even a small telescope reveals that it goes through phases like the Moon. Galileo discovered that Venus displays a full range of phases, and he used this as an argument to show that Venus must circle the Sun and not the Earth (see Figure 1.16). The planet's actual surface is not visible because it is shrouded by dense clouds that reflect about 70 percent of the sunlight that falls on them, frustrating efforts to study the underlying surface, even with cameras in orbit around the planet (Figure 9.1).

In contrast, Mars is more tantalizing as seen through a telescope (Figure 9.2). The planet is distinctly red, due (as we now know) to the presence of iron oxides in its soil. This color may account for its association with war (and bloodletting) in the legends of early cultures. The best resolution obtainable from telescopes on the ground is about 100 km, or about the same as what we can see on the Moon with the unaided eye. At this resolution, no hint of topographic structure can be detected—no mountains, no valleys, not even impact craters. On the other hand, bright polar ice caps can be seen easily, together with dusky surface markings that sometimes change in outline and intensity from season to season. Of all the planets, only Mars has a surface that can be clearly made out from the Earth.

For a few decades around the turn of the 20th century, some astronomers believed they saw evidence of an intelligent civilization on Mars. The controversy began in 1877, when the Italian astronomer Giovanni Schiaparelli announced that he could see long, faint, straight lines on Mars that he called *canale,* or channels. In English-speaking countries, the term was mistakenly translated as *canals,* implying an artificial origin.

Even before Schiaparelli's observations, astronomers had watched the bright polar caps change size with the seasons and had seen variations in the dark surface features. With a little imagination, it was not difficult to picture the

NASA/ARC

■ **FIGURE 9.1**

Venus as Photographed by the Pioneer Venus Orbiter This ultraviolet image shows upper-atmosphere cloud structure that would be invisible at visible wavelengths. Note that there is not a glimpse of the planet's surface.

Steve Larson, University of Arizona

■ **FIGURE 9.2**

Mars Seen from the Earth's Surface These are among the best Earth-based photos of Mars, taken in 1988 when the planet was exceptionally close to the Earth. The polar caps and dark surface markings are evident, but not the topographic features.

Percival Lowell: Dreaming of an Inhabited Mars

Percival Lowell (1855–1916) was born into the well-to-do Massachusetts family about whom John Bossidy made the famous toast:

> And this is good old Boston,
> The home of the bean and the cod,
> Where the Lowells talk to the Cabots
> And the Cabots talk only to God.

Percival's brother Lawrence became president of Harvard University, and his sister Amy, who liked to dress in men's clothing and smoke cigars, became a distinguished poet. Percival was already interested in astronomy as a boy; he made observations of Mars at age 13. His undergraduate thesis at Harvard dealt with the origin of the solar system, but he did not pursue this interest immediately. Instead he entered the family business and traveled extensively in Asia. In 1892, however, he decided to dedicate himself to carrying on Schiaparelli's work and solving the mysteries of the martian canals.

In 1894, with the help of astronomers at Harvard but using his own funds, Lowell built an observatory on a high plateau in Flagstaff, Arizona, where he hoped the seeing would be clear enough to show him Mars in unprecedented detail. He and his assistants quickly accumulated a tremendous number of drawings and maps, purporting to show a vast network of martian canals (see Figure 9.2). He elaborated his ideas about the inhabitants of the red planet in several books, including *Mars* (1895)

Percival Lowell in about 1910, observing with his 24-inch telescope at Flagstaff, Arizona.

and *Mars and Its Canals* (1906), and in hundreds of articles and speeches. As Lowell put it:

> A mind of no mean order would seem to have presided over the system we see—a mind certainly of considerably more comprehensiveness than that which presides over the various departments of our own public works. Party politics, at all events, have had no part in them; for the system is planet-wide. . . . Certainly what we see hints at the existence of beings who are in advance of, not behind us, in the journey of life.

Lowell's views captured the public imagination and inspired many novels and stories, the most famous of

which was H. G. Wells' *War of the Worlds* (1897). In this famous "invasion" novel, the thirsty inhabitants of a dying planet Mars (based entirely on Lowell's ideas) come to conquer the Earth with advanced technology.

Although the Lowell Observatory first became famous for its work on the martian canals, both Lowell and the observatory eventually turned to other projects as well. As we will see in Chapter 11, Lowell became interested in the search for a ninth (and then undiscovered) planet in the solar system. In 1930 Pluto was found at the Lowell Observatory, and it is not a coincidence that the name selected for the new planet starts with Lowell's initials. It was also at the Lowell Observatory that the first measurements were made of the great speed at which galaxies are moving away from us, observations that would ultimately lead to our modern view of an expanding universe.

Lowell continued to live at his observatory, marrying at age 53 and publishing extensively. He relished the debate his claims about Mars caused far more than the astronomers on the other side, who often complained that Lowell's work was making planetary astronomy a less respectable field. At the same time, the public fascination with the planets fueled by Lowell's work (and its interpreters) may, several generations later, have helped fan support for the space program and the many missions whose results grace the pages of our text.

canals as crops bordering irrigation ditches that brought water from the melting polar ice to the parched deserts of the red planet. (They assumed the polar caps were composed of water ice, which isn't exactly true, as we will see shortly).

Until his death in 1916, the most effective proponent of intelligent life on Mars was Percival Lowell, a self-made American astronomer and member of the wealthy Lowell family of Boston (see the Voyagers in Astronomy box). A skilled author and speaker, Lowell made what seemed to the public a convincing case for intelligent Martians, who had constructed the huge canals to preserve their civilization in the face of a deteriorating climate (Figure 9.3).

The argument for a race of Martians, however, hinged on the reality of the canals, a matter that remained in serious

■ **FIGURE 9.3**

Lowell's Mars Globe One of the remarkable globes of Mars prepared by Percival Lowell, showing a network of dozens of canals, oases, and triangular water reservoirs that he claimed were visible on the red planet.

Lowell Observatory

than 200 years, a period during which tens of thousands of martian days have passed. As a result, the rotation period can be calculated to within a few hundredths of a second.

The rotational axis of Mars has a tilt of about 25°, similar to the tilt of the Earth's axis. Thus Mars experiences seasons very much like those on Earth. Because of the longer martian year (almost two Earth years), however, each season there lasts about six of our months.

Since no surface detail can be seen through Venus' clouds, its rotation period can be found only by bouncing radar signals off the planet (as explained for Mercury in the last chapter). The first radar observations of the planet's rotation were made in the early 1960s. Later, topographical surface features were identified on the planet that showed up in the reflected radar signals. The rotation period of Venus, precisely determined from the motion of such "radar features" across its disk, is 243 days. Even more surprising than how *long* Venus takes to rotate is the fact that it spins in a backward or retrograde direction (east to west).

Stop for a moment, and think about how odd this slow rotation makes the calendar on Venus. The planet takes 225 Earth days to orbit the Sun and 243 Earth days to spin on its axis. As a result, the time the Sun takes to return to the same place in the sky turns out to be 117 Earth days. (If you say "See you tomorrow" on Venus, you'll have a long time to wait.) Although we do not know the reason for Venus' slow backward rotation, we can guess that it may have suffered one or more extremely powerful collisions during the formation process of the solar system (see Chapter 8).

9.1.3 Basic Properties of Venus and Mars

Before discussing each planet individually, let us compare some of their basic properties with each other and with the Earth (Table 9.1). Venus is in many ways the Earth's twin, with a mass 0.82 times the mass of the Earth and an almost identical density. The amount of geological activity is also relatively high, almost as high as on Earth. On the other hand, with a surface pressure nearly 100 times greater than ours, Venus' atmosphere is not at all like that of Earth. The surface of Venus is also remarkably hot, with a temperature of 730 K (over 850°F), hotter than the self-cleaning cycle of your oven. One of the major challenges presented by Venus is to understand why the atmosphere and surface environment of this twin have diverged so sharply from those of our own planet.

Mars, by contrast, is rather small, with a mass only 0.11 times the mass of the Earth. It is larger than either the Moon or Mercury, however, and unlike them it retains a thin atmosphere. Mars is also large enough to have supported considerable geological activity in the past. But the most fascinating thing about Mars is that long ago it probably had a thick atmosphere and seas of liquid water; the conditions we associate with the development of life. There is even a chance that some form of life persists today in protected environments below the martian surface.

dispute among astronomers. The canal markings were always difficult to study, glimpsed only occasionally because atmospheric conditions caused the tiny image of Mars to shimmer in the telescope. Lowell saw canals everywhere (even a few on Venus), but many other observers could not see them at all and remained unconvinced of their existence. When larger telescopes than Lowell's failed to confirm the presence of canals, the skeptics felt vindicated.

Astronomers had given up the idea of martian canals by the 1930s, although it persisted in the public consciousness until the first spacecraft photographs clearly showed that no canals existed. Now it is generally accepted that the straight lines were an optical illusion, the result of the human mind's tendency to see order in random features that are glimpsed dimly at the limits of the eye's resolution. When we see small, dim dots of surface markings, our minds tend to connect those dots into straight lines.

9.1.2 Rotation of the Planets

Astronomers have determined the rotation period of Mars with great accuracy by watching the motion of permanent surface markings; its sidereal day is 24 h 37 m 23 s, just a little longer than the rotation period of the Earth. This high precision is not obtained by watching Mars for a single rotation, but by noting how many turns it makes over a long period of time. Good observations of Mars date back more

TABLE 9.1 *Properties of Earth, Venus, and Mars*

	Earth	Venus	Mars
Semimajor axis (AU)	1.00	0.72	1.52
Period (year)	1.00	0.61	1.88
Mass (Earth = 1)	1.00	0.82	0.11
Diameter (km)	12,756	12,102	6,790
Density (g/cm^3)	5.5	5.3	3.9
Surface gravity (Earth = 1)	1.00	0.91	0.38
Escape velocity (km/s)	11.2	10.4	5.0
Rotation period (hours or days)	23.9 h	243 d	24.6 h
Surface area (Earth = 1)	1.00	0.90	0.28
Atmospheric pressure (bar)	1.00	90	0.007

9.2 THE GEOLOGY OF VENUS

Since Venus has about the same size and composition as the Earth, we might expect its geology to be similar. This is partly true, but Venus does not exhibit the same kind of *plate tectonics* as the Earth, and we will see that its lack of erosion results in a very different surface appearance.

9.2.1 Spacecraft Exploration of Venus

More spacecraft have been sent to Venus than to any other planet. Although the 1962 U.S. Mariner 2 flyby was the first, the Soviet Union launched most of the subsequent missions to Venus. In 1970 Venera 7 became the first probe to land and broadcast data from the surface of Venus. It operated for 23 minutes before succumbing to the high surface temperature. Additional Venera probes and landers followed, photographing the surface and analyzing the atmosphere and soil.

To understand the geology of Venus, however, we needed to make a global study of its surface, a task made very difficult by the perpetual cloud layers surrounding the planet. The problem resembles the challenge facing air traffic controllers on many an evening at San Francisco airport, when the weather is so foggy that they can't locate the incoming planes visually. The solution is similar in both cases: Use a radar instrument to probe through the obscuring layer.

The first crude global radar map was made by the U.S. Pioneer Venus orbiter in the late 1970s, followed by better maps from the twin Soviet Venera 15 and 16 radar orbiters in the early 1980s. However, most of our information on the geology of Venus is derived from the U.S. Magellan spacecraft, which mapped Venus from orbit between 1991 and 1993. With its powerful *imaging radar*, Magellan was able to study the surface at a resolution of 100 m, much

higher than that of previous missions, yielding our first detailed look at the surface of our sister planet (Figure 9.4). (The Magellan spacecraft returned more data to Earth than all previous planetary missions combined; each 100 minutes of data transmission from the spacecraft provided enough information, if translated into characters, to fill two 30-volume encyclopedias.)

Consider for a moment how good Magellan's resolution of 100 m really is. It means the radar images from Venus can show anything on the surface larger than a football field. Suddenly a whole host of topographic features on Venus became accessible to our view. As you look at the radar images throughout this chapter, bear in mind that these are constructed from radar reflections, not from visible-light photographs. For example, bright features on these images are an indication of rough terrain, while darker regions are smoother.

JPL/USGS

■ **FIGURE 9.4**
Radar Map of Venus Global radar image of Venus centered on 180° longitude. This composite image has a resolution of a few kilometers. Colors have been added to indicate elevation, with blue meaning low and red high. The large continent of Aphrodite stretches around the equator, where the bright (therefore rough) surface has been deformed by tectonic forces in the crust of Venus.

9.2.2 Probing Through the Clouds of Venus

The radar maps of Venus reveal a planet that looks much the way the Earth might look if our planet's surface were not constantly being changed by erosion and deposition of sediment. Because there is no water or ice on Venus and the surface wind speeds are low, almost nothing obscures or erases the complex geological features produced by the movements of Venus' crust and by volcanic eruptions. Having finally penetrated below the clouds of Venus, we find its surface to be naked, revealing the history of hundreds of millions of years of geological activity. Venus is in many ways a geologist's dream planet.

About 75 percent of the surface of Venus consists of lowland lava plains. Superficially, these plains resemble the basaltic ocean basins of the Earth, but they were not produced in quite the same way. There is no evidence of subduction zones (see Chapter 7) on Venus, indicating that, unlike the Earth, this planet never experienced plate tectonics. Although *convection* (the rising of hot materials) in its mantle generated great stresses in the crust of Venus, they did not start great continental plates moving. The formation of the lava plains of Venus more nearly resembles that of the lunar maria. Both were the result of widespread lava eruptions without the crustal spreading associated with plate tectonics.

Rising above the lowland lava plains are two full-scale continents of mountainous terrain. The largest continent on Venus, called Aphrodite, is about the size of Africa (you can see it stand out in Figure 9.4). Aphrodite stretches along the equator for about one third of the way around the planet. Next in size is the northern highland region Ishtar, which is about the size of Australia. Ishtar contains the highest region on the planet, the Maxwell Mountains, which rise 11 km above the surrounding lowlands. (The Maxwell Mountains are the only feature on Venus named after a man. They commemorate James Clerk Maxwell, whose theory of electromagnetism led to the invention of radar. All other features are named for women, historic or mythological.)

9.2.3 Craters and the Age of the Venusian Surface

One of the first questions astronomers addressed with the high-resolution Magellan images was the age of the surface of Venus. Remember that the age of a surface is rarely the age of the world it is on. A young age merely implies an active geology; the planet has forces at work that can resurface the world in short order. As described in the previous chapter, such ages can be derived from counting impact craters (Figure 9.5a is an example of what these craters look like on the Venus radar images). The more densely cratered the surface, the greater its age. The largest crater on Venus (called Mead) is 275 km in diameter, slightly larger than the largest known terrestrial crater (Chicxulub) but much smaller than the lunar impact basins.

(a)

(b)

NASA/JPL

■ FIGURE 9.5

Impact Craters on Venus (a) Large impact craters in the Lavinia region of Venus. Because they are rough, the crater rims and ejecta appear brighter on these radar images than do the smoother surrounding lava plains. The largest of these craters has a diameter of 50 km. (b) The small, complex crater named after writer Gertrude Stein. The triple impact was caused by the breaking apart of the incoming asteroid during its passage through the thick atmosphere of Venus. The projectile had an initial diameter of between 1 and 2 km.

You might think that the thick atmosphere of Venus would protect the surface from impacts, burning up the projectiles long before they could reach the surface. But this is the case for only smaller projectiles. Crater statistics show very few craters less than 10 km in diameter, indicating that projectiles smaller than about 1 km (the size that typically produces a 10-km crater) were stopped by the atmosphere. Those craters with diameters from 10 to 30 km are frequently distorted or multiple, apparently because the incoming projectile broke apart in the atmosphere before it could strike the ground (Figure 9.5b). If we limit ourselves to impacts that produce craters with diameters of 30 km or larger, however, then crater counts are as useful on Venus for measuring surface age as they are on airless bodies such as the Moon.

There are typically only about 15 percent as many craters in the venusian plains as on the lunar maria, indicating a surface age only about 15 percent as great, or roughly 500 million years old. These results indicate that Venus is indeed a planet with persistent geological activity, intermediate between that of the Earth's ocean basins (which are younger and more active) and that of its continents (which are older and less active).

Almost all of the craters on Venus look fresh, with little degradation or filling in by either lava or windblown dust. This is one way we know that the rates of erosion or sediment deposition are very low. We have the impression that relatively little has happened since the venusian plains were last resurfaced by large-scale volcanic activity. Apparently Venus experienced some sort of planetwide volcanic convulsion about 500 million years ago, a mysterious event that is unlike anything we know of in terrestrial history.

9.2.4 Volcanoes on Venus

Like the Earth, Venus is a planet with widespread volcanism, and we can see many different types of volcanic features. In the lowland plains, volcanic eruptions are the principal way the surface is renewed, with large flows of highly fluid lava destroying old craters and generating a fresh surface. In addition, numerous younger volcanic mountains and other structures are associated with surface hot spots—places where convection in the planet's mantle transports the interior heat (and hot magma) to the surface.

The largest individual volcano on Venus, called Sif Mons, is about 500 km across and 3 km high—broader but lower than the Hawaiian volcano Mauna Loa. At its top is a volcanic crater or *caldera* about 40 km across, and its slopes show individual lava flows up to 500 km long. Thousands of smaller volcanoes dot the surface, down to the limit of visibility of the Magellan images, which correspond to cones or domes about the size of a shopping-mall parking lot. Most of these seem similar to terrestrial volcanoes. Other volcanoes have unusual shapes, such as the "pancake domes" illustrated in Figure 9.6.

All of this volcanism is the result of eruption of lava onto the surface of the planet. But the hot magma rising from the interior of a planet does not always make it to the surface. On

NASA/JPL

■ FIGURE 9.6

Pancake-Shaped Volcanoes on Venus These remarkable circular domes, each about 25 km across and about 2 km tall, are the result of eruptions of highly viscous (sludgy) lava.

both the Earth and Venus, this upwelling magma can collect to produce bulges in the crust. Many of the granite mountain ranges on Earth, such as the Sierra Nevada in California, involve subsurface collections of magma that push up on the material above them. Such subsurface bulges are common on Venus, where they produce large circular or oval features called *coronae* (singular: *corona*) (Figure 9.7).

NASA/JPL

■ FIGURE 9.7

The "Miss Piggy" Corona Fotla Corona, located in the plains to the south of Aphrodite Terra. This circular feature, about 200 km across, is a region where magma from deeper inside Venus is pushing up on the surface. Curved fracture patterns clearly show where the material beneath has put stress on the surface. A number of pancake and dome volcanoes are also visible. Fotla was a Celtic fertility goddess. Some students see a resemblance between this corona and Miss Piggy of the Muppets (her right ear is the pancake volcano in the upper center of the image).

9.2.5 Tectonic Activity

Convection currents in the mantle of Venus do more than bring magma to the surface. As on the Earth, these convection currents push and stretch the crust itself. Such forces are called **tectonic,** and the geological features that result from these forces are called *tectonic features*. On Venus' lowland plains, tectonic forces have broken the lava surface to create remarkable patterns of ridges and cracks (Figure 9.8). In a few places the crust has even torn apart to gen-erate rift valleys. The circular features associated with co-ronae are tectonic ridges and cracks, and most of the mountains of Venus also owe their existence to tectonic forces.

The Ishtar continent, which has the highest elevations on Venus, is the most dramatic product of these tectonic forces. Ishtar and its tall Maxwell Mountains resemble the Tibetan Plateau and Himalayan Mountains on the Earth. Both are the product of compression of the crust, and both are maintained by the continuing forces of mantle convection.

9.2.6 On Venus' Surface

The successful Venera landers of the 1970s found them-selves on an extraordinarily inhospitable planet, with a sur-face pressure of 90 bars and a temperature hot enough to melt lead and zinc. Despite these unpleasant conditions, the spacecraft were able to photograph their surroundings and collect surface samples for chemical analysis before their instruments gave out.

The probes found that the rock in the landing areas is igneous, primarily basalts. Examples of the Venera photo-graphs are shown in Figure 9.9. Each picture shows a flat, desolate landscape with a variety of rocks, some of which may be ejecta from impacts. Other areas show flat, layered lava flows.

NASA/JPL

■ FIGURE 9.8

Ridges and Cracks This region of the Lakshmi Plains on Venus has been fractured by tectonic forces to produce a cross-hatched grid of cracks and ridges. Be sure to notice the fainter linear features that run perpendicular to the brighter ones! This image shows a region 40 km across. Lakshmi is a Hindu goddess of prosperity.

9.3 THE MASSIVE ATMOSPHERE OF VENUS

The atmosphere of Venus produces the high surface tem-perature and shrouds the surface in a perpetual red twi-light. The Sun does not shine directly through the heavy clouds, but the surface is fairly well lit by diffuse light—about the same as the light on Earth under a heavy over-cast. The weather at the bottom of this deep atmosphere remains perpetually hot and dry, with calm winds. Because of the heavy blanket of clouds and atmosphere, one spot on the surface of Venus is similar to any other as far as weather is concerned.

USSR Academy of Science and Carle Pieters, Brown University

■ FIGURE 9.9

The Surface of Venus Views of the surface of Venus from the Venera 13 spacecraft. The upper images show the scene as re-corded by the spacecraft cam-eras; everything is orange be-cause the thick atmosphere of Venus absorbs the bluer colors of light. Computer processing has transformed the bottom images to show what the scenes would look like under Earth lighting condi-tions. The horizon is visible in the upper corner of each image.

TABLE 9.2	*Atmospheric Compositions of Earth, Venus, and Mars (in %)*		
Gas	Earth	Venus	Mars
Carbon dioxide (CO_2)	0.03	96	95.3
Nitrogen (N_2)	78.1	3.5	2.7
Argon (Ar)	0.93	0.006	1.6
Oxygen (O_2)	21.0	0.003	0.15
Neon (Ne)	0.002	0.001	0.0003

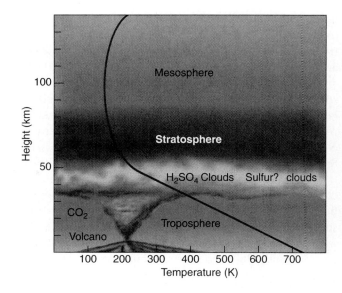

■ FIGURE 9.10

Venus' Atmosphere The layers of the massive atmosphere of Venus, based on data from the Pioneer and Venera entry probes. Height is measured along the left axis, the bottom scale shows temperatures, and the red line allows you to read off the temperature at each height. Notice how steeply the temperature rises below the clouds thanks to the planet's huge greenhouse effect.

9.3.1 Composition and Structure of the Atmosphere

The most abundant gas on Venus is carbon dioxide (CO_2), which accounts for 96 percent of the atmosphere. The second most common gas is nitrogen. The predominance of carbon dioxide over nitrogen is not surprising when you recall that the Earth's atmosphere would also be mostly carbon dioxide if this gas were not locked up in marine sediments (see Section 7.3).

Table 9.2 compares the compositions of the atmospheres of Venus, Mars, and the Earth. Expressed in this way, as percentages, the proportions of the major gases are very similar for Venus and Mars, but in total quantity their atmospheres are dramatically different. With its surface pressure of 90 bars, the venusian atmosphere is more than 10,000 times more massive than its martian counterpart. Overall, the atmosphere of Venus is very dry; absence of water is one of the important ways that Venus differs from the Earth.

The venusian atmosphere has a huge troposphere that extends up to at least 50 km above the surface (Figure 9.10). Within the troposphere, the gas is heated from below and circulates slowly, rising near the equator and descending over the poles. Being at the base of the atmosphere of Venus is something like being a kilometer or more below the ocean surface on the Earth. There the mass of water evens out temperature variations and results in a uniform environment—the same effect the thick atmosphere has on Venus.

In the upper troposphere, between 30 and 60 km above the surface, a thick cloud layer is composed primarily of sulfuric acid droplets. Sulfuric acid (H_2SO_4) is formed from the chemical combination of sulfur dioxide (SO_2) and water (H_2O). In the atmosphere of the Earth, sulfur dioxide is one of the primary gases emitted by volcanoes, but it is quickly diluted and washed out by rainfall. In the dry atmosphere of Venus, this unpleasant substance is apparently stable. Below 30 km the venusian atmosphere is clear of clouds.

9.3.2 Surface Temperature on Venus

The high surface temperature of Venus was discovered by radio astronomers in the late 1950s and confirmed by the Mariner and Venera probes. How can our neighbor planet be so hot? Although Venus is somewhat closer to the Sun

than is the Earth, its surface is hundreds of degrees hotter than you would expect from the extra sunlight it receives. Scientists wondered what could be heating the surface of Venus to a temperature above 700 K. The answer turned out to be the *greenhouse effect*.

The greenhouse effect works on Venus just as it does on the Earth (see Section 7.3). But since Venus has so much more CO_2—almost a million times more—the effect is much stronger. Sunlight that diffuses through the atmosphere heats the surface just as on Earth, but the thick CO_2 acts as a blanket, making it very difficult for the infrared radiation from the ground to get back into space. As a result, the surface heats up until eventually it is emitting enough to balance the energy it receives from the Sun.

Has Venus always had such a massive atmosphere and high surface temperature, or might it have evolved to such conditions from a climate that was once more nearly Earthlike? The answer to this question is of particular interest to us as we look at the increasing levels of CO_2 in the Earth's atmosphere, as discussed in Section 7.4. Are we in any danger of transforming our own planet into a hellish place like Venus?

Let us try to reconstruct the possible evolution of Venus from an Earth-like beginning to its present state. We imagine that Venus began with moderate temperatures, water oceans, and much of its CO_2 dissolved in the ocean or chemically combined with the surface rocks, as is the case on Earth today. Then we allow for modest additional heating—by a small rise in the energy output of the Sun, for example, or by an increase in atmospheric CO_2. When we make models of how Venus' atmosphere would respond to such effects, it turns out that even a small amount of extra heat can lead to increased evaporation from the oceans and the release of gas from surface rocks.

This in turn means a further increase in the atmospheric CO_2 and H_2O, gases that would amplify the greenhouse effect. That would lead to still more heating and the release of further CO_2 and H_2O. Unless some other processes intervene, the temperature thus continues to rise. Such a situation is called the *runaway greenhouse effect.*

We want to emphasize that the runaway greenhouse effect is not just a larger greenhouse effect; it is an evolutionary process. The atmosphere evolves from having a small greenhouse effect, such as on the Earth, to a situation where greenhouse warming is a major factor, as we see today on Venus. Once the larger greenhouse conditions develop, the planet establishes a new, much hotter equilibrium near its surface. Reversing the situation is difficult because water vapor in the planet's atmosphere is not stable in the presence of ultraviolet light. Ultraviolet light from the Sun tends to break down the molecules of H_2O into their constituent parts—oxygen and hydrogen. The light element hydrogen can escape from the atmospheres of the terrestrial planets, leaving the oxygen behind to combine chemically with surface rock. The loss of water is therefore an *irreversible* process; once the water is gone, it cannot be restored. There is evidence that this is just what happened to the water once present on Venus.

Although we do not know the point at which a stable greenhouse effect breaks down and turns into a runaway greenhouse effect, Venus stands as clear testament to the fact that a planet cannot continue heating indefinitely without a major change in its oceans and atmosphere. It is a conclusion that we and our descendants will surely want to pay close attention to.

9.4 THE GEOLOGY OF MARS

Mars is more interesting to most people than Venus because it is more hospitable. Even from the distance of the Earth, we can see surface features on Mars and follow the seasonal changes in its polar caps (Figure 9.11). Although the surface today is dry and cold, evidence collected by spacecraft suggests that Mars once had blue skies and lakes of liquid water. Even today, it is the sort of place we can imagine astronauts visiting and perhaps even setting up permanent bases.

9.4.1 Spacecraft Exploration of Mars

Mars has been intensively investigated by spacecraft, and many additional missions are planned. The first visitor was the U.S. Mariner 4, which flew by Mars in 1965 and transmitted 22 photos to Earth. These pictures showed an ap-

J. Bell, Cornell University; NASA/STScI

■ FIGURE 9.11

Mars Photographed by the Hubble Space Telescope   This is one of the best photos of Mars taken from our planet, obtained by the HST in June 2001 when Mars was only 68 million km away. The resolution is about 20 km—much better than can be obtained with ground-based telescopes but still insufficient to reveal the underlying geology of Mars.

parently bleak planet with abundant impact craters. In those days craters were unexpected; some people who were romantically inclined still hoped to see canals or something like them. In any case, newspaper headlines sadly announced that Mars was a "dead planet."

In 1971 Mariner 9 became the first spacecraft to orbit another planet, mapping the entire surface of Mars at a resolution of about 1 km and discovering a great variety of geological features, including volcanoes, huge canyons, intricate layers on the polar caps, and channels that appeared to have been cut by running water. Geologically, Mars didn't look so dead after all. Next came the Viking spacecraft, which were among the most ambitious and successful of all planetary missions.

Two Viking *orbiters* surveyed the planet and served to relay communications for two *landers* on the surface. After an exciting and sometimes frustrating search for a safe landing spot, the Viking 1 lander touched down on the surface of Chryse Planitia (the Plains of Gold) on July 20, 1976, exactly seven years after Neil Armstrong's historic first step on the Moon. Two months later, Viking 2 landed with equal success in another plain farther north, called Utopia. The last message from the Viking landers arrived on November 5, 1982, after an enormously successful mission of studying the red planet.

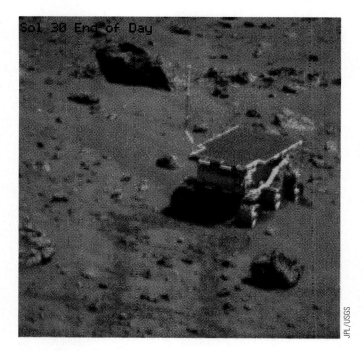

Sol 30 End of Day

JPL/USGS

■ **FIGURE 9.12**

Surface View from Mars Pathfinder The scene from the Pathfinder lander shows a windswept plain, sculpted long ago when water flowed out of the martian highlands and into the depression where the spacecraft landed. The Sojourner rover, the first wheeled vehicle on Mars, is about the size of a microwave oven. The flat top contains solar cells that provide electricity to run the vehicle.

Mars then languished unvisited for two decades after Viking. Two more spacecraft were launched toward Mars, by NASA and the Russian Space Agency, but both failed before reaching the planet. Then in the 1990s NASA began a new exploration program using smaller and less expensive spacecraft. The first of the new missions, appropriately called Pathfinder, landed a small rover on the martian surface on July 4, 1997 (Figure 9.12). An orbiter called Mars Global Surveyor arrived a few months later and began high-resolution photography of the surface. These were followed by a second orbiter in 2002, and both a European lander and a pair of NASA surface rovers are being built for 2004. Ultimately, many NASA planners hope that humans will fly to Mars on piloted missions, just as they did to the Moon.

9.4.2 Global Properties of Mars

Mars has a diameter of 6790 km, just over half the diameter of the Earth, giving it a total surface area very nearly equal to the continental area of our planet. Its density of 3.9 g/cm^3 suggests a composition consisting primarily of silicates but with a small metal core. The planet has no global magnetic field, although there are areas of strong surface magnetization that indicate a global field billions of years ago. Apparently the red planet has no liquid material in its core today that would conduct electricity.

Thanks to the Mars Global Surveyor, we have mapped the entire planet (Figure 9.13). A laser altimeter on board made millions of separate measurements of the surface topography to a precision of a few meters—good enough to show even the annual deposition and evaporation of the polar caps. Like the Earth, Moon, and Venus, the surface of Mars has continental or highland areas as well as widespread volcanic plains. The total range in elevation from the top of the highest mountain (Olympus Mons) to the bottom of the deepest basin (Hellas) is 31 km.

Approximately half the planet consists of heavily cratered highland terrain, found primarily in the southern hemisphere. The other half, which is mostly in the north, contains younger, lightly cratered volcanic plains at an average elevation about 5 km lower than the highlands. Remember that we saw a similar pattern on Earth, the Moon, and Venus. A geological division into older highlands and younger lowland plains seems to be characteristic of all the terrestrial planets except Mercury.

Lying across the north–south division of Mars is an uplifted continent the size of North America. This is the 10-km-high Tharsis bulge, a volcanically active region crowned by four great volcanoes that rise still higher into the martian sky.

9.4.3 Volcanoes on Mars

The lowland plains of Mars look very much like the lunar maria, and they have about the same density of impact craters. Like the lunar maria, they probably formed between 3 and 4 billion years ago. Apparently Mars experienced extensive volcanic activity at about the same time the Moon did, producing similar basaltic lavas.

The largest volcanic mountains of Mars are found in the Tharsis area (you can see them in Figure 9.13), although smaller volcanoes dot much of the surface. The most dramatic volcano on Mars is Olympus Mons (Mount Olympus), with a diameter larger than 500 km and a summit that towers more than 20 km above the surrounding plains (Figure 9.14). The volume of this immense volcano is nearly 100 times greater than that of Mauna Loa in Hawaii. Placed on the Earth's surface, Olympus would more than cover the entire state of Washington or Missouri.

Images taken from orbit allow scientists to search for impact craters on the slopes of these volcanoes in order to estimate their age. Many of the volcanoes show fair numbers of such craters, suggesting that they ceased activity a billion years or more ago. However, Olympus Mons has very, very few impact craters. Its present surface cannot be more than about 100 million years old; it may even be much younger. Some of the fresh-looking lava flows might have been formed a hundred years ago, or a thousand, or a million, but geologically speaking they are quite young. This leads geologists to the conclusion that Olympus Mons probably remains intermittently active today—something future Mars land developers may want to keep in mind.

JPL/NASA

Altitude [km]

■ **FIGURE 9.13**

Mars Globe from Laser Ranging   These globes are highly precise topographic maps, reconstructed from millions of individual elevation measurements made with the Mars Global Surveyor. Color is used to indicate elevation. The hemisphere on the left includes Olympus Mons, the highest mountain on Mars; the hemisphere on the right includes the Hellas basin, which has the lowest elevation on Mars.

Computer graphic by artist Kees Veenenbos

■ **FIGURE 9.14**

Olympus Mons   The largest volcano on Mars, and probably the largest in the solar system, is Olympus Mons, illustrated in this computer-generated rendering based on data from the Mars Global Surveyor's Laser Altimeter. Placed on Earth, the base of Olympus Mons would completely cover the state of Missouri; the caldera, the circular opening at the top, is 65 km across, about the size of Los Angeles.

206   CHAPTER 9   EARTH-LIKE PLANETS: VENUS AND MARS

■ **FIGURE 9.15**

Martian Landslides This Viking orbiter image shows one section of the Valles Marineris canyon system. The canyon walls are about 100 km apart here. Look carefully and you can see enormous landslides whose debris is piled up underneath the cliff walls, which tower some 10 km above the canyon floor.

9.4.4 Martian Cracks and Canyons

The Tharsis bulge has many interesting geological features in addition to its huge volcanoes. In this part of the planet the surface itself has bulged upward, forced by great pressures from below, resulting in extensive tectonic cracking of the crust. Among the most spectacular tectonic features are the canyons called the Valles Marineris (or Mariner Valleys, named after Mariner 9, which first revealed them to us). They extend for about 5000 km (nearly a quarter of the way around Mars) along the slopes of the Tharsis bulge. If it were on Earth, this canyon system would stretch all the way from Los Angeles to Washington, D.C. The main canyon is about 7 km deep and up to 100 km wide, large enough for the Grand Canyon of the Colorado River to fit comfortably into one of its side canyons. The opening image for this chapter shows a spectacular overview of part of Valles Marineris.

The term *canyon* is somewhat misleading here because the Valles Marineris canyons have no outlets and were not cut by running water. They are basically tectonic cracks, produced by the same crustal tensions that caused the Tharsis uplift. However, water has played a later role in shaping the canyons, primarily by seeping from deep springs and undercutting the cliffs. This undercutting led to landslides that gradually widened the original cracks into the great valleys we see today (Figure 9.15). Today the primary form of erosion in the canyons is probably wind.

While the Tharsis bulge and Valles Marineris are impressive, in general we see fewer tectonic structures on Mars than on Venus. In part, this may reflect a lower general level of geological activity, as would be expected for a smaller planet. But it is also possible that evidence of widespread faulting has been buried by wind-deposited sediment over much of Mars. Like the Earth, Mars may have hidden part of its geological history under a cloak of soil.

9.4.5 The View on the Martian Surface

Three spacecraft have landed successfully on Mars: Vikings 1 and 2 and Mars Pathfinder. All sent back photos that showed a desolate but strangely beautiful landscape, including numerous angular rocks interspersed with dune-like deposits of fine-grained, reddish soil (Figure 9.16).

Mars Pathfinder was targeted to land in the outwash plain below the mouth of the ancient martian river called Ares Vallis. Presumably, the most recent floods took place when the Ares Vallis river flowed more than 3 billion years ago, and the landscape has been little changed since that time. Pathfinder carried a small, solar-powered rover named Sojourner, about the size of a microwave oven (see Figure 9.12). Sojourner, operating on the surface for three months, traveled a total distance of about 100 m and measured the composition of 8 rocks, as well as several soil samples. Instruments on all three of the landers found that the soil consisted of clays and iron oxides, as had long been expected from the red color of the planet.

Although the Pathfinder found itself in the river mouth as planned, scientists were disappointed to find that all the

NASA/JPL/USGS

■ FIGURE 9.16

Three Martian Landing Sites The Mars landers—Viking 1 in Chryse, Viking 2 in Utopia, and Pathfinder in Ares Valley—all photographed their immediate surroundings. It is apparent from the similarity of these three photos that each spacecraft touched down on a flat, windswept plain littered with rocks ranging from tiny pebbles up to meter-size boulders. It is probable that most of Mars looks like this on the surface.

rocks measured were of volcanic origin and had roughly the same composition. They had hoped that a variety of rocks would be present, including perhaps some *sedimentary* samples—rocks deposited by wind or water. But this mission, at least, did not turn up any such examples.

The earlier Viking landers were bigger and carried a wider range of instruments. They peered at their surroundings through color stereo cameras, sniffed the atmosphere with a variety of analytical tools, and poked at nearby rocks and soil with their mechanical arms. As part

of its mission of searching for martian life, each lander collected soil samples and brought them on board for analysis (as we will describe in more detail shortly). The Viking weather stations operated for several years, providing a perspective on martian weather.

The temperatures they measured varied greatly due to the absence of moderating oceans and clouds. Typically, the summer maximum at Viking 1 was 240 K ($-33°C$), dropping to 190 K ($-83°C$) at the same location just before dawn. The lowest air temperatures, measured farther north by Viking 2, were about 173 K ($-100°C$). During the winter, Viking 2 also photographed water frost deposits on the ground (Figure 9.17). We make a point of saying "water frost" here because at some locations on Mars, it gets cold enough for carbon dioxide (dry ice) to freeze out of the atmosphere as well.

Most of the winds measured at the three landing sites were low to moderate, only a few kilometers per hour. However, Mars is capable of great windstorms that can shroud the entire planet with windblown dust. Such high winds can strip the surface of some of its loose, fine covering, leaving the rocks exposed. The Pathfinder lander, which NASA renamed the Carl Sagan Memorial Station to honor the well-known astronomer who died while the spacecraft was on the way to Mars (see the box in Chapter 6), found that each sunny afternoon the atmosphere became turbulent as heat waves rose off the surface. This turbulence generated dust devils, which play an important role in lifting the fine dust into the atmosphere. As the dust devils strip off the top layer of light dust and expose darker material underneath, they can produce fantastic patterns on the ground (Figure 9.18).

In spite of the thin martian air, wind plays an important role in redistributing surface material. Figure 9.19 shown a beautiful area of dark sand dunes on top of lighter material. Much of the material stripped out of the martian canyons has been dumped in extensive dune fields like this, which are concentrated at high latitudes. The constant redistribution of surface material gave scientists some confidence in interpreting the results from the three landers—presumably they were thus able to obtain a globally representative sample of dust from each site.

The Pathfinder measurements continued for 83 martian days, at which point the lander succumbed to the cold temperatures and fell permanently silent. It is possible the rover continued to operate for several more days, but without the lander we could not communicate with it. We can imagine the little rover still moving in the red martian sand and plaintively calling for instructions, cut off forever from its makers on Earth.

9.4.6 Martian Samples

Much of what we know of the Moon, including the circumstances of its origin, comes from studies of lunar samples, but spacecraft have not yet returned martian samples to Earth for laboratory analysis. It is with great interest, therefore, that scientists have discovered that samples of

NASA/JPL

■ **FIGURE 9.17**

Water Frost in Utopia Surface frost photographed at the Viking 2 landing site during late winter.

Malin Space Science Systems/NASA

■ **FIGURE 9.18**

Tracks of Dust Devils This high-resolution photo from the Mars Global Surveyor shows the dark tracks of several dust devils that have stripped away a thin coating of light-colored dust. This view is of an area about 3 km across. Dust devils are one of the most important ways that dust gets redistributed by the martian winds.

NASA/JPL/Malin Space Science Systems

■ **FIGURE 9.19**

Sand Dunes on Mars These dark dunes in the interior of Proctor Crater overlay a lighter sandy surface. Each dune in this high-resolution view is about 1 km across.

FIGURE 9.20

Martian Meteorite A fragment of basalt, ejected from Mars in a crater-forming impact, that eventually arrived on the Earth's surface.

martian material are nevertheless already here on Earth, available for study. About two dozen of these martian rocks exist, all members of a rare class of *meteorites* (Figure 9.20), rocks that have fallen from space (see Chapter 13 for more on meteorites).

How would rocks have escaped from Mars? Many impacts have occurred on the red planet, as shown by its heavily cratered surface. The larger impacts can eject debris at so great a speed that some fragments actually attain escape velocity from Mars. A long time later (typically a few million years), a very small fraction of these fragments collide with the Earth and survive their passage through our atmosphere, just like other meteorites. (By the way, several rocks from the Moon have also reached our planet as meteorites, although we were able to demonstrate their lunar origin only after the Apollo missions brought back samples directly from the lunar surface.)

Most of the martian meteorites are volcanic basalts; most of them are also relatively young, about 1.3 billion years old. We know from details of their composition that they are not from the Earth or the Moon. Besides, there was no volcanic activity on the Moon to form them as recently as 1.3 billion years ago. It would be very difficult for ejecta from impacts on Venus to escape through its thick atmosphere. By the process of elimination, the only reasonable origin seems to be Mars, where the Tharsis volcanoes were certainly active at that time.

The martian origin of these meteorites was confirmed by analysis of tiny gas bubbles trapped inside several of them. These bubbles match the atmospheric properties of Mars as measured directly by Viking. It appears that some atmospheric gas was trapped in the rock by the shock of the impact that ejected it from Mars and started it on its way toward Earth.

One of the most exciting results from analysis of these martian samples has been the discovery of both water and organic (carbon-based) compounds in them, which suggests that Mars may once have had oceans and perhaps even life on its surface. As we have already hinted, there is other evidence for the presence of flowing water on Mars in the remote past.

9.5 MARTIAN POLAR CAPS AND CLIMATE

The atmosphere of Mars today has an average surface pressure of only 0.007 bar, less than 1 percent that of the Earth. (This is how thin the air is about 30 km above the Earth's surface.) Martian air is composed primarily of carbon dioxide (95 percent), with about 3 percent nitrogen and 2 percent argon. The proportions of different gases are similar to those in the atmosphere of Venus (see Table 9.2), but a lot less of each gas is found on Mars.

9.5.1 Clouds on Mars

Several types of clouds can form in the martian atmosphere. First there are dust clouds, raised by winds, which can sometimes grow to produce global-scale dust storms. Second are water-ice clouds similar to those on Earth. These often form around mountains, just as happens on our planet. Finally, the CO_2 of the atmosphere can itself condense at high altitudes to form hazes of dry ice crystals. The CO_2 clouds have no counterpart on Earth, since on our planet temperatures never drop low enough (down to about 150 K or about $-125°C$) for this gas to condense.

Although the atmosphere contains water vapor and occasional clouds of water ice, liquid water is not stable under present conditions on Mars. Part of the problem is the low temperatures on the planet. But even if the temperature on a sunny summer day rises above the freezing point, the low pressure means that liquid water still cannot exist on the surface, except at the lowest elevations. At a pressure of less than 0.006 bar, only the solid and vapor forms of water are possible. In effect, the boiling point is as low or lower than the freezing point, and water changes directly from solid to vapor without an intermediate liquid state.

9.5.2 The Polar Caps

Through a telescope, the most prominent surface features on Mars are the bright polar caps, which change with the seasons, similar to the seasonal snow cover on Earth. We do not usually think of the winter snow in northern latitudes as a part of our polar caps, but seen from space, the thin snow merges with the Earth's thick, permanent ice caps to create an impression much like that seen on Mars.

The *seasonal caps* on Mars are composed not of ordinary snow but of frozen CO_2 (dry ice). These deposits condense directly from the atmosphere when the surface tem-

46103
600 m

NASA/JPL/Main Space Science Systems

■ FIGURE 9.21

Layers at the Martian North Pole   The small inset in the left image shows a map of the residual north polar cap of Mars, which is about 1000 km across and composed of water ice. The small black box in the middle of the map shows the area covered in the tilted Viking orbiter image at left. The box in that image shows the area of the Mars Global Surveyor high-resolution image at right. On the right image, we see a slope on the edge of the permanent north polar cap, with dozens of layers visible—some thinner than 10 m.

perature drops below about 150 K. The caps develop during the cold martian winters and extend down to about 50° latitude by the start of spring.

Quite distinct from these thin seasonal caps of CO_2 are the *permanent* or *residual caps* that are always present near the poles (Figure 9.21). The southern permanent cap has a diameter of 350 km and is composed of frozen CO_2 deposits together with a great deal of water ice. Throughout the southern summer, it remains at the freezing point of CO_2, 150 K, and this cold reservoir is thick enough to survive the summer heat intact.

The northern permanent cap is different. It is much larger, never shrinking to a diameter less than 1000 km, and is composed of water ice. Summer temperatures in the north are too high for the frozen CO_2 to be retained. Measurements from the Mars Global Surveyor have established the exact elevations in the north polar region of Mars, showing that it is a large basin about the size of our own Arctic Ocean basin. The ice cap itself is about 3 km thick, with a total volume of about 10 million km^3 (similar to that of the Earth's Mediterranean Sea). If Mars ever had extensive liquid water, this north polar basin would have contained a shallow sea. There is some indication of ancient shorelines visible, but better images will be required to verify this suggestion.

The polar caps represent a huge reservoir of water compared with the very small amounts of water vapor in the atmosphere. However, the caps appear to be only the tip of the iceberg. Recent measurements, made by the Mars Odyssey orbiter, indicate that about one third of the surface crust (in the top meter) at high latitudes is composed of water ice or permafrost. Unlike Venus or the Moon, Mars is not dry—it is just too cold for liquid water to be present.

Images taken from orbit also show a distinctive type of terrain surrounding the permanent polar caps, as shown in Figure 9.21. At latitudes above 80° in both hemispheres, the surface consists of recent layered deposits that cover the older cratered ground below. Individual layers are typically ten to a few tens of meters thick, marked by alternating light and dark bands of sediment. Probably the material in the polar deposits is dust carried by wind from the equatorial regions of Mars.

What do these terraced layers tell us about Mars? Some cyclic process is depositing dust and ice over long periods of time. The time scales represented by the polar layers are tens of thousands of years. Apparently the martian climate experiences periodic changes at intervals similar to those between ice ages on the Earth. Calculations indicate that the causes are probably also similar: The gravitational pull of the other planets produces variations in Mars' orbit and tilt as the great clockwork of the solar system goes through its paces.

NASA /from Mars Digital Image Map, processing by Brian Fessler, LPI

■ FIGURE 9.22

Runoff Channels These runoff channels in the old martian highlands are interpreted as the valleys of ancient rivers fed by either rain or underground springs. The width of this image is about 200 km.

9.5.3 Channels and Gullies on Mars

Although no liquid water exists on Mars today, evidence has accumulated that long ago rivers flowed on the red planet. Two kinds of geological features appear to be remnants of ancient watercourses, while a third class—smaller gullies—suggests intermittent outbreaks of liquid water even today. We will examine each of these features in turn.

In the highland equatorial plains there are multitudes of small, sinuous (twisting) channels—typically a few meters deep, some tens of meters wide, and perhaps 10 or 20 km long (Figure 9.22). They are called *runoff channels* because they look like what geologists would expect from troughs that carried the surface runoff of ancient rainstorms. These runoff channels seem to be telling us that the planet had a very different climate long ago. To estimate the age of these channels, we look at the cratering record. Crater counts show that this part of the planet is more cratered than the lunar maria but less cratered than the lunar highlands. Thus the runoff channels are probably older than the lunar maria, presumably about 4 billion years old.

The second set of water-related features we see are *outflow channels* (Figures 9.23 and 9.24), much larger than the runoff channels. The largest of these, which drain into the Chryse basin where Pathfinder landed, are 10 km or more wide and hundreds of kilometers long. Many features of these outflow channels have convinced geologists that they were carved by huge volumes of running water, far too great to be produced by ordinary rainfall. Where could such floodwater have come from on Mars?

As far as we can tell, the regions where the outflow channels originate contained abundant water frozen in the soil as permafrost. Some local source of heating must have

NASA / Malin Space Science Systems

■ FIGURE 9.23

Evidence of Liquid Water on Mars This intriguing channel, called Nanedi Valles, resembles Earth riverbeds in some (but not all) ways. The tight curves and terraces seen in the channel certainly suggest the sustained flow of a fluid like water. The channel is about 2.5 km across, and the entire Mars Global Surveyor image is 10 km wide.

released this water, leading to a period of rapid and catastrophic flooding. Perhaps this heating was associated with the formation of the volcanic plains on Mars, which date back to roughly the same time as the outflow channels.

Note that neither the runoff channels nor the outflow channels are wide enough to be visible from Earth, nor do they follow straight lines. They could not have been the "canals" Percival Lowell imagined seeing on the red planet.

The third type of water feature, the smaller *gullies*, was discovered by the camera on the Mars Global Surveyor. These images achieved a resolution of a few meters, good enough to see something as small as a truck or bus on the surface (Figures 9.25 and 9.26). On the steep walls of valleys and craters at high latitudes, there are often erosional

NASA/JPL

Malin Space Science Systems/NASA

◼ FIGURE 9.24

Outflow Channels   Here we see a region of large outflow channels, photographed by Viking. These features appear to have been formed in the distant past from massive floods of water. The width of this image is about 150 km.

◼ FIGURE 9.26

Stratification in the Martian Crust   As many as a hundred layers can be seen in this high-resolution photo of a wind-eroded mesa within the Valles Marineris canyons. Many geologists interpret this photo as evidence for layers of sediment deposited in an ancient martian sea. The width of the image is only 1.5 km.

Malin Space Science Systems/NASA

(a)

NASA/JPL/Malin Space Science Systems

(b)

◼ FIGURE 9.25

Recent Gullies on Mars   (a) Gullies on a cliff near the south polar cap. (b) Gullies on the wall of Newton Crater. Each image is about 2 km across.

Astronomy and Pseudoscience: The "Face on Mars"

People like human faces. We humans have developed great skill in recognizing people and interpreting facial expressions. We also have a tendency to see faces in many natural formations, from clouds to the Man in the Moon. One of the curiosities that emerged from the Viking orbiters' global mapping of Mars was the discovery of a strangely shaped mesa in the Cydonia region that resembled a human face. Despite later rumors of a cover-up, the "Face on Mars" was, in fact, recognized by Viking scientists and included in one of the early mission press releases. At the low resolution and oblique lighting under which the Viking image was obtained, the mile-wide mesa has something of a Sphinx-like appearance.

Unfortunately, a small band of individuals decided that this formation was an artificial, carved sculpture of a human face placed on Mars by an ancient civilization that thrived there hundreds of thousands of years ago. A band of "true believers" grew around the face and tried to deduce the nature of the "sculpture" and who made it. This group linked the face to a variety of other pseudoscientific phenomena such as crop circles (patterns in fields of grain, mostly in Britain, now known to be the work of pranksters).

Members of this group accused NASA of covering up evidence of intelligent life on Mars, and they received a great deal of help in publicizing their perspective from tabloid media. Some of the believers picketed JPL at the time of the failure of the Mars Observer spacecraft, circulating stories that the "failure" of the Mars Observer was itself a fake, and that its true (now secret) mission was to photograph the face.

The high-resolution Mars Observer Camera (MOC) was reflown on the Mars Global Surveyor mission, which arrived at Mars in 1997. On April 5, 1998, in Orbit 220, the MOC obtained an oblique image of the face at a resolution of 4 m/pixel, a factor of ten improvement in resolution over the Viking image. Another image in 2001 had even higher resolution. Immediately released by NASA, the new images showed a low mesa-like hill cut crossways by several roughly linear ridges and depressions, which were misidentified in the 1976 photo as the eyes and mouth of a face. Only with a large dose of imagination can any resemblance to a face be seen in the new images, demonstrating how dramatically our interpretation of geology can change with large improvements in resolution.

After 20 years of promoting pseudoscientific interpretations and various conspiracy theories, will the Face on Mars believers now accept reality? Probably they will find a way out, perhaps by accusing NASA of faking the new picture. One suggestion they have already made is that the secret mission of the Mars Observer included a nuclear bomb used to destroy the face before it could be photographed in greater detail by the Mars Global Surveyor.

Space scientists find these suggestions incredible. NASA is spending increasing sums for research on life in the universe, and a major objective of current and upcoming Mars missions is to search for evidence of past microbial life on Mars. Conclusive evidence of extraterrestrial life would be one of the great discoveries of science and incidentally might well lead to increased funding for NASA. The idea that NASA or other government agencies would (or could) mount a conspiracy to suppress such welcome evidence is truly bizarre.

Alas, the Face on Mars story is only one example of a whole series of conspiracy theories that are kept before the

---

features that look like gullies carved by flowing water. These gullies are very young; not only are there no superimposed impact craters, but in some instances the gullies seem to cut across recent wind-deposited dunes. Perhaps there is liquid water underground that can occasionally break out to produce short-lived surface flows before the water can freeze or evaporate. These gullies are one of the most exciting and mysterious recent discoveries on Mars.

9.5.4 Climate Change on Mars

The evidence discussed so far suggests that billions of years ago martian temperatures must have been warmer and the atmosphere must have been much more substantial than it is today. That way, ancient Mars could have had river channels and even great lakes filled with water; but what could have changed the climate on Mars so dramatically?

We presume that, like the Earth and Venus, Mars formed with a much thicker atmosphere than it has now, and that this atmosphere maintained a higher surface temperature thanks to the greenhouse effect. But Mars is a smaller planet, and its lower gravity means that gases could escape more easily than from Earth and Venus. As more and more of the atmosphere escaped into space, the temperature on the surface gradually fell.

Eventually Mars became so cold that water froze out of the atmosphere, further reducing its ability to retain heat. The planet experienced a sort of *runaway refrigerator ef-*

public by dedicated believers, by people out to make a fast buck, and by irresponsible media attention. Others include the "urban legend" that the Air Force has the bodies of extraterrestrials at a secret base, the widely circulated report that UFOs crashed near Roswell, New Mexico (actually it was a balloon carrying scientific instruments to find evidence of Soviet nuclear tests), or the notion that alien astronauts helped build the Egyptian pyramids and many other ancient monuments because our ancestors were too stupid to do it alone.

In response to the increase in publicity given to these "fiction science" ideas, a group of scientists, educators, scholars, and magicians (who know a good hoax when they see one) have formed the Committee for the Scientific Investigation of Claims of the Paranormal (CSICOP). Two of the authors of your book are active on the committee. For more information about its work delving into the rational explanations for paranormal claims, see their excellent magazine, *The Skeptical Inquirer*, or check out their Web site at www.csicop.org/.

The "Face on Mars" is seen at the left in low resolution from Viking and to the right with 20 times better resolution from the Mars Global Surveyor.

fect, just the opposite of the runaway greenhouse effect that occurred on Venus. Probably this loss of atmosphere took place within less than a billion years after Mars formed. The result is the cold, dry Mars we see today.

Conditions a few meters below the martian surface, however, may be much different. There liquid water might persist, kept warm by the internal heat of the planet. Even on the surface, long climate cycles may be caused by the changing orbit and tilt of the planet. At times, one or both of the polar caps might melt, releasing a great deal of water vapor into the atmosphere. Perhaps an occasional impact by a comet might produce a temporary atmosphere that is thick enough to permit liquid water on the surface for a few weeks or month. Some have even suggested that future technology

might allow us to *terraform* Mars—that is, to engineer its atmosphere and climate in ways that might make the planet more hospitable for long-term human habitation.

9.5.5 The Search for Life on Mars

If there was running water on Mars in the past, perhaps there was life as well; could life, in some form, remain in the martian soil today? Testing this possibility, however unlikely, was one of the primary objectives of the Viking landers. These landers carried, in addition to the instruments already discussed, miniature biological laboratories to test for microorganisms in the martian soil. They looked for evidence of *respiration* by living animals, *absorption of*

Planetary Protection

When scientists begin to search for life on another planet, they must make sure that we do not contaminate the other world with life carried from the Earth. At the very beginning of spacecraft exploration on Mars, an international agreement specified that all landers were to be carefully sterilized to avoid accidentally transplanting terrestrial microbes to Mars. In the case of Viking, we know the sterilization was successful. Viking's failure to detect martian organisms also implies that these experiments did not detect hitchhiking terrestrial microbes.

As we have learned more about the harsh conditions on the martian surface, the sterilization requirements have been somewhat relaxed. It is evident that no terrestrial microbes could grow on the martian surface, with its low temperature, absence of water, and intense ultraviolet radiation. Microbes might survive in a dormant, desiccated state, but they cannot grow and proliferate on Mars.

The problem of contaminating Mars will become more serious, however, as we begin to search for life below the surface, where temperatures are higher and no ultraviolet light penetrates. The situation will be even more daunting if we consider human flights to Mars. Any humans will carry with them a multitude of terrestrial microbes of all kinds, and it is hard to imagine how we can effectively keep the two biospheres isolated from each other if Mars has indigenous life. Perhaps the best situation could be one in which the two life-forms are so different that each is effectively invisible to the other—not recognized on a chemical level as living or as potential food.

The most immediate issue of public concern is not with the contamination of Mars but with any dangers associated with returning Mars samples to the Earth. NASA is committed to the complete biological isolation of returned samples until they are demonstrated to be safe. Even though the chances of contamination are extremely low, it is better to be safe than sorry.

Most likely there is no danger, even if there is life on Mars and alien microbes hitch a ride to Earth inside some of the returned samples. In fact, Mars is sending samples to Earth all the time in the form of the Mars meteorites. Since some of these bugs (if they exist) could probably survive the trip to Earth, we may have been exposed many times over to martian microbes. Either they do not interact with our terrestrial life, or in effect our planet has already been inoculated against such alien bugs.

---

nutrients offered to organisms that might be present, and an *exchange of gases* between the soil and its surroundings for any reason whatsoever. Martian soil was scooped up by the spacecraft's long arm and placed into the experimental chambers, where it was isolated and incubated in contact with a variety of gases, radioactive isotopes, and nutrients to see what would happen. A fourth instrument pulverized the soil and analyzed it carefully to determine what organic (carbon-bearing) material it contained.

The Viking experiments were so sensitive that, had one of the spacecraft landed anywhere on Earth (with the possible exception of Antarctica), it would easily have detected life. But, to the disappointment of many scientists and members of the public, no life was detected on Mars. The soil tests for absorption of nutrients and gas exchange did show some activity, but this was most likely caused by chemical reactions that began as water was added to the soil and had nothing to do with life. In fact, these experiments showed that martian soil seems much more chemically active than terrestrial soils because of its exposure to solar ultraviolet radiation (since Mars has no ozone layer).

The organic chemistry experiment showed no trace of organic material, which is apparently destroyed by the sterilizing effect of this ultraviolet light. While the possibility of life on the surface has not been eliminated, most experts consider it negligible. Although Mars has the most Earth-like environment of any planet in the solar system, the sad fact is nobody seems to be home today, at least on the surface.

However, there is no reason to think that life could not have begun on Mars about 4 billion years ago, at the same time it started on Earth. The two planets had very similar surface conditions then. Thus the attention of scientists has shifted to the search for *fossil* life on Mars. One of the primary questions to be addressed by future spacecraft is whether Mars once supported its own life-forms and, if so, how this martian life compared with that on our own planet. In 1996, several scientists reported that one of the martian meteorites found in the Antarctic contained evidence of fossil microorganisms, but further examination of this evidence has left most scientists unconvinced.

Future missions will include the return of martian samples selected (we hope) from sedimentary rocks at sites (such as ancient lake beds or hot springs) that once held water and thus perhaps ancient life. The most powerful searches for martian life (past or present) will thus be carried out in our laboratories here on Earth.

9.6 DIVERGENT PLANETARY EVOLUTION

Venus, Mars, and our own planet Earth form a remarkably diverse triad of worlds. Although all three orbit at roughly the same distance from the Sun and all apparently started with about the same chemical mix of silicates and metals, their evolutionary paths have diverged. As a result, Venus became hot and dry, Mars became cold and dry, and only the Earth ended up with what we consider a hospitable climate.

We have discussed the runaway greenhouse effect on Venus and the runaway refrigerator effect on Mars. But we do not understand exactly what started these two planets down these evolutionary paths. Was the Earth ever in danger of a similar fate? Or might it still be diverted onto one of these paths, perhaps due to stress on the atmosphere generated by human pollutants? One of the reasons for studying Venus and Mars is to seek insight into these questions.

Some people have even suggested that if we understood the evolution of Mars and Venus better, we could possibly reverse their evolution and restore more Earth-like environments. While it seems unlikely that humans could ever make either Mars or Venus into a replica of the Earth, considering such possibilities is a useful part of our more general quest to understand the delicate environmental balance that distinguishes our planet from its two neighbors. In Chapter 13 we will return to the comparative study of the terrestrial planets and their divergent evolutionary histories.

SURFING THE WEB

General Venus sites:

- *The Nine Planets Site:*
  seds.lpl.arizona.edu/nineplanets/nineplanets/venus.html

- *Views of the Solar System Site:*
  www.solarviews.com/eng/venus.htm

- *Planetary Sciences Site:*
  nssdc.gsfc.nasa.gov/planetary/planets/venuspage.html

- *Magellan Mission to Venus:*
  www.jpl.nasa.gov/magellan/
  Magellan radar images, the news releases, and mission background, as well as technical information and links.

General Mars sites:

- *The Nine Planets Site:*
  seds.lpl.arizona.edu/nineplanets/nineplanets/mars.html

- *Views of the Solar System Site:*
  www.solarviews.com/eng/mars.htm

- *Planetary Sciences Site:*
  nssdc.gsfc.nasa.gov/planetary/planets/marspage.html

- *Center for Mars Exploration Home Page:*
  cmex-www.arc.nasa.gov/
  A fine site with background information, maps and images, Mars mission summaries, educational resources, and lots of links.

- *Mars Missions Home Page: mars.jpl.nasa.gov*
  Images, mission overviews, updates, and educational activities.

- *Mars Explorer for Armchair Astronauts:*
  www-pdsimage.wr.usgs.gov/PDS/public/mapmaker
  Browse through a map of Mars made by the cartographers at the U.S. Geological Survey, zoom in on an area of your choice, and let the mapmaker program create a detailed map for you.

- *Face on Mars:*
  science.nasa.gov/headlines/y2001/ast24may_1.htm
  All the information on the new images of "the face," with background and even links to a climbing trail for the "face" mesa.

Planetary Protection Information:
www.spaceref.com/Directory/Astrobiology_and_Life_Science/planetary_protection/
A private site full of information and articles about all aspects of space science; this section has a good introduction to the ideas of protecting the Earth and other planets from contamination.

9.1 Venus, the nearest planet, is a great disappointment through the telescope because of its impenetrable cloud cover. Mars is more tantalizing, with dark markings and polar caps. Early in the 20th century it was widely believed that the "canals" of Mars indicated intelligent life there. Mars has only 11 percent the mass of the Earth, but Venus is nearly our twin in size and mass. Mars rotates in 24 hours and has seasons like the Earth; Venus has a retrograde rotation period of 243 days. Both planets have been extensively explored by spacecraft.

9.2 Venus has been mapped by radar, especially with the Magellan spacecraft. Its crust consists of 75 percent lowland lava plains, numerous volcanic features, and many large coronae, which are the expression of subsurface volcanism. The planet has been modified by widespread **tectonics** driven by mantle convection, forming complex patterns of ridges and cracks and building high continental regions such as Ishtar. The surface is extraordinarily inhospitable, with pressure of 90 bars and temperature of 730 K, but several Russian Venera landers investigated it successfully.

9.3 The atmosphere of Venus is 96 percent CO_2. Thick clouds at altitudes of 30 to 60 km are made of sulfuric acid, and a CO_2 greenhouse effect maintains the high surface temperature. Venus presumably reached its current state from more Earth-like initial conditions as a result of a runaway greenhouse effect, which included the loss of large quantities of water.

9.4 Much of what we know about Mars is derived from six spacecraft: two Viking orbiters and the Mars Global Surveyor in orbit, and two Viking landers and the Mars Pathfinder operating on the surface. Mars has heavily cratered highlands in its southern hemisphere but younger, lower volcanic plains over much of its northern half. The Tharsis bulge, as big as North America, includes several huge volcanoes; Olympus Mons is more than 20 km high and 500 km in diameter. The Valles Marineris canyons are tectonic features widened by erosion. The Viking and Pathfinder landers revealed barren, windswept plains at Chryse and Utopia. Currently there is great interest in martian meteorites, which are samples of the martian crust and may preserve information from the time that Mars was warmer and wetter and could even have supported life.

9.5 The martian atmosphere has a surface pressure of less than 0.01 bar and is 95 percent CO_2. It has dust clouds, water clouds, and carbon dioxide (dry ice) clouds. Liquid water is not possible today, but there is subsurface permafrost at high latitudes. Seasonal polar caps are made of dry ice, but the northern residual cap is water ice. Evidence of a very different climate in the past is found in water erosion features, both runoff channels and outflow channels, the latter carved by catastrophic floods. Even more mysterious are the gullies that seem to indicate small but recent episodes of water erosion. The Viking landers searched for martian life in 1976, with negative results, but life might have flourished long ago. The search for fossil life is one of the prime motivations for future sample return missions.

9.6 Earth, Venus, and Mars have diverged in their evolution from what may have been similar beginnings. We need to understand why if we are to protect the environment of the Earth.

INTER-ACTIVITY

A Your group has been asked by high NASA officials to start planning the first human colony on Mars. Begin by making a list of what sorts of things humans would need to bring along to be able to survive for years on the surface of the red planet.

B As a publicity stunt, the mayor of Venus, Texas (there really is such a town), proposes that NASA fund a mission to Venus with humans on board. Clearly, the good mayor neglected to take an astronomy course in college. Have your group assemble a list of as many reasons as possible why it is unlikely that humans will soon land on the surface of Venus.

C Even if humans would have trouble surviving on the surface of Venus, this does not mean we could not learn a lot more about our veiled sister planet. Have your group brainstorm a series of missions (pretend cost is no object) that would provide us with more detailed information about the Venus atmosphere, surface, and interior.

D Some time late in the 21st century, when travel to Mars has become somewhat routine, a very wealthy couple asks you to plan a honeymoon tour of Mars that includes the most spectacular sights on the red planet. Constitute your group as the Percival Lowell

Memorial Tourist Agency, and come up with a list of not-to-be missed tourist stops on Mars. (If you have time to complete this assignment outside of class and have access to the Web, be sure to check out some of the Mars Web sites as part of your research.)

E. Astronomers have been puzzled and annoyed about the extensive media publicity that was given the small group of "true believers" who claimed the "Face on Mars" was not a natural formation (see Making Connections: *Astronomy and Pseudoscience*). Have your group make a list of the reasons many of the media were so enchanted by this story. What do you think astronomers could or should do to get the skeptical, scientific perspective about such issues before the public?

F. Assume your group is a Long-Term Planetary Environment Planning Committee at the Environmental Protection Agency during a future administration. Make a list of ways in which human activity is affecting the Earth's climate, and discuss what might be done about it.

G. Your group is a special committee of scientists set up by the United Nations to specify how any Mars samples should be returned to Earth so that possible martian microbes do not harm Earth life. What precautions would you recommend, starting at Mars and going all the way to the labs that analyze the martian samples back on Earth?

REVIEW QUESTIONS

1. List several ways that Venus, Earth, and Mars are similar, and several ways that they are different.
2. Although neither Venus nor Mars has oceans today, both are considered to have "continents" somewhat like those on the Earth. Describe the continents on each planet. What fraction of the surface area do continents take up on each world?
3. Compare the current atmospheres of Earth, Venus, and Mars in terms of composition, thickness (and pressure at the surface), and the greenhouse effect.
4. How might Venus' atmosphere have evolved to its present state through a runaway greenhouse effect?

5. The Magellan mission has taught us the most about Venus, while the Viking and Mars Global Surveyor missions have been the source of the largest amount of information about Mars. Compare the instruments on each mission, and describe how each helped us understand the planet it surveyed.
6. Describe the current atmosphere on Mars. What evidence suggests that it must have been different in the past?
7. Explain the runaway refrigerator effect and the role it may have played in the evolution of Mars.
8. What evidence do we have that there was running (liquid) water on Mars?

THOUGHT QUESTIONS

9. What are the advantages of using radar imaging rather than ordinary cameras to study the topography of Venus? What are the relative advantages of these two approaches to mapping the Earth or Mars?

10. Venus and Earth are nearly the same size and distance from the Sun. What are the main differences in the geology of the two planets? What might be some of the reasons for these differences?

11. Why is there so much more carbon dioxide in the atmosphere of Venus than in that of the Earth? Why so much more than on Mars?

12. If the Viking missions were such a rich source of information about Mars, why have we sent the Pathfinder, Global Surveyor, and other more recent missions to Mars?

Make a list of questions about Mars that still puzzle astronomers.

13. Compare Mars with Mercury and the Moon in terms of overall properties. What are the main similarities and differences?

14. Contrast the mountains on Mars and Venus with those on Earth and the Moon.

15. We believe that all of the terrestrial planets had similar histories when it comes to impacts from space. Explain how this idea can be used to date the formation of the martian highlands, the martian basins, and the Tharsis volcanoes. How certain are the ages derived for these features?

16. Is it likely that life ever existed on either Venus or Mars? Justify your answer in each case.

17. Suppose that, decades from now, astronauts are sent to Mars and Venus. In each case, describe what kind of protective gear they would have to carry, and what their chances for survival would be if their spacesuits ruptured.

18. We believe that Venus, Earth, and Mars all started with a significant supply of water. Explain where that water is now for each planet.

FIGURING FOR YOURSELF

It is interesting to estimate the amount of water (in the form of ice) on Mars and to compare this with the amount of water on Earth. In each case, we can find the total volume of a layer on a sphere by multiplying the area of the sphere $(4\pi r^2)$ by the thickness of the layer. For the Earth, the ocean water is equivalent to a layer 3 km thick spread over the entire planet, and the radius of the Earth is 6.378×10^6 m (see Appendix 6). The volume of this water is therefore $4\pi(6.378 \times 10^6)^2 = 5.1 \times 10^{14}$ m^2 times the thickness of 3000 m. This gives 1.5×10^{18} cubic meters of water. Since water has a density of 1 ton per cubic meter, the mass is 5.1×10^{18} tons.

19. For Mars most of the water we are sure of is in the form of ice near the poles. We can calculate the amount of ice in one of the residual polar caps if it is (for example) 2 km thick and has a radius of 400 km (the area of a circle is $\pi \times$ radius^2). Estimate how much water might be present in the polar caps (using the dimensions given above).

Compare this with the amount of water in the Earth's oceans calculated above.

20. Estimate the amount of water there could be in a global region of subsurface permafrost on Mars (assuming two permafrost thicknesses, 1 and 10 km, and a concentration of ice in the permafrost of 10 percent by volume). Compare the two results you get with the amount of water in the Earth's oceans calculated above.

21. At its nearest approach, Venus comes within about 40 million km of the Earth. How distant is it at its farthest?

22. If you weigh 150 lb on the surface of the Earth, how much would you weigh on Venus? On Mars?

23. Calculate the relative land areas—that is, the amounts of the surface not covered by liquids—of the Earth, Moon, Venus, and Mars. (Assume that 70 percent of the Earth is covered with water.)

SUGGESTIONS FOR FURTHER READING

On Venus
Cooper, H. *The Evening Star: Venus Observed.* 1993, Farrar, Straus and Giroux. Good primer on Venus exploration, with a focus on the Magellan mission.

Grinspoon, D. "Venus Unveiled" in *Astronomy*, May 1997, p. 44.

Grinspoon, D. *Venus Revealed: A New Look Below the Clouds of Our Mysterious Twin Planet.* 1997, Addison Wesley. Excellent overview of Venus, including the fascinating geology revealed by the Magellan radar images.

Kargel, J. "Rivers of Venus" in *Sky & Telescope*, Aug. 1997, p. 32. On lava channels.

Kasting, J. et al. "How Climate Evolved on the Terrestrial Planets" in *Scientific American*, Feb. 1988. Compares Venus with the other inner planets.

Saunders, S. "Venus: A Hellish Place Next Door" in *Astronomy*, Mar. 1990, p. 18.

Robinson, C. "Magellan Reveals Venus" in *Astronomy*, Feb. 1995, p. 32.

Stofan, E. "The New Face of Venus" in *Sky & Telescope*, Aug. 1993, p. 22.

On Mars
Barile, S. "2002: A Martian Odyssey" in *Mercury* (the magazine of the Astronomical Society of the Pacific), Sept./Oct. 2002, p. 30. On the first year of observations with the spacecraft's instruments.

Bell, J. "Mars Pathfinder: Better Science?" in *Sky & Telescope*, July 1998, p. 36. Good review of mission results.

Bergreen, L. *Voyage to Mars: NASA's Search for Life Beyond Earth.* 2000, Riverhead. History of NASA's Mars exploration over the last decade or so.

Goldsmith, D. *The Hunt for Life on Mars.* 1997, Dutton. Popular discussion of life on Mars, with particular attention to the story of Mars rock ALH 84001 and its possible microbial fossils.

Golombeck, M. "The Mars Pathfinder Mission" in *Scientific*

American, July 1998, p. 40. Good review by the mission's chief scientist.

Hartmann, W. "Red Planet Renaissance" in *Astronomy,* July 2000, p. 36. Review of our view of Mars by a noted astronomer.

Hoyt, W. *Lowell and Mars.* 1976, U. of Arizona Press. Probably the best study of the canal controversy.

Kargel, J. and Strom, R. "The Ice Ages of Mars" in *Astronomy,* Dec. 1992, p. 40.

Kerr, R. "Requiem for Life on Mars? Support for Microbes [in Meteorites] Fades" in *Science,* vol. 282, p. 1398 (Nov. 20, 1998).

Morton, O. *Mapping Mars: Science, Imagination, and the Birth of a New World.* 2002, Picador. Enjoyable and authoritative ac-count of current Mars studies from a well-known science journalist.

Newcott, W. "Return to Mars" in *National Geographic,* Aug. 1998. Report on the Pathfinder mission with great color images, some in 3D.

Pendick, D. "The Real Men in Black" in *Astronomy,* July 1999, p. 36. Profile of John Rummel, NASA's Planetary Protection Officer, whose task it is to see that microorganisms from other worlds don't contaminate the Earth.

Raeburn, P. and Golombek, M. *Mars: Uncovering the Secrets of the Red Planet.* 2000, Econo-Clad books. The story of the Pathfinder mission from the perspective of the project scientist.

The Jovian Planets This montage shows the four jovian planets to scale. Each image was recorded by one of the Voyager spacecraft that flew by these giant outer worlds.

NASA/JPL

What do we learn about the Earth by studying the planets? Humility.

**Andrew Ingersoll
in discussing the
Voyager missions
in 1986**

THINKING AHEAD

Are you a surface chauvinist? You should be! You grew up on a planet with a solid surface and probably take it for granted that a world must have a surface. Yet, in the outer solar system, four planets do not have solid surfaces. You cannot land *on* Jupiter, for example, but only *in* it. What would it be like to visit such a world?

Beyond Mars and the asteroid belt we encounter a new region of the solar system: the realm of the giants. Here planets are much larger, the distances between them are vastly increased, and worlds are accompanied by extensive systems of satellites and rings. From many perspectives the outer solar system is where the action is, and the giant planets are the more important members of the Sun's family. When compared to these outer giants, the little cinders of rock and metal that orbit closer to the Sun can seem like insignificant afterthoughts.

10.1 EXPLORING THE OUTER PLANETS

Five planets—Jupiter, Saturn, Uranus, Neptune, and Pluto—are found in the outer solar system. The first four of these are the giant or jovian planets, the subject of this chapter. In contrast, Pluto is a small world, by far the smallest in the planetary system. Pluto is physically similar to the satellites of the giant planets and is therefore discussed with them in Chapter 11.

10.1.1 Composition and Chemistry

As we saw in Chapter 6, most of the mass in the planetary system is in the outer solar system. Jupiter alone exceeds the mass of all the other planets combined (Figure 10.1). The chemistry of the outer solar system is also different, with hydrogen rather than oxygen dominating. When the planets and the Sun were forming, it was cooler in the parts of the solar nebula farther from the Sun. This allowed water ice and other volatile compounds to condense, whereas these materials remained as gas (and eventually dissipated) in the inner solar system.

As a result, the outer planets accumulated a significantly greater mass of solid material. This meant that the core bodies for the giant planets had more gravity to capture the gases hydrogen and helium, which were abundant

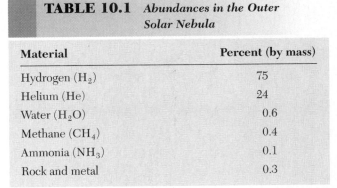

TABLE 10.1	*Abundances in the Outer Solar Nebula*
Material	**Percent (by mass)**
Hydrogen (H_2)	75
Helium (He)	24
Water (H_2O)	0.6
Methane (CH_4)	0.4
Ammonia (NH_3)	0.1
Rock and metal	0.3

in the solar nebula. Hence, the outer planets were able to grow significantly bigger than those closer to the Sun. Table 10.1 lists the main materials that were available in the outer solar nebula to form planets, based on the elements we observe making up the Sun.

With so much hydrogen available, the chemistry of the outer solar system became *reducing* (see Chapter 6). Most of the oxygen combined with hydrogen to make H_2O and was thus unavailable to form oxidized compounds with other elements. As a result, the compounds detected in the atmospheres of the giant planets are hydrogen-based gases such as methane (CH_4) and ammonia (NH_3), or more complex hydrocarbons (combinations of hydrogen and carbon) such as ethane (C_2H_6) and acetylene (C_2H_2).

NASA/JPL/University of Arizona

■ FIGURE 10.1

Jupiter Jupiter as seen by the Cassini spacecraft on its way to Saturn. The storm system called the Great Red Spot is visible to the lower right. Unlike many spacecraft photos, the colors here are only slightly enhanced.

TABLE 10.2 *Missions to the Outer Solar System*

Planet	Spacecraft	Encounter Date
Jupiter	Pioneer 10	Dec. 1973 (flyby)
	Pioneer 11	Dec. 1974 (flyby)
	Voyager 1	Mar. 1979 (flyby)
	Voyager 2	July 1979 (flyby)
	Ulysses °	Dec. 1991 (flyby)
	Galileo	Dec. 1995 (orbiter and probe)
	Cassini	Dec. 2002 (flyby)
Saturn	Pioneer 11	Sept. 1979 (flyby)
	Voyager 1	Nov. 1980 (flyby)
	Voyager 2	Aug. 1981 (flyby)
	Cassini	July 2004 (orbiter)
Uranus	Voyager 2	Jan. 1986 (flyby)
Neptune	Voyager 2	Aug. 1989 (flyby)

° The Ulysses spacecraft was designed to study the polar regions of the Sun. To get away from the ecliptic, it needed a gravity boost from flying by Jupiter.

10.1.2 Exploration of the Outer Solar System

Seven spacecraft, six from the United States and one from Europe, have penetrated beyond the asteroid belt into the realm of the giants. The challenges of probing so far away from Earth are considerable. Flight times to the giant planets are measured in years to decades rather than the few months required to reach Venus or Mars. Spacecraft must be highly reliable and capable of a fair degree of independence and autonomy. This is because, even at the speed of light, messages take several hours to pass between Earth and the spacecraft. If a problem develops near Saturn, for example, the spacecraft computer must deal with it directly. To wait hours for the alarm to reach Earth and instructions to be routed back to the spacecraft could spell disaster.

These spacecraft also must carry their own electrical energy sources, since sunlight is too weak to supply energy through solar cells. Heaters are required to keep instruments at proper operating temperatures, and spacecraft must have powerful radio transmitters and large antennas if their precious data are to be transmitted to receivers on Earth a billion kilometers or more away. Table 10.2 summarizes the spacecraft missions to the outer solar system.

The first spacecraft to investigate the regions past Mars were Pioneers 10 and 11, launched in 1972 and 1973 as pathfinders to Jupiter. One of their main objectives was simply to determine whether a spacecraft could actually navigate through the belt of asteroids that lies beyond Mars without colliding with small particles in the belt. Another was to measure the radiation hazards in the enormous *mag-*

NASA/JPL

■ **FIGURE 10.2**

The Voyager Spacecraft The camera (at bottom) is on an arm that swivels. To the left of the white main antenna for sending data back to Earth, you can see a gold disk; this is the cover for an audiovisual recording of the sights and sounds of Earth, included for the remote possibility that some other civilization might find the capsule someday. The entire spacecraft weighs about a ton on Earth.

netosphere (or zone of magnetic influence) of Jupiter. Both spacecraft passed through the asteroids without incident, but the energetic particles in Jupiter's magnetic field nearly wiped out their electronics, providing information necessary for the safe design of subsequent missions.

Pioneer 10 flew past Jupiter in 1973, after which it sped outward toward the limits of the solar system. Pioneer 11 undertook a more ambitious program, using the gravity of Jupiter during its 1974 encounter to divert it toward Saturn, which it reached in 1979. The two small craft eventually became the first human-built objects to leave the realm of the planets.

Much more advanced were Voyagers 1 and 2, launched in 1977 (Figure 10.2). The Voyagers each carried 11 scientific instruments, including cameras and spectrometers, as well as devices to measure the characteristics of planetary magnetospheres. Voyager 1 reached Jupiter in 1979 and used a gravity assist from that planet to take it on to Saturn in 1980 (see Figure 2.12). Voyager 2 arrived at Jupiter four months later and then followed a different path to complete a grand tour of the outer planets, reaching Saturn in 1981, Uranus in 1986, and Neptune in 1989. This trajectory was made possible by the approximate alignment of the four giant planets on the same side of the Sun. About once every 175 years, these planets are in a position that allows a single spacecraft to visit them all by using gravity-assisted flybys to adjust its course for each next encounter. Because of the alignment, every planet in the outer solar system except Pluto has been visited by spacecraft.

All of these missions were flybys of the giant planets: One quick look and the spacecraft moved on. For a more detailed study, we must send spacecraft that can go into orbit around a planet. This was the mission of Galileo, launched toward Jupiter in 1989, and Cassini, sent to Saturn in 1997.

10.1.3   Galileo and Cassini

The Galileo spacecraft arrived at Jupiter in 1995, beginning its investigations by deploying an entry probe into the planet for the first direct studies of Jupiter's gaseous outer layers. Like a flaming meteor, the probe plunged at a shallow angle into the jovian atmosphere, traveling at a speed of 50 kilometers per second. (That's more than 160,000 kilometers per hour—fast enough to fly from New York to San Francisco in 100 seconds!)

Atmospheric friction slowed the probe within two minutes, reaching a maximum deceleration of nearly 230 g and producing temperatures at the front of its heat shield as high as 15,000°C. This high entry speed was a result of acceleration by the strong gravitational attraction of Jupiter. As the probe speed dropped to 2500 km/h, the remains of the glowing heat shield were jettisoned, and a parachute was deployed to lower the instrumented probe spacecraft more gently into the atmosphere (Figure 10.3). The data from the probe instruments were relayed to Earth via the Galileo main spacecraft.

The probe continued to operate for nearly an hour, descending 200 km into the atmosphere. A few minutes later the polyester parachute melted, and within a few hours the main aluminum and titanium structure of the probe vaporized to become a part of Jupiter itself. About two hours after the receipt of the final probe data, the main spacecraft fired its retro-rockets so it could be captured into orbit around the planet, where its primary objectives were to study Jupiter's large and often puzzling moons (see Chapter 11).

The Cassini mission to Saturn (Figure 10.4), a cooperative venture between NASA and the European Space Agency, is similar to Galileo in its twofold approach. Arriving in 2004, Cassini will go into orbit around Saturn and begin studying its rings and satellites. In January 2005 Cassini will also deploy an entry probe. In this case the probe will not explore Saturn itself, but its intriguing moon Titan, whose surface is veiled by a thick, opaque atmosphere (see Chapter 11).

NASA/ARC

■ FIGURE 10.3

Galileo Probe Fall into Jupiter   This artist's impression shows the Galileo probe descending into the clouds on its parachute just after the protective heat shield separated. The probe made its measurements of Jupiter's atmosphere on December 7, 1995.

NASA/JPL

■ FIGURE 10.4

Cassini Mission Drops Its Probe   In this artist's conception, the Huygens probe drops from the Cassini orbiter into the hazy clouds of Saturn's moon Titan.

Engineering and Space Science: Teaching an Old Spacecraft New Tricks

By the time Voyager 2 arrived at Neptune in 1989, 12 years after its launch, the spacecraft was beginning to show almost human signs of old age. The arm on which the camera and other instruments were located was "arthritic"—it could no longer move easily in all directions. The communications system was "hard of hearing"—part of its radio receiver was not operating anymore. The "brains" had significant "memory loss"—some of the computer memory on board had failed. And the whole spacecraft was beginning to run out of energy—its radioactive generators had begun showing serious signs of wear.

To make things even more of a challenge, the craft's mission at Neptune was in many ways the most difficult of all four flybys. For example, since sunlight at Neptune is 900 times weaker than at Earth, the on-board camera had to take longer exposures in this light-starved environment. This was a nontrivial requirement, given that the spacecraft was shooting by Neptune at ten times the speed of a rifle bullet, or 42,000 mi/h (at this speed, it could have gone around the Earth's equator in only 36 min).

The solution was to swivel the camera backward at a rate that would exactly compensate for the forward motion of the spacecraft. But since the arm on which the camera was mounted could not move with the ease and grace it once had (and because instructions from Earth took about 4 hours to reach Neptune), engineers had to preprogram the ship's computer to execute an incredibly complex series of maneuvers for each image. Some of the beautiful illustrations you see in this chapter are a testament to their ingenuity.

Another problem that confronted the mission scientists was the sheer distance of the craft from its controllers on Earth. Voyager 2 communicated its results (including the pattern of dark and light dots that make up each image) by means of its on-board radio transmitter. By the time the radio beam arrived at the Earth, it had traveled (and spread out over) about 4.8 billion km. The power that reached us was approximately 10^{-16} W, or 20 billion times less power than it takes to operate a digital watch. Thirty-eight different antennas on four continents were used by NASA to collect the faint signals from the spacecraft and decode the precious information about Neptune that they contained.

10.2 THE JOVIAN PLANETS

Let us now examine the four jovian planets in more detail. Our approach is not just to catalog their characteristics but to compare them with each other, noting their similarities and differences, and attempting to relate their properties to their differing masses and distances from the Sun.

10.2.1 Basic Characteristics

The giant planets are much farther from the Sun than we are. The distance of Jupiter from the Sun is 778 million km—5.2 times the Earth's distance, or 5.2 AU; thus Jupiter takes just under 12 years to circle the Sun. Saturn is about twice as far away as Jupiter and takes very nearly the standard human generation of 30 years to complete one revolution.

As we saw in Chapter 2, Uranus and Neptune were not discovered until larger telescopes came into regular use. Uranus orbits at 19 AU with a period of 84 years, while Neptune at 30 AU requires 165 years for each circuit of the Sun. Not until 2010 will a full Neptune "year" have passed since its discovery in 1845.

Jupiter and Saturn have many similarities in composition and internal structure, although Jupiter is nearly four times more massive. In contrast, Uranus and Neptune are smaller worlds that differ in composition and structure from their larger siblings. Some of the main properties of these four planets are summarized in Table 10.3.

Jupiter, the giant among giants, has enough mass to make 318 Earths. Its diameter is about 11 times the Earth's (and about one tenth of the Sun's). Jupiter's average density is 1.3 g/cm^3, much lower than that of any of the terrestrial planets. (Recall that water has a density of 1 g/cm^3.) Jupiter's material is spread out over a volume so large that more than 1400 Earths could fit within it.

Saturn's mass is 95 times that of the Earth, and its average density is only 0.7 g/cm^3—the lowest of any planet. Since this is less than the density of water, Saturn would be light enough to float—if a bathtub large enough to contain it existed (but then imagine the ring it would leave)!

Uranus and Neptune each have a mass about 15 times that of the Earth and hence only 5 percent as great as

TABLE 10.3 *Basic Properties of the Jovian Planets*

Planet	Distance (AU)	Period (years)	Diameter (km)	Mass (Earth = 1)	Density (g/cm^3)	Rotation (hours)
Jupiter	5.2	11.9	142,800	318	1.3	9.9
Saturn	9.5	29.5	120,540	95	0.7	10.7
Uranus	19.2	84.1	51,200	14	1.3	17.2
Neptune	30.1	164.8	49,500	17	1.6	16.1

Jupiter. Their densities of 1.3 g/cm^3 and 1.6 g/cm^3, respectively, are much higher than that of Saturn, in spite of the reduced internal compression that results from a weaker gravity than that of Saturn. This tells us that their composition must be fundamentally different, consisting for the most part of heavier materials than hydrogen and helium, the primary constituents of Jupiter and Saturn. We will return to this question of their composition shortly.

10.2.2 Appearance and Rotation

When we look at the giant planets, we see only their atmospheres, composed primarily of hydrogen and helium gas (see the image that opens this chapter). The uppermost clouds of Jupiter and Saturn, the part we see when looking down at these planets from above, are composed of ammonia crystals. On Neptune the upper clouds are made of methane. On Uranus we see no obvious cloud layer at all, but only a deep and featureless haze.

Seen through a telescope, Jupiter is a colorful and dynamic planet. Distinct details in its cloud patterns allow us to determine the rotation rate of its atmosphere at the cloud level, although such atmospheric rotation may have little to do with the spin of the underlying planet. Much more fundamental is the rotation of the mantle and core; these can be determined by periodic variations in radio waves coming from Jupiter, which are controlled by its magnetic field. Since the magnetic field originates deep inside the planet, it shares the rotation of the interior. This rotation period of 9 h 56 m gives Jupiter the shortest "day" of any planet. In the same way, we can measure that the underlying rotation period of Saturn is 10 h 40 m. Uranus and Neptune have slightly longer rotation periods of about 17 hours, also determined from the rotation of their magnetic fields.

Remember from Chapters 3 and 9 that the Earth and Mars have seasons because their spin axes, instead of "standing up straight," are tilted relative to the Sun's orbital plane. This means that, as the Earth revolves around the Sun, sometimes one hemisphere and sometimes the other "leans into" the Sun.

What are the seasons like for the giant planets? The spin axis of Jupiter is tilted by only 3°, so there are no seasons to speak of. Saturn, however, does have seasons, since its spin axis is inclined at 27° to the perpendicular to its orbit. Neptune has about the same tilt as Saturn (29°); therefore, it experiences similar seasons (only more slowly). The strangest seasons of all are on Uranus, which has a spin axis tilted by 98° with respect to the north direction. Practically speaking, we can say that Uranus orbits on its side, and its ring and satellite system follows along, orbiting about Uranus' equator (Figure 10.5).

We don't know what caused Uranus to be tipped over like this, but one possibility is a collision with a large planetary body when our system was first forming. Whatever the cause, this unusual tilt creates dramatic seasons. When Voyager 2 arrived at Uranus, its south pole was facing directly into the Sun. The southern hemisphere was having a roughly 21-year sunlit summer, while during that same period the northern hemisphere was plunged into darkness. For the next 21-year season, the Sun shines on Uranus' equator, and both hemispheres go through cycles of light and dark (Figure 10.6). Then there are 21 years of an illuminated northern hemisphere and a dark southern hemisphere. If you were to install a floating platform at the south pole of Uranus, for example, it would experience 42 years of light and 42 years of darkness. Any future astronauts crazy enough to set up camp there could spend most of their lives without ever seeing the Sun.

10.2.3 Composition and Structure

Although we cannot see into these planets, astronomers are confident that the interiors of Jupiter and Saturn are composed primarily of hydrogen and helium. Of course, these gases have been measured only in their atmospheres, but calculations first carried out 50 years ago showed that these two light gases are the only possible materials out of which a planet with the observed masses and densities of Jupiter and Saturn could be constructed.

Having said that much, we should add that the precise internal structure of the two planets is not so easy to predict. This is mainly because the planets are so big that the hydrogen and helium in their centers become tremendously compressed and behave in ways that these gases can never behave on Earth. The best theoretical models we have of Jupiter's structure predict a central pressure greater than 100 million bars and a central density of about

Belinda
Puck
Rosalind   Epsilon Ring
Porta
Bianca
Pole   Equator
Desdemona
Cressida
Juliet

Erich Karkoschka, U. of Arizona and NASA

■ **FIGURE 10.5**

Infrared Image of Uranus   The infrared camera on the Hubble Space Telescope took these false-color images of the planet Uranus, its ring system, and satellites in 1997. The south pole of the planet (marked with a + on the right image) faces the Sun; its green color shows a strong local haze. The two images were taken 90 min apart, and during that time the five reddish clouds can be seen to rotate around the parallel to the equator. The rings (which are very faint in visible light, but prominent in infrared) and eight satellites can be seen around the equator.

31 g/cm^3. (By contrast, the Earth's core has a central pressure of 4 million bars and a central density of 17 g/cm^3.)

At the pressures inside the giant planets, familiar materials can take on strange forms. A few thousand kilometers below Jupiter and Saturn's visible clouds, pressures become so great that hydrogen changes from a gaseous to a liquid state. Still deeper, this liquid hydrogen is further compressed and begins to act like a metal, something it never does on Earth. (In a metal, electrons are not firmly attached to their parent nuclei but can wander around. This is why metals are such good conductors of electricity.) On Jupiter, the greater part of the interior is liquid metallic hydrogen.

Because Saturn is less massive, it has only a small volume of metallic hydrogen, but most of its interior is liquid. Uranus and Neptune are too small to reach internal pressures sufficient to liquefy hydrogen. We will return to the question of the metallic hydrogen layers when we examine the magnetic fields of the giant planets.

Each of these planets has a core composed of heavier materials, as demonstrated by detailed analyses of their gravitational fields. Presumably these cores are the original rock-and-ice bodies that formed before the capture of gas from the surrounding nebula (see Chapters 6 and 13). The cores exist at pressures of tens of millions of bars. While scientists speak of the giant planet cores being composed of rock and ice, we can be sure that neither rock nor ice assumes any familiar forms at such pressures and the accompanying high temperatures. What is really meant by *rock* is any material made up primarily of iron, silicon, and oxygen. The term *ice* denotes materials composed primarily of the elements carbon, nitrogen, and oxygen in combination with hydrogen.

Figure 10.7 illustrates the interior structures of the four jovian planets. It appears that all four have similar cores of rock and ice. On Jupiter and Saturn, the cores constitute only a few percent of the total mass, consistent with the initial composition of raw materials shown in Table 10.1. However, most of the mass of Uranus and Neptune resides in these cores, demonstrating that the two planets were unable to attract massive quantities of hydrogen and helium when they were first forming.

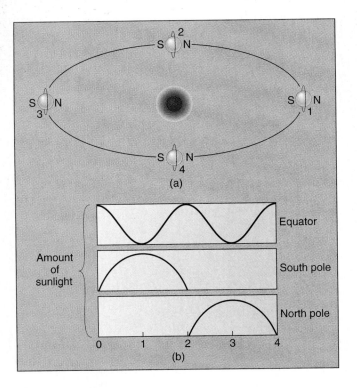

(a)

(b)

■ FIGURE 10.6

The Strange Seasons on Uranus (a) The orbit of Uranus as seen from above. At the time Voyager 2 arrived (position 1), the south pole was facing the Sun. As we move counterclockwise in the diagram, we see the planet 21 years later at each step. (b) The amount of sunlight seen at the poles and the equator of Uranus over the course of its 84-year revolution around the Sun.

10.2.4 Internal Heat Sources

Because of their large sizes, all of the giant planets were strongly heated during their formation by the collapse of surrounding material onto their cores. Jupiter, being the largest, was the hottest. Some of this primordial heat can still remain inside such large planets. In addition, it is possible for giant, largely gaseous planets to generate heat after formation by slowly contracting. (With so large a mass, even a minuscule amount of shrinking can generate significant heat.) The effect of these internal energy sources is to raise the temperatures in the interiors and atmospheres of the planets higher than we would expect from the heating effect of the Sun alone.

Jupiter has the largest internal energy source, amounting to 4×10^{17} watts (W); that is, it is glowing with the equivalent of 4 million billion 100-W light bulbs. This is about the same as the total solar energy absorbed by Jupiter. The atmosphere of Jupiter is therefore something of a cross between a normal planetary atmosphere (like the Earth's), which obtains most of its energy from the Sun, and the atmosphere of a star, which is entirely heated by an internal energy source. Most of the internal energy of Jupiter is primordial heat, left over from the formation of the planet 4.5 billion years ago.

Saturn has an internal energy source about half as large as that of Jupiter, which means (since its mass is only about one quarter as great) that it is producing twice as much energy per kilogram of material as does Jupiter. Since Saturn is expected to have much less primordial heat, there must be another source at work generating most of this 2×10^{17} W of power. This source is the separation of helium from hydrogen in Saturn's interior. In the liquid hydrogen mantle, the heavier helium forms drops that sink toward the core, releasing gravitational energy. In effect, Saturn is still differentiating—letting lighter material rise and heavier material fall.

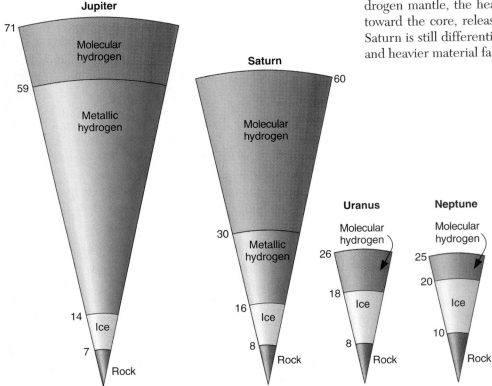

■ FIGURE 10.7

Internal Structures of the Jovian Planets Jupiter and Saturn are composed primarily of hydrogen and helium, but Uranus and Neptune consist, in large part, of compounds of carbon, nitrogen, and oxygen. (The diagrams are drawn to scale; numbers show radius in thousands of km.)

FIGURE 10.8

Jupiter in Radio Waves This false-color image of Jupiter was made with the Very Large Array (of radio telescopes) in New Mexico. We see part of the magnetosphere, brightest in the middle because the largest number of charged particles are in the equatorial zone of Jupiter. The planet itself is slightly smaller than the circular green region in the center. Different colors are used to indicate different intensities of synchrotron radiation.

Imke dePater/NRAO

Uranus and Neptune are different. Neptune has a small internal energy source, while Uranus does not emit a measurable amount of internal heat. As a result, these two planets have almost the same temperature, in spite of Neptune's greater distance from the Sun. No one knows why these two planets differ in their internal heat.

10.2.5 Magnetic Fields

Each of the giant planets has a strong magnetic field, generated by electric currents in its rapidly spinning interior. Associated with the magnetic fields are the planets' *magnetospheres*, which are defined as the regions around the planet within which the planet's own magnetic field dominates over the general interplanetary magnetic field. The magnetospheres of these planets are their largest features, extending millions of kilometers into space.

In the late 1950s, astronomers discovered that Jupiter was a source of radio waves that got more intense at longer rather than at shorter wavelengths—just the reverse of what is expected from thermal radiation (radiation caused by the normal vibrations of particles within all matter; see Chapter 4). Such behavior is typical, however, of the radiation emitted when high-speed electrons are accelerated by a magnetic field. We call this **synchrotron radiation** because it was first observed on Earth in particle accelerators, called synchrotrons. This was our first hint that Jupiter must have a strong magnetic field.

Later observations showed that the radio waves are coming from a region surrounding Jupiter with a diameter several times that of the planet itself (Figure 10.8). The evidence suggested, therefore, that a vast number of charged atomic particles must be circulating around Jupiter, spiraling around the lines of force of a magnetic field associated with the planet. This is just what we observe happening, but on a smaller scale, in the Van Allen belt around the Earth (see Chapter 7). The magnetic fields of Saturn, Uranus, and Neptune, discovered by the spacecraft that first passed close to these planets, work in a similar way.

Inside each magnetosphere, charged particles are trapped into spiraling around in the magnetic field; as a result they can be accelerated to high energies. The charged particles come from the Sun or from the neighborhood of the planet itself. In Jupiter's case, Io, one of its moons,

turns out to have a constant series of volcanic eruptions that propel charged particles into space.

The axis of Jupiter's magnetic field (the line that connects the magnetic north pole with the magnetic south pole) is not aligned exactly with the axis of rotation of the planet; rather, it is tipped by about 10°. Uranus and Neptune have even greater magnetic tilts of 60° and 55°, respectively. Saturn's field, on the other hand, is perfectly aligned with its rotation axis. Why different planets have such different magnetic tilts is not well understood.

The physical processes around the jovian planets turn out to be milder versions of what astronomers find in many distant objects, from the remnants of dead stars to the puzzling distant powerhouses we call quasars. One reason to study the magnetospheres of the giant planets and the Earth is that they provide nearby accessible analogues of more energetic and challenging cosmic processes.

10.3 ATMOSPHERES OF THE GIANT PLANETS

The atmospheres of the jovian planets are the parts we can observe or measure directly. Since these planets have no solid surfaces, their atmospheres are more representative of their general compositions than is the case with the terrestrial planets. These atmospheres also present us with some of the most dramatic examples of weather patterns in the solar system. As we will see, storms on these planets can grow bigger than the entire planet Earth.

10.3.1 Atmospheric Composition

When sunlight reflects from the atmospheres of the giant planets, the atmospheric gases leave their "fingerprints" in the spectrum of light (as we saw in Chapter 4). Spectroscopic observations of the jovian planets began in the 19th century, but for a long time astronomers were not able to interpret the spectra they observed. As late as the 1930s, the most prominent features photographed in these spectra remained unidentified. Then better spectra revealed the presence of methane (CH_4) and ammonia (NH_3) in the atmospheres of Jupiter and Saturn.

James Van Allen: Several Planets Under His Belt

The career of physicist James Van Allen has spanned the birth and growth of the space age; in fact he played a major role in its development. Born in Iowa in 1914, Van Allen received his PhD from the University of Iowa. He then worked for several research institutions and served in the Navy during World War II. In 1951 he was appointed Professor of Physics at the University of Iowa, where he has remained ever since.

After the war, Van Allen and his collaborators began using the V2 rockets captured from the Germans, as well as rockets built in the United States, to explore cosmic radiation in the Earth's outer atmosphere. Van Allen designed a technique, nicknamed "the rockoon," in which a small rocket is lifted up by a balloon before being launched, enabling it to reach a higher altitude.

Over dinner one night in 1950, Van Allen and several colleagues came up with the idea of the International Geophysical Year (IGY), an opportunity for scientists around the world to coordinate their investigations of the physics of the Earth, especially research done at high altitudes. Stretched to 18 months, the IGY took place between July 1957 and December 1958. In 1955, the United States and the Soviet Union each committed themselves to launching an Earth-orbiting satellite during IGY, a competition that began what came to be known as the space race.

When the Soviet Union won the first lap of the race by launching Sputnik 1 in October 1957, the U.S. government spurred its scientists and engineers to even greater efforts to get something into space and thereby maintain the country's prestige. However, the main U.S. satellite program, called Vanguard, was running into all sorts of difficulties, and each of its early launches crashed or exploded. In the meantime, a second team of

James Van Allen

D. Hokanson

■ ■ ■ ■ ■ ■ ■ ■ ■ ■ ■ ■ ■

After WWII, Van Allen began using captured V2 rockets to probe nearby space.

■ ■ ■ ■ ■ ■ ■ ■ ■ ■ ■ ■ ■

rocket engineers and scientists had quietly been working on a military launch vehicle called Jupiter-C. Van Allen spearheaded the design of the instruments aboard the small satellite, called Explorer I, that this rocket would carry. On January 31, 1958, Explorer I became the first U.S. satellite in space.

Unlike Sputnik, Explorer I was equipped to make scientific measurements. Van Allen had placed an instrument aboard Explorer I that could detect high-energy charged particles above the atmosphere, and it was this instrument that discovered what we now call the Van Allen belts around the Earth. This was the first scientific discovery of the space program and made Van Allen's name a household word around the world.

Van Allen and his colleagues continued making measurements of the magnetic and particle environment around our planet with more sophisticated spacecraft and instruments. When Pioneers 10 and 11 made the first exploratory surveys of the Jupiter and Saturn environments, they again carried instruments designed by Van Allen. Some scientists therefore refer to the charged-particle zones around those planets as Van Allen belts as well. (Once, when Van Allen was giving a lecture at the University of Arizona, the graduate students in planetary science asked him if he would leave his belt at the school. It is now proudly displayed as the "Van Allen belt.")

Van Allen has continued to be a strong supporter of space science and has been an eloquent senior spokesman for the American scientific community, warning NASA not to put all its efforts into human exploration but to remember how productive and (relatively) inexpensive its robot probes and instruments have always been in helping us learn more about the solar system.

At first astronomers thought that methane and ammonia might be the main constituents of these atmospheres, but now we know that hydrogen and helium are actually the dominant gases. The confusion arose because neither hydrogen nor helium possesses easily detected spectral features. It was not until the Voyager spacecraft measured the far-infrared spectra of Jupiter and Saturn that a reliable abundance for the elusive helium could be found. The compositions of the two atmospheres are generally similar, except that on Saturn there is less helium—the result of the precipitation of helium that contributes to Saturn's internal energy source. The most precise measurements of

NASA/JPL

■ FIGURE 10.9

Jupiter's Colorful Clouds
Also shown in this Voyager photo are two of Jupiter's satellites: Io (*left*) and Europa (*right*). The colors have been somewhat exaggerated to emphasize the differences among the zones and bands. The actual colors are a bit more muted, as shown in Figure 10.1.

composition were made on Jupiter by the Galileo entry probe in 1995, and as a result, we know the abundances of some elements in the jovian atmosphere even better than those in the Sun.

10.3.2 Clouds and Atmospheric Structure

The clouds of Jupiter are among the most spectacular sights in the solar system, much beloved by makers of science-fiction films. They range in color from white to orange to red to brown, swirling and twisting in a constantly changing kaleidoscope of patterns (Figure 10.9). Saturn shows similar but much more subdued cloud activity: Instead of vivid colors, its clouds have a nearly uniform butterscotch hue (Figure 10.10).

Different gases freeze at different temperatures. At the temperatures and pressures of the upper atmospheres of Jupiter and Saturn, methane remains a gas but ammonia can condense and freeze. (Similarly, water vapor condenses high in the Earth's atmosphere to produce clouds of ice crystals.) The primary clouds that we see around these planets,

Space Telescope Science Institute/NASA

■ FIGURE 10.10

Saturn Over Four Years These beautiful images of Saturn were recorded by the Hubble Space Telescope between 1996 and 2000. Since Saturn is tilted by 27°, we see the orientation of the rings around its equator change as the planet moves along its orbit. Note the horizontal bands in the atmosphere.

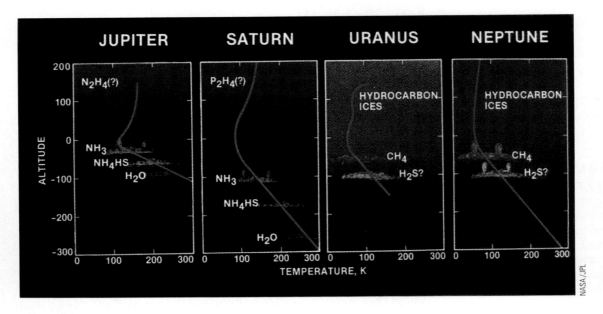

■ **FIGURE 10.11**

Atmospheric Structure for the Jovian Planets In each diagram, the yellow line shows how the temperature (see the scale on the bottom) changes with altitude. The location of the main layers on each planet is also shown.

whether from a spacecraft or through a telescope, are composed of frozen ammonia crystals. The ammonia clouds mark the upper edge of the planets' *troposphere;* above that lies the *stratosphere,* the coldest part of the atmosphere. (These layers were defined for the Earth in Chapter 7.)

Figure 10.11 is a diagram showing the structure and clouds in the atmospheres of all four jovian planets. On both Jupiter and Saturn, the temperature near the cloud tops is about 140 K (only a little cooler than the polar caps of Mars). On Jupiter this cloud level is at a pressure of about 0.1 bar, but on Saturn it occurs lower in the atmosphere, at about 1 bar. Because the ammonia clouds lie so much deeper on Saturn, they are more difficult to see, and the overall appearance of the planet is much more bland than is Jupiter.

Within the troposphere of these planets, the temperature and pressure both increase with depth. Through breaks in the ammonia clouds, we can see tantalizing glimpses of other cloud layers that can form in these deeper regions of the atmosphere—regions that were sampled directly by the Galileo probe for Jupiter.

As it descended to a pressure of 5 bars, the probe should have passed into a region of frozen water clouds, then below that into clouds of liquid water droplets, perhaps similar to the common clouds of the terrestrial troposphere. At least this is what scientists expected. But the probe saw no water clouds, and it measured a surprisingly low abundance of water vapor in the atmosphere. It soon became clear to the Galileo scientists that the probe happened to descend through an unusually dry, cloud-free region of the atmosphere—a giant downdraft of cool, dry gas.

Andrew Ingersoll of Caltech, a member of the Galileo team, called this entry site the "desert" of Jupiter. It's a pity that the probe did not enter in a more representative region, but that's the luck of the cosmic draw. The probe continued to make measurements to a pressure of 22 bars but found no other cloud layers before its instruments stopped working. It also detected lightning storms, but only at great distances, suggesting that the probe itself was in a region of clear weather.

Above the visible ammonia clouds in Jupiter's atmosphere, we find the clear stratosphere, which reaches a minimum temperature near 120 K. At still higher altitudes temperatures rise again, just as they do in the upper atmosphere of the Earth, because here the molecules absorb ultraviolet light from the Sun. This input of energy causes chemical reactions among the atmospheric gases—a process we call **photochemistry.** In Jupiter's upper atmosphere, photochemical reactions create a variety of fairly complex compounds of hydrogen and carbon that form a thin layer of smog far above the visible clouds. We show this smog as a fuzzy orange region in Figure 10.11; however, this thin layer does not block our view of the clouds beneath it.

We must mention one puzzle of the main jovian clouds: the mystery of their colors. The ammonia-condensation clouds identified on the planet should be white, like water clouds on Earth, yet we see beautiful and complex patterns of pastel red, orange, and brown. Some additional chemical or chemicals must be present to create such colors, but we do not know what they are. Various photochemically produced organic compounds have been suggested, as well as sulfur and red phosphorus. But there are no firm identifications, nor any immediate prospects of solving this mystery.

The atmospheric structure of Saturn is similar to that of Jupiter. Temperatures are somewhat colder, however, and the atmosphere is more extended, due to Saturn's lower surface gravity. Thus the layers are stretched out over a longer distance (as you can see in Figure 10.11). Overall, though, the same atmospheric regions, condensation clouds, and photochemical reactions that we see on Jupiter should be present on Saturn.

Unlike Jupiter and Saturn, Uranus is almost entirely featureless as seen at wavelengths that range from the ultraviolet to the infrared (see its rather boring image in the opening collage for this chapter). Calculations indicate that the basic atmospheric structure of Uranus should resemble that of Jupiter and Saturn, although its upper clouds (at the 1-bar pressure level) are composed of methane rather than ammonia. However, the absence of an internal heat source suppresses up-and-down movement and leads to a very stable atmosphere with little visible structure.

Neptune differs from Uranus in its appearance, although their basic atmospheric temperatures are similar. The upper clouds are composed of methane, which forms a thin cloud layer near the top of the troposphere at a temperature of 70 K and a pressure of 1.5 bars. Most of the atmosphere above this level is clear and transparent, with less haze than is found on Uranus. The scattering of sunlight by gas molecules lends Neptune a pale blue color similar to that of the Earth's atmosphere (Figure 10.12). Another cloud layer, perhaps composed of hydrogen sulfide ice particles, exists below the methane clouds at a pressure of 3 bars.

Unlike Uranus, Neptune has an atmosphere in which *convection currents*—vertical drafts of gas—emanate from the interior, powered by the planet's internal heat source. These currents carry warm gas above the 1.5-bar cloud level, forming additional clouds at elevations about 75 km higher. These high-altitude clouds form bright white patterns against the blue planet beneath. Voyager photographed distinct shadows on the methane cloud tops, permitting the altitudes of the high clouds to be calculated. (See Figure 10.13, a remarkable close-up of Neptune's outer layers that could never have been obtained from Earth.)

10.3.3   Winds and Weather

The atmospheres of the jovian planets have many regions of high pressure (where there is more air) and low pressure (where there is less). Just as it does on Earth, air flows between these regions, setting up wind patterns that are then distorted by the rotation of the planet. By observing the changing cloud patterns on the jovian planets, we can measure wind speeds and track the circulation of their atmospheres.

The atmospheric motions we see on these planets are fundamentally different from those on the terrestrial planets. The giants spin faster, and their rapid rotation tends to smear out the circulation of the air into horizontal (east–

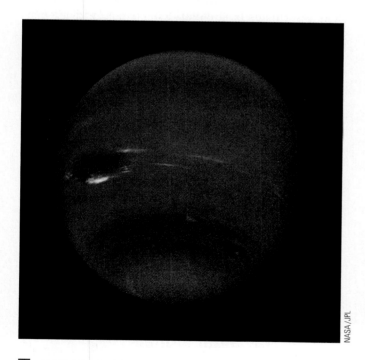

■ FIGURE 10.12

Neptune   The planet Neptune is seen here as photographed by Voyager in 1989. The blue color, exaggerated with computer processing, is caused by the scattering of sunlight in the upper atmosphere.

■ FIGURE 10.13

High Clouds in the Atmosphere of Neptune
These bright, narrow cirrus clouds are made of methane ice crystals. From the shadows they cast on the thicker cloud layer below, we can measure that they are about 75 km higher than the main clouds.

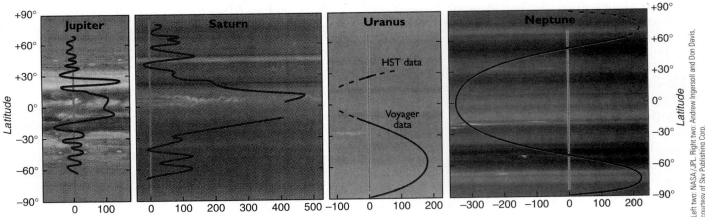

Left two: NASA/JPL. Right two: Andrew Ingersoll and Don Davis, courtesy of Sky Publishing Corp.

■ **FIGURE 10.14**

Winds on the Giant Planets   The speed and direction of the east–west winds on Jupiter, Saturn, Uranus, and Neptune, as measured by Voyager, are shown on these photographic maps. Latitude is shown on the left column of each chart, with the equator being 0°. The dark line shows the wind speed at each latitude (measured relative to the planet's core). Notice that Saturn has very strong winds at the equator.

west) patterns parallel to the equator. In addition, there is no solid surface below the atmosphere against which the circulation patterns can rub and lose energy (which is how tropical storms on Earth ultimately die out when they come over land). As we have seen, on all the giants except Uranus, heat from the inside contributes about as much energy to the atmosphere as sunlight from the outside. This means that deep convection currents of rising hot air and falling cooler air circulate throughout the atmospheres of the planets in the vertical direction.

The main features of Jupiter's visible clouds (see Figures 10.1 and 10.9), for example, are alternating dark and light bands that stretch around the planet parallel to the equator. These bands are semipermanent features, although they shift in intensity and position from year to year. Consistent with the small tilt of Jupiter's axis, the pattern does not change with the seasons.

More fundamental than these bands are underlying east–west wind patterns in the atmosphere, which do not appear to change at all, even over many decades. These are displayed in Figure 10.14, which indicates how strong the winds are at each latitude for the giant planets. At Jupiter's equator, a jet stream flows eastward with a speed of about 90 m/s (300 km/h), similar to the speed of jet streams in the Earth's upper atmosphere. At higher latitudes there are alternating east- and west-moving streams, with each hemisphere an almost perfect mirror image of the other. Saturn shows a similar pattern, but with a much stronger equatorial stream flowing at a speed of 1300 km/h.

The light zones on Jupiter are regions of upwelling air capped by white ammonia cirrus clouds. They apparently represent the tops of upward-moving convection currents. The darker belts are regions where the cooler atmosphere

moves downward, completing the convection cycle; they are darker because fewer ammonia clouds mean we can see deeper into the atmosphere, perhaps down to a region of ammonium hydrosulfide (NH_4SH) clouds. The Galileo probe sampled one of the clearest of these dry downdrafts.

In spite of the strange seasons induced by the 98° tilt of its axis, Uranus' basic circulation is parallel with its equator, as is the case on Jupiter and Saturn. The mass of the atmosphere and its capacity to store heat are so great that the alternating 42-year periods of sunlight and darkness have little effect. In fact, Voyager measurements show that the atmospheric temperatures are even a few degrees higher on the dark winter side than on the hemisphere facing the Sun. Here is another indication that the behavior of such giant planet atmospheres is a complex problem: We simply do not understand the seasonal effects on Uranus in detail.

Neptune's weather is characterized by strong east–west winds generally similar to those observed on Jupiter and Saturn. The highest wind speeds near its equator reach 2100 km/h (600 m/s), nearly twice as fast as the peak winds on Saturn. The Neptune equatorial jet stream actually approaches supersonic speeds (faster than the speed of sound in Neptune's air).

10.3.4   Giant Storms on Giant Planets

Superimposed on the regular atmospheric circulation patterns we have just described are many local disturbances—weather systems or *storms*, to borrow the term we use on Earth. The most prominent of these are large, oval-shaped, high-pressure regions on both Jupiter and Neptune.

■ FIGURE 10.15

The Great Red Spot This is the largest storm system on Jupiter, as seen during the Voyager spacecraft flyby. Below and to the right of the Red Spot is one of the white ovals, which are similar but smaller high-pressure features. The white oval is roughly the size of planet Earth, to give you a sense of the huge scale of the weather patterns we are seeing. The colors on the Jupiter image have been somewhat exaggerated here so astronomers (and astronomy students) can study their differences more effectively. See Figure 10.1 to get a better sense of the colors your eye would actually see near Jupiter.

The largest and most famous storm on Jupiter is the Great Red Spot, a reddish oval in the southern hemisphere. The Red Spot changes slowly; it was 25,000 km long when Voyager arrived in 1979, but it had shrunk to 20,000 km by the end of the Galileo mission in 2000 (Figure 10.15). The giant storm has persisted in Jupiter's atmosphere ever since astronomers were first able to observe it after the invention of the telescope, more than 300 years ago.

In addition to its longevity, the Red Spot differs from terrestrial storms in being a high-pressure region; on our planet such storms are regions of lower pressure. The Red Spot's counterclockwise rotation has a period of six days. Three similar but smaller disturbances (about as big as the Earth) formed on Jupiter in the 1930s. They look like white ovals and one can be seen clearly below and to the right of the Great Red Spot in Figure 10.15. In 1998 the Galileo spacecraft watched as two of these ovals collided and merged into one (Figure 10.16).

We don't know what causes the Great Red Spot or the white ovals, but we do understand how they can last so long once they form. On Earth, the lifetime of a large oceanic hurricane or typhoon is typically a few weeks, or even less when it moves over the continents and encounters friction with the land. Jupiter has no solid surface to slow down an atmospheric disturbance, and furthermore the sheer size of the disturbances lends them stability. We can calculate that on a planet with no solid surface, the lifetime of anything as large as the Red Spot should be measured in centuries, while lifetimes for the white ovals should be measured in decades, which is pretty much what we have observed.

Despite Neptune's smaller size and different cloud composition, the 1989 Voyager images showed that it had an atmospheric feature surprisingly similar to the jovian Great Red Spot. Neptune's Great Dark Spot was nearly 10,000 km long (see Figure 10.12). On both planets, the giant storms formed at latitude 20°S, had the same shape, and took up about the same fraction of the planet's diameter. The Great Dark Spot rotated with a period of 17 days, versus about 6 days for the Great Red Spot. When the repaired Hubble Space Telescope examined Neptune in the mid 1990s, however, astronomers could find no trace of the Great Dark Spot on their images.

Although many of the details of the weather on the jovian planets are not yet understood, it is clear that if you are a fan of dramatic weather, these worlds are the place to look. We study the features in these atmospheres not only for what they have to teach us about conditions in the jovian planets, but also because we hope they can help us understand the weather on Earth just a bit better.

(a)

(c)

(b)

■ **FIGURE 10.16**

Storms on Jupiter   Three examples of storms on Jupiter, illustrating the use of enhanced color and contrast to bring out faint features. (a) We show two versions of the same Galileo view of the large white ovals that are the most long-lived storms after the Great Red Spot. The top image shows how the scene might appear to the human eye, while the bottom is in false color (and includes infrared information). The left white oval is about three-fourths the size of the Earth. Note the cyclone-shaped storm between the two top ovals. (b) This 1996 Galileo image shows an enhanced view of a large hole in the clouds, similar to the place where the Galileo probe entered the atmosphere. The image is about 44,000 km across, and the dark hole (which is free of clouds) is nearly as wide as the Earth. (c) Here we see a thunderhead on Jupiter—a plume of water clouds carried to high altitudes by a strong updraft. The white plume is 1000 km across and about 25 km high. Such clouds occasionally form in regions of high atmospheric turbulence, such as this site near the outer edge of the Great Red Spot.

SURFING THE WEB

General Jupiter sites:

- *The Nine Planets Site:*
  seds.lpl.arizona.edu/nineplanets/nineplanets/jupiter.html

- *Views of the Solar System Site:*
  www.solarviews.com/eng/jupiter.htm

- *Planetary Sciences Site:*
  nssdc.gsfc.nasa.gov/planetary/planets/jupiterpage.html

General Saturn sites:

- *The Nine Planets Site:*
  seds.lpl.arizona.edu/nineplanets/nineplanets/saturn.html

- *Views of the Solar System Site:*
  www.solarviews.com/eng/saturn.htm

- *Planetary Sciences Site:*
  nssdc.gsfc.nasa.gov/planetary/planets/saturnpage.html

General Uranus sites:

- *The Nine Planets Site:*
  seds.lpl.arizona.edu/nineplanets/nineplanets/uranus.html

- *Views of the Solar System Site:*
  www.solarviews.comsolar/eng/uranus.htm

- *Planetary Sciences Site:*
  nssdc.gsfc.nasa.gov/planetary/planets/uranuspage.html

General Neptune sites:

- *The Nine Planets Site:*
  seds.lpl.arizona.edu/nineplanets/nineplanets/neptune.html

- *Views of the Solar System Site:*
  www.solarviews.com/eng/neptune.htm

- *Planetary Sciences Site:*
  nssdc.gsfc.nasa.gov/planetary/planets/neptunepage.html

Galileo Mission Site: www.jpl.nasa.gov/galileo
All the new Galileo images and results can be found here, together with educational activities and background information.

Voyager Project Home Page: voyager.jpl.nasa.gov
In 1998, Voyager 1 became the most distant human-made object in space. This site chronicles the planetary flybys of the two Voyagers and gives updates on their continuing mission to explore deep space and search for the heliopause. There is a good science summary for each planet the Voyagers investigated.

Cassini Mission Site: saturn.jpl.nasa.gov/index.cfm
Information about the mission and its engineering and science objectives. There are images of mission equipment and goals and several pages for educators.

SUMMARY

10.1 The outer solar system contains the four jovian (or giant) planets: Jupiter, Saturn, Uranus, and Neptune. Their chemistry is generally reducing, and Jupiter and Saturn have overall compositions similar to that of the Sun. These planets have been explored by the Pioneer, Voyager, Galileo, and Cassini spacecraft. Voyager 2, perhaps the most successful of all space-science missions, explored Jupiter (1979), Saturn (1981), Uranus (1986), and Neptune (1989)—a grand tour of the giant planets. Galileo successfully deployed a probe into the atmosphere of Jupiter in 1995 before beginning a 5-year orbital mission.

10.2 Jupiter is 318 times more massive than the Earth. Saturn is about 25 percent, and Uranus and Neptune only 5 percent, as massive as Jupiter. All four have deep atmospheres and opaque clouds, and all rotate quickly (with periods from 10 to 17 hours). Jupiter and Saturn have exten-

sive mantles of liquid hydrogen. Uranus and Neptune are depleted in hydrogen and helium relative to Jupiter and Saturn (and the Sun). Each giant planet has a core of "ice" and "rock" of about 10 Earth masses. Jupiter, Saturn, and Neptune have major internal heat sources, obtaining as much (or more) energy from their interiors as by radiation from the Sun. Uranus has no measurable internal heat. Jupiter has the strongest magnetic field and largest magnetosphere of any planet, first discovered by radio astronomers from observations of **synchrotron radiation.**

10.3 The four giant planets have generally similar atmospheres, composed mostly of hydrogen and helium. Their atmospheres contain small quantities of methane and ammonia gas, both of which also condense to form clouds. Deeper (invisible) cloud layers consist of water and possibly ammonium hydrosulfide (Jupiter and Saturn) and hydrogen sulfide (Neptune). In the upper atmospheres, hydrocarbons and other trace compounds are produced by **photochemistry.** We do not know what colors the clouds of Jupiter. Atmospheric motions on the giant planets are dominated by east–west circulation. Jupiter displays the most active cloud patterns, with Neptune second. Saturn is generally bland, and Uranus is featureless (perhaps due to its lack of an internal heat source). Large storms (oval-shaped high-pressure systems such as the Great Red Spot on Jupiter and the Great Dark Spot on Neptune) can be found in some of the giant planet atmospheres.

INTER-ACTIVITY

 A A new member of Congress has asked your group to investigate why the Galileo probe launched into the Jupiter atmosphere in 1995 survived only 57 minutes. Make a list of all the reasons the probe did not last longer, and why it was not made more durable. (Remember that the probe had to hitch a ride to Jupiter!)

B Select one of the jovian planets and get your group to write a script for an evening news weather report for the planet you chose. Be sure you specify roughly how high in the atmosphere the region lies for which you are giving the report.

C What does your group think should be the next step to learn more about the giant planets? Put cost considerations aside for a moment: What kind of mission would you recommend to NASA to learn more about these giant worlds? Which world or worlds should get the highest priority and why?

 D Suppose that an extremely dedicated (and slightly crazy) astronomer volunteers to become a human probe into Jupiter (and somehow manages to survive the trip through Jupiter's magnetosphere alive). As he enters the upper atmosphere of Jupiter, would he fall faster or slower than he would fall doing the same suicidal jump into the atmosphere of the solid Earth? Groups that have some algebra background could even calculate the force he would feel compared to the force on Earth. (*Bonus question:* If he were in a capsule, falling into Jupiter feet first, and the floor of the capsule had a scale, what would the scale show as his weight?)

REVIEW QUESTIONS

1. Describe the differences in the chemical makeup of the inner and outer parts of the solar system. What is the relationship between what the planets are made of and the temperature where they formed?
2. How did the giant planets grow to be so large?
3. Why is it difficult to drop a probe like Galileo's gently into the atmosphere of Jupiter? How did the Galileo engineers solve this problem?
4. What are the visible clouds on the four giant planets composed of, and why are they different from each other?
5. Describe the seasons on the planet Uranus.
6. Compare the atmospheric circulation (weather) for the four giant planets.
7. What are the main atmospheric heat sources of each of the giant planets?
8. How do the clouds on the four giant planets differ?

THOUGHT QUESTIONS

9. Jupiter is denser than water, yet composed for the most part of two light gases, hydrogen and helium. What makes Jupiter as dense as it is?

10. Would you expect to find free oxygen gas in the atmospheres of the giant planets? Why or why not?

11. Why would a tourist brochure (of the future) describing the most dramatic natural sights of the giant planets have to be revised more often than one for the terrestrial planets?

12. The water clouds believed to be present on Jupiter and Saturn exist at temperatures and pressures similar to those in the clouds of the terrestrial atmosphere. What would it be like to visit such a location on Jupiter or Saturn? In what ways would the environment differ from that in the clouds of Earth?

13. Describe the different processes that lead to substantial internal heat sources for Jupiter and Saturn. Since these two objects generate much of their energy internally, should they be called stars instead of planets? Justify your answer.

14. Use sources in your college library to learn more about the Galileo mission. What technical problems occurred between the mission launch and the arrival of the craft in the jovian system, and how did the mission engineers deal with them? (Good sources of information include *Astronomy* and *Sky & Telescope* magazines, plus the mission web site.)

FIGURING FOR YOURSELF

The wind speeds in circular storm systems can be formidable on both Earth and the giant planets. First think about our big terrestrial hurricanes. If you watch their behavior in satellite images shown on the weather channels, you will see that they require about one day to rotate. If a storm has a diameter of 400 km and rotates once in 24 hours, what is the wind speed? Speed equals distance divided by time. The distance in this case is the circumference ($\pi \times$ diameter), or approximately 1200 km, and the time is 24 hours, so the speed at the edge of the storm is 50 km/h, or about 35 mi/h. Toward the center of the storm, of course, the wind speeds can be much higher.

15. Jupiter's Great Red Spot rotates in 6 days and has a circumference equivalent to a circle with radius 10,000 km. Calculate the wind speed at the outer edge of the spot.

16. Calculate the wind speed at the edge of Neptune's Great Dark Spot, which was 2500 km wide and rotated in 17 days.

17. Calculate how many Earths would fit into the volumes of Saturn, Uranus, and Neptune.

18. As the Voyager spacecraft penetrated into the outer solar system, the illumination from the Sun declined. Relative to the situation at Earth, how bright is the sunlight at each of the jovian planets?

19. The ions in the inner parts of the jovian magnetosphere rotate with the same period as Jupiter. Calculate how fast they are moving at the orbit of Jupiter's moon Io (see Appendix 8). Will these ions strike Io from behind or in front as it moves about Jupiter?

SUGGESTIONS FOR FURTHER READING

Beatty, J. "Into the Giant" in *Sky & Telescope*, Apr. 1996, p. 20. On the Galileo probe.

Beebe, R. "Queen of the Giant Storms" in *Sky & Telescope*, Oct. 1990, p. 359. Excellent review of the Red Spot.

Cowling, T. "Big Blue: The Twin Worlds of Uranus and Neptune" in *Astronomy*, Oct. 1990, p. 42. Nice long review of the two planets.

Hanlon, M. *The Worlds of Galileo.* 2001, St. Martin's Press. Good overview of the Galileo mission and its discoveries.

Harland, D. *Jupiter Odyssey: The Story of NASA's Galileo Mission.* 2000, Praxis/Springer-Verlag. Another good mission summary.

Littmann, M. *Planets Beyond: Discovering the Outer Solar System,* 2nd ed. 1989, John Wiley. Good introduction to Uranus, Neptune, and Pluto.

Lunine, J. "Neptune at 150" in *Sky & Telescope*, Sept. 1996, p. 38. Nice review.

McAnally, J. "A Jupiter Observing Guide" in *Sky & Telescope*, Oct. 2000, p. 124. A good introduction for serious amateurs.

Miner, E. *Uranus.* 1990, Ellis Horwood/Simon & Shuster.

Miner, E. & Wessen, R. *Neptune: The Planet, Rings, and Satellites.* 2001, Springer Verlag.

Nellis, E. "Making Metallic Hydrogen" in *Scientific American*, May 2000, p. 84. On experiments on Earth to duplicate what is found naturally in Jupiter.

Rogan, J. "Bound for the Ringed Planet" in *Astronomy*, Nov. 1997, p. 36. On the Cassini mission.

Sagan, C. *Pale Blue Dot.* 1994, Random House. Chapters 6–9 explore the outer solar system.

Smith, B. "Voyage of the Century" in *National Geographic*, Aug. 1990, p. 48. Beautiful summary of the Voyager mission to all four outer planets.

This montage, assembled from individual Galileo and Voyager images, shows a "family portrait" of Jupiter (with its giant red spot) and its four large satellites. From top to bottom, we see Io, Europa, Ganymede, and Callisto. The colors are exaggerated by image processing to emphasize contrasts. See Figure 10.1 for a better sense of the true colors of Jupiter's atmosphere, for example.

11 · Rings, Moons, and Pluto

Our imaginations always fall short of anticipating the beauty we find in nature.

Geologist Laurence Soderblom, discussing the 1989 Voyager encounter with Neptune's moons

THINKING AHEAD

On an active world, a map is always temporary. The boundaries of towns change, new roads are built, rivers slowly alter their channels. Over geological time, even the Earth's mountains, seas, and continents change shape and form. But what kind of map would be worth making of a world where the surface landforms changed substantially in only a decade? Such a world can be found orbiting the planet Jupiter.

All of the giant planets are accompanied by moons or satellites (astronomers use the two terms interchangeably), which orbit about them like planets in a miniature solar system. More than 100 satellites are known in the outer solar system, too many to name individually or discuss in any detail. The characteristics of the more important moons are summarized in Appendix 8. Astronomers expect that many other small satellites will be found in the future.

Before the Voyager missions to the outer planets, even the largest of these satellites were mere points of light in our telescopes. Suddenly, in less than a decade, we had close-up images (and much nonvisual information as well); they then became individual worlds for us, each with unique features and peculiarities. The Galileo mission added greatly to our knowledge of the satellites of Jupiter, and the Cassini mission will, we hope, do the same for the Saturn system.

In this chapter we focus our attention on a few of the more interesting satellites and on the tiny planet Pluto, which is smaller than a number of the outer planets' moons. Then we examine the ring systems that gird each of the four giant planets. The processes we see at work in these rings have applications ranging from the formation of planetary systems to the spiral structure of galaxies.

11.1 RING AND SATELLITE SYSTEMS

The rings and satellites of the outer solar system are not composed of the same materials as the mostly rocky objects in the inner solar system. We should expect this, since they formed in regions of lower temperature, cool enough so that large quantities of water ice were available as building materials. Most of these objects also contain dark, organic compounds mixed with their ice and rock. The presence of this dark, primitive material often means that the objects reflect very little light. Don't be surprised, therefore, to find many objects in the ring and satellite systems that are both icy and black.

Most of the satellites in the outer solar system are in *direct* or regular orbits; that is, they revolve about their parent planet in a west-to-east direction and in the plane of the planet's equator. In addition, a number of irregular satellites orbit in a *retrograde* (east-to-west) direction or else have orbits of high eccentricity (more elliptical than circular) or high inclination (moving in and out of the planet's equatorial plane). These are usually small satellites, located relatively far from their planet; they were probably formed far away and subsequently captured by the planet they now orbit. (Perhaps the fact that they were not born locally will excuse their ill-mannered behavior.)

11.1.1 The Jupiter System

Jupiter has 52 known satellites (that's the number as we write; probably more will have been discovered by the time you read this sentence) and a faint ring. The satellites include four large moons—Callisto, Ganymede, Europa, and Io (see the image that opens the chapter)—discovered in 1610 by Galileo and therefore often called the *Galilean satellites*. The smaller of these, Europa and Io, are about the size of our Moon, while the larger, Ganymede and Callisto, are bigger than the planet Mercury. We will discuss the Galilean satellites individually in a moment and try to understand the interesting differences among them.

Most of the jovian moons are much smaller. The majority are in retrograde orbits more than 20 million km from Jupiter; these are very likely small captured asteroids.

11.1.2 The Saturn System

Saturn has at least 30 known satellites in addition to a magnificent set of rings (Figure 11.1). The largest of the satellites, Titan, is almost as big as Ganymede in the jovian system, and

■ **FIGURE 11.1**

Saturn Family Portrait This NASA montage of Voyager images shows a perspective view of Saturn and several of its satellites. The satellites are not to scale: Giant Titan is the tiny orange ball toward the back. The moon in the foreground is called Enceladus. The color of Saturn is intensified by the image processing.

NASA/JPL

it is the only satellite with a substantial atmosphere. We will discuss it in detail in the next section. Saturn has six other large regular satellites with diameters between 400 and 1600 km, a handful of small moons orbiting in or near the rings, and a collection of captured strays similar to those of Jupiter.

The rings of Saturn, one of the most impressive sights in the solar system, are broad and flat, with only a few gaps. They are not solid, but rather a huge collection of icy fragments, all orbiting the equator of Saturn in a traffic pattern that makes rush hour in a big city look simple by comparison. Individual ring particles are composed of water ice and are typically the size of ping-pong, tennis, and basketballs.

11.1.3   The Uranus System

The ring and satellite system of Uranus is tilted at 98°, just like the planet itself. It consists of 11 rings and more than 20 known satellites. The five largest satellites are similar in size to the six regular satellites of Saturn, with diameters of 500 to 1600 km, while most of the smaller satellites and the ring particles are very dark, in contrast to those of Saturn.

Discovered in 1977, the rings of Uranus are narrow ribbons of material with broad gaps in between—very different from the broad rings of Saturn. Astronomers suppose that the ring particles are confined to these narrow paths by the gravitational effects of numerous small satellites, most of which we have not yet glimpsed. (We will discuss how satellites can do this in Section 11.4.4.)

11.1.4   The Neptune System

Neptune has eight known satellites: six regular satellites close to the planet, and two irregular satellites farther out. The most interesting of these is Triton, a relatively large moon in a retrograde orbit. Triton has an atmosphere, and active volcanic eruptions were discovered there by Voyager in its 1989 flyby. We will compare it in detail with the planet Pluto, which it resembles, later in this chapter.

The rings of Neptune are narrow and faint. Like those of Uranus, they are composed of dark materials and are thus not easy to see.

11.2   THE GALILEAN SATELLITES AND TITAN

From 1996 to 1999 the Galileo spacecraft careered through the jovian system on a complex but carefully planned trajectory that provided repeated close encounters with the Galilean satellites. Although scientists had thought they understood the basics of these satellites from the two Voyager flybys of 1979, the new data brought many surprises. And, to the delight of scientists and the public, the high-resolution Galileo cameras offered stunning views that far exceeded the Voyager images in their scientific value. This wealth of new information provides us the opportunity, quite literally, to rewrite the textbooks on the Galilean satellites. Beginning in 2004, we can expect a similar bonanza of information about Titan, obtained from the Cassini spacecraft.

We discuss each of these five large satellites in turn, but to compare them with each other, we first list some basic facts about them (and our own Moon) in Table 11.1.

11.2.1   Callisto: An Ancient, Primitive World

We begin our discussion of the Galilean satellites with the outermost one, Callisto—not because it is remarkable but because it is not. This makes it a convenient object with which other, more active, worlds can be compared. Its distance from Jupiter is about 2 million km, and it orbits the planet in 17 days. Like our own Moon, Callisto rotates in the same period as it revolves, so it always keeps the same face toward Jupiter. Callisto's day thus equals its month: 17 days. Its noontime surface temperature is only 130 K (about 140°C below freezing), so that water ice is stable on its surface the year around.

Callisto has a diameter of 4820 km, almost the same as the planet Mercury (Figure 11.2). Yet its mass is only one third as great, which means its density (the mass divided by the volume) must be only one third as great as well. This tells us that Callisto has far less of the rocky and metallic materials found in the inner planets and must instead be an

TABLE 11.1 *The Largest Satellites*			

Name	Diameter (km)	Mass (Moon = 1)	Density (g/cm^3)	Reflectivity (%)
Moon	3476	1.0	3.3	12
Callisto	4820	1.5	1.8	20
Ganymede	5270	2.0	1.9	40
Europa	3130	0.7	3.0	70
Io	3640	1.2	3.5	60
Titan	5150	1.9	1.9	20

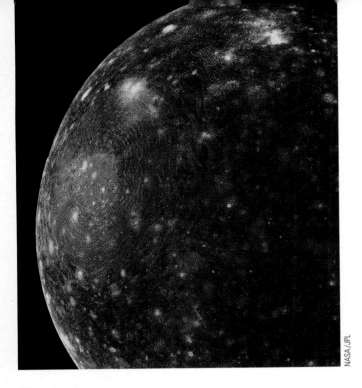

NASA/JPL

■ **FIGURE 11.2**

Callisto Jupiter's outermost large satellite shows a heavily cratered surface of dark ice in this Voyager image. The multi-ringed feature at center left is the giant impact basin called Valhalla.

icy body through much of its interior. Such a satellite can show us how the geology of an icy object compares with one made primarily of rock.

Unlike the worlds we have studied so far, Callisto has not fully *differentiated* (separated into layers of different density materials). We can tell that it lacks a dense core

from the details of its gravitational pull on the Galileo spacecraft during several very close flybys. This fact surprised scientists, who expected that all the big icy moons would be differentiated. It should be easier for an icy body to differentiate than for a rocky one because the melting temperature of ice is so low. Only a little heating will soften the ice and get the process started, allowing the rock and metal to sink to the center while the slushy ice floats to the surface. Yet Callisto seems to have frozen solid before the process of differentiation was complete.

The surface of Callisto is covered with impact craters, like the lunar highlands. The survival of these craters tells us that an icy object can retain impact craters in its surface. An especially impressive impact feature, called Valhalla (after the home of the gods in Norse mythology), can be seen as a bull's-eye on Figure 11.2. The light material brought up by the impact in the center is as large as the state of Colorado, and the rings surrounding it have a diameter almost as large as the continental United States. Callisto is unique among the planet-sized objects of the solar system in the absence of interior forces to drive geological change. You might say that the satellite was stillborn, and it has remained geologically dead for more than 4 billion years.

In thinking about ice so far from the Sun, we must take care not to judge its behavior from the much warmer ice we know and love on Earth. At the temperatures of the outer solar system, ice on the surface is nearly as hard as rock and it behaves similarly. Ice on Callisto does not deform or flow like ice in glaciers on the Earth.

In general form, the icy craters of Callisto look very much like the rocky craters of the Moon and Mercury. Yet when we look at the close-up photographs from Galileo, the similarity with lunar craters evaporates (Figure 11.3). Some process has been at work to break down crater walls and bury parts of the surface. Although the results of this

Arizona State U./JPL/NASA

■ **FIGURE 11.3**

Worn Craters on Callisto
This pair of Galileo images shows an area about 30 km across in the Valhalla region of Callisto. Details as small as 10 m across can be seen. Sunlight is illuminating the surface from the left. Note that the impact craters are eroded even though Callisto has no atmosphere, and that white frost seems to have condensed on the highest parts of the eroded crater rims.

erosion look rather like the effect of windblown dust on Earth and Mars, this explanation cannot be correct because Callisto has no atmosphere. Instead, it must be the slow process of *sublimation* of ice (turning from solid directly into gas due to heating by the Sun) that breaks down the craters over billions of years and redistributes the dusty (non-icy) surface material.

11.2.2 Ganymede, the Largest Satellite

Ganymede, the largest satellite in the solar system, also shows a great deal of cratering (Figure 11.4). About one quarter of its surface seems to be as old and heavily cratered as that of Callisto; the rest formed more recently, as we can tell by the sparse covering of impact craters as well as the relative freshness of the craters. If we judge from crater counts, this fresher terrain on Ganymede is somewhat younger than the lunar maria or the martian volcanic plains, perhaps 2 to 3 billion years old.

The differences between Ganymede and Callisto are more than skin deep. Ganymede is a differentiated world, like the terrestrial planets. Measurements of its gravity field tell us that the rock sank to form a core about the size of our Moon, with a mantle and crust of ice "floating" above it. In

JPL/NASA

■ FIGURE 11.5

Close-up of Ganymede   Details as small as 75 m across can be seen in this image of Jupiter's largest moon, taken by the Galileo spacecraft cameras. We see an area 55 by 35 km in the Uruk Sulcus region, with a complex assortment of ridges, grooves, craters, and evidence of crustal shifting. The linelike sets of features are parallel sunlit ridges. We can see signs that the icy surface has pulled apart and slid sideways in shaping the landscape.

Galileo Imagining Team; JPL/NASA

■ FIGURE 11.4

Ganymede   This global view of Ganymede, the largest moon in the solar system, was assembled from images taken by the Galileo spacecraft. The large circular dark region at the upper right is called Galileo Regio. Such darker places are older, more heavily cratered regions; the lighter areas are younger. Really bright spots are sites of geologically recent impacts.

addition, the Galileo spacecraft discovered that Ganymede has a magnetic field—the sure signature of a partially molten interior. There is very likely liquid water trapped within the interior. Thus, Ganymede is not a dead world but rather a place of continuing geological activity powered by an internal heat source. Some surface features could be as young as the surface of Venus (half a billion years).

The younger terrain is the result of tectonic and volcanic forces (Figure 11.5). In some places the crust apparently cracked, flooding many of the craters with water from the interior. Extensive mountain ranges were formed from compression of the crust, forming long ridges with parallel valleys spaced a few kilometers apart. In some places older impact craters were split and pulled apart. There are even indications of large-scale crustal movements that are similar to the plate tectonics of the Earth.

Why is Ganymede so different from Callisto? Possibly the small difference in size and internal heating between the two led to this divergence in their evolution. But more likely the gravity of Jupiter is to blame for Ganymede's continuing geological activity. Ganymede is close enough to Jupiter that *tidal forces* from the giant planet may have episodically heated its interior and triggered major convulsions on its crust.

As we discussed in Section 3.6, a tidal force results from the unequal gravitational pull on two sides of a body. In a complex kind of modern dance, the large satellites of Jupiter are caught in the varying gravity grip of both the giant planet and each other. This leads to gravitational flexing or kneading in their centers, which can heat them up. (A fuller explanation is given in the section on Io, below.) We will see as we move inward to Europa and Io that the role of jovian tides becomes more critical for satellites close to the planet.

11.2.3 Europa, a Satellite with an Ocean

Europa and Io, the inner two Galilean satellites, are not icy worlds like most of the satellites of the outer planets. With densities and sizes similar to our Moon, they appear to be predominantly rocky objects. How did they fail to acquire a majority share of the ice that must have been plentiful in the outer solar system at the time of their formation?

The most probable cause is Jupiter itself, which was hot enough to radiate a great deal of infrared energy during the first few million years after its formation. This infrared radiation would have heated the disk of material near the planet that would eventually coalesce into the closer moons. Thus any ice near Jupiter was vaporized, leaving Europa and Io with compositions more appropriate to bodies in the inner solar system.

Despite its mainly rocky composition, Europa has an ice-covered surface, as astronomers have long known from examining spectra of sunlight reflected from it (Figure 11.6). In this it resembles the Earth, which also has a layer of water on its surface, but in Europa's case most of the water may be frozen. There are very few impact craters, indicating that the surface of Europa is in a continual state of geological self-renewal. Judging from crater counts, the surface must be no more than a few million years old, and perhaps substantially less. In terms of its ability to erase impact craters, Europa is more geologically active than the Earth!

When we look at close-up photos of Europa, we see a strange, complicated surface (Figures 11.7 and 11.8). For the most part the satellite is extremely smooth, but it is crisscrossed with cracks and low ridges that often stretch for thousands of kilometers across the icy plains. Some of these long lines are single, but most are double or

■ **FIGURE 11.6**

Europa's Intriguing Surface Jupiter's moon Europa is the only place in the solar system beyond Earth that appears to have liquid water—an ocean beneath a thick crust of ice. Both the contrast and the color are highly exaggerated in this wide-angle view of a region about 1200 km across. Scientists enhance the color to bring out subtle differences in composition between the different types of terrain. Europa's crust is nearly pure water ice, with very few hills or other topographic features.

■ **FIGURE 11.7**

Evidence for an Ocean on Europa A close-up of an area called Conamara Chaos is shown here with enhanced color. This view is 70 km wide in its longer dimension. It appears that Conomara is a region where Europa's icy crust is (or recently was) relatively thin and there is easier access to the possible liquid or slushy ocean beneath. Not anchored to solid crust underneath, many of the ice blocks here seem to have slid or rotated from their original positions. In fact, the formations seen here look similar to views of floating sea-ice and icebergs in the Earth's Arctic Ocean.

FIGURE 11.8

A Faded Crater We know that the surface of Europa is geologically young because there are so few impact craters. One of the largest craters on Europa is Pwyll, about 26 km in diameter. Even though Pwyll is the result of a recent impact, its crater rim is nearly gone, fading away as the icy crust bends and readjusts over time. The darker material in and outside the crater was probably dredged up from deeper inside Europa by the force of the impact.

multiple, looking rather like the remnants of a colossal freeway system (Figure 11.9 and the first panel of Figure 11.11).

It is very difficult to make straight lines on a planetary surface. In Chapter 9, we explained that when Percival Lowell saw what appeared to him to be straight lines on Mars (the so-called martian "canals"), he attributed them to the engineering efforts of intelligent beings. We now know the lines on Mars were optical illusions, but the lines on Europa are real. If the crust consists of ice floating without much friction on an ocean of liquid water, then such long cracks would be expected. To examine this exciting possibility more closely, NASA redirected the Galileo spacecraft during the third and fourth years of its mission to pass nearer to Europa and send back very high-resolution images of its intriguing surface (Figure 11.10, for example).

The Galileo images appear to confirm the existence of a global ocean. In many places the surface of Europa looks just as we would expect for a thick layer of ice that was broken up into giant icebergs and ice flows and then refrozen in place. When the ice breaks, water or slush from below may be able to seep up through the cracks and make the ridges and multiple-line features we observe. Many episodes of ice cracking, shifting, rotating, and refreezing are required to explain the complexity we see. The icy crust might vary in thickness from a kilometer or so up to 20 km.

FIGURE 11.9

Europa's Complex Geological History In this high-resolution view, the ice is wrinkled and crisscrossed by long ridges. Where these ridges intersect, we can see which ones are older and which younger; the younger ones cross over the older ones. To some observers, this system of ridges resembles a giant freeway system on Europa, but the ridges are much wider than our freeways and are a natural result of the flexing of the moon.

FIGURE 11.10

Detail of an Ice Ridge A very high-resolution Galileo image of one young double ridge. The area in this picture is only 17 km across, and the ridge itself is about 3 km wide and 300 m high. It appears to have formed when viscous icy material was forced up through a long straight crack in the crust.

Europa   Ganymede   Callisto

DLR, the German Aerospace Center; JPL/NASA

■ FIGURE 11.11

Comparing Three Icy Crusts   These Galileo images show the surfaces of Europa, Ganymede, and Callisto (from left to right) with the same resolution (detail). Note that the number of craters (and thus the age of the surface we see) increases as we go from Europa to Ganymede to Callisto. The Europa image is one of those where the system of cracks and ridges resembles a freeway system.

In some places the ice appears to be free-floating, in others it seems to be grounded against the rocky ocean floor. Confirmation that this liquid ocean is present today came from measurements of the small magnetic field induced by Europa's interactions with the magnetosphere of Jupiter. The "magnetic signature" of Europa is that of a liquid water ocean, not of ice or rock.

If Europa really has a large ocean of liquid water under its ice, then it is the only place in the solar system, other than Earth, with such an environment. To remain liquid, this ocean must be warmed by heat escaping from the interior of Europa. Hot (or at least warm) springs might be active there, analogous to those we have discovered in the deep oceans of the Earth. However, to maintain large quantities of liquid water, there must be substantial heating from Europa's interior—more than can be explained by the energy from radioactive rocks (the mechanism that heats the interiors of the terrestrial planets).

Such heat can be generated by the tidal heating mechanism we mentioned earlier, but the heating will vary at different times during Europa's orbit. Scientists still don't know how much tidal energy is produced inside Europa and whether this heating is steady enough to maintain a global ocean at all times. At some times the water may be in the form of an icy slush, depending on subtle changes in the orbit of Europa.

What makes the idea of an ocean with possible warm springs especially exciting is the discovery in *Earth's oceans* of large ecosystems of primitive life around subterranean hot springs. Such life derives all its energy from the mineral-laden water and thrives independent of the sunlight shining on the Earth's surface. Is it possible that similar ecosystems could exist today under the ice of Europa? Some people have suggested that Europa is the most likely place beyond the Earth to find life in the solar system. This possibility ensures that Europa will be a continuing target of planetary missions in the future.

11.2.4   Io, a Volcanic Satellite

Io, the innermost of Jupiter's Galilean satellites, is a close twin of our Moon, with nearly the same size and density. We might therefore expect it to have experienced a similar history. Its appearance, as photographed from space, tells us another story, however (Figure 11.12). Instead of being a dead, cratered world, Io turns out to have the highest level of volcanism in the solar system, greatly exceeding that of the Earth.

Io's active volcanism was discovered by the Voyager spacecraft. Eight volcanoes were actually seen erupting when Voyager 1 passed in March 1979, and six of these were still active four months later when Voyager 2 passed.

NASA/JPL/USGS

■ FIGURE 11.12

Two Sides of Io   A composite of Voyager images assembled by the U.S. Geologic Survey shows both sides of the volcanically active moon Io. The orange deposits are sulfur snow; the white is sulfur dioxide. (Carl Sagan once quipped that Io looks like it desperately needs a shot of penicillin.)

With the improved instruments carried by the Galileo spacecraft, more than 50 eruptions were found during the calendar year 1997 alone. Many of the eruptions produce graceful plumes that extend hundreds of kilometers out into space (Figure 11.13).

The Galileo data show that most of the volcanism on Io consists of hot silicate lava, like the volcanoes on Earth. Sometimes the hot lava encounters frozen deposits of sulfur and sulfur dioxide. When these icy deposits are suddenly heated, the result is great eruptive plumes far larger than any ejected from terrestrial volcanoes. As the rising plumes cool, the sulfur and sulfur dioxide recondense as solid particles, which fall back to the surface in colorful "snowfalls" that extend as much as a thousand kilometers from the vent. Planetary scientists, comparing Galileo images to earlier Voyager maps, have found at least a dozen areas on Io larger than the entire state of Connecticut that have been resurfaced by volcanic activity in the years between the two missions. Major new surface features were even seen to appear between Galileo orbits, as shown in Figure 11.14.

As the Galileo mission drew to a close, controllers were willing to take risks in getting close to Io. Approaching this satellite is always a dangerous maneuver because the belts of atomic particles trapped in the magnetic environment of Jupiter (see Chapter 10) are at their most intense near Io's orbit. Indeed, in its very first pass by Io, the spacecraft absorbed damaging radiation beyond its design levels. To

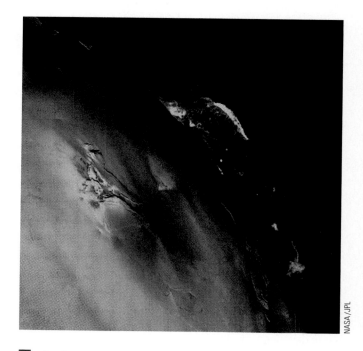

NASA/JPL

■ FIGURE 11.13

Volcanic Plume on Io   The large plume rising about 250 km above Io's surface (seen in white above the edge of the moon) erupted from the center of an enormous volcanic feature that takes up most of this image. It is named Pele, after the Hawaiian goddess of fire, and is about the size of the state of Alaska.

NASA/JPL/U. of Arizona

■ **FIGURE 11.14**

Volcanic Changes on Io These three images were taken of the same 1700-km-square region of Io in April 1997, September 1997, and July 1999. The dark volcanic center called Pillan Patera experienced a huge eruption, producing a dark deposit some 400 km across. In the right image, however, some of it is being covered by the reddish material from the volcano Pele (see Figure 11.13), indicating that both volcanic vents probably continue to be active. Also, a small unnamed volcano to the right of Pillan has erupted since 1997, and some of its dark deposit and a yellow ring around it are visible on the right image. The color range is exaggerated in these images.

keep the system working at all, controllers had to modify or disable various fault-protection software routines in the on-board computers. In spite of these difficulties, the space-craft achieved four successful Io flybys, obtaining photos and spectra of the surface with unprecedented resolution.

Our new maps of Io reveal more than 100 recently active volcanoes on its surface (Figure 11.15). For more that 50 of these, infrared measurements have been made of the hot lava. Some of these eruptions show temperatures as high as 1500°C, hotter than all but a few terrestrial volcanoes. Huge flows spread out from these vents, covering about 25 percent of the moon's total surface with still-warm lava. From these measurements, it seems clear that the bright surface colors that first attracted attention to Io are the result of a thin veneer of sulfur compounds. The underlying volcanism is driven by eruptions of molten silicates, just like on the Earth (Figure 11.16).

11.2.5 Tidal Heating

How can Io maintain such ongoing volcanism in spite of its small size? The answer, as we hinted earlier, lies in gravitational heating of the satellite by Jupiter. Io is about the same distance from Jupiter as our Moon is from the Earth. Yet Jupiter is more than 300 times more massive than Earth, causing forces that pull the satellite into an elongated shape, with a several-kilometer-high bulge extending toward Jupiter.

If Io always kept exactly the same face turned toward Jupiter, this bulge would not generate heat. However, Io's orbit is not exactly circular due to gravitational perturbations (tugs) from Europa and Ganymede. In its slightly ec-

centric orbit, Io twists back and forth with respect to Jupiter, at the same time moving nearer and farther from the planet on each revolution. The twisting and flexing heat Io, much as repeated flexing of a wire coat-hanger heats the wire.

After billions of years, this constant flexing and heating have taken their toll on Io, driving away water and carbon dioxide and other gases, until now sulfur and sulfur compounds are the most volatile materials remaining. The moon's inside is entirely melted, and the crust itself is constantly recycled by volcanic activity.

In moving inward toward Jupiter from Callisto to Io, we have encountered more and more evidence of geological activity and internal heating. Callisto has a surface more than 4 billion years old; Ganymede has been active as recently at the past billion years; Europa has an icy crust at most a few million years old; and parts of the visible surface of Io were deposited in the brief 16-year interval between the Voyager and Galileo missions. (Three of these surfaces are compared in Figure 11.11.) Just as the character of the planets in our solar system depends in large measure on their distance from the Sun (and on the amount of heat they receive), so it appears that distance from a giant planet like Jupiter can play a large role in the composition and evolution of its satellites (at least partly due to differences in internal heating by Jupiter's unrelenting tides).

11.2.6 Titan, a Satellite with Atmosphere

Titan, discovered in 1655 by the Dutch astronomer Christian Huygens, is the largest satellite of Saturn. It was the first satellite discovered after Galileo saw the four large moons

NASA/JPL/U. of Arizona

■ **FIGURE 11.15**

Giant Lava Flow on Io   This mosaic of Galileo images shows a region 500 by 200 km that contains the longest active lava flow yet discovered in the solar system. We see a patchwork quilt of dark lava interacting with the colored snows of sulfur compounds. The most recent lava eruptions are the darkest because they have not yet had a chance to be covered by the sulfur dioxide plumes the lava flows cause. It is not always clear where the liquid lava is coming up from Io's interior because the fresh lava can travel under older lava deposits and emerge a long distance—hundreds of kilometers—from its point of origin.

NASA/JPL

■ **FIGURE 11.16**

Lava Fountains on Io   Galileo captured a number of eruptions along the chain of huge volcanic calderas (or pits) on Io called Tvashtar Catena. In this false-color image, infrared and visible-light information have been combined. The orange-yellow areas at left are places where fresh, hot lava is erupting. The rounded ribbon of lava is about 60 km long.

NASA/JPL

■ **FIGURE 11.17**

Titan's Thick Atmosphere   To the Voyager cameras, Titan looked like a big fuzzy orange ball (see Figure 11.1). In this enhanced close-up view, we see the main orange haze layer that blocks our view of the surface and also the blue haze that lies higher in the moon's atmosphere.

of Jupiter. Titan has roughly the same diameter, mass, and density as Callisto or Ganymede. Presumably it also has a similar composition—about half ice and half rock, although we don't know whether this satellite has differentiated. What is most remarkable about Titan is the presence of a substantial atmosphere, discovered by Gerard Kuiper in 1944. He noticed that spectra of Titan showed absorption features indicating the presence of methane gas (Figure 11.17).

The 1980 Voyager flyby of Titan determined that the surface density of its atmosphere is four times greater than that on the Earth. The atmospheric pressure on this satellite is 1.6 bars, higher than that on any other moon and, remarkably, even higher than that of the terrestrial planets Mars and Earth. The atmospheric composition is primarily nitrogen, an important way in which Titan's atmosphere resembles Earth's.

A variety of additional compounds have been detected in Titan's upper atmosphere, including carbon monoxide (CO), hydrocarbons (compounds of hydrogen and carbon) such as methane (CH_4), ethane (C_2H_6), and propane (C_3H_8), and nitrogen compounds such as hydrogen cyanide (HCN), cyanogen (C_2N_2), and cyanoacetylene (HC_3N). Their presence indicates an active chemistry, in which sunlight interacts with atmospheric nitrogen and methane to create a rich mix of organic molecules. This is very much the way we believe organic compounds were formed on the Earth when it still had its original atmosphere. The discovery of HCN on Titan is particularly interesting, since this molecule is the starting point for the formation of some components of DNA, the fundamental genetic molecule essential to life on Earth.

Titan has multiple layers of clouds (Figure 11.18). The lowest are within the bottom 10 km of the atmosphere; these are condensation clouds probably composed of methane. Much higher is a dark reddish haze or smog, consisting of complex organic chemicals created by the energy

of sunlight falling on the molecules of Titan's atmosphere. Formed at an altitude of several hundred kilometers, this aerosol slowly settles downward to the surface, where it may have built up a deep layer of tarlike organic chemicals.

Much to the disappointment of the Voyager scientists, the spacecraft revealed that the reddish haze completely blankets Titan, hiding its surface from our view. Still the Voyager data allow us to make models of what it might be like to land on this intriguing world. For example, we know that Titan's surface temperature is about 90 K, held uniform by the blanketing atmosphere. At such a low temperature, methane plays the same role that water does on Earth: The gas is only a minor constituent of the atmosphere, but it can exist in gas, liquid, or solid form. Methane condenses to form clouds, and there may even be seas or lakes of liquid methane and ethane, as well as solid methane ice, on the surface.

Organic compounds are chemically stable in Titan's cold environment, unlike those on the warmer, oxidizing Earth. Titan's surface, therefore, probably records a chemical history that goes back billions of years. Many scientists believe that this satellite will provide more insights into the early history of Earth's atmosphere, and even into the origin of life, than any other object in the solar system. We hope to know more about the Titan environment when the Huygens probe carried by the Cassini spacecraft parachutes down to the surface early in 2005.

11.3 TRITON AND PLUTO

In this section we discuss two apparently similar objects: Neptune's satellite Triton and the planet Pluto. Seeking to understand the ways in which these two worlds are similar to each other is an example of the comparative approach to studying astronomy. Pluto is the only planet not visited by a spacecraft; therefore comparing it with Triton, which was studied by Voyager in 1989, is one way we can learn more about it.

11.3.1 Triton and Its Volcanoes

Triton (don't get it confused with Titan) has a diameter of 2720 km and a density of 2.1 g/cm^3, indicating that it's probably composed of 75 percent rock mixed with 25 percent water ice. Measurements indicate that Triton's surface has the coldest temperature of any of the worlds our robot representatives have visited. Because its reflectivity is so high (about 80 percent), Triton reflects most of the solar energy that falls on it, resulting in a surface temperature between 35 and 40 K.

The surface material of Triton is made of frozen water, nitrogen, methane, and carbon monoxide. Methane and nitrogen exist as gas in most of the solar system, but they are frozen at Triton's temperatures. Only a small quantity of ni-

■ FIGURE 11.18

The Structure of Titan's Atmosphere Titan has the most Earth-like atmosphere in the solar system, although it is much colder than our planet.

text rotated on right side of image readsNASA/JPL/Image processing at USGS

■ **FIGURE 11.19**

Neptune's Moon Triton   This mosaic of Voyager images of Triton shows a wide range of surface features. The pinkish area at the bottom is Triton's large southern polar cap. The south pole of Triton faces the Sun here, and the slight heating effect is driving some of the material northward, where it is colder.

trogen vapor persists to form an atmosphere. Although the surface pressure of this atmosphere is only 16 millionths of a bar, this is sufficient to support thin haze or cloud layers.

Triton's surface, like that of many other satellites in the outer solar system, reveals a long history of geological evolution (Figure 11.19). Although some impact craters are found, many regions have been flooded fairly recently by the local version of "lava" (perhaps water or water–ammonia mixtures). There are also mysterious regions of jumbled or mountainous terrain.

The Voyager flyby of Triton took place at a time when the satellite's southern pole was tipped toward the Sun, allowing this part of the surface to enjoy a period of relative warmth. (Remember that "warm" on Triton is still outrageously colder than anything we experience on Earth.) A polar cap covers much of Triton's southern hemisphere, apparently evaporating along the northern edge. This polar cap may consist of frozen nitrogen that was deposited during the previous winter. Remarkably, the Voyager images showed that the evaporation of this polar cap generates geysers or volcanic plumes of nitrogen gas. (Fountains of such gas rose about 10 km high, visible in the thin atmosphere because dust from the surface rose with them and

colored them dark.) These plumes differ from the volcanic plumes of Io in their composition and also in that they derive their energy from sunlight warming the surface rather than from internal heat.

11.3.2   Discovery of Pluto

Now we turn to Pluto, a small planet that resembles Triton in many ways and is quite different from the giant planets in the outer solar system. Pluto was discovered through a careful, systematic search, unlike Neptune, whose position was calculated from gravitational theory. Nevertheless, the history of the search for Pluto began with indications that Uranus and Neptune had slight departures from their predicted orbits (you may want to review the last section of Chapter 2 for more on what such departures mean). Early in the 20th century several astronomers—most notably Percival Lowell, then at the peak of his fame as an advocate of intelligent life on Mars (see Chapter 9)—became interested in this problem.

At the time Lowell made his calculations, Neptune had moved such a short distance since its discovery that its orbit could not be used effectively to search for perturbations by an unknown ninth planet. Therefore Lowell and his contemporaries based their calculations primarily on the tiny remaining irregularities in the motion of Uranus. Lowell's computations indicated two possible locations for a perturbing planet; the more likely of the two was in the constellation of Gemini. He predicted a mass for the planet intermediate between those of the Earth and Neptune (his calculations gave about 6.6 Earth masses). Other astronomers, however, obtained other solutions from the tiny orbital irregularities, even including one model that indicated two planets beyond Neptune.

At his Arizona observatory, Lowell searched without success for the unknown planet from 1906 until his death in 1916. Subsequently, his brother donated to the observatory a 33-cm, wide-field photographic telescope that could record a 12° x 14° area of the sky on a single photograph. The new camera went into operation in 1929.

In February 1930, a young observing assistant named Clyde Tombaugh (see the Voyagers in Astronomy box), comparing photographs he made on January 23 and 29 of that year, found an object whose motion appeared to be about right for a planet far beyond the orbit of Neptune (Figure 11.20). It was within 6° of the position Lowell had predicted for the unknown planet. The new planet was named for Pluto, the Roman god of the underworld, who dwelt in remote darkness, just like the new planet. The choice of this name, among hundreds suggested, was helped by the fact that the first two letters were Percival Lowell's initials.

Although at the time the discovery of Pluto appeared to be a vindication of gravitational theory similar to the triumph of Adams and Leverrier in predicting the position of Neptune, we now know that Lowell's calculations were wrong. When the mass of Pluto was finally measured, it was found to be much less than that of the Moon. Pluto could not possibly have exerted any measurable pull on either Uranus or Neptune. Astronomers are now convinced that the reported small anomalies in the motions of Uranus and Neptune are not, and never were, real.

11.3.3   Pluto's Motion and Satellite

Pluto's orbit has the highest inclination to the ecliptic (17°) of any planet, meaning that the plane of its orbit is significantly tilted relative to the plane in which the other planets move. Pluto also has the largest eccentricity (0.248) in our solar system, so that its orbit deviates quite a bit from a perfect circle. Its average distance from the Sun is 40 AU, or 5.9 billion km, but its perihelion distance is less than 4.5 billion km, within the orbit of Neptune. Even though the orbits of these two planets cross, there is no danger of collision because of the high inclination of Pluto's orbit. (By the way, many of the giant planets found orbiting other stars have an even greater orbital eccentricity than Pluto.)

Pluto completes its orbital revolution in a period of 248.6 years; since its discovery in 1930, it has traversed less than one quarter of its long path around the Sun. But almost 50 years after that discovery, astronomers found that it is not alone in its distant travels. In 1978, astronomers found that Pluto has a satellite, which was named Charon after the boatman who ferried the dead into the realm of Pluto in mythology. The exact nature of Charon's orbit was not confirmed until 1985, when the system had turned enough that the satellite and the planet began to go behind each other as seen from the Earth. These observations showed that the satellite was in a retrograde orbit and had a diameter of about 1200 km, more than half the size of Pluto itself. This makes Charon the satellite whose size is the largest fraction of its parent planet (Figure 11.21). Seen from Pluto, Charon would be as large as eight full moons on Earth.

Furthermore, Pluto orbits "on its side" like Uranus, and Charon circles in its equatorial plane around its rotation axis. The gravitational interaction between the two worlds has locked them in the tightest of embraces. Not only does Charon take the same time to rotate and revolve, thus keeping the same face toward Pluto, but this period—the month on Pluto—is the same length as Pluto's day. This means that if you were on the Charon-facing side of Pluto,

■ **FIGURE 11.20**

Pluto's Motion Portions of the two photographs by which Clyde Tombaugh discovered Pluto in 1930. The left one was taken on January 23 and the right on January 29. Note that Pluto, indicated by an arrow, has moved among the stars during those six nights. If we hadn't put an arrow next to it, you probably would never have spotted the dot that moved.

Lowell Observatory

Clyde Tombaugh: From the Farm to Fame

Clyde Tombaugh discovered Pluto when he was 24 years old, and his position as staff assistant at the Lowell Observatory was his first paying job. Tombaugh had been born on a farm in Illinois, but when he was 16 his family moved to Kansas. There, with his uncle's encouragement, he observed the sky through a telescope the family had ordered from the Sears catalog. Tombaugh later constructed a larger telescope on his own and devoted his nights (when he wasn't too tired from farm work) to making detailed sketches of the planets.

In 1928, after a hailstorm ruined the crop, Tombaugh decided he needed a job to help support his family. Although he had only a high school education, he thought of becoming a telescope builder. He sent his planet sketches to the Lowell Observatory, seeking advice about whether such a career choice was realistic. By a wonderful twist of fate, his query arrived just when the Lowell astronomers realized that a renewed search for a ninth planet would require a very patient and dedicated observer.

The large photographic plates (pieces of glass with photographic emulsion on them) that Tombaugh

Clyde Tombaugh at the Lowell Observatory at the time of his discovery of Pluto in 1930.

was hired to take at night and search during the day contained an average of about 160,000 star images each. How to find Pluto among them? The technique involved taking two photographs about a week apart. During that week, a planet would move a tiny bit, while the stars remained in the same place relative to each other. A new instrument called a "blink comparator" could quickly alternate the two images in an eyepiece. The stars,

being in the same position on the two plates, would not appear to change as the two images were "blinked." But a moving object would appear to wiggle back and forth as the plates were alternated.

After examining more than 2 million stars (and many false alarms), Tombaugh found his planet on February 18, 1930. The astronomers at the observatory checked his results carefully, and the find was announced on March 13, the 149th anniversary of the discovery of Uranus. Congratulations and requests for interviews poured in from around the world. Visitors descended on the observatory by the scores, wanting to see the place where the first new planet in almost a century had been discovered, as well as the person who had discovered it.

In 1932 Tombaugh took leave from Lowell, where he had continued to search and blink, to get a college degree. Eventually he received a master's degree in astronomy and taught navigation for the Navy during World War II. In 1955, after working to develop a rocket-tracking telescope, he became a professor at New Mexico State University, where he helped found the astronomy department. He died in 1997.

FIGURE 11.21

Pluto and Charon An image of Pluto and its satellite Charon, taken with the Faint Object Camera on the Hubble Space Telescope. The clarity of the image allowed astronomers to measure the diameters of the two worlds to an accuracy of 1 percent.

you would see the moon hover in the same place all the time, while from Pluto's opposite hemisphere Charon would never be visible.

11.3.4 The Nature of Pluto

Pluto has not been visited by spacecraft, and it is so faint that the world's largest telescopes are required to study it. The diameter of Pluto is 2190 km, only 60 percent that of the Moon. From the diameter and mass, we find a density of 2.1 g/cm^3, suggesting that Pluto is a mixture of rocky materials and water ice in about the same proportions as in Triton.

Pluto's surface is highly reflective, and its spectrum demonstrates the presence of frozen methane, carbon monoxide, and nitrogen. The surface temperature ranges from about 50 K when Pluto is farthest from the Sun to 60 K when it's closest. Even this small difference is enough to cause a partial sublimation of the methane and nitrogen ice to generate an atmosphere when Pluto is close to the Sun. In fact, for the past few years Pluto has been in its warmest period, near perihelion, and its atmosphere is accordingly near maximum size and density. Observations of a distant star seen through that thin atmosphere suggest that the surface pressure is about a ten-thousandth of the Earth's. Because Pluto is a few degrees warmer than Triton, its atmospheric pressure is about ten times greater.

In February 1999, Pluto crossed the orbit of Neptune and became the ninth planet once more (it had been eighth from the Sun since 1979). Now it will slowly swing farther and farther from the Sun, and more and more of its atmosphere will slowly freeze out.

We have a long way to go to understand how Pluto arrived at its unusual place in the outer solar system. In some ways Pluto is like the odd cousin that everyone hopes will not show up at the next family reunion. Neither its path around the Sun nor its size resembles the jovian planets. Pluto is much more similar to a moon such as Triton than to a planet such as Saturn. We now know that Pluto is the largest example of a whole family of distant asteroid- or cometlike objects in what is called the Kuiper belt, which we will discuss in Chapter 12.

11.4 PLANETARY RINGS

All four of the giant planets have rings, with each ring system consisting of billions of small particles or moonlets orbiting close to their planet. Each of these rings displays a complicated structure that seems related to interactions between the ring particles and the larger satellites. However, the four ring systems are very different from each other in mass, structure, and composition (Table 11.2).

Saturn's large ring system is made up of icy particles spread out into several vast, flat rings containing a great deal of fine structure. The Uranus and Neptune ring systems, on the other hand, are nearly the reverse of Saturn's; they consist of dark particles confined to a few narrow rings

TABLE 11.2 *Properties of Ring Systems*

Planet	Outer Radius (km)	Outer Radius (R_{planet})	Mass (kg)	Reflectivity (%)
Jupiter	128,000	1.8	10^{10}(?)	?
Saturn	140,000	2.3	10^{19}	60
Uranus	51,000	2.2	10^{14}	5
Neptune	63,000	2.5	10^{12}	5

with broad empty gaps in between. Jupiter's ring and at least one of Saturn's are merely transient dust bands, constantly renewed by erosion of dust grains from small satellites. (Figure 11.22 is a diagram comparing the four ring systems.) In this section we focus on the two most massive ring systems, those of Saturn and Uranus.

11.4.1 What Causes Rings?

A ring is a collection of vast numbers of particles, each like a tiny moon obeying Kepler's laws as it follows its own orbit around the planet. Thus the inner particles revolve faster than those farther out, and the ring as a whole does not rotate as a solid body. In fact, it is better not to think of a ring rotating at all, but rather to consider the revolution of its individual moonlets.

If the ring particles were widely spaced, they would move independently, like separate small satellites. However, in the main rings of Saturn and Uranus the particles are close enough to each other to exert mutual gravitational influence, and occasionally even to rub together or bounce off each other in low-speed collisions. Because of these interactions, we see phenomena such as waves that move across the rings the way water waves move over the surface of the ocean.

There are two basic theories of how such rings come to be. First is the breakup hypothesis, which suggests that the rings are the remains of a shattered satellite. A passing comet or asteroid might have collided with the satellite, breaking it into pieces. Tidal forces then inhibited the fragments from coming together again, and they dispersed into a disk.

The second hypothesis, which takes the reverse perspective, suggests that the rings are made of particles that were unable to come together to form a satellite in the first place. In either theory, the gravity of the planet plays an important role. Close to the planet, tidal forces can tear bodies apart or inhibit loose particles from coming together. The rings of both Saturn and Uranus lie within the region in which large satellites are probably not stable against such forces (the edge of this region is sometimes called the *Roche limit*, after the French mathematician who first worked out the theory). We do not know which explanation holds for the rings, although many scientists have concluded that at least a few of the rings are relatively young and must therefore be the result of breakup.

Four Ring Systems A diagram showing the locations of the ring systems of the four giant planets. The left axis represents the planet's surface. The dotted vertical line is the limit inside which gravitational forces can break up satellites. Notice that each planet's system is drawn *to a different scale*, so that this stability limit lines up for all four of them. The black dots are the inner satellites of each planet on the same scale as its rings. Notice that only really small bodies survive inside the stability limit.

11.4.2 Rings of Saturn

The rings of Saturn are one of the most beautiful sights in the solar system (Figures 10.10 and 11.23). From outer to inner, the three brightest rings are labeled with the extremely romantic names of A, B, and C Rings. Table 11.3 gives the dimensions of the rings in both kilometers and units of the radius of Saturn, R_s. The B Ring is the brightest and has the most closely packed particles, whereas the A and C Rings are translucent. The total mass of the B Ring, which is probably close to the mass of the entire ring system, is about equal to that of an icy satellite 250 km in diameter (thus it could have originated in the breakup of such a satellite). Between the A and B Rings is a wide gap named the Cassini Division after Gian Domenico Cassini, who first glimpsed it in 1675.

The rings of Saturn are very broad and very thin. The width of the main rings is 70,000 km, yet their thickness is only about 20 m. If we made a scale model of the rings out of paper the thickness of the pages in this book, we would have to make them 1 km across—about eight city blocks. On this scale, Saturn itself would loom as high as an 80-story building.

The ring particles are composed primarily of water ice, and they range from grains the size of sand up to house-sized boulders. An insider's view of the rings would probably resemble a bright cloud of floating snowflakes and hailstones, with a few snowballs and larger objects, many of these loose aggregates of smaller particles (Figure 11.24).

In addition to the broad A, B, and C Rings, Saturn has a handful of very narrow rings no more than 100 km wide. The most substantial of these, which lies just outside the A Ring, is called the F Ring; its surprising appearance is discussed below. In general, Saturn's narrow rings resemble the rings of Uranus and Neptune.

TABLE 11.3 *Rings of Saturn*

Ring Name	Outer Edge (R_s)	Outer Edge (km)	Width (km)
F	2.324	140,180	90
A	2.267	136,780	14,600
Cassini Division	2.025	122,170	4,590
B	1.949	117,580	25,580
C	1.525	92,000	17,490

NASA/JPL

■ **FIGURE 11.23**

Saturn's Rings from Underneath Since you are looking up at sunlight filtering through the rings in this Voyager image, the brightest parts of the ring system are the gaps and the least dense regions. The denser B Ring is not transmitting much light.

■ **FIGURE 11.24**

Ring Pieces An artist's idealized version of the sizes of typical particles inside the A Ring of Saturn. The ball at the center is roughly the size of an adult's head. In the real rings, the particles are not spherical but are likely to have irregular shapes.

11.4.3   Rings of Uranus and Neptune

The rings of Uranus are narrow and black, making them almost invisible from the Earth. The nine main rings were discovered in 1977 from observations made of a star as Uranus passed in front of it. We call the passage of one astronomical object in front of another an *occultation*. During the 1977 occultation, astronomers expected the star's light to disappear as the planet moved across it. But in addition the star dimmed briefly several times before Uranus reached it, as each narrow ring passed between the star and the telescope. Thus the rings were mapped out in detail even though they could not be seen or photographed directly, like counting the number of cars in a train at night by watching the blinking of a light as the cars successively pass in front of it. When Voyager approached Uranus in 1986, it was able to study the rings at close range; the spacecraft also photographed two new rings (Figure 11.25).

The outermost and most massive of the rings of Uranus is called the Epsilon Ring. It is only about 100 km wide and probably no more than 100 m thick (similar to the F Ring of Saturn). The Epsilon Ring encircles Uranus at a distance of 51,000 km, about twice the radius of Uranus. This ring probably contains as much mass as all of Uranus' other ten rings combined; most of them are narrow ribbons less than 10 km wide—just the reverse of the broad rings of Saturn.

The individual particles in the uranian rings are nearly as black as lumps of coal. While astronomers do not understand the composition of this material in detail, it seems to consist in large part of black carbon and hydrocarbon compounds. Organic material of this sort is rather common in the outer solar system. Many of the asteroids and comets (discussed in Chapter 12) also are composed of dark, tarlike materials. In the case of Uranus, its ten small inner satel-

■ **FIGURE 11.25**

The Narrow Rings of Uranus The Voyager team had to expose this image for a long time to get a glimpse of these dark rings; you can see the grainy structure of "noise" in the electronics of the camera in the picture background.

lites have a similar composition, suggesting that one or more satellites might have broken up to make the rings.

The rings of Neptune are generally similar to those of Uranus but even more tenuous (Figure 11.26). There are only four of them, and the particles are not uniformly distributed along their lengths. Because these rings are so difficult to investigate from the Earth, it will probably be a long time before we understand them very well.

11.4.4   Interactions Between Rings and Satellites

Much of our fascination with planetary rings is a result of their intricate structures, most of which owe their existence to the gravitational effect of satellites. Without satellites the rings would be flat and featureless. Indeed, it is becoming clear that without satellites there would probably be no rings at all because, left to themselves, thin disks of matter gradually spread and dissipate.

Most of the gaps in Saturn's rings, and also the location of the outer edge of the A Ring, result from gravitational resonances with small inner satellites. A **resonance** takes place

■ FIGURE 11.26

Neptune's Rings A long exposure of the rings of Neptune, as photographed by Voyager in 1989. Note the two denser regions of the outer ring.

■ FIGURE 11.27

Saturn's F Ring and Its Shepherds This Voyager image shows the narrow, complex F Ring of Saturn, with its two shepherd satellites Pandora and Prometheus.

when two objects have orbital periods that are exact ratios of each other, such as one-to-two or one-to-three. For example, any particle in the gap at the inner side of the Cassini Division of Saturn's rings would have a period equal to one-half that of Saturn's satellite Mimas. Such a particle would be nearest Mimas in the same part of its orbit every second revolution. The repeated gravitational tugs of Mimas, acting always in the same direction, would perturb it, forcing it into a new orbit outside the gap and therefore no longer representing a resonance. In this way, the Cassini Division became depleted of ring material over long periods of time.

One of the most interesting rings of Saturn is the narrow F Ring, which contains several apparent ringlets within its 90-km width. In places the F Ring breaks up into two or three parallel strands that sometimes show bends or kinks. Most of the rings of Uranus and Neptune are also narrow ribbons like the F Ring of Saturn. Clearly, the gravity of some objects must be keeping the particles in these rings from spreading out.

The best hypothesis is that the rings are controlled gravitationally by small satellites orbiting very close to them. This certainly seems to be the case for Saturn's F Ring: Voyager showed that it is bounded by the orbits of two satellites, now called Pandora and Prometheus (Figure 11.27). These two small objects (each about 100 km in diameter) are referred to as *shepherd satellites*, since their gravitation serves to "shepherd" the ring particles and keep them confined to a narrow ribbon. A similar situation applies to the Epsilon Ring of Uranus, which is shepherded by the satellites Cordelia and Ophelia. These two shepherds, each about 50 km in diameter, orbit about 2000 km inside and outside the ring.

Theoretical calculations suggest that the other narrow rings in the uranian and neptunian systems should also be controlled by shepherd satellites, but none has been located. The calculated diameter for such shepherds—about 10 km—was just at the limit of detectability for the Voyager cameras, so it is impossible to say whether they are present or not. (Given all the narrow rings we see, many scientists still hope to find another, more satisfactory mechanism for keeping them confined.)

Just recently, astronomers have found a relatively narrow ring of dust around another star (called HR 4796A) and have applied the same kind of analysis to suggest that there must be shepherding *planets* on either side of the ring to keep it so narrow. (At the distance of the star, such planets cannot be seen with our present telescopes—the same situation we face for the smaller shepherd moons around the outer planets.) This is a wonderful example of how techniques developed in one branch of astronomy often find application in another.

The main rings of Saturn include several narrow gaps that are the result of satellites, which clear lanes in a manner similar to that of the shepherd satellites. As each small satellite moves through its gap, it produces waves in the surrounding ring material like the wake left by a moving ship. One of these satellites, called Pan, was discovered in 1991 from its wake; calculations based on the waves it produced pinpointed its location, and when the appropriate ten-year-old Voyager picture was examined, there it was!

NOTE: Many of the sites about planets and planetary missions listed for Chapters 6 and 10 also include good information about the satellites of the planets. For example, the Nine Planets site has separate sections for many of the main moons in the solar system.

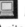 *Galileo Mission Moons Page:*
www.jpl.nasa.gov/moons/moons.html
A fine summary of what the Galileo spacecraft found out about the Galilean moons, with some of the great images from the mission.

 General Pluto sites:

- *The Nine Planets Site:*
  seds.lpl.arizona.edu/nineplanets/nineplanets/pluto.html
- *Views of the Solar System Site:*
  www.solarviews.com/eng/pluto.htm
- *Planetary Sciences Site:*
  nssdc.gsfc.nasa.gov/planetary/planets/plutopage.html

 The Clyde Tombaugh Web Site:
www.klx.com/clyde/index.html
The discoverer of Pluto passed away in January 1997 at age 90. Jeff Faust at MIT has brought together articles on Tombaugh and his work, plus information about Pluto.

Pluto Home Page: dosxx.colorado.edu/plutohome.html
A site from astronomer Fran Bagenal at the University of Colorado with good images, diagrams (with captions), brief information, and links about Pluto.

Pluto Portal: www.plutoportal.net
Another good site with information about Pluto.

New Horizons Mission: pluto.jhuapl.edu
A site describing a planned mission to Pluto.

The Planetary Rings Node: ringside.arc.nasa.gov/
Mark Showalter at NASA's Ames Research Center maintains this somewhat technical site devoted to the rings. It includes lots of images, movies, links, and information.

SUMMARY

11.1 The four jovian planets are accompanied by impressive systems of satellites and rings. More than 100 satellites have been discovered in the outer solar system. Most of these satellites are composed in part of water ice, with varying amounts of rock and dark primitive material mixed in. Of the four ring systems, Saturn's is the largest; Uranus and Neptune have narrow rings of dark material, and Jupiter has a tenuous ring of dust.

11.2 The largest satellites are Ganymede, Callisto, and Titan, all low-density objects that are composed of more than half water ice. Callisto has an ancient cratered surface, while Ganymede shows evidence of extensive tectonic and volcanic activity, persisting until perhaps a billion years ago. Voyager was not able to see Titan's surface because it is unique among jovian-planet moons in having a thick atmosphere (surface pressure of 1.6 bar) with an opaque haze layer. Io and Europa are denser and smaller, each about the size of our Moon. Europa is the only other object in the solar system thought to have a global ocean of liquid water (under an ice crust). Io is the most volcanically active object in the solar system; its eruptions are powered by tidal interactions with Jupiter.

11.3 Triton, Neptune's retrograde satellite, and Pluto resemble each other in size, composition, and temperature. Triton has a very thin atmosphere, and we can see eruptions of nitrogen when surface layers of its extensive polar cap evaporate while facing the Sun. The planet Pluto, discovered by Clyde Tombaugh in 1930, has never been visited by spacecraft. Observations indicate the presence of frozen methane, carbon monoxide, and nitrogen on its surface, and a thin (perhaps temporary) atmosphere. Pluto's moon Charon has a diameter more than half of Pluto's, and its and Pluto's motions are gravitationally locked.

11.4 Rings are composed of vast numbers of individual particles (ranging in size) orbiting so close to a planet that its gravitational forces could have broken larger pieces apart or kept pieces from gathering together. Saturn's rings are broad, flat, and nearly continuous, except for a handful of gaps. The particles are mostly water ice, with typical dimensions of a few centimeters. The rings of Uranus are narrow ribbons separated by wide gaps and contain much less mass. They were discovered in 1977 during a stellar occultation. Neptune's rings are similar but contain even less material. Much of the complex structure of the rings is due to waves and **resonances** induced by satellites within the rings or orbiting outside them. Voyager found that several thin rings were accompanied by shepherd moons that kept the ring material confined.

INTER-ACTIVITY

A Imagine it's the distant future and humans can now travel easily among the planets. Your group is a travel agency, with the task of designing a really challenging tour of the Galilean satellites for a group of sports enthusiasts. What kind of activities are possible on each world? How would rock climbing on Ganymede, for example, differ from rock climbing on Earth? (If you design an activity for Io, you had better bring along very strong radiation shielding. Why?)

B In the same spirit as activity A, have your agency design a tour that includes the seven most spectacular sights among any and all of the satellites or rings covered in this chapter. What are the "not to be missed" destinations that future tourists will want to visit and why? Which of the sights you pick are going to be spectacular if you are on the moon's surface or inside the ring, and which would look interesting only from far away in space?

C In this chapter we could cover only a few of the dozens of satellites in the outer solar system. Using the Web (see the Surfing the Web sections in this and the previous chapters) or your college library, organize your group into a research team and find out more about one of the moons we did not cover in any detail. Our favorites include Uranus' Miranda with its jigsaw puzzle surface, Saturn's Mimas with a "knockout" crater called Herschel, and Saturn's Iapetus whose two hemispheres differ significantly. Prepare a report to attract tourists to the world you selected.

D In a novel entitled *2010*, science fiction writer Arthur C. Clarke, inspired by the information coming back from the Voyager spacecraft, had fun proposing a lifeform under the ice of Europa that was evolving toward intelligence. Suppose future missions do indeed find some sort of life (not necessarily intelligent but definitely alive) under the ice of Europa—life that evolved completely independently from life on Earth. Have your group discuss what effect such a discovery would have on humanity's view of itself. What should be our attitude toward such a life-form? Do we have an obligation to guard it against contamination by our microbes and viruses? Or, to take an extreme position, should we wipe it out before it becomes competitive with Earth life or contaminates our explorers with microorganisms we are not prepared to deal with? Who should be in charge of making such decisions?

E Work together to make a list of all the reasons it is hard to send a mission to Pluto. How would you design a mission that will get there despite your list of difficulties?

F Your group has been asked by NASA to come up with one or more missions to learn about Europa. Review what we know about this satellite so far and then design a robotic mission that would answer some of the questions we have. You can assume that budget is not a factor, but your instruments have to be realistic. (Bear in mind that Europa is cold and far from the Sun.)

G Imagine your group is the first landing party on Pluto (let's hope you remembered to bring long underwear!). You land in a place where Charon is visible in the sky and you observe Charon for one Earth week. Describe what Charon will look like during that week. Now you move your camp to the opposite hemisphere of Pluto. What will Charon look like there during the course of a week?

REVIEW QUESTIONS

1. What are the satellites of the outer planets made of, and why is their composition different from that of the Moon?
2. Compare the geology of Callisto, Ganymede, and Titan. Why does Titan have an atmosphere while the other two large satellites do not?
3. What is the evidence for a liquid water ocean on Europa, and why is this interesting to scientists searching for extraterrestrial life?
4. Explain the energy source that powers the volcanoes of Io.
5. Compare the properties of Titan's atmosphere with those of the Earth's atmosphere.
6. How was Pluto discovered? Why did it take so long to find it?
7. How are Triton and Pluto similar?
8. Describe and compare the rings of Saturn and Uranus, including their possible origins.

9. Why do you think the outer planets have such extensive systems of rings and satellites, while the inner planets do not?

10. Which would have the longer orbital period: a satellite 1 million km from the center of Jupiter, or a satellite 1 million km from the center of Earth? Why?

11. Ganymede and Callisto were the first icy objects to be studied from a geological point of view. Summarize the main differences between their geology and that of the rocky terrestrial planets.

12. Compare the properties of the volcanoes on Io with those of terrestrial volcanoes. Give at least two similarities and two differences.

13. Would you expect to find more impact craters on Io or Callisto? Why?

14. Why is it unlikely that humans will be traveling to Io? (*Hint:* Review the information about Jupiter's magnetosphere in Chapter 10.)

15. Where did the nitrogen in Titan's atmosphere come from? Compare its origin with that of the nitrogen in our atmosphere.

16. Do you think there are many impact craters on the surface of Titan? Why or why not?

17. Why do you suppose the rings of Saturn are made of bright particles, whereas the particles in the rings of Uranus and Neptune are black?

18. Suppose you miraculously removed all of Saturn's satellites. What would happen to its rings?

FIGURING FOR YOURSELF

Using the occultations of stars by the rings of Uranus, astronomers have been able to measure details in the ring structure to a resolution of 10 km. This is a much higher resolution than can be obtained in a conventional photo of the rings. Let's figure out what angular resolution (in arcsec) a space telescope in Earth orbit would have to achieve to obtain equal resolution. To solve this problem, we use the "small-angle formula" to relate angular and linear diameter in the sky. For angles in the sky that are small, the formula is usually written as

$$\frac{\text{angular diameter}}{206{,}265} = \frac{\text{linear diameter}}{\text{distance}}$$

where angular diameter is expressed in arcsec. The distance of Saturn near opposition is about 9 AU = 1.4×10^9 km. Substituting in the above formula and solving for the angular resolution, we get $(206{,}265 \times 10)/1.4 \times 10^9$, which is about 10^{-3} arcsec, or a milliarcsec. This is not possible for our telescopes to achieve. For comparison, the best resolution from either the HST or ground-based telescopes is about 0.1 arcsec, or 100 times worse than what we would need. This is why such occultation measurements are so useful for astronomers.

19. How close to Uranus would a spacecraft have to get to obtain the same resolution as in our example above with a camera that has an angular resolution of 2 arcsec?

20. How close to Saturn would a spacecraft have to be to make out detail in its rings as small as 20 km, if its camera has an angular resolution of 5 arcsec?

21. If the HST has an angular resolution of 0.1 arcsec and Pluto is at its closest distance to the Sun, what is the smallest detail the telescope could make out on its surface? Does this tell you why astronomers really hope NASA or some other space agency will send a spacecraft to the ninth planet?

22. Saturn's A, B, and C Rings extend about 75,000 to 137,000 km from the center of the planet. Use Kepler's third law to calculate the difference between how long the inner edge and the outer edge of the three-ring system take to revolve about the planet.

23. Use the information in Appendix 8 to calculate what you would weigh on Titan, Io, and Uranus' satellite Miranda.

SUGGESTIONS FOR FURTHER READING

On Satellites

Carroll, M. "Europa: Distant Ocean, Hidden Life?" in *Sky & Telescope*, Dec. 1997, p. 50.

Elliot, J. "The Warming Wisps of Triton" in *Sky & Telescope*, Feb. 1999, p. 42.

Harland, D. M. *Jupiter Odyssey: The Story of NASA's Galileo Mission;* 2000, Springer-Praxis. Good mission overview.

Johnson, T. "The Galileo Mission to Jupiter and Its Moons" in *Scientific American,* Feb. 2000, p. 40. Ten-page review of mission results.

Johnson, T., et al. "The Moons of Uranus" in *Scientific American,* Apr. 1987.

Lorenz, R. and Mitton, J. *Lifting Titan's Veil: Exploring the Giant Moon of Saturn.* 2002, Cambridge U. Press.

Mackenzie, D. "Is There Life Under the Ice?" in *Astronomy,* Aug. 2001, p. 32. On future exploration of Europa.

Rothery, D. *Satellites of the Outer Planets: Worlds in Their Own Right.* 1999, Oxford U. Press.

Spencer, J. "Galileo's Closest Look at Io" in *Sky & Telescope,* May 2001, p. 40.

Talcott, R. "The Violent Volcanoes of Io" in *Astronomy,* May 1993, p. 41.

On Pluto

Binzel, R. "Pluto" in *Scientific American,* June 1990.

Burnham, R. "At the Edge of Night: Pluto and Charon" in *Astronomy,* Jan. 1994, p. 41.

Harrington, R. and Harrington, B. "The Discovery of Pluto's Moon" in *Mercury,* Jan./Feb. 1979, p. 1.

Hoyt, W. *Planets X and Pluto.* 1980, U. of Arizona Press. Superb history of the search for Pluto and other outer planets.

Levy, D. *Clyde Tombaugh: Discoverer of Pluto.* 1991, U. of Arizona Press.

Stern, S. "Journey to the Farthest Planet" in *Scientific American,* May 2002, p. 56. On possible missions to Pluto and the Kuiper belt.

Stern, S. and Mitton, J. *Pluto and Charon.* 1998, Wiley.

Tombaugh, C. "The Discovery of Pluto" in *Mercury,* May/June 1986, p. 66, and July/Aug. 1986, p. 98.

On Rings

Burns, J., et al. "Bejeweled Worlds" in *Scientific American,* Feb. 2002, p. 64. On rings throughout the solar system.

Elliot, J., et al. "Discovering the Rings of Uranus" in *Sky & Telescope,* June 1977, p. 412.

Esposito, L. "The Changing Shape of Planetary Rings" in *Astronomy,* Sept. 1987, p. 6.

Sobel, D. "Secrets of the Rings" in *Discover,* Apr. 1994, p. 86. Discusses the outer planet ring systems.

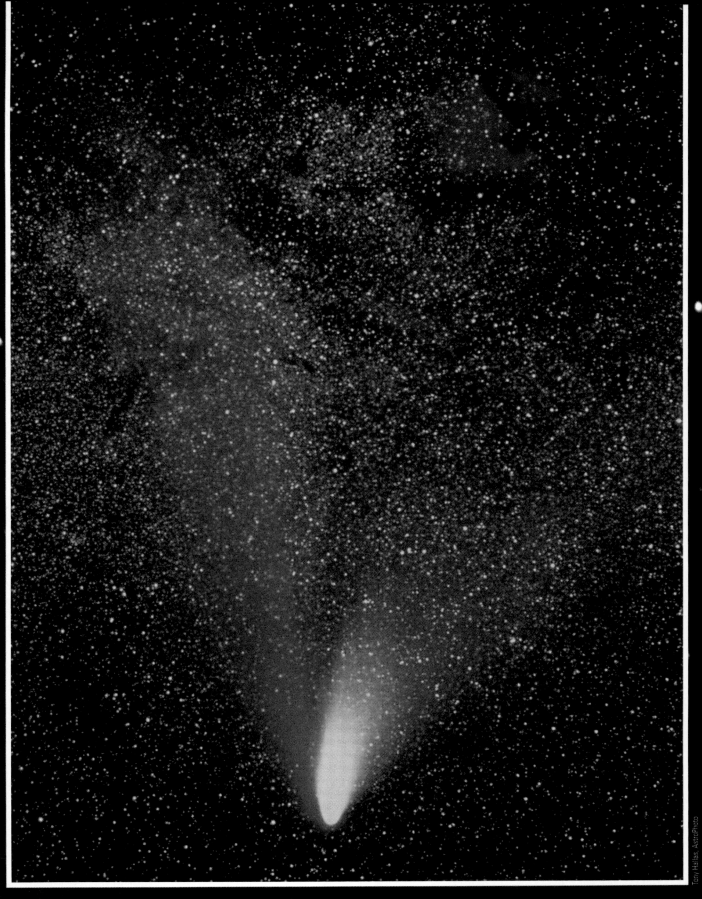

Comet Hale–Bopp was one of the most attractive and easily visible comets of the 20th century. It is shown here as it appeared in the sky on March 8, 1997, near the much more distant North America Nebula (the reddish glowing cloud of gas). You can see the comet's long blue ion tail and the shorter white dust tail.

Comets and Asteroids: Debris of the Solar System

The day will yet come when posterity will be amazed that we remained ignorant of things that will to them seem so plain. . . . Men will some-day be able to demon-strate in what regions comets have their paths [and] what is their size and constitution. Let us be satisfied with what we have discovered, and leave a little truth for our descendants to find out.

Lucius Annaeus Seneca, in *Natural Questions* (first century)

THINKING AHEAD

Hundreds of smaller members of the solar system, called asteroids and comets, are known to have crossed the Earth's orbit in the past, and many others will do so in the centuries ahead. What could we do if we knew a few years in advance that one of these bodies (with significant mass) would hit the Earth?

To understand the early history of life on Earth, scientists need to study ancient fossils. To reconstruct the early history of the solar system, we need cosmic fossils—material that formed when our system was very young. When it comes to answering our questions about origins, the planets them-selves are largely mute. Melted, battered by giant impacts, twisted by tectonic forces, they retain little evidence of their births. Reconstructing their early history is almost as difficult as determining the circumstances of human birth by merely looking at an adult. Instead, we must turn to clues provided by the surviving remnants of the creation process—ancient but smaller objects with which we share our cosmic neighborhood.

Astronomers have rather arbitrarily divided these objects into two cate-gories. **Asteroids** are rocky or metallic and contain little volatile (easily evap-orated) material. They differ from the terrestrial planets primarily in size and are sometimes even called minor planets. **Comets** are icy objects that contain frozen water and other volatile materials (but with solid grains mixed in). Comets are most easily noticed when they get near the Sun and form a visible

Virtual Laboratories

Kuiper Belt Objects

(but temporary) atmosphere and an extended tail of gas and dust.

12.1 ASTEROIDS

The asteroids are mostly found in the broad space between Mars and Jupiter, a region of the solar system called the *asteroid belt.* Asteroids are too small to be seen without a telescope; the first of them was not discovered until the beginning of the 19th century.

12.1.1 Discovery and Orbits of the Asteroids

In the late 1700s, many astronomers were hunting for an additional planet they thought should exist in the gap between the orbits of Mars and Jupiter. Some German astronomers even organized themselves into a group they called "the celestial police" to search for this "missing" member of the solar system. However, it was another astronomer who discovered the first asteroid as a by-product of his search for new comets.

In January 1801, the Sicilian astronomer Giovanni Piazzi initially thought he had found this missing planet when he discovered the first asteroid, which he named Ceres, orbiting at 2.8 astronomical units (AU) from the Sun. His discovery was followed the next year by that of another little planet in a similar orbit, and two more were found in 1804 and 1807. Clearly, there was not a single missing planet between Mars and Jupiter but rather a whole group of objects, each no more than 1000 km across.

By 1890 more than 300 had been discovered by sharp-eyed observers. In that year, Max Wolf of Heidelberg introduced astronomical photography to the search for asteroids, greatly accelerating the discovery of additional objects (Figure 12.1). More than 20,000 asteroids now have well-determined orbits. The pace of discovery is so fast that, although it required 197 years to find the first 10,000 asteroids, only 3 years were needed to double this number to 20,000.

Asteroids are given both a number (corresponding to the order of discovery) and a name. Originally, the names of asteroids were chosen from goddesses in Greek or Roman mythology. After exhausting these and other female names (including, later, those of wives, friends, flowers, cities, and others), astronomers turned to the names of colleagues (and other people of distinction) whom they wished to honor. For example, asteroids 2410 and 4859 are named Morrison and Fraknoi, for two authors of this text.

It would be a formidable task to discover, determine orbits for, and catalog all the asteroids bright enough to be photographed with modern telescopes. Nevertheless, the total number of such objects can be estimated by systematically sampling regions of the sky. These studies indicate that approximately a million (10^6) asteroids have diameters larger than 1 km.

FIGURE 12.1

Asteroid Trails Time exposure showing trails left by asteroids (arrows). The telescope moved backward to compensate for the rotation of the Earth; thus the stars remained fixed. But in the time the image was taken, the asteroids moved relative to the stars and left a trail.

The largest asteroid is Ceres, with a diameter just less than 1000 km. Two, Pallas and Vesta, have diameters of about 500 km, and about 15 more are larger than 250 km (Table 12.1). The number of asteroids increases rapidly with decreasing size; there are about 100 times more objects 10 km across than there are 100 km across. The total mass in all the asteroids, which probably represent just a tiny fraction of the original asteroid population, is less than that of the Moon. The orbits of the four largest asteroids were illustrated in Figure 2.10.

TABLE 12.1 *The Largest Asteroids*

Name	Year of Discovery	Semimajor Axis (AU)	Diameter (km)	Class°
Ceres	1801	2.77	940	C
Pallas	1802	2.77	540	C
Vesta	1807	2.36	510	†
Hygeia	1849	3.14	410	C
Interamnia	1910	3.06	310	C
Davida	1903	3.18	310	C
Cybele	1861	3.43	280	C
Europa	1868	3.10	280	C
Sylvia	1866	3.48	275	C
Juno	1804	2.67	265	S
Psyche	1852	2.92	265	M
Patientia	1899	3.07	260	C
Euphrosyne	1854	3.15	250	C

° C = carbonaceous; S = stony; M = metallic.
† Vesta has a very unusual (once thought unique) basaltic surface.

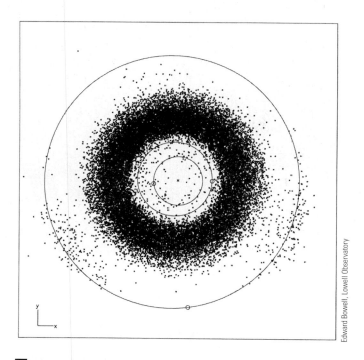

Edward Bowell, Lowell Observatory

■ **FIGURE 12.2**

Asteroid Positions The positions of more than 6000 asteroids in February 1990, seen from above. Also shown are the planets Earth, Mars, and Jupiter. Notice the two groups of Trojan asteroids in approximately the same orbit as Jupiter.

The asteroids all revolve about the Sun in the same west-to-east direction as the planets, and most of their orbits lie near the plane in which the Earth and other planets circle. The majority of asteroids lie in a zone we call the **asteroid belt,** defined as the region that contains all asteroids with semimajor axes in the range 2.2 to 3.3 AU (Figure 12.2). (See Chapter 2 if you need a reminder of the terms that apply to orbits.) Asteroids at these distances take 3.3 to 6 years to orbit the Sun. Although more than 75 percent of the known asteroids are in the main belt, they are not closely spaced. The volume of the belt is actually very large, and the typical spacing between objects (down to 1 km in size) is several million kilometers. (This is fortunate for spacecraft like Galileo and Cassini that need to travel through the belt without a collision.)

Still, over the long history of our solar system, there must have been a good number of collisions among the asteroids themselves. In 1917, the Japanese astronomer Kiyotsuga Hirayama found that some asteroids fall into *families*, groups with similar orbital characteristics. He hypothesized that each family may have resulted from a breakup of a larger body or, more likely, from the collision of two asteroids. Slight differences in the speeds with which the various fragments left the collision scene account for the small spread in orbits now observed for the different asteroids in a given family. Several dozen such families exist, and observations have shown that individual members of the larger ones are physically similar, just as we would expect if they were fragments of a common parent.

12.1.2 Composition and Classification

Asteroids are as different as black and white. The majority are very dark, with reflectivities of only 3 to 4 percent, like a lump of coal. However, another large group has typical reflectivities of 15 to 20 percent, a little greater than that of the Moon, and still others have reflectivities as high as 60 percent. To understand more about these differences and how they are related to chemical composition, astronomers study the spectra of the light reflected from asteroids for clues about their composition.

The dark asteroids are revealed from spectral studies to be *primitive* bodies (chemically little changed since the beginning of the solar system) composed of silicates mixed with dark, organic carbon compounds. Two of the largest asteroids, Ceres and Pallas, are primitive, as are almost all of the objects in the outer third of the belt. Most of the primitive asteroids are classed as C asteroids, where C stands for carbonaceous or carbon-rich.

The second most populous asteroid group is the S asteroids, where S stands for a stony or silicate composition. Here the dark carbon compounds are missing, resulting in higher reflectivity and clearer spectral signatures of silicate minerals. At first astronomers wondered if this was evidence of subsequent heating or chemical alteration. However, it appears that most of the S-type asteroids are also primitive; their different composition indicates they were formed in a different location from the C-type asteroids.

Asteroids of a third class, much less numerous than those of the first two, are composed primarily of metal and are called M asteroids. Spectroscopically, the identification of metal is difficult, but for at least the largest M asteroid, Psyche, this identification has been confirmed by radar. Since a metal asteroid, like an airplane or ship, is a much better reflector of radar than is a stony object, Psyche appears bright when we aim a radar beam at it.

How did such metal asteroids come to be? We suspect that each came from a parent body large enough for its molten interior to settle out or differentiate. The heavier metals sank to the center. When this parent body shattered in a later collision, the fragments from the core were rich in metals. There is enough metal in even a 1-km M-type asteroid to supply the world with iron and most other industrial metals for the foreseeable future, if only we could bring one safely to Earth.

In addition to the M asteroids, a few other asteroids show signs of early heating and differentiation. These have basaltic surfaces like the volcanic plains of the Moon and Mars; the large asteroid Vesta (discussed in a moment) is in the latter category. Why only a small percentage of the total number of asteroids have these characteristics, we do not know.

The different classes of asteroids are grouped together at different distances from the Sun (Figure 12.3). By tracing how their compositions vary with distance from the Sun, we can reconstruct some of the properties of the solar nebula from which they originally formed.

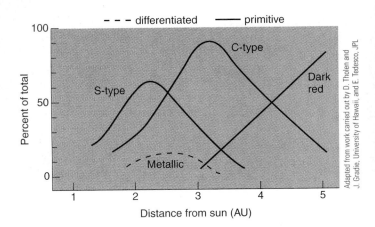

Adapted from work carried out by D. Tholen and J. Gradie, University of Hawaii, and E. Tedesco, JPL

FIGURE 12.3

Where Different Types of Asteroids Are Found
Asteroids of different composition are distributed at different distances from the Sun. The S type and C type are both primitive; the M type consists of cores of differentiated parent bodies.

FIGURE 12.4

A Piece of Vesta   Photograph of one of the meteorites believed to be a volcanic fragment from the crust of asteroid Vesta.

12.1.3   Vesta: A Volcanic Asteroid

Vesta is one of the most interesting of the asteroids. It orbits the Sun with a semimajor axis of 2.4 AU in the inner part of the main belt. Its relatively high reflectivity of almost 30 percent makes it the brightest of the main-belt objects, visible to the unaided eye if you know just where to look. But its real claim to fame is that its surface is covered with basalt, indicating that Vesta was once volcanically active in spite of its small size (about 500 km in diameter).

You may be surprised to learn that samples of Vesta's surface are available for direct study in the laboratory (and we thus know a great deal about this asteroid). Scientists have long suspected that *meteorites*—chunks of solar system material that survived their trip to the Earth's surface (see Chapter 13)—come from the asteroids, but there is generally no way to identify the particular source of a given meteorite that strikes the Earth. However, because of Vesta's unusual surface composition, the identification of its fragments seems fairly firm. (As we discussed in Chapter 9, we are also able to determine which meteorites come from Mars because we find pockets of trapped Mars gases in them.)

The meteorites believed to come from Vesta are a group of about 30 basaltic meteorites very similar in composition (Figure 12.4). Chemical analysis has shown that they cannot have come from the Earth, Moon, or Mars. On the other hand, their spectra (measured in the laboratory) perfectly match the spectrum of Vesta obtained telescopically. The age of the lava flows from which these meteorites derived has been measured at 4.4 to 4.5 billion years, very soon after the formation of the solar system. This age is consistent with what we might expect for Vesta; whatever process heated such a small object was probably intense and short-lived.

12.1.4   Asteroids Up Close

On the way to its 1995 encounter with Jupiter, the Galileo spacecraft was targeted to fly close to two main-belt asteroids called Gaspra and Ida. Gaspra, the first target, is an S-type asteroid. Astronomers knew from variations in its brightness that it had a rotation period of 7 hours. (Its name, incidentally, comes from a resort where 19th-century Russians enjoyed vacationing; the U.S. equivalent would be to call an asteroid Niagara or Mazatlan.) The Galileo camera revealed that Gaspra is 16 km long and highly irregular, as befits a fragment from a catastrophic collision (Figure 12.5).

The detailed images have allowed us to count the craters on Gaspra and to estimate the length of time its surface has been exposed to collisions. From the rather sparse number of craters, the Galileo scientists have concluded that Gaspra is only about 200 million years old (that is, the collision that formed Gaspra and other members of its family took place about 200 million years ago). Calculations suggest that an asteroid the size of Gaspra can expect another catastrophic collision sometime in the next billion years, at which time it will be disrupted to form another generation of still-smaller fragments.

The second target was Ida, a larger S-type asteroid 56 km long. The greatest surprise of the Galileo flyby of Ida, and the most important result scientifically, was the discovery of a satellite—named Dactyl—in orbit about the asteroid (Figure 12.6). Although only 1.5 km in diameter, smaller than many college campuses, Dactyl provides scientists with something otherwise beyond their reach—a measurement of the mass and density of Ida using Kepler's laws. The satellite's distance of about 100 km and its orbital period of about 24 hours indicate that Ida has a density of approximately 2.5 g/cm^3, which matches the density of primitive rocks. More recently, both large visible-light telescopes and high-powered planetary radar have discovered other asteroidal satellites, so that we are now able to accumulate valuable data on asteroid masses and densities.

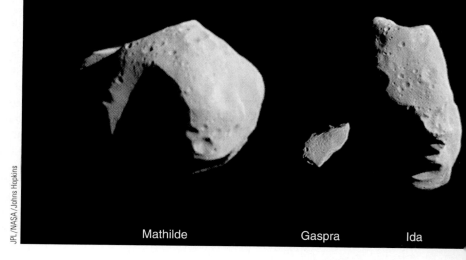

JPL/NASA/Johns Hopkins

Mathilde   Gaspra   Ida

■ **FIGURE 12.5**

Mathilde, Gaspra, and Ida The first three asteroids photographed from spacecraft flybys, printed to the same scale. Gaspra and Ida are S type and were investigated by the Galileo spacecraft; Mathilde is C type and was a flyby target for the NEAR-Shoemaker spacecraft.

We might mention that Phobos and Deimos, the two satellites of Mars, are probably captured asteroids. They were first studied at close range by the Viking orbiters in 1977 and later by the Mars Global Surveyor. Both are irregular, somewhat elongated, and heavily cratered. Their largest dimensions are about 25 km and 13 km, respectively.

In 1996, NASA launched the Near Earth Asteroid Rendezvous (NEAR) spacecraft, designed to match orbit with an asteroid rather than merely flying past at high speed. On the way to Eros, its prime target (see Section 12.2.4), the NEAR spacecraft flew past a much larger C-type asteroid in the main belt, called Mathilde. It passed so close that the mass of the asteroid could be measured without a satellite, just from its influence on the path of the spacecraft. To our surprise, the density turned out to be very low—about 1.6 g/cm^3.

Since no rock has a density this low, Mathilde must be a loosely bound collection of smaller rocks with substantial spaces between them—a sort of orbiting gravel bank. Another clue came from the fact that Mathilde has huge craters (as you can see in Figure 12.5). The impacts that formed them would have shattered a denser body, but a porous collection of pieces might well have survived such hits.

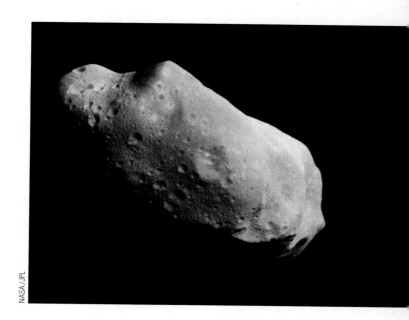

NASA/JPL

■ **FIGURE 12.6**

Ida and Dactyl Asteroid Ida and its moon Dactyl photographed by the Galileo spacecraft in 1993. Irregularly shaped Ida is 56 km in its longest dimension, while Dactyl is about 1.5 km across.

12.2 ASTEROIDS FAR AND NEAR

Not all asteroids are in the main asteroid belt. In this section we consider some special groups of asteroids that stray outside the belt's boundaries.

12.2.1 The Trojans

The Trojan asteroids are located far beyond the main belt, orbiting the Sun at 5.2 AU, about the same distance as Jupiter. The gravity of the giant planet makes most orbits of asteroids near it unstable, but calculations first carried out in the 18th century show that there are two points in the or-

bit of Jupiter near which an asteroid can remain almost indefinitely. These two points make equilateral triangles with Jupiter and the Sun (Figure 12.7). Between 1906 and 1908, four such asteroids were found; the number has now increased to several hundred. These asteroids are named for heroes from Homer's epic poem, the *Iliad,* and are collectively called the Trojans.

Measurements of the reflectivities and spectra of the Trojans show that they are dark, primitive objects like those in the outer part of the asteroid belt. They appear faint because they are so dark and far away, but actually the larger Trojans are quite sizable. Four of them—Hektor, Diomedes, Agamemnon, and Patroclus—have diameters between 150 and 200 km.

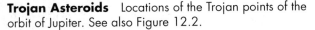

■ **FIGURE 12.7**

Trojan Asteroids Locations of the Trojan points of the orbit of Jupiter. See also Figure 12.2.

12.2.2 Asteroids in the Outer Solar System

There may be many asteroids with orbits that carry them far beyond Jupiter, but they are difficult to detect and only a few have been discovered. The largest of these mysterious objects is Chiron, with a path that carries it from just inside the orbit of Saturn at its closest approach to the Sun out to almost the distance of Uranus. The diameter of Chiron is estimated to be about 200 km.

In 1992, a still-more-distant object named Pholus was discovered with an orbit that takes it 33 AU from the Sun, beyond the orbit of Neptune. Pholus has the reddest surface of any object in the solar system, indicating a strange (and still unknown) surface composition. As more objects are discovered in these distant reaches, astronomers have decided that they too will be given the names of centaurs from classical mythology; this is because the centaurs were half human, half horse, and these new objects display some of the properties of both asteroids and comets.

In 1988, as it was reaching its closest approach to the Sun, Chiron brightened by about a factor of two. This is not the sort of behavior we expect from self-respecting asteroids. Astronomers realized that Chiron (and perhaps the other centaurs) likely contained abundant volatile materials such as water ice or carbon monoxide ice. As the object gets closer to the Sun, these materials begin to evaporate and produce bright reflective atmospheres of gas and dust. As we will see in the next section, this is precisely the behavior we associate with *comets*. But Chiron and Pholus are much larger than any known comet. What a spectacular show they would put on if either was diverted into the inner solar system! Chiron has been shown to be in an unstable orbit, so this may yet happen sometime in the distant future.

12.2.3 Earth-Approaching Asteroids

Asteroids that stray far *outside* the main belt are of interest mostly to astronomers. But asteroids that come *inward*, especially those with orbits that come close to or cross the or-

bit of the Earth, are of interest to political leaders, military planners, and all the readers of this text. Some of these asteroids briefly become the closest celestial objects to us. In 1994 a 2-km asteroid passed within 1.1 million km of the Earth, and later that same year a 1-km object was picked up passing closer than the Moon. Some of these small asteroids have collided with the Earth in the past, and others will surely do so in the future, as we saw in Chapter 7. Together with any comets that come close to our planet, such asteroids are known collectively as **near-Earth objects (NEOs).** As the dinosaurs found out 65 million years ago, the collision of a significant-sized NEO could be a catastrophe for life on our planet.

Astronomers have been urging for some time that the first step in protecting the Earth from future impacts must be to learn what potential impactors are out there. Not until 1998, however, did NASA formally adopt what has been called the Spaceguard Survey Goal: to discover and track 90 percent of the Earth-approaching asteroids larger than 1 km in diameter within ten years—that is, by 2008. By the end of 2002, more than 640 of these 1-km NEOs had been discovered, as well as many smaller ones. Figure 12.8 shows how the pace of discoveries has been increasing over the last years.

How will astronomers know when they have discovered 90 percent of these asteroids? There are several ways to estimate the total number even before they have been individually located. One way is to look at the numbers of larger craters on the Moon. Remember that these craters were made by impacts just like the ones we are considering; it's just that the Moon preserves billions of years of impacts on its airless surface, while the Earth soon erases the imprints of past impacts. Thus the number of larger craters on the Moon allows us to estimate how often impacts have occurred on the Moon and Earth over the past several billion years. The number of impacts is directly related to the

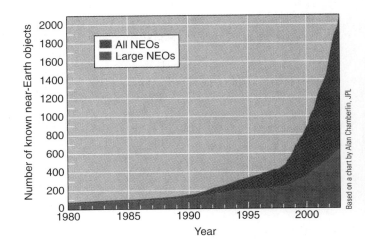

■ **FIGURE 12.8**

Discovery of Near-Earth Asteroids The accelerating rate of discovery of NEOs is illustrated in this graph, which shows both the total number of known NEOs and the number of those 1 km in diameter or larger, the size that dominates the current impact risk on Earth.

number of asteroids and comets that are typically orbiting in the inner solar system.

Another approach is to see how often the surveys (which are mostly automated searches for points of light that move among the stars) rediscover a previously known asteroid. At the beginning of a survey, all the objects it finds will be new. But as the survey becomes more complete, more and more of the moving points the survey cameras record will be rediscoveries. Applying such reasoning, astronomers calculate that there are probably between 1000 and 1200 such Earth-approaching asteroids larger than 1 km, and we have already discovered more than half of them.

The best news is that none of the NEOs discovered so far is on a trajectory that will impact the Earth within the foreseeable future. However, we can't speak for the 500 or so larger than 1 km that have not yet been found or the much more numerous smaller ones. Researchers who work with asteroid orbits estimate that for asteroids we are not yet tracking, we will have about a 5-second warning that one is going to hit the Earth—we won't see it until it enters the atmosphere! Clearly, this estimate gives us a lot of motivation to continue the surveys and track as many of the larger asteroids as possible.

Though entirely predictable over times of a few centuries, the orbits of Earth-approaching asteroids are unstable over long time spans. These objects will meet one of two fates: Either they will impact one of the terrestrial planets, or they will be ejected gravitationally from the inner solar system due to a near-encounter with a planet. The probabilities of these two outcomes are about the same. The timescale for impact or ejection is only about a hundred million years, very short compared with the age of the solar system. Calculations show that approximately one quarter of the current Earth-approaching asteroids will eventually end up colliding with the Earth itself. The larger of these impacts will generate environmental catastrophes for our planet as described in Section 7.5.

If the current population of Earth-approaching asteroids will be removed by impact or ejection in a hundred million years, there must be a continuing source of new objects to replenish our supply. Some of them probably come from the asteroid belt, where collisions between asteroids can eject fragments into Earth-crossing orbits. Others may be dead comets that have exhausted their volatile materials (see the next section). So far we have no way of distinguishing between asteroidal fragments and the solid remnants of former comets.

If an NEO comes close enough, it can be studied with radar as well as with visible-light telescopes. Radar can even be used to produce an image of the asteroid. Steven Ostro and his colleagues at the NASA Jet Propulsion Laboratory have succeeded in imaging more than a dozen with sufficient resolution to establish their shapes and sizes. Their greatest success was with a 5-km-long NEO called Toutatis, which approached to within 3 million km of the Earth in 1992. Several radar images of Toutatis are shown in Figure 12.9, obtained with the NASA radar system in California. It appears to be a double object, consisting of two irregular lumps, with diameters of 3 km and 2 km, squashed together.

NASA/JPL, courtesy of Steven Ostro

■ FIGURE 12.9

Radar Images of Toutatis Reconstruction of the size and shape of asteroid Toutatis obtained from radar data collected during its close flyby of Earth in December 1992. Toutatis appears to consist of two irregular, lumpy bodies rotating in contact with each other. Its maximum dimension is about 5 km.

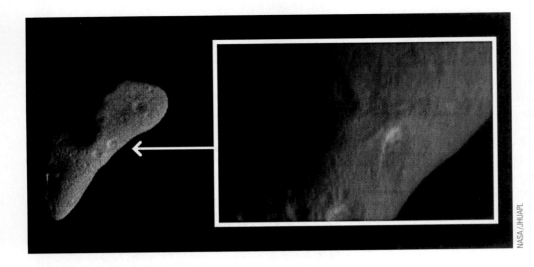

NASA/JHUAPL

■ **FIGURE 12.10**

Eros in Color The left image shows Eros from a distance of about 1700 km, taken two days before the NEAR-Shoemaker spacecraft went into orbit around the asteroid. The asteroid's surface is heavily cratered, indicating that the surface is old. The image in the box at right was taken from a distance of 283 km and thus shows much more detail.

12.2.4 A Rendezvous with Eros

On Valentine's Day 1999, the first spacecraft to explore an Earth-approaching asteroid went into orbit around the asteroid Eros, and one year later it landed on the surface. The arrival month was dictated by the position of the asteroid in the solar system, but the specific arrival date was selected to highlight the asteroid's name, which honors the Greco-Roman god of love. Eros is highly elongated: 34 km long but only 11 km wide (Figure 12.10). It is the largest of the Earth-approaching asteroids—somewhat larger than the impacting body that killed off the dinosaurs when it collided with our planet.

Getting into an intimate relationship with Eros was the culmination of the NEAR mission, launched in 1996 as part of NASA's Discovery series. On its way to Eros, the NEAR spacecraft was renamed after planetary geologist Eugene Shoemaker, a pioneer in our understanding of craters and impacts, who died in a tragic car accident in 1997.

For a year the NEAR–Shoemaker spacecraft orbited the little asteroid at various altitudes, measuring its surface and interior composition as well as mapping Eros from all sides (Figure 12.11). An additional chemical analysis of the surface was performed after the landing. The new information shows that Eros is a fractured chip off a larger body, made of some of the most primitive materials in the solar system. Several other asteroids have been revealed as loosely bound rubble, but not Eros. Its uniform density (about the same as that of Earth's crust) and extensive global-scale grooves and ridges show that it is a cracked but solid rock.

Although it is solid inside, Eros does seem to have a good deal of loose surface material sliding down toward lower elevations. In some places the surface rubble layer is 100 m deep. The top of this rubble layer is dotted with scattered, half-buried boulders. There are so many of these boulders that at the closest distances they are more numerous than the craters (Figure 12.12). Of course, with the gravity so low on this small world, a visiting astronaut would find loose boulders rolling toward her pretty slowly and could easily leap high enough to avoid being hit.

One reason scientists are interested in the interior structure of Earth-approaching asteroids like Eros is that humans will probably need to defend themselves against an asteroid impact someday. If we ever found one of these as-

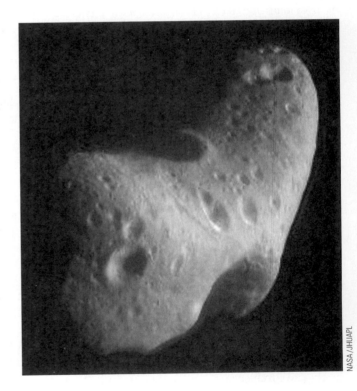

NASA/JHUAPL

■ **FIGURE 12.11**

Looking Down on the North Pole of Eros Constructed from six images taken from an altitude of 200 km, this mosaic shows Eros from a different angle. The large crater at the top has been named Psyche (after the maiden who was Eros' lover in classical mythology) and is about 5.3 km wide. A saddle-shaped region can be seen directly below it. Craters of many different sizes are visible.

FIGURE 12.12

Craters and Boulders Taken from an altitude of 43 km, this scene on Eros is about 1.3 km across. You can see many boulders, some dramatically displayed on the black skyline. Because of the orientation of the spacecraft, the viewer gets the impression that the cliff on the left side of the image is steep, but in reality the slope is gentle.

NASA/JHUAPL

teroids on a collision course, we would certainly try to deflect it so it would miss the Earth. With today's technology, such deflection would be done by exploding a nuclear bomb near the asteroid and nudging it off course. But to achieve a successful deflection, we need to know about the density and interior structure of the asteroid.

A nearby explosion would have a much greater effect on a solid rocky asteroid such as Eros than on a loose rubble pile such as Mathilde. Think of climbing a sand dune compared to climbing a rocky hill with the same slope. On the dune, much of our energy is absorbed in the slipping sand, so the climb is much more difficult. In a comparable way, a nuclear explosion would have little effect on the orbit of a loosely bound asteroid.

12.3 THE "LONG-HAIRED" COMETS

Comets differ from asteroids primarily in their icy composition, a difference that also makes them become much brighter as they approach the Sun. In some early cultures, these "hairy stars" were considered omens of disaster. Today we no longer fear the comets but eagerly anticipate those that come close enough to us to put on a good sky show.

12.3.1 Appearance of Comets

Comets have been observed from the earliest times; accounts of spectacular comets are found in the histories of virtually all ancient civilizations. The typical comet is not spectacular, however, but has the appearance of a rather faint, diffuse spot of light somewhat smaller than the Moon

and many times less brilliant. Comets are relatively small chunks of icy material that develop an atmosphere as they get closer to the Sun. Later there may be a very faint, nebulous **tail** extending several degrees away from the main body of the comet.

Like the Moon and planets, comets appear to wander among the stars, slowly shifting their positions in the sky from night to night. Unlike the planets, however, most comets appear at unpredictable times, which perhaps explains why they have frequently inspired fear and superstition. Comets typically remain visible for periods that vary from a couple of weeks to several months.

12.3.2 Comet Orbits

The study of comets as members of the solar system dates from the time of Isaac Newton, who first suggested that their orbits were extremely elongated ellipses. Newton's colleague Edmund Halley (see the Voyagers in Astronomy box) developed these ideas, and in 1705 he published calculations of 24 cometary orbits. In particular, he noted that the orbits of the bright comets appearing in the years 1531, 1607, and 1682 were so similar that the three could well be the same comet, returning to perihelion at average intervals of 76 years. If so, he predicted that the object should next return about 1758. Although Halley had died, when the comet appeared as he predicted, it was given the name Comet Halley (rhymes with "valley") in honor of the man who first recognized it as a permanent member of our solar system.

We now know from historical records that Comet Halley has actually been observed and recorded on every passage near the Sun at intervals from 74 to 79 years since 239 B.C.E. The period of its return varies somewhat

Edmund Halley: Astronomy's Renaissance Man

Halley, a brilliant astronomer who made contributions in many fields of science and statistics, was by all accounts a generous, warm, outgoing man. In this he was quite the opposite of his good friend Isaac Newton, whose great work, the *Principia* (see Chapter 2), Halley encouraged, edited, and helped pay to publish. Halley himself published his first scientific paper at age 20, while still in college. As a result, he was given a royal commission to go to Saint Helena (a remote island off the coast of Africa where Napoleon would later be exiled) to make the first telescopic survey of the southern sky. After returning, he received the equivalent of a master's degree and was elected to the prestigious Royal Society in England, all at the age of 22.

In addition to his work on comets, Halley was the first astronomer to recognize that the so-called "fixed" stars move relative to each other, by noting that several bright stars had changed their positions since the ancient Greek catalogs published by Ptolemy. He wrote a paper on the possibility of an infinite universe, proposed that some stars may be variable, and discussed the nature and size of *nebulae* (glow-

Edmund Halley (1656–1742)

Yerkes Observatory, University of Chicago

■ ■ ■ ■ ■ ■ ■ ■ ■ ■ ■ ■

Halley was the first astronomer to recognize that so-called "fixed" stars move relative to each other.

■ ■ ■ ■ ■ ■ ■ ■ ■ ■ ■ ■

ing cloudlike structures visible in telescopes). While in Saint Helena, Halley observed the planet Mercury going across the face of the Sun and devel-

oped the mathematics of how such *transits* could be used to establish the scale of the solar system.

In other fields, Halley published the first table of human life expectancies (the precursor of life-insurance statistics); wrote papers on monsoons, trade winds, and tides (charting the tides in the English Channel for the first time); laid the foundations for the systematic study of the Earth's magnetic field; studied evaporation and how inland waters become salty; and even designed an underwater diving bell. He served as a British diplomat, advising the emperor of Austria and squiring the future czar of Russia around England (avidly discussing, we are told, both the importance of science and the quality of local brandy).

For a brief time Halley was a deputy comptroller of the British Mint (where he uncovered some local graft) and served as a Navy captain for scientific voyages (once he had to deal with a mutiny). In 1703 he became a professor of geometry at Oxford and in 1720 was appointed Astronomer Royal of England. He continued observing the Earth and the sky and publishing his ideas for another 20 years, until death claimed him at age 85.

---

because of orbital changes produced by the pull of the giant planets. In 1910 the Earth was brushed by the comet's tail, causing much needless public concern. Comet Halley last appeared in our skies in 1986 (Figure 12.13), when it was met by several spacecraft that gave us a wealth of information about its makeup; it will return in 2061.

Only a few comets return in a time measurable in human terms, like Comet Halley. Most of the short-period comets have had their orbits changed by coming too close to one of the giant planets—most often Jupiter (they are thus sometimes called Jupiter-family comets). Most comets have long periods and will take thousands of years to return, if they return at all. As we will see later in this chapter, most Jupiter-family comets also come from a different source than Halley and the long-period comets.

Observational records exist for thousands of comets. Today, many new comets are discovered each year, not just by terrestrial observers but by spacecraft instruments as well. Most never become conspicuous and are visible only on photographs made with large telescopes. Every few years, however, a comet may appear that is bright enough to be seen easily with the unaided eye. Table 12.2 lists some well-known comets.

We are fortunate to have been visited by two bright comets in recent years; perhaps you had the chance to see at least one of them. First, in March 1996, came Comet Hyakutake, with a very long tail. A year later Comet Hale–Bopp appeared; it was as bright as the brightest stars and remained visible for several weeks, even in urban areas (see the image that opens this chapter).

12.3.3 The Comet's Nucleus

When we look at a comet, all we see is its temporary atmosphere of gas and dust illuminated by sunlight. This atmosphere is called the comet's head, or *coma*. Since the gravity of such small bodies is very weak, the atmosphere is rapidly escaping all the time; it must be replenished by new material, which has to come from somewhere. The source is the small, solid **nucleus** inside, usually hidden by the glow from the much larger atmosphere surrounding it. The nucleus is the real comet, the fragment of ancient material responsible for the atmosphere and the tail (Figure 12.14).

The modern theory of the physical and chemical nature of comets was first proposed by Harvard astronomer Fred Whipple in 1950. Before Whipple's work, many astronomers thought that a comet's nucleus might be a loose aggregation of solids, the sort of orbiting "gravel bank" model we discussed for the asteroid Mathilde. Whipple proposed instead that the nucleus is a solid object a few kilometers across, composed in substantial part of water ice (but with other ices as well) mixed with silicate grains and dust. This proposal became known as the "dirty snowball" model.

The water vapor and other volatiles that escape from the nucleus when it is heated can be detected in the comet's head and tail, and we can use spectra to analyze what atoms and molecules the nucleus ice consists of. We are somewhat less certain of the non-icy component, however. No large fragments of solid matter from a comet have ever survived passage through the Earth's atmosphere.

Some very fine, microscopic grains of comet dust have been collected in the Earth's upper atmosphere, however, and have been studied in the laboratory (Figure 12.15). The spacecraft that encountered Comet Halley in March 1986 also carried dust detectors. From these various investigations it seems that much of the "dirt" in the dirty snowball

Anglo-Australian Observatory

■ FIGURE 12.13

Comet Halley This composite of three images (one in red, one in green, one in blue) shows Comet Halley as seen with a large telescope in Australia in 1985. During the time the three images were taken in sequence, the comet moved among the stars. The telescope was moved to keep the image of the comet steady, causing the stars to become colored streaks in the background.

TABLE 12.2 *Some Interesting Comets*

Name	Period	Significance
Great Comet of 1577	Long	Tycho Brahe showed it was beyond the Moon
Great Comet of 1843	Long	Brightest recorded comet; visible in daytime
Daylight Comet of 1910	Long	Brightest comet of the 20th century
West	Long	Nucleus broke into pieces (1976)
Hyakutake	Long	Passed within 15 million km of Earth (1996)
Hale–Bopp	Long	Brightest recent comet (1997)
Swift–Tuttle	133 years	Parent comet of Perseid meteor shower
Halley	76 years	First comet found to be periodic; explored by spacecraft in 1986
Borrelly	6.8 years	Flyby by Deep Space 1 spacecraft (2000)
Biela	6.7 years	Broke up in 1846 and not seen again
Wild 2	6.4 years	Target of Stardust sample return mission (2004)
Tempel 1	5.7 years	Target of Deep Impact mission (2005)
Encke	3.3 years	Shortest known period

Donald Brownlee, University of Washington

■ **FIGURE 12.14**

Parts of a Comet This schematic illustration shows the main parts of a comet. Note that the different structures are not to scale.

is tiny bits of dark, primitive hydrocarbons and silicates, rather like the material thought to be present on the dark, primitive asteroids.

Since the nuclei of comets are small and dark, they are difficult to study from the Earth. Two direct measurements of a comet nucleus have been obtained, however: one in 1986, when three spacecraft swept past Comet Halley at

■ **FIGURE 12.15**

Captured Comet Dust A particle that is believed to be a tiny fragment of cometary dust, collected in the upper atmosphere of the Earth.

close range, and one in 2001, when a NASA spacecraft made a close flyby of the much fainter Comet Borrelly. Although the European spacecraft Giotto came the closest, passing just 600 km from the Halley nucleus, the best photos of a comet nucleus were obtained for Borrelly from a distance of 2000 km (Figure 12.16). In spite of their much different brightness and level of activity, both Halley and Borrelly turn out to be about the same size: 10 km and 8 km long, respectively. Both are also exceedingly dark, with reflectivity less than 2 percent, which makes these chunks of dirty ice darker than coal.

Max Planck Institut für Aeronomie, and Ball Aerospace Corporation, courtesy of Harold Reitsema

JPL/NASA

(a)

(b)

■ **FIGURE 12.16**

Close-ups of Comets Halley and Borrelly

(a) This historic photograph of the black, irregularly shaped nucleus of Comet Halley was obtained by the Giotto spacecraft from a distance of about 1000 km. The bright areas are jets of material escaping from the surface. The length of the nucleus is 10 km, and details as small as 1 km can be made out. (b) Even more detail is visible in this photo of Comet Borrelly taken at a range of 2000 km. The nucleus is 8 km long and very dark, with an average reflectivity less than 2 percent.

12.3.4 The Comet's Atmosphere

The spectacular activity that allows us to see comets is caused by the evaporation of cometary ices heated by sunlight. Beyond the asteroid belt, where comets spend most of their time, these ices are solidly frozen. But as a comet approaches the Sun, it begins to warm up. If water (H_2O) is the dominant ice, significant quantities vaporize as temperatures rise toward 200 K, which happens somewhat beyond the orbit of Mars. The evaporating H_2O in turn releases the dust that was mixed with the ice. Since the comet's nucleus is so small, its gravity cannot hold back either the gas or the dust, both of which flow away into space at speeds of about 1 km/s.

The comet continues to absorb energy as it approaches the Sun. A great deal of this energy goes into the evaporation of its ice, as well as into heating the surface. However, observations of many comets indicate that the evaporation is not uniform and that most of the gas is released in sudden spurts, perhaps confined to a few areas of the surface. Such jets were observed directly on the surfaces of Comets Halley and Borrelly (see Figure 12.16). Most of the comet's surface is apparently inactive, with the ice buried under a layer of black silicates and carbon compounds.

Expanding into space at a speed of about 1 km/s, the comet's atmosphere can reach an enormous size. The diameter of a comet's head is often as large as Jupiter, and it can approach a diameter of a million kilometers (Figure 12.17).

Measurements show that the composition of the gas is primarily H_2O (about 80 percent in the case of Comet Halley) plus a few percent each of carbon dioxide (CO_2) and carbon monoxide (CO), along with small quantities of many additional gases, including hydrocarbons. Ultraviolet light from the Sun can break up the molecules of H_2O into oxygen and hydrogen. Modern telescopes have revealed huge hydrogen clouds around comets, up to tens of millions of kilometers across, making them, temporarily, the largest objects in the solar system.

Many comets develop tails as they approach the Sun. A comet's tail is an extension of its atmosphere, consisting of the same gas and dust that make up its head. As early as the 16th century, observers realized that comet tails always point away from the Sun (Figure 12.18), not back along the comet's orbit. Newton proposed that comet tails are formed by a repulsive force of sunlight driving particles away from the head—an idea close to our modern view. In addition to sunlight, however, we now know that cometary gas is also carried outward by streams of ions (charged particles) emitted by the Sun (Figure 12.19).

12.3.5 The Kuiper Belt

If we define a "comet" by its chemical composition (lots of water ice) rather than by its appearance (possessing an atmosphere and tail), then there is another group of "comets" that are now being found in large numbers at the edge of the solar system, beyond the orbit of Neptune. The first of these trans-Neptunian objects was discovered in 1992 by David Jewitt and Jane Luu of the University of Hawaii (Figure 12.20). They are dark (low reflectivity) and can be significantly larger than any of the comets ever seen in the inner solar system. Because of their great distance, as-

University of Arizona, courtesy of Uwe Fink

■ **FIGURE 12.17**

Head of Comet Halley   Here we see the cloud of gas and dust that make up the head or coma of Comet Halley on January 20, 1986. On this scale, the nucleus (hidden inside the cloud) would be a dot too small to see.

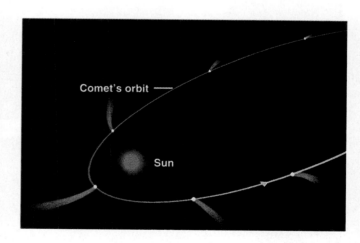

■ **FIGURE 12.18**

Comet Tail Points Away from the Sun   The orientation of a typical comet tail changes as the comet passes perihelion. Approaching the Sun, the tail is behind the incoming comet head, but on the way out, the tail precedes the head.

■ **FIGURE 12.19**

Comet Tails Comet Mrkos was photographed in 1957 with a wide-field telescope at Palomar Observatory. Note the smoother tail of dust curving to the right as individual dust particles spread out along the comet's orbit, and the straight ion tail pushed outward from the Sun by its wind of charged particles.

Caltech/Palomar Observatory

tronomers can detect are only the largest—generally more than 100 km in diameter.

More than 500 of these objects have now been found, and it is expected that larger survey telescopes currently under discussion could detect thousands. If any of them

ever wondered into the inner solar system, they would probably develop a spectacular atmosphere and tail, but in the bitter cold beyond Neptune they are solid as a rock.

The existence of this vast reservoir of primitive icy objects—related to the comets and asteroids but not really part of either group—had been predicted theoretically 50 years earlier by the astronomer Gerard Kuiper. Today astronomers call this region beyond Neptune the **Kuiper belt** and its "inhabitants" Kuiper belt objects, or **KBOs.** We will discuss the KBOs in the next section when we look at the origin of comets.

12.4 ORIGIN AND EVOLUTION OF COMETS AND THEIR KIN

The comets and their close relatives, the KBOs, are mostly remnant ice fragments from the origin of the solar system. They are probably the most primitive objects we can study, preserved unchanged for billions of years in the deep freeze of the outer solar system.

12.4.1 The Oort Comet Cloud

Although comets are part of the solar system, observations show that most of them come initially from very great distances. By following their orbits, we can calculate that the *aphelia* (points farthest from the Sun) of new comets typi-

■ **FIGURE 12.20**

Kuiper Belt Object Two images of object 1992QB taken in 1992 with the 3.5-m New Technology Telescope at the European Southern Observatory. The object, circled on each image as it moved among the stars, was the first member of the Kuiper belt to be found.

cally have values near 50,000 AU (more than a thousand times farther than Pluto). This clustering of aphelion distances was first noted by Dutch astronomer Jan Oort, who in 1950 proposed an idea for the origin of those comets that is still accepted today.

It is possible to calculate that a star's gravitational *sphere of influence*—the distance within which it can exert sufficient gravitation to hold onto orbiting objects—is about one third of the distance to the nearest other stars. Stars in the vicinity of the Sun are spaced in such a way that the Sun's sphere of influence extends only a little beyond 50,000 AU, or about 1 light year (LY). At such distances, however, objects in orbit about the Sun can be perturbed by the gravity of passing stars. Some of the perturbed objects can then take on orbits that bring them much closer to the Sun.

Oort suggested, therefore, that the new comets were examples of objects orbiting the Sun near the edge of its sphere of influence, whose orbits had been disturbed by nearby stars, eventually bringing them close to the Sun where we can see them. The reservoir of ancient icy objects from which such comets are derived is now called the **Oort comet cloud.**

Astronomers estimate that there are about a trillion (10^{12}) comets in the Oort cloud. In addition, we estimate that about ten times this number of potential comets could be orbiting the Sun in the volume of space between the planets and the Oort cloud at 50,000 AU. These objects remain undiscovered because they are too faint to be seen directly and their orbits are too stable to permit any of them to be deflected inward close to the Sun. The total number of cometary objects could thus be on the order of 10 trillion (10^{13}), a very large number indeed.

What is the mass represented by 10^{13} comets? We can make an estimate if we assume something about comet sizes and masses (see the Figuring for Yourself section at the end of the chapter). Let us suppose that the nucleus of Comet Halley is typical. Its observed volume is about 600 km^3. If the primary constituent is water ice with a density of about 1 g/cm^3, then the total mass of Halley's nucleus must be about 6×10^{14} kg. This is about one ten billionth (10^{-10}) of the mass of the Earth.

If our estimate is reasonable and there are 10^{13} comets with this mass out there, their total mass would be equal to about 1000 Earths—greater than the mass of all the planets put together! Therefore, cometary material could be the most important constituent of the solar system after the Sun itself.

12.4.2 KBOs and Plutinos

The Oort cloud spawns comets we can see because they are deflected into the inner solar system by passing stars. But we now know of a second source of comets. The Kuiper belt represents a relatively nearby reservoir of cometary material, in the form of a flattened disk that starts just beyond Neptune. Smaller KBOs are occasionally ejected gravitationally into the inner solar system, where they can

be captured by Jupiter and thereby contribute to the short-period or Jupiter-family comets. The Centaur asteroids are probably intermediate objects that are in transit from the Kuiper belt to the inner solar system. (You can see how awkward the traditional terminology is, when the same object is called a KBO for most of its life, then a Centaur asteroid, and finally a short-period comet.)

The biggest KBOs discovered so far have impressive diameters, approaching 1000 km. Once we have orbits and distances for such KBOs, we can estimate their diameters from how bright they look. The bigger the object, the more surface it has to reflect the rays of the Sun. But to make this estimate, we must make some reasonable assumptions about how reflective their surfaces are. Our estimates of their size are only as good as these assumptions. The two largest known KBOs, called Varuna and Quaoar (which was discovered in 2002), have diameters similar to the largest asteroid (Ceres) and Pluto's moon, Charon. That's significantly larger than any comet that has been seen approaching the Sun.

Many of the more than 500 known KBOs have orbits like that of Pluto, with periods less than 300 years. These KBOs are sometimes called *plutinos*. Some astronomers have suggested that Pluto can simply be considered the largest member of the Kuiper belt, thus explaining why it is so different from the other outer planets. In any case, it appears that the Kuiper belt is heavily influenced by the gravity of Neptune, which confines most of its members to a distance from the Sun between 35 and 50 AU.

The objects in the Oort cloud and the Kuiper belt probably had different origins, and they may therefore have different compositions. Astronomers are therefore very interested in making detailed measurements of the comets that are derived from these two source regions. Most of the bright comets that have been studied in the past (Halley, Hyakutake, Hale–Bopp) are Oort-cloud comets, but several of the comets targeted for spacecraft measurements in the next decade are Jupiter-family comets from the Kuiper belt (see Table 12.2).

12.4.3 The Fate of Comets

Any comet we see today will have spent nearly its entire existence in the Oort cloud or the Kuiper belt at a temperature near absolute zero. But once a comet enters the inner solar system, its previously uneventful life history begins to accelerate. It may, of course, survive its initial passage near the Sun and return to the cold reaches of space where it spent the previous 4.5 billion years. At the other extreme, it may impact the Sun or come so close that it is destroyed on its first perihelion passage (several such impacts have been observed with space telescopes that monitor the Sun). Sometimes, however, the new comet does not come that close to the Sun but instead interacts with one or more of the planets.

A comet that comes within the gravitational influence of a planet has three possible fates. It can (1) impact the planet, ending the story at once; (2) be speeded up and

Comet Hunting as a Hobby

When amateur astronomer David Levy, the codiscoverer of Comet Shoemaker–Levy 9, found his first comet, he had already spent 928 fruitless hours searching through the dark night sky. But the discovery of that first comet only whetted his appetite. Since then he has found 7 others on his own and 13 more working with others. Despite this impressive record, he ranks only fourth in the record books for number of comet discoveries. But David is still young and hopes to break the record someday.

All around the world, dedicated amateur observers spend countless nights scanning the sky for new comets. Astronomy is one of the very few fields of science where amateurs can still make a meaningful contribution, and the discovery of a comet is one of the most exciting ways they can establish their place in astronomical history. Don Machholz, a California amateur (and comet hunter) who has been making a study of comet discoveries, reported that between 1975 and 1995, 38 percent of all comets discovered were found by amateurs. Those 20 years yielded 67 comets for amateurs, or almost four per year. That might sound pretty encouraging to new comet hunters, until they learn that the average number of hours the typical amateur spent searching for a comet before finding one was about 420. Clearly, this is not an activity for impatient personalities.

What do comet hunters do if they think they have found a new comet? First, they must check the object's location in an atlas of the sky to make sure it really is a comet. Since the first sighting of a comet usually occurs when it is still far from the Sun and before it sports a significant tail, it will look

ejected, leaving the solar system forever; or (3) be perturbed into an orbit with a shorter period. In the last case, its fate is sealed. Each time it approaches the Sun, it loses part of its material and also has a significant chance of collision with a planet. Once the comet is in a short-period orbit, its lifetime starts being measured in thousands, not billions, of years. Comet Halley is the best-known example of a comet that has been reoriented in this way.

A few comets end their lives catastrophically by breaking apart (sometimes for no apparent reason; Figure 12.21).

■ **FIGURE 12.21**

Breakup of Comet LINEAR A ground-based view and a much more detailed photo with the HST showing the multiple fragments of the nucleus of Comet LINEAR. The comet disintegrated in July 2000 for no apparent reason.

Comet LINEAR
Fragments
August 5, 2000

HST

University of Hawaii

Hubble Space Telescope
WFPC2

NASA and H. Weaver (Johns Hopkins University)
STScI-PRC00-27

University of Hawaii & STScI/NASA

like only a small, fuzzy patch. And, through most amateur telescopes, so will nebulae (clouds of cosmic gas and dust) and galaxies (distant groupings of stars). Next they must check that they have not come across a comet that is already known, in which case they will only get a pat on the back instead of fame and glory. Then they must reobserve or rephotograph it some time later to see whether its motion in the sky is appropriate for comets.

Often comet hunters who think they have made a discovery get an-

Amateur astronomer David Levy

other comet hunter elsewhere in the country to confirm it. If everything checks out, the place they contact is the Central Bureau for Astronomical Telegrams at the Harvard–Smithsonian Center for Astrophysics in Cambridge, Massachusetts (Web site cfa-www. harvard.edu/cfa/ps/cbat.html). If the discovery is confirmed, the bureau will send the news out to astronomers and observatories around the world. One of the unique rewards of comet hunting is that the discoverer's name becomes associated with the new comet—a bit of cosmic fame that few hobbies can match.

Especially spectacular was the fate of a faint comet called Shoemaker–Levy 9, which broke into about 20 pieces when it passed close to Jupiter in July 1992. The fragments of Shoemaker–Levy were actually captured into a very elongated, two-year orbit around Jupiter, more than doubling the number of known jovian satellites. This was only a temporary enrichment of Jupiter's family, however, because in July 1994 the comet fragments crashed into Jupiter, releasing energy equivalent to millions of megatons of TNT.

As each cometary fragment streaked into the jovian atmosphere at a speed of 60 km/s, it disintegrated and ex-

ploded, producing a hot fireball that carried the comet dust as well as atmospheric gases to high altitudes. These fireballs were clearly visible in profile, with the actual point of impact just beyond the jovian horizon as viewed from the Earth (Figure 12.22). As each explosive plume fell back into Jupiter, a region of the upper atmosphere larger than the Earth was heated to incandescence and glowed brilliantly for about 15 min, a glow we could detect with infrared-sensitive telescopes.

Dark clouds of debris settled into the stratosphere of Jupiter, producing long-lived "bruises" (each larger than

■ **FIGURE 12.22**

Comet Impact on Jupiter
This sequence of images shows the last fragment of Comet Shoemaker–Levy 9 to hit Jupiter. Although the initial impact is on the edge of Jupiter as seen from Earth, the giant planet's rotation soon brings the fireball caused by the impact into view. The time of each observation and the filter through which it was taken are shown in each box.

NASA/STScI

the Earth) that could be easily seen through even small telescopes (Figure 12.23). Millions of people all over the world peered at Jupiter through telescopes or followed the event via television and the Internet. Seeing these impact explosions on Jupiter helps us to appreciate the disaster that would happen to our planet if we were hit by a comet or asteroid.

For comets that do not meet so dramatic an end, measurements of the amount of gas and dust in their atmospheres permit us to estimate the total losses during one orbit. Typical loss rates are up to a million tons per day from an active comet near the Sun, adding up to some tens of millions of tons per orbit. At that rate, a typical comet will be gone after a few thousand orbits. This will probably be the fate of Comet Halley in the long run.

■ **FIGURE 12.23**

Impact Dust Cloud on Jupiter   One group of features resulting from the impact of Comet Shoemaker–Levy 9 with Jupiter, seen with the Hubble Space Telescope 105 minutes after the impact that produced the dark rings (the compact black dot came from another fragment). The inner edge of the diffuse outer ring is about the same size as the Earth. Later, the winds on Jupiter blended these features into a broad spot that remained visible for more than a month.

SURFING THE WEB

 Asteroid and Comet Impact Hazards Page: impact.arc.nasa.gov/
Informative site about the dangers of asteroids and comets hitting the Earth and what we might do about it.

 Near-Earth Asteroid Rendezvous Mission: near.jhuapl.edu/
Background information and great images from the mission that went by Mathilde and landed on Eros.

Kuiper Belt Home Page: www.ifa.hawaii.edu/~jewitt/kb.html
David Jewitt of the University of Hawaii keeps track of the objects that have been discovered in the Kuiper belt and provides useful background material.

Deep Space 1 Mission: nmp.jpl.nasa.gov/ds1/
This was the mission that flew by Comet Borrelly.

Stardust Mission Home Page: stardust.jpl.nasa.gov/
This NASA mission will encounter Comet Wild 2 in 2004, collecting comet dust and returning it to Earth in 2006.

Deep Impact Mission: www.ss.astro.umd.edu/deepimpact
A mission to fly by and impact a probe on comet Tempel 1 in 2005.

Information About Quaoar, the Largest Known KBO:

• *At Caltech:*
http://www.gps.caltech.edu/~chad/quaoar/

• *From the Hubble:*
http://hubblesite.org/newscenter/archive/2002/17/

Sites About Asteroids in General:

• *Nine Planets Site:*
seds.lpl.arizona.edu/billa/tnp/asteroids.html

• *Views of the Solar System Site:*
www.solarviews.com/eng/asteroid.htm

SUMMARY

The solar system includes many objects that are much smaller than the planets and their larger satellites. The rocky ones are called **asteroids,** and the icy ones are called **comets.**

12.1 Ceres is the largest asteroid; about 15 are larger than 250 km and 100,000 are larger than 1 km. Most are in the **asteroid belt,** between 2.2 and 3.3 AU from the Sun. The presence of asteroid families in the belt indicates that many asteroids are the remnants of asteroid collisions and fragmentation. The asteroids include both primitive and differentiated objects. Most asteroids are classed as C type, meaning they are composed of carbonaceous materials. Dominating the inner belt are S-type (stony) asteroids, with a few M-type (metallic) ones. Vesta is rare in having a volcanic (basaltic) surface. We now have spacecraft images of several asteroids and have detected several asteroid satellites, making it possible to measure asteroid masses and densities.

12.2 The Trojan asteroids are dark, primitive objects gathered into two swarms controlled by Jupiter's gravity at 5.2 AU from the Sun. Of great interest are the Earth-approaching asteroids, called **near-Earth objects (NEOs),** estimated to number about 1100 (down to 1 km in diameter). These are on unstable orbits, and on timescales of 10^8 years they will either impact one of the terrestrial planets or be ejected. Most of them probably come from the asteroid belt, but some may be dead comets. Radar images provide us a close view of some of these without a spacecraft encounter.

12.3 Halley first showed that some comets are on closed orbits and return periodically to the Sun. The heart of a comet is its **nucleus,** a few kilometers in diameter and composed of volatiles (primarily H_2O) and solids (including both silicates and carbonaceous materials). Whipple first suggested this "dirty snowball" model in 1950; it has been confirmed by spacecraft studies of Comets Halley and Borrelly. As the nucleus approaches the Sun, its volatiles evaporate (perhaps in localized jets or explosions) to form the comet's head or atmosphere, which escapes at about 1 km/s. The atmosphere streams away from the Sun to form a long **tail.**

12.4 Oort proposed in 1950 that comets are derived from what we now call the **Oort comet cloud,** which surrounds the Sun out to about 50,000 AU (near the limit of the Sun's gravitational sphere of influence) and contains between 10^{12} and 10^{13} comets. Comets also come from the **Kuiper belt,** a disk-shaped region beyond the orbit of Neptune. Modern telescopes enable us to discover and study the largest of these **KBOs** directly. Comets are primitive bodies left over from the formation of the outer solar system. Once a comet is diverted into the inner solar system, it typically survives no more than a few thousand perihelion passages before losing all its volatiles. Some comets die spectacular deaths: Shoemaker–Levy 9, for example, broke into 20 pieces before colliding with Jupiter in 1994.

INTER-ACTIVITY

A Your group is a congressional committee charged with evaluating the funding for an effort to find all the NEOs (near-Earth objects) that are larger than 0.5 km across. Make a list of reasons it would be useful to humanity to find such objects. What should we (could we) do if we found one that will hit the Earth in a few years?

B Many cultures considered comets bad omens. Many legends associate comets with the death of kings, losses in war, ends of dynasties. Did any members of your group ever hear about such folktales? Discuss why comets may have gotten this bad reputation.

C Because asteroids have a variety of compositions and a low gravity that makes the removal of materials quite easy, some people have suggested that mining asteroids may be a way to get needed resources in the future. Make a list of materials in asteroids (and comets that come to the inner solar system) that may be valuable to a space-faring civilization. What are the pros and cons of undertaking mining operations on these small worlds?

D As discussed in the last box of the chapter, amateur comet hunters typically spend more than 400 hours scanning the skies with their telescopes to find a comet. That's a lot of time to spend (usually alone, usually far from city lights, usually in the cold, and always in the dark). Discuss with members of your group whether you can see yourself being this dedicated. Why do people undertake such quests? Do you envy their dedication?

 E The largest Kuiper belt objects known are estimated to be about half the diameter of Pluto and are sometimes called miniplanets. All the planets in our solar system have so far been named after mythological gods. Have your group discuss whether they think we should continue this naming tradition with miniplanets. Why or why not? (The group that discovered the largest KBO so far decided to go beyond Greek and Roman mythology. They named it Quaoar, after a creation god of the Tongva tribe, the Native Americans who lived in the region of Southern California where Caltech, the university where the work was done, is located.)

F The discovery of the miniplanets in the Kuiper belt again raises the issue of the nature of Pluto. Have your group review the information about Pluto and Triton (see Chapter 11) and the large Kuiper belt objects (in this chapter), and then discuss whether you think Pluto should be considered a planet, the largest KBO, or something else. Give reasons for your choice.

REVIEW QUESTIONS

1. Why are asteroids and comets important to our understanding of solar system history?
2. Describe the main differences between C-type and S-type asteroids.
3. Compare asteroids of the asteroid belt with the Earth-approaching asteroids. What are the main differences between the two groups?
4. Describe the nucleus of a typical comet, and compare it with an asteroid of similar size.
5. Describe the origin and eventual fate of the comets we see from the Earth.
6. What evidence do we have for the existence of the Kuiper belt? What kind of objects are found there?
7. How did Comet Shoemaker–Levy 9 end its life in 1994?

THOUGHT QUESTIONS

8. Give at least two reasons today's astronomers are so interested in the discovery of additional Earth-approaching asteroids.

9. Suppose you were designing a spacecraft that would match course with an asteroid and follow along its orbit. What sorts of instruments would you put on board, and what would you like to learn?

10. Suppose you were designing a spacecraft that would match course with a comet and move with it for a while. What sorts of instruments would you put on board, and what would you like to learn?

11. Suppose a comet were discovered approaching the Sun, one whose orbit would cause it to collide with the Earth 20 months later, after perihelion passage. (This is approximately the situation described in the science-fiction novel

Lucifer's Hammer by Larry Niven and Jerry Pournelle.) What could we do? Would there be any way to protect ourselves from a catastrophe?

12. We believe that chains of comet fragments like Comet Shoemaker–Levy 9's have collided not only with the jovian planets, but occasionally with their satellites. What sort of features would you look for on the outer planet satellites to find evidence of such collisions? (As an extra bonus, can you find any images of such features on a satellite like Callisto? You can use a Web site of planetary images, such as *Planetary Photojournal*, at: photojournal.jpl.nasa.gov.)

13. Why have we found so many KBOs in the last decade and not before?

14. Why is it hard to give exact diameters for even the larger KBOs?

FIGURING FOR YOURSELF

Suppose the Oort comet cloud contains 10^{12} comets with an average diameter of 10 km each. How can we estimate the mass of the total Oort comet cloud? We can start by assuming that typical comets are about the size of Halley and Borrelly, with a diameter of 10 km and a density appropriate to water ice, which is 1 g/cm^3 or 1000 kg/m^3. We know that density = mass/volume, the volume of a sphere = $\frac{4}{3}\pi R^3$, and the radius $R = \frac{1}{2}D$. Therefore, for each comet,

$$\text{mass} = \text{density} \times \text{volume} = \text{density} \times \frac{4}{3}\pi\left(\frac{1}{2}D\right)^3$$

Given that 10 km = 10^4 m, each comet's mass is

$$1000 \text{kg/m}^3 \times \frac{4}{3} \times 3.14 \times \frac{1}{8} \times (10^4)^3 \text{m}^3 \approx 10^{15} \text{kg}$$
$$= 10^{12} \text{ tons}$$

To calculate the total mass of the cloud, we multiply this typical mass for one comet by the number of comets:

$$10^{15} \times 10^{12} = 10^{27} \text{ kg}$$

15. How does the total mass we calculated above compare to the mass of Jupiter? To the mass of the Sun? (Give a numerical answer.)

16. How would the calculation change if a typical comet in the Oort cloud is only 1 km in diameter?

17. How would the calculation change if a typical comet in the Oort cloud is larger—say, 50 km in diameter?

18. The calculation above refers to the known Oort cloud, the source for most of the comets we see. If, as some astronomers suspect, there are ten times this many cometary objects in the solar system, how does the total mass of cometary matter compare with the mass of Jupiter?

19 If the Oort comet cloud contains 10^{12} comets and ten new comets are discovered coming close to the Sun each year, what percentage of the comets have been used up since the beginning of the solar system?

20. The mass of the asteroids is found mostly in the larger asteroids, so to estimate the total mass we need to consider only the larger objects. Suppose the three largest asteroids—Ceres (1000 km), Pallas (500 km), and Vesta (500 km)—account for half the total mass. Assume that each of these three asteroids has a density of 3 g/cm^3 and calculate their total mass. Multiply your result by 2 to obtain an estimate for the mass of the total asteroid belt. How does this compare with the mass of the Oort comet cloud?

21. Make a similar estimate for the mass of the Kuiper belt assuming that Pluto (about 2000 km), Quauar, and Varuna (each 1000 km) are the three largest (they are the biggest found so far) and that all three have Pluto's density of 2 g/cm^3. How does this mass compare with the mass of the main asteroid belt? (In fact there are probably many undiscovered large KBOs, so the true mass could be as much as ten times larger than in this example.)

22. What is the period of revolution about the Sun for an asteroid with a semimajor axis of 3 AU in the middle of the asteroid belt?

23. What is the period of revolution for a comet with aphelion at 5 AU and perihelion at the orbit of the Earth?

SUGGESTIONS FOR FURTHER READING

On Asteroids

Asphang, E. "The Small Planets" in *Scientific American,* May 2000, p. 46. On asteroids and NEAR results.

Beatty, J. "NEAR Falls for Eros" in *Sky & Telescope,* May 2001, p. 34. On the landing, with an album of photos.

Bell, J. and Mitton, J., eds. *Asteroid Rendezvous: NEAR Shoemaker's Adventures at Eros.* 2002, Cambridge U. Press.

Binzel, R. "A New Century for Asteroids" in *Sky & Telescope,* July 2001, p. 44.

Binzel, R. "Asteroids Come of Age" in *Science,* Sept. 2000, vol. 289, p. 2065.

Burnham, R. "Here's Looking at Ida" in *Astronomy,* Apr. 1994, p. 38.

Kowal, C. *Asteroids: Their Nature and Utilization,* 2nd ed. 1996, John Wiley/Praxis. Introductory book by the astronomer who discovered Chiron.

Zimmerman, R. "Ice Cream Sundaes and Mashed Potatoes" in *Astronomy,* Feb. 1999, p. 54. On the NEAR mission.

On Comets

Aguirre, E. "The Great Comet of 1997" in *Sky & Telescope,* July 1997, p. 50. On Comet Hale–Bopp.

Benningfield, D. "Where Do Comets Come From?" in *Astronomy,* Sept. 1990, p. 28.

Bortle, J. "A Halley Chronicle" in *Astronomy,* Oct. 1985, p. 98. A chronology of each pass.

Brandt, J. and Chapman, R. *Rendezvous in Space.* 1992, W. H. Freeman. Introduction by two leading comet experts.

Davies, J. *Beyond Pluto: Exploring the Outer Limits of the Solar System.* 2001, Cambridge U. Press. Discovery of minor planets in the Kuiper belt.

Gore, R. "Halley's Comet '86: Much More Than Met the Eye" in *National Geographic,* Dec. 1986, p. 758.

Levy, D. "My Life and Hard Times as a Comet Hunter" in *Sky & Telescope,* Oct. 2002, p. 72.

Levy, D. *The Quest for Comets.* 1994, Plenum Press. Personal story of comet discovery and comet science by an amateur astronomer who has found many comets.

Spencer, J. and Mitton, J., eds. *The Great Comet Crash: The Collision of Comet Shoemaker–Levy 9 and Jupiter.* 1995, Cambridge U. Press. Good nontechnical summary.

Stern, A. "Chiron: Interloper from the Kuiper Disk?" in *Astronomy,* Aug. 1994, p. 26.

Talcott, R. "Comet Borrelly's Dark Nature" in Astronomy, Apr. 2002, p. 42. Results of the Deep Space 1 flyby.

Weissman, P. "The Oort Cloud" in *Scientific American,* Sept. 1998, p. 84.

Collisions between planetesimals during the formative stages of the solar system. Many meteorites are fragments of these planetesimals, dating back to the origin of our planetary system.

13

Cosmic Samples and the Origin of Planetary Systems

I *could more easily believe that two Yankee professors would lie than that stones would fall from the heavens.*

Thomas Jefferson, responding to a report of a Connecticut meteorite fall in 1807

THINKING AHEAD

Imagine you are a scientist examining a sample of rock that had fallen from space a few days earlier and you find within it some of the chemical building blocks of life. How could you determine whether those "organic" materials were extraterrestrial or merely the result of earthly contamination?

We conclude this survey of our solar system with a discussion of its origin and evolution. Some of these ideas were introduced in Chapter 6; we now return to them, using the information we have learned about individual planets and smaller objects. In addition, astronomers have a new source of data to consider: Planets recently found around more than a hundred other stars provide us with a perspective that extends beyond our own particular (and perhaps atypical) solar system.

First, we look at another crucial way that astronomers learn about the ancient history of the solar system—by examining samples of primitive matter, the debris of the processes that formed the solar system some 4.5 billion years ago. Unlike the Apollo Moon rocks, these samples of cosmic material come to us free of charge; they literally fall from the sky. We call this material cosmic dust and meteorites.

13.1 METEORS

As we saw in the last chapter, the ices in comets evaporate when they get close to the Sun, spraying millions of tons of rock and dust into the inner solar system. The Earth is surrounded by this material, sweeping through thin clouds and streams of dust in its orbit about the Sun. As each of the larger dust or rock particles enters the Earth's atmosphere, it creates a brief fiery trail; this is often called a *shooting star*, but it is properly known as a **meteor**.

13.1.1 Observing Meteors

Meteors are tiny solid particles that enter the Earth's atmosphere from interplanetary space (Figure 13.1). Since the particles move at speeds of many kilometers per second, friction with the air vaporizes them at altitudes between 80 and 130 km. The resulting flashes of light fade out within a few seconds. These "shooting stars" got their name because at night their luminous vapors look like stars moving rapidly across the sky. To be visible, a meteor must be within about 200 km of the observer. On a typical dark, moonless night, an alert observer can see half a dozen meteors per hour. Over the entire Earth, the total number of meteors bright enough to be visible is about 25 million per day.

The typical meteor is produced by a particle with a mass of less than 1 gram—no larger than a pea. The light you see comes from the much larger region of glowing gas surrounding this little grain of interplanetary material. Because of its high speed, the energy in a pea-sized meteor is as great as that of an artillery shell fired on Earth, but it is all dispersed high in the Earth's atmosphere.

If a particle the size of a golf ball strikes our atmosphere, it produces a much brighter trail called a fireball (Figure 13.2). A piece as large as a bowling ball has a fair

■ **FIGURE 13.1**

Meteor Trails Two meteors can be seen against a starry background that includes the Milky Way.

chance of surviving its fiery entry if its approach speed is not too high. The total mass of meteoric material entering the Earth's atmosphere is estimated to be about 100 tons per day (which seems like a lot, if you imagine it all falling

■ **FIGURE 13.2**

Fireball When a larger piece strikes the Earth's atmosphere, it can make a bright fireball. Note that during the long exposure needed to catch a fireball, the Earth's motion has made the stars into streaks.

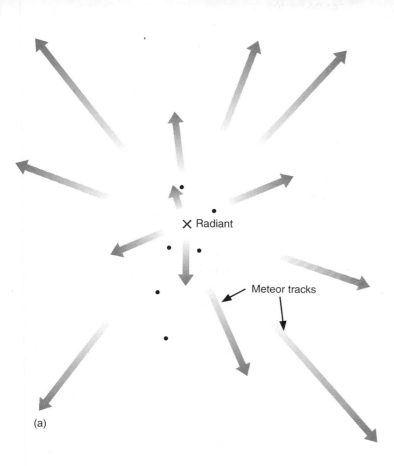

X Radiant

Meteor tracks

(a)

(b)

◼ FIGURE 13.3

The Radiant of a Meteor Shower The tracks of the meteors diverge from a point in the distance, just as long, straight railroad tracks appear to do.

on your home, but is not very significant when spread out over the whole planet).

13.1.2 Meteor Showers

Many—perhaps most—of the meteors that strike the Earth can be associated with specific comets. Some of these periodic comets still return to our view; others have long ago fallen apart, leaving only a trail of dust behind them. The many dust particles from a given comet retain approximately the orbit of their parent and continue to move together through space (spreading out over the orbit with time). When the Earth, in its travels around the Sun, crosses such a dust stream, we see a sudden burst of meteor activity that usually lasts several hours; these events are called **meteor showers.**

The dust particles that produce meteor showers are moving together in space before they encounter the Earth. Thus, as we look up at the atmosphere, their parallel paths seem to diverge from a place in the sky called the *radiant*. This is the direction in space from which the meteor stream is moving at us, just as long railroad tracks seem to diverge from a single spot on the horizon (Figure 13.3). Meteor showers are often designated by the constellation in which this radiant is located: For example, the Perseid meteor shower has its radiant in the constellation of Perseus. The characteristics of some of the more famous meteor showers are summarized in Table 13.1.

TABLE 13.1 *Major Annual Meteor Showers*

Shower Name	Date of Maximum	Associated Comet	Comet's Period (years)
Quadrantid	January 3	Unknown	—
Lyrid	April 21	Thatcher	415
Eta Aquarid	May 4	Halley	76
Delta Aquarid	July 30	Unknown	—
Perseid	August 11	Swift–Tuttle	105
Draconid	October 9	Giacobini–Zinner	7
Orionid	October 20	Halley	76
Taurid	October 31	Encke	3
Leonid	November 16	Tempel–Tuttle	33
Geminid	December 13	Phaethon°	1.4

° Phaethon is an Earth-approaching asteroid, not a comet.

The meteoric dust is not always evenly distributed along the orbit of the comet, so during some years more meteors are seen when the Earth intersects the dust stream, and in other years fewer are seen. A very clumpy distribution is associated with the Leonid meteors, which in 1833 and again in 1866 (after an interval of 33 years—the period

FIGURE 13.4

A Meteor Storm A woodcut of the great meteor shower or storm of 1833, shown here with considerable artistic license.

of the comet) yielded the most spectacular showers (sometimes called *meteor storms*) ever recorded (Figure 13.4). During the Leonid storm on November 17, 1966, up to a hundred meteors were observed per second in some locations. The Leonid shower of 2001 was not so intense, but it peaked at nearly a thousand meteors per hour—one every few seconds—observable from any dark viewing site. Perhaps some of our readers saw this "once in a lifetime" event.

The most dependable meteor display is the Perseid shower, which appears each year for about three nights near August 11. In the absence of bright moonlight, you can see meteors at a rate of one every few minutes during a typical Perseid shower. Astronomers estimate that the total combined mass of the particles in the Perseid swarm is nearly a billion tons; the comet that gave rise to the particles, called Swift–Tuttle, must originally have had at least that much mass. However, if its initial mass were comparable to the mass measured for Comet Halley (see Chapter 12), then

Swift–Tuttle would have contained several hundred billion tons, suggesting that only a very small fraction of the original cometary material survives in the meteor stream.

No shower meteor has ever survived its flight through the atmosphere and been recovered for laboratory analysis. However, there are other ways to investigate the nature of these particles and thereby to gain additional insight into the comets from which they are derived. Analysis of the flight paths of meteors shows that most of them are very light or porous, with densities typically less than 1.0 g/cm^3. If you placed a fist-sized lump of meteor material on a table in the Earth's gravity, it might well fall apart under its own weight.

Such light particles break up very easily in the atmosphere, accounting for the failure of even relatively large shower meteors to reach the ground. Comet dust is apparently fluffy, rather inconsequential stuff, as we can also infer from the tiny comet particles recovered in the Earth's atmosphere (see Figure 12.15). This fluff, by its very nature, cannot reach the Earth's surface intact. However, more substantial fragments from asteroids do make it into our laboratories, as we will see in the next section.

 METEORITES: STONES FROM HEAVEN

Any fragment of interplanetary debris that survives its fiery plunge through the Earth's atmosphere is called a **meteorite.** Meteorites fall only rarely in any one locality, but over the entire Earth thousands fall each year. These rocks from the sky carry a remarkable record of the formation and early history of the solar system.

13.2.1 Extraterrestrial Origin of Meteorites

Although occasional meteorites have been found throughout history, their extraterrestrial origin was not accepted by scientists until the beginning of the 19th century. Before that, these strange stones were either ignored or considered to have a supernatural origin.

The falls of the earliest recovered meteorites are lost in the fog of mythology. A number of religious texts mention stones from heaven, which sometimes arrived at opportune moments to smite the enemies of the authors of those texts. At least one sacred meteorite has apparently survived in the form of the Ka'aba, the holy black stone in Mecca that is revered by Islam as a relic from the time of the Patriarchs—although understandably no chip from this sacred stone has been subject to detailed chemical analysis.

The modern scientific history of the meteorites begins in the late 18th century, when a few scientists suggested that some strange-looking stones had such peculiar composition and structure that they were probably not of terrestrial origin. The general acceptance that indeed "stones fall

Showering with the Stars

Observing a meteor shower is one of the easiest and most enjoyable astronomy activities for beginners. The best thing about it is that you don't need a telescope or binoculars—in fact, they would positively get in your way! What you do need is a site far from city lights with an unobstructed view of as much sky as possible. Although the short, bright lines in the sky made by individual meteors could, in theory, be traced back to a radiant point (as shown in Figure 13.3), the quick blips of light that represent the end of the meteor could happen anywhere above you.

So the trick of observing meteor showers is not to restrict your field of view, but to lie back and scan the sky alertly. Try to select a good shower (see the list in Table 13.1) and a night when the Moon will not be bright at the time you are observing. The Moon, street lights, car headlamps, and bright flashlights will all get in the way of your seeing the faint meteor streaks.

■■■■■■■■■■■■■■■■■■■■

The trick of observing meteor showers is not to restrict your field of view, but to lie back and scan the sky alertly.

■■■■■■■■■■■■■■■■■■■■

You will also see more meteors after midnight, when you are on the hemisphere of the Earth that faces forward—in the direction of the Earth's revolution around the Sun. Before midnight, you are observing from the "back side" of the Earth, and the only meteors you see will be those that traveled fast enough to catch up with the Earth's orbital motion.

When you've gotten away from all the lights, give your eyes about 15 minutes to get "dark adapted"—that is, for the pupil of your eye to open up as much as possible. (This adaptation is the same thing that happens in a dark movie theater. When you first enter, you can't see a thing, but eventually, as your pupils open wider, you can see pretty clearly by the faint light of the screen.)

Seasoned meteor observers find a hill or open field and make sure to bring warm clothing, a blanket, and a thermos of hot coffee or chocolate with them. (It's also nice to take along someone with whom you enjoy sitting in the dark.) Don't expect to see fireworks or a laser show; meteor showers are subtle phenomena, best approached with a patience that reflects the fact that some of the dust you are watching burn up may first have been gathered into its parent comet more than 4.5 billion years ago, as the solar system was just forming.

---

from the sky" occurred after a scientific team led by French physicist Jean Baptiste Biot investigated a well-observed fall in 1803. As indicated by the opening quote, however, it took a few more years for the news to reach the young republic on the western shores of the Atlantic.

Meteorites sometimes fall in groups or showers. A fall occurs when a single larger object breaks up during its violent passage through the atmosphere. A fall of this kind happened outside of Chicago in March 2003, with more than a hundred fragments recovered from streets, backyards, and even inside homes on the first day alone. It is important to remember that such a *shower of meteorites* has nothing to do with a *meteor shower*. No meteorites have ever been recovered in association with meteor showers. Whatever the ultimate source of the meteorites, they do not appear to come from the comets or their associated particle streams.

13.2.2 Meteorite Falls and Finds

Meteorites are found in two ways. First, sometimes bright meteors (fireballs) are observed to penetrate the atmosphere to very low altitudes. If we search the area beneath the point where the fireball burned out, we may find one or more remnants that reached the ground. Observed *meteorite falls*, in other words, may lead to the recovery of fallen meteorites. (A few meteorites have even hit buildings or, very rarely, people—see the Making Connections box, *Some Striking Meteorites.*) But there are many false alarms. For example, every few months the newspapers report that a meteorite has been implicated in the start of a fire. Such stories have always proved to be wrong. The meteorite is ice-cold in space, and most of its interior remains cold even after its brief fiery plunge through the atmosphere. A freshly fallen meteorite is more likely to acquire a coating of frost than to start a fire.

Second, people sometimes discover unusual-looking rocks that turn out to be meteoritic; these are termed *meteorite finds*. Now that the public has become meteorite-conscious, many unusual fragments, not all of which turn out to be from space, are sent to experts each year. Some scientists divide these objects into two categories: "meteorites" and "meteorwrongs." Outside Antarctica (see the next paragraph), genuine meteorites turn up at an average rate of 25 or so per year. Most of these end up in natural

MAKING CONNECTIONS

Some Striking Meteorites

Although meteorites fall regularly onto the Earth's surface, few of them have much impact(!) on human civilization. There is so much water and uninhabited land on our planet that rocks from space typically fall where no one even sees them come down. But given the number of meteorites that land each year, you may not be surprised that a few have struck buildings, cars, and even people.

In November 1982, for example, Robert and Wanda Donahue of Wethersfield, Connecticut, were watching *M*°*A*°*S*°*H* on television when a 6-lb meteorite came thundering through their roof, making a hole in the living room ceiling. After bouncing back up into the attic, it finally came to rest under their dining room table.

Eighteen-year-old Michelle Knapp of Peekskill, New York, got quite a surprise one morning in October 1992. She had just purchased her very first car, her grandmother's 1980 Chevy Malibu. But she awoke to find its rear end mangled and a crater in the family driveway—thanks to a 3-lb meteorite. Michelle was not sure whether to be devastated by the loss of her car or thrilled by all the media attention.

In June 1994, José Martín and his wife were driving from Madrid, Spain, to a golfing vacation when a fist-sized meteorite crashed through the windshield of their car, bounced off the dashboard, broke José's little finger, and then landed in the back seat. Before Martín, the most recent person known to have been struck by a meteorite was Annie Hodges of Sylacauga, Alabama. In November 1954, she was napping on a couch when a meteorite came through the roof, bounced off a large radio set, and hit her first on the arm and then on the leg.

In a recent compilation, amateur astronomer and meteorite historian Christopher Spratt found 61 recorded instances between 1790 and 1990 of meteorites striking human habitats or machinery (and we presume other cases that have gone unreported or unrecorded). Although his

© 1982, 1992 Dan Haar/ The Hartford Courant

John McAuliffe, deputy chief of the Wethersfield, Connecticut, Fire Department, examines the hole made by a meteorite that crashed into the home of Robert and Wanda Donahue on November 8, 1982.

compilation makes for an impressive list, two centuries are a long time and the Earth is a big place. The chance of you or your property being hit by a meteorite is so small that it is certainly not worth losing sleep over.

---

history museums (Figure 13.5) or specialized meteoritic laboratories throughout the world.

Since the 1980s, a new source from the Antarctic has dramatically increased our knowledge of meteorites. More than ten thousand meteorites have been recovered from the Antarctic as a result of the motion of the ice in some parts of that continent (Figure 13.6). Meteorites that fall in regions where ice accumulates are buried and then carried slowly to other areas where the ice is gradually worn away. After thousands of years, the rock again finds itself on the surface, along with other meteorites carried to these same locations. The ice thus concentrates the meteorites that have fallen both in a large area and over a long period of time.

13.2.3 Meteorite Classification

The meteorites in our collections have a wide range of compositions and histories, but traditionally they have been placed into three broad classes. First are the **irons,** composed of nearly pure metallic nickel–iron. Second are the **stones,** the term used for any silicate or rocky meteorite. Third are the much rarer **stony-irons,** made (as the name implies) of mixtures of stone and metallic iron (Figure 13.7).

Of these three types, the irons and stony-irons are the most obviously extraterrestrial because of their metallic content. Pure iron almost never occurs naturally on Earth; it is generally found here as an oxide (chemically combined

FIGURE 13.5

Meteorite Find This 1906 photo shows a 15-ton iron meteorite found in the Willamette Valley in Oregon. Although known to Native Americans in the area, it was "discovered" by an enterprising local farmer in 1902, who proceeded to steal it and put it on display. It was eventually purchased for the American Museum of Natural History and is now on display in the Hayden Planetarium in New York City, the largest iron meteorite in the United States.

with oxygen) or other mineral ore. Therefore, if you ever come across a chunk of metallic iron, it is sure to be either man-made or a meteorite.

The stones are much more common than the irons but more difficult to recognize. Often laboratory analysis is required to demonstrate that a particular sample is really of extraterrestrial origin, especially if it has lain on the ground

for some time and been subject to weathering. The most scientifically valuable stones are those collected immediately after they fall, or the Antarctic samples preserved in a nearly pristine state by the ice.

Table 13.2 summarizes the frequencies of occurrence of the different classes of meteorites among the fall, find, and Antarctic categories.

FIGURE 13.6

Antarctic Megteorite An iron meteorite lying on the Antarctic ice just before being added to our collection.

FIGURE 13.7

A Variety of Meteorite Types At the top left is a piece of the Allende carbonaceous meteorite, with white inclusions that may date back to before the formation of the solar nebula. At the top right is a fragment of the iron meteorite responsible for the formation of Meteor Crater in Arizona (see Chapter 12). At bottom left is a piece of the Imilac stony-iron meteorite, a beautiful mixture of green olivine crystals and metallic iron. At bottom right is part of the Mern primitive stony meteorite.

TABLE 13.2 *Frequency of Occurrence of Meteorite Classes*

Class	Falls	Finds	Antarctic
Primitive stones	88%	51%	85%
Differentiated stones	8	1	12
Irons	3	42	2
Stony-irons	1	5	1

13.2.4 Ages and Compositions of Meteorites

It was not until the ages of meteorites were measured and their compositions analyzed in detail that scientists appreciated their true significance. The meteorites include the oldest and most primitive materials available for direct study in the laboratory. The ages of stony meteorites can be determined from the careful measurement of radioactive isotopes and their decay products, as described in Chapter 6. Almost all meteorites have radioactive ages between 4.50 and 4.56 billion years, as old as any ages we have measured in the solar system. The few exceptions are igneous rocks that have been ejected from cratering events on the Moon or Mars.

The average age for the most primitive meteorites, calculated using the most accurate values now available for radioactive half-lives, is 4.55 billion years, with an uncertainty of less than 0.01 billion years. This value (which we round off to 4.5 billion years in this book) is taken to represent the *age of the solar system*—the time since the first solids condensed and began to form into larger bodies.

The traditional classification of meteorites into irons, stones, and stony-irons is easy to use because it is obvious from inspection which category a meteorite falls into (although it may be much more difficult to distinguish a meteoritic stone from a terrestrial rock). Much more significant, however, is the distinction between *primitive* and *differentiated* meteorites. The differentiated meteorites are fragments of larger parent bodies that were molten before they broke up, allowing the denser materials (such as metals) to sink to their centers. Like many rocks on Earth, they have been subject to a degree of chemical reshuffling, with the different materials sorted according to density. Differentiated meteorites include the irons, which come from the metal cores of their parent bodies, and the stony-irons, which probably originate in regions between a metal core and a stony mantle.

13.2.5 The Most Primitive Meteorites

For information on the *earliest* history of the solar system, we turn to the most primitive meteorites—those made of materials that have not been subject to great heat or pressure since their formation. A comparison of their spectra indicates that their parent bodies are almost certainly asteroids. Recall from Chapter 12 that asteroids are believed to be fragments left over from the formation process of the solar system; thus it makes sense that they should be the parent bodies of the primitive meteorites.

The great majority of the meteorites that reach the Earth are primitive stones. Many of them are composed of light-colored gray silicates with some metallic grains mixed in, but there is also an important group of darker stones called **carbonaceous meteorites.** As their name suggests, these meteorites contain carbon as well as various complex organic compounds—chemicals based on carbon, which on Earth are the simple building blocks of life. In addition, some of them also contain chemically bound water, and many are depleted in metallic iron. The carbonaceous meteorites probably originate among the dark, carbonaceous C-type asteroids concentrated in the outer part of the asteroid belt. Among the most useful of these meteorites have been the Allende meteorite that fell in Mexico (see Figure 13.7) and the Murchison meteorite that fell in Australia (both in 1969), and the much more recent Tagish Lake meteorite that landed in a winter snowdrift on Tagish Lake, Canada, in 2000. (The fragile bits of dark material from the Tagish Lake meteorite were readily visible against the white snow, although at first they were mistaken for wolf droppings!)

The Murchison meteorite is known for the variety of organic chemicals it has yielded. Most of the carbon compounds in carbonaceous meteorites are complex, tarlike substances that defy exact analysis. Murchison also contains 16 amino acids (the building blocks of proteins), 11 of which are rare on Earth. The most remarkable thing about them is that they include equal numbers with right-handed and left-handed molecular symmetry. Amino acids can have either kind of symmetry, but all life on Earth has evolved using only the left-handed versions to make proteins. The presence of both kinds of amino acids clearly demonstrates that the ones in the meteorite had a nonterrestrial origin.

These naturally occurring amino acids and other complex organic compounds in Murchison—formed without the benefit of the sheltering environment of planet Earth—show that a great deal of interesting chemistry must have taken place when the solar system was forming. Perhaps some of the molecular building blocks of life on Earth were first delivered by primitive meteorites and comets. This is an interesting idea because our planet was probably much too hot for any organic materials to survive its earliest history. But after the Earth's surface cooled, the asteroid and comet fragments that pelted it could have refreshed its supply of organic materials.

13.3 FORMATION OF THE SOLAR SYSTEM

The comets, asteroids, and meteorites are surviving remnants from the processes that formed the solar system. (The planets and the Sun, of course, also are the products of the formation process, although the material in them has undergone a wide range of changes.) We are now ready to put together the information from the previous chapters in order to discuss what is known about the origin of the solar system.

13.3.1 Observational Constraints

There are certain basic properties of the planetary system that any theory of its formation must explain. These may be summarized under three categories: motion constraints, chemical constraints, and age constraints. We call them *constraints* because they place restrictions on our theories; unless a theory can explain the observed facts, it will not survive in the competitive marketplace of ideas that characterizes the endeavor of science.

There are many regularities to the motions in the solar system. We saw that the planets all revolve around the Sun in the same direction and approximately in the plane of the Sun's own rotation. In addition, most of the planets rotate in the same direction as they revolve and most of the satellites also move in counterclockwise orbits (when seen from the north). With the exception of the comets, the motions of the system members define a disk or frisbee shape. On the other hand, a full theory must be prepared to deal with the exceptions to these trends, such as the retrograde rotation of Venus.

In the realm of chemistry, we saw that Jupiter and Saturn have approximately the same composition—dominated by hydrogen and helium—as the Sun and the stars. These are the two largest planets, with sufficient gravity to hold on to any gas present when and where they formed; thus we might expect them to be representative of the original material out of which the solar system formed. Each of the other members of the planetary system is, to some degree, lacking in the light elements. A careful examination of the composition of solid solar-system objects shows a striking progression from the metal-rich inner planets, through those made predominantly of rocky materials, out to objects with ice-dominated compositions in the outer solar system. The comets in the Oort cloud and the Kuiper belt are also icy objects, whereas the asteroids represent a transitional rocky composition with abundant dark, carbon-rich material.

As we saw in Chapter 6, this general chemical pattern can be interpreted as a temperature sequence, with the inner parts of the system strongly depleted in materials that could not condense (form a solid) at the high temperatures found near the Sun. Again, however, there are important exceptions to the general pattern. For example, it is difficult to explain the presence of water on Earth and Mars if these planets formed in a region where the temperature was too hot for ice to condense, unless the ice or water was brought in later from cooler regions.

As far as age is concerned, we discussed that radioactive dating demonstrates that some rocks on the surface of the Earth have been present for at least 3.8 billion years, and that certain lunar samples are 4.4 billion years old. In addition, the primitive meteorites all have radioactive ages near 4.5 billion years. The age of these unaltered building blocks is considered the age of the planetary system. The similarity of the measured ages tells us that planets formed and their crusts cooled within a few hundred million years (at most) of the beginning of the solar system. Further, detailed examination of primitive meteorites indicates that they are made primarily from material that condensed or coagulated out of a hot gas; few identifiable fragments appear to have survived from before this hot-vapor stage 4.5 billion years ago.

13.3.2 The Solar Nebula

All of the foregoing constraints are consistent with the general idea introduced in Chapter 6—that the solar system formed 4.5 billion years ago out of a rotating cloud of hot vapor and dust called the solar nebula, with an initial composition similar to that of the Sun today. As the solar nebula collapsed under its own gravity, material fell toward the center, where things became more and more concentrated and hot. Increasing temperatures in the shrinking nebula vaporized most of the solid material that was originally present.

At the same time, the collapsing nebula began to rotate faster and faster through the conservation of angular momentum (see Chapter 2). Near the poles of the spin, the nebular material fell directly into the center, while faster moving material collapsed into a flat disk revolving around the central object (Figure 13.8). The existence of this disk-shaped rotating nebula explains the primary motions in the solar system just discussed.

Picture the solar nebula at the end of the collapse phase, when it was at its hottest. With no more gravitational energy (from material falling in) to heat it, most of the nebula began to cool. The material in the center, however, where it was hottest and most crowded, formed a *star* that was able to maintain high temperatures in its immediate neighborhood by producing its own energy. Turbulence and magnetic fields within the disk drained away angular momentum and allowed some material to continue to fall into the growing star, while the rest of the disk gradually stabilized.

The temperature within the disk decreased with increasing distance from the Sun, much as the planets' temperatures vary with position today. As the disk cooled, the gases interacted chemically to produce compounds; eventually these compounds condensed into liquid droplets or solid grains. This is similar to the process by which raindrops on Earth condense from moist air as it rises over a mountain.

Let's look in more detail at how material condensed at different places in the maturing disk (Figure 13.9). The first materials to form grains were the metals and various rock-forming silicates. As the temperature dropped, these were joined throughout much of the solar nebula by sulfur compounds and by carbon- and water-rich silicates, such as those now found abundantly among the asteroids. However, in the inner parts of the disk, the temperature never dropped low enough for such materials as ice or carbona-

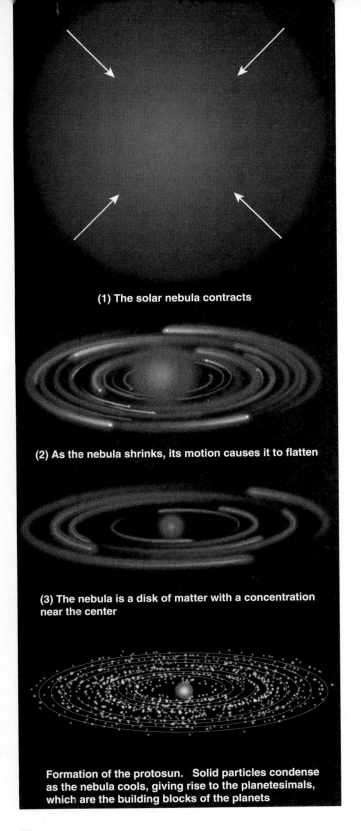

(1) The solar nebula contracts

(2) As the nebula shrinks, its motion causes it to flatten

(3) The nebula is a disk of matter with a concentration near the center

Formation of the protosun. Solid particles condense as the nebula cools, giving rise to the planetesimals, which are the building blocks of the planets

■ **FIGURE 13.8**

Steps in Forming the Solar System An artist has drawn a very rough sketch of the steps in the formation of the solar system from the solar nebula. As the nebula shrinks, its rotation causes it to flatten into a disk. Much of the material is concentrated in the hot center, which will ultimately become a star. Away from the center, solid particles can condense as the nebula cools, giving rise to planetesimals, the building blocks of the planets and satellites.

■ **FIGURE 13.9**

The Chemical Condensation Sequence in the Solar Nebula The scale at left shows temperature; at right are the materials that would condense out at each temperature under the conditions expected to prevail in the nebula.

ceous organic compounds to condense, so they are lacking on the innermost planets.

Far from the Sun, cooler temperatures allowed the oxygen to combine with hydrogen and condense in the form of water (H_2O) ice. Beyond the orbit of Saturn, carbon and nitrogen combined with hydrogen to make ices such as methane (CH_4) and ammonia (NH_3). This chemical condensation sequence explains the basic chemical composition differences among various regions of the solar system.

13.3.3 Formation of the Terrestrial Planets

The grains that condensed in the solar nebula rather quickly joined into larger and larger aggregates, until most of the solid material was in the form of *planetesimals* a few kilometers to a few tens of kilometers in diameter (see Section 6.4). Some planetesimals still survive today as the comets and asteroids. Others have left their imprint on the cratered surfaces of many of the worlds we studied in earlier chapters. A substantial step up in size is required, however, to go from planetesimals to planets.

Some planetesimals were large enough to attract their neighbors gravitationally and thus to grow by the process called **accretion.** Although the intermediate steps are not well understood, ultimately several dozen centers of accretion seem to have grown in the inner solar system. Each of these attracted surrounding planetesimals until it had acquired a mass similar to that of Mercury or Mars. At this stage we may think of these objects as *protoplanets*—"not quite ready for prime time" planets.

Each of these protoplanets continued to grow by the accretion of planetesimals. Every incoming planetesimal was accelerated by the gravity of the protoplanet, striking with great energy—sufficient to melt both the projectile and a part of the impact area. Soon the entire protoplanet was heated to above the melting temperature of rocks. The result was *planetary differentiation*, with heavier metals sinking toward the core and lighter silicates rising toward the surface. As they were heated, the protoplanets lost some of their more volatile constituents (the lighter gases), leaving more of the heavier elements and compounds behind.

13.3.4 Formation of the Giant Planets

In the outer solar system, where the raw materials included ices as well as silicates, the protoplanets grew to be much larger, with masses ten times that of the Earth. These protoplanets of the outer solar system were so large that they were able to attract and hold the surrounding gas. As the hydrogen and helium rapidly collapsed onto their cores, the giant planets were heated by the energy of contraction. Such heating also begins the process of forming stars. But these giant planets were far too small to raise their central temperatures and pressures to the point where nuclear reactions could begin (and it is such reactions that give us our definition of a star). After glowing dull red for a few thousand years, the giant planets gradually cooled to their present state.

The collapse of nebular gas onto the cores of the giant planets explains how these objects acquired nearly the same hydrogen-rich composition as the Sun itself. The process was most efficient for Jupiter and Saturn; hence their compositions are most nearly "cosmic." Much less gas was captured by Uranus and Neptune, which is why these two planets have compositions dominated by the icy and rocky building blocks that made up their large cores, rather than by hydrogen and helium gas.

13.3.5 Further Evolution of the System

All of the processes we have just described, from the collapse of the solar nebula to the formation of protoplanets, occurred within a few million years. However, the story of the formation of the solar system was not complete at this stage; there were many planetesimals and other debris that did not initially accumulate to form the planets. What was their fate?

Recall from Chapter 12 that we have evidence that the comets visible to us are merely the tip of the cosmic iceberg (if you'll pardon the pun). Most comets are believed to be in the Oort cloud, far from the region of the planets. Additional comets are in the Kuiper belt, which stretches beyond the orbit of Neptune. Our models show that the Oort cloud comets probably formed near the present orbits of Uranus and Neptune, but they were ejected from their initial orbits by the gravitational influence of the giant planets.

In the inner parts of the system, remnant planetesimals and perhaps several dozen protoplanets continued to whiz about. Over the vast span of time we are discussing, collisions among these objects were inevitable. Giant impacts at this stage probably stripped Mercury of part of its mantle and crust, reversed the rotation of Venus, and broke the Earth apart to create the Moon.

Smaller-scale impacts also added mass to the inner protoplanets. Because the gravity of the giant planets could "stir up" the orbits of the planetesimals, the material striking the inner protoplanets could have come from almost anywhere within the solar system. In contrast to the previous stage of accretion, therefore, this new material did not represent just a narrow range of compositions (from "local" planetesimals).

Much of the debris striking the inner planets was ice-rich material that had condensed in the outer part of the solar nebula. As this cometlike bombardment progressed, the Earth accumulated the water and various organic compounds that would later be critical to the formation of life. Mars and Venus probably also acquired water and organic materials from the same source.

Gradually, as the planets swept up or ejected the remaining debris, most of the leftover planetesimals disappeared. In two regions, however, stable orbits are possible where leftover planetesimals could avoid impacting the planets or being ejected from the system. These regions are the asteroid belt between Mars and Jupiter and the Kuiper belt beyond Neptune. The planetesimals (and their fragments) that survive in these special locations are what we now call the asteroids and KBOs.

13.4 PLANETS BEYOND THE SOLAR SYSTEM

One of the most exciting recent developments in astronomy is the discovery of planets circling other stars—a discovery that makes it possible to begin to use the comparative approach to studying planetary systems, just as we have been doing for the study of individual planets. But planets outside the solar system are extremely difficult to detect. Recall that we see planets in our own system only because they reflect sunlight and are close by. When we look to the other stars, we find that the amount of light a planet reflects is a depressingly tiny fraction of the light its star gives off. Furthermore, from a distance, planets are lost in the glare of their much brighter parent stars.

You might compare a planet orbiting a star to a mosquito flying around one of those giant spotlights at shopping center openings. From close up, you might spot the mosquito. But now imagine viewing the scene from some distance away—say, from an airplane. You could see the spotlight just fine, but what are your chances of catching the light reflected from the mosquito?

Even though astronomers have not yet succeeded in *seeing* a planet around another star, they have detected planets elsewhere *indirectly*—by studying the disk of material from which planets might be condensing and by observing a planet's effects on its parent star.

13.4.1 Embryonic Planetary Systems

It is a lot easier to detect the spread-out raw material from which planets might be assembled than to detect planets after they are fully formed. From our study of the solar system, we understand that planets form by the gathering together of gas and dust particles in a circumstellar disk spinning around a young star. Each dust particle is heated by the young protostar and radiates in the infrared region of the spectrum. Before any planets form, we can detect such radiation from all of the individual dust particles that are destined to become part of planets. We can also detect the silhouette of the disk if it blocks bright light coming from a source behind it (Figure 13.10).

However, once the dust particles gather together and form a few planets (and maybe some moons), the overwhelming majority of the dust is hidden in the interiors of the planets where we cannot see it. All we can detect is the radiation from the outside surfaces, which cover a drastically smaller area than the huge dusty disk from which they formed. The amount of infrared radiation is therefore greatest before the dust particles combine into planets.

Many of the circumstellar disks show internal structure, although the details are beyond the resolving power of current telescopes. It seems as if these disks are donut-shaped, with one or more gaps close to the star. Such gaps may indicate that the gas and dust in the disk has already collapsed to form a planet (Figure 13.11). The planets themselves are far too small and faint to be seen directly, but the depletion of raw materials hints at the presence of something invisible in the inner part of the circumstellar disk—and that something may well be one or more planets.

The presence of a disk of gas and dust appears to be an essential part of star formation. Observations show that nearly all very young protostars have disks and that the disks range in size from 10 to 1000 AU. (For comparison, the average diameter of the orbit of Pluto—that is, the rough size of our own planetary system—is 80 AU.) The mass contained in these disks is typically 1 to 10 percent of the mass of our own Sun, which is similar to the mass we have been discussing for our own solar nebula. Observations therefore demonstrate that a large fraction of stars begin their lives with enough material in the right place to form a planetary system. Older stars generally do not have such circumstellar disks; this observation is consistent with the rather rapid dissipation of the solar nebula as discussed above. These comparisons are encouraging, and they suggest that planet formation may be common. But this information still does not give us a way to detect the planets themselves.

13.4.2 Search for Orbital Motion

Astronomers have tried a variety of techniques to search for evidence of planets around mature stars that have lost their disks, with dramatic recent success. Since planets are

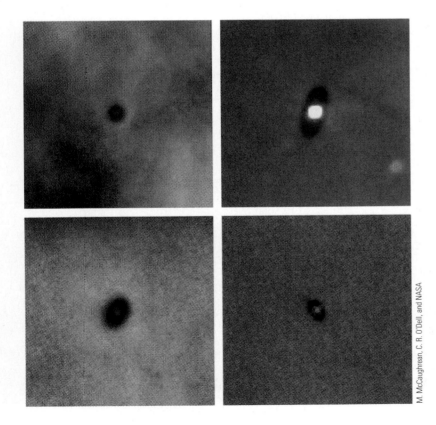

M. McCaughrean, C. R. O'Dell, and NASA

■ FIGURE 13.10

Disks Around Protostars These Hubble Space Telescope images show four disks around young stars in the Orion Nebula. The dark, dusty disks are seen silhouetted against the bright backdrop of the glowing gas in the nebula. The size of each image is about 30 times the diameter of our planetary system; this means the disks we see here range in size from two to eight times the orbit of Pluto. The red glow at the center of each disk is a young star, no more than a million years old.

■ **FIGURE 13.11**

Model of a Disk with Embedded Plane Model for a circumstellar disk. This computer simulation shows the empty lane and spiral density waves that result as one giant planet is forming within the disk.

Geoffrey Bryden/JPL/NASA

so difficult to image, the kinds of experiments possible with current technology look for the effects of planets *on their stars*. In particular, astronomers are measuring subtle changes in the star's motion through space caused by the pull of surrounding planets.

To understand how this approach works, consider a single Jupiter-like planet in orbit about a star. Both the planet and the star in such systems actually revolve about their *common center of mass*. Remember from our discussion in Chapter 2 that gravity is a mutual attraction. The star and the planet each exert a force on the other, and we can find a stable point between them about which both objects move. The smaller the mass of a body in such a system, the larger the orbit. So a massive star barely swings around the center of mass, while a low-mass planet makes a much larger "tour."

Suppose the planet is like Jupiter and has a mass about one-thousandth that of the star; in this case, the size of the star's orbit is one-thousandth the size of the planet's. To get a sense of how difficult observing such motion might be, let's see how hard Jupiter would be to detect from the distance of a nearby star. Consider alien astronomers trying to observe our own system from Alpha Centauri, the closest star system to our own (about 4 LY away).

Even if they had the best telescopes we presently use, these alien astronomers could not detect Jupiter directly. But there are two ways these alien astronomers could try to detect the orbital motion of the Sun. One way is to look for changes in its position on the sky. The second is to use the Doppler effect to look for changes in its velocity.

The diameter of Jupiter's apparent orbit viewed from Alpha Centauri is 10 seconds of arc (arcsec), and the diameter of the Sun's orbit is 0.010 arcsec. (A second of arc is

1/3600 degree.) If the astronomers could measure the apparent position of the Sun (which is bright and easy to detect) to sufficient precision, they would see it describe an orbit of diameter 0.010 arcsec with a period equal to that of Jupiter, which is 12 years. In other words, if they watched the Sun for 12 years, they would see it wiggle back and forth in the sky by this minuscule fraction of a degree. From the observed motion and the period of the "wiggle," they could deduce the mass of Jupiter and its distance using Kepler's laws.

Measuring positions in the sky this accurately is extremely difficult, and so far Earth-based astronomers have not made any confirmed detections of planets using this technique. New NASA missions are being planned, however, which will someday make such measurements from space.

As the star and planet orbit each other, part of their motion will be in our line of sight (toward us or away from us). Such motion (as discussed in Chapter 4) can be measured using the *Doppler effect* and the star's spectrum. As the star moves in orbit around the system's center of mass in response to the gravitational tug of an orbiting planet, the lines in its spectrum will shift back and forth (Figure 13.12).

Let's again consider the example of the Sun. Its *radial velocity* (motion toward or away from us) changes by about 13 m/s with a period of 12 years because of the gravitational pull of Jupiter. This corresponds to about 30 mi/h, roughly the speed at which many of us drive around town. Detecting motion at this level in a star's spectrum presents an enormous technical challenge, but several groups of astronomers around the world, using spectrometers designed for this purpose, have recently succeeded.

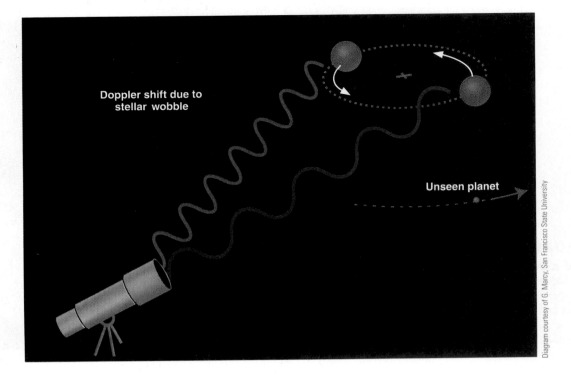

Doppler shift due to
stellar wobble

Unseen planet

Diagram courtesy of G. Marcy, San Francisco State University

■ FIGURE 13.12

Doppler Method of Detecting Planets The motion of a star and a planetary companion around a common center of mass can be detected as a cyclical change in the Doppler shift of the star. When the star is moving away from us, the lines in its spectrum show a tiny redshift; when it is moving toward us, they show a tiny blueshift. The change in color (wavelength) has been exaggerated here for educational purposes. In reality, the Doppler shifts we measure are extremely small and require sophisticated equipment to be detected.

13.4.3 Discovery of Planets

Michel Mayor and Didier Queloz of the Geneva Observatory used precise Doppler measurements of velocity to discover the first planet in 1995 (Figure 13.13). This planet orbits a star resembling our Sun called 51 Pegasi, about 40 LY away. (The star can be found in the sky near the great square of Pegasus, the flying horse of Greek mythology, one of the easiest-to-find star patterns.) To everyone's surprise, the planet takes a mere 4.2 days to orbit around the star. Remember that Mercury, the innermost planet in our solar system, takes 88 days to go around once.

This means the planet must be very close to 51 Pegasi, circling it about 7 million km away (Figure 13.14). At that distance, the energy of the star should heat the planet's surface to a temperature of more than a thousand degrees C (a bit hot for future tourism). From its motion, astronomers calculate that this planet has at least half the mass of Jupiter, making it clearly a jovian and not a terrestrial-type planet.

Since that initial discovery, the rate of progress has been breathtaking. As we write this paragraph, more than 100 giant planets have been discovered orbiting other stars. Most of the discoveries are due to Geoffrey Marcy, Paul Butler, Steven Vogt, and their collaborators, primarily based

M. Mayor and D. Queloz

■ FIGURE 13.13

European Planet-Hunting Team Michel Mayor and Didier Queloz of the Geneva Observatory were the first to discover a giant planet around a Sun-like star. They are continuing their work using telescopes in Europe and in Chile and have found several other planets since 1995.

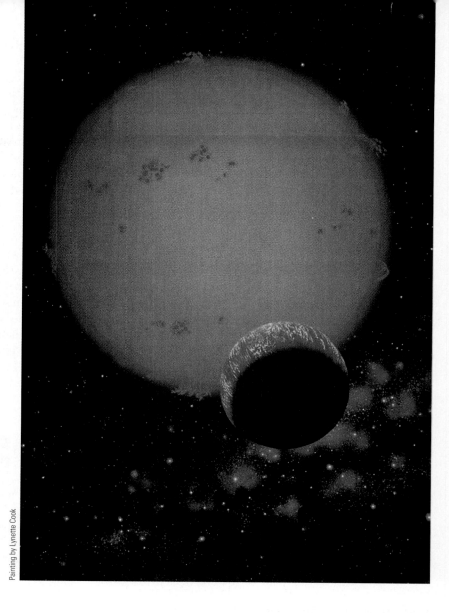

FIGURE 13.14

Giant Planet Close to a Sun-Like Star
This artist's concept shows what the giant
planet discovered orbiting the star 51 Pegasi
might look like close up. As explained in the
chapter, this planet was the first of more than
a dozen jovian planets found around other
stars whose orbits turned out to be smaller
than the orbit of Mercury in our own system.
The planet around 51 Pegasi is only about
7 million km from its star, taking a mere
4.2 days to complete its orbit. At that distance, the surface temperature of the planet
would be so hot that (as astronomer Geoff
Marcy has quipped) "A roast chicken might
be done in a microsecond." The artist has
shown prominences and sunspots on 51
Pegasi, evidence of an active atmosphere that
might extend a significant way to the giant
planet. She also shows the planet with bands
like Jupiter, although our measurements can
allow us to estimate only the mass of the
planet, not its density, and thus we have
no idea what sorts of materials the planet
is made of.

Painting by Lynette Cook

on observations with the 3-m telescope at the Lick Observatory in California and the 10-m Keck telescope in Hawaii
(Figure 13.15). There is about one new discovery each
month, and so we recommend that you refer to the Web
sites listed at the end of this chapter for up-to-date information. But stop for a minute and think about what we are
discussing. Less than a decade after the first discovery, we
now know ten times as many planets outside our solar system than within it!

Many of the initial planets discovered are massive
(Jupiter-mass or larger) planets orbiting close to their stars;
astronomers have called these "hot Jupiters." The existence
of giant planets so close to their stars was a surprise, and
these observations have forced us to re-think our ideas
about how planetary systems form (something we'll discuss
in a moment). But for now, bear in mind that the Doppler-
shift method—which relies on the pull of a planet making
its star "wiggle" back and forth around the center of
mass—is most effective at finding planets that are close to
their stars and massive. These cause the biggest "wiggles"
in the motion of their stars and take the shortest time to

San Francisco State University

FIGURE 13.15

American Planet-Hunting Team   Paul Butler and Geoff
Marcy were both at San Francisco State University when they
confirmed the discovery of the planet around 51 Pegasi and
went on to discover most of the planets that have been found
around other stars so far. Later, they separated, with Butler
going to the Anglo-Australian Observatory to look for planets
in the Southern Hemisphere and Marcy continuing the search
in the Northern Hemisphere.

complete a full orbit. So if such planets exist, we would expect to be finding them first. (Scientists call this a *selection effect*—where our technique of discovery selects certain kinds of objects as "easy finds.") As we spend longer times watching target stars and as our ability to measure smaller Doppler shifts improves, this technique can reveal more distant and less massive planets, too. It is only very recently that astronomers have found a handful of solar systems like our own—that is, systems with Jupiter-sized planets taking several years to orbit their stars.

Figure 13.16 and Table 13.3 summarize the characteristics of the planets discovered through 2002. (If some of the terms used to describe orbits in the table are not familiar, refer back to Chapter 2.) The masses of the planets found so far range from about a tenth the mass of Jupiter to more than 13 Jupiter masses. Astronomers know only the *minimum* mass of the planets. To determine the exact mass using the Doppler shift and Kepler's laws, we must also know the angle at which the planet's orbit is oriented to our view—something we don't have an independent way of knowing under most circumstances. Still, if the minimum mass is as large as the ones listed in the table, then we know for sure that we are dealing with giant planets, many of which have significantly more mass than our Jupiter.

13.4.4 Other Planetary Systems

The first system of multiple planets orbiting a solar-type star was discovered orbiting around Upsilon Andromedae, a star that is about 3 billion years old and about 44 LY away. The first planet discovered in this system has an orbital period of the planet of 4.6 days, and its mass is at least 70 percent the mass of Jupiter. However, the velocity variations of the star do not follow the smooth curve that would be expected if its motion were influenced by a single planet. The perturbations are caused by two other planets, both of which are more massive than Jupiter. The orbit of the most distant one is slightly larger than that of Mars. A few other systems with multiple planets are now known and the data contain strong hints that more such systems will be found as time goes on.

Another interesting planet is the one orbiting the star HD 209458, also in the constellation of Pegasus. It turns out this planet passes directly in front of its parent star for about 3 hours every 3.5 days as we view it from Earth. A careful measurement of the tiny bit of starlight that is blocked as the planet crosses the face of the star then tells us how large the planet is. Using the Hubble Space Telescope, astronomers are now able to rule out any moons around this planet larger than 1.2 Earths, since such a moon would also have caused some measurable dimming.

The planet around HD 209458 has about 70 percent the mass of Jupiter but its radius is about 35 percent larger than Jupiter's. Perhaps the planet is bloated because it is so hot. Orbiting in less than 4 days, it is so close to the star that the temperature of the planet is predicted to be more than 1500° Celsius. (Indeed, observations with the Hubble Space Telescope indicate that the outer atmosphere of the planet is actually evaporating into space, forming a giant "coma" that extends over 200,000 km.) In any case, these measurements confirm that the planet is a gas giant. When the planet passes in front of HD 209458, the atoms in the planet's atmosphere absorb starlight. Observations of this absorption were first made at the wavelengths of the yellow sodium lines, and show that there is sodium in the atmosphere of the planet. Searches for other elements, such as methane, water vapor, and potassium, will be made in the future.

As more Doppler observations are made and more time passes, astronomers expect to find planets with still lower masses, at larger distances from their parent stars, and hence with periods more like those of the giant planets in our own solar system. In order to detect a planet with an orbit and a mass like that of Jupiter, astronomers would have to make measurements with an accuracy of about 3 m/s for more than a decade, and most of the 2000 stars currently being surveyed for planets have been monitored for fewer than 10 years.

13.5 COMPARISON WITH OTHER PLANETARY SYSTEMS

Until the middle 1990s, the study of the origin of planets focused on our single known example—the solar system. Although there had been a great deal of speculation about planets circling other stars, none had actually been detected. Logically enough, in the absence of data most scientists assumed that our own system was likely to be typical. As we saw in the previous section, however, most of the newly discovered systems are quite different from our own. This unexpected result requires us to reexamine the theory of planetary formation.

13.5.1 The Configurations of Other Planetary Systems

Let's summarize the data on the first hundred external planets, all discovered between 1995 and 2002, to see what they can tell us about the way other planetary systems are configured. As we described in the previous section, these hundred planets were found from a careful survey of about 700 solar-type stars. Thus roughly 15 percent of these stars had systems such as we will describe below. The other 85 percent either have no planets or (more likely) have planetary systems that do not include hot Jupiters. It is important to remember that we still cannot tell whether our solar system is typical or highly unusual. Certainly the first hundred external planets are very different from our own system, but this is at least partly a consequence of the limitations of current techniques, which could not detect a

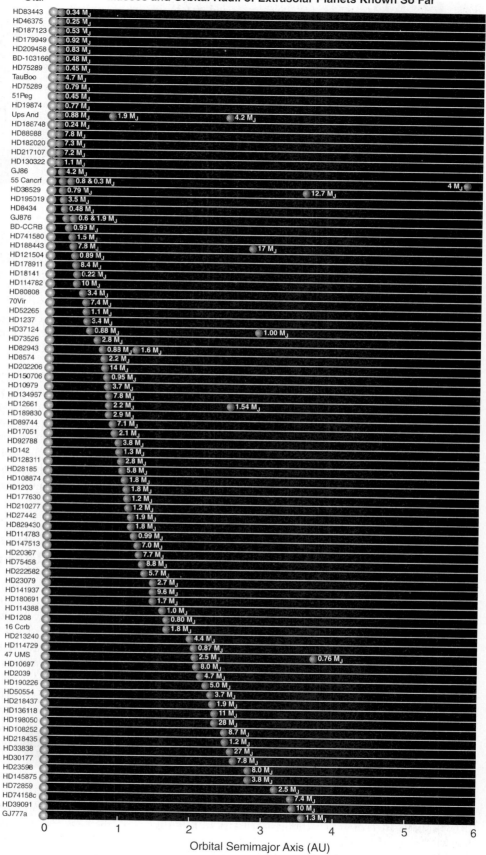

FIGURE 13.16

Masses and Orbital Radii of Known Extrasolar Planets A plot showing the planetary systems discovered by summer 2002, along with their masses in terms of the mass of Jupiter and their sizes of their orbits measured in astronomical units. Note that four of the stars have more than one planetary companion.

TABLE 13.3	*Some Properties of the First Extrasolar Planets Discovered*	
Number Found	**Characteristic**	
1	System of triple planets	
8	Systems containing two planets	
16	Planets with periods shorter than ten days	
33	Planets closer to their stars than Mercury	
40	Planets closer to their stars than 1 AU, which is the semimajor axis of Earth's orbit	
49	Planets with periods longer than 1 year	
54	Planets with eccentricities larger than Pluto, which is the largest in the solar system	
6	Planets with minimum mass less than that of Saturn; all are closer to their Sun than Mercury	

■ FIGURE 13.17

Masses of External Planets This graph shows the masses of the first hundred extrasolar planets. Note that few, if any, have masses larger than about 12 times Jupiter; that is, all are real planets, not failed stars. The number of planets seems to be increasing toward small masses, in spite of the bias against finding small planets by the Doppler method.

system like our own in which the only giant planets have periods of 12 years or longer.

One third of these systems have true "hot Jupiters"—that is, planets of jovian mass that are closer than the orbit of Mercury. Fully half of them have giant planets with periods of less than 1 year—that is, lying inside the Earth's orbit. Nothing like this was expected from the theory of formation of our solar system as presented in Section 13.3, in which giant planets could form only in a region that was cold enough for ice to condense in the circumstellar disk.

Roughly half of the systems studied so far include one or more giant planets with eccentric orbits—eccentricities greater than that of Pluto (0.25), which has the most irregular planetary orbit in our solar system. Orbits of high eccentricity are not expected for planets that form in a spinning disk where the building blocks have nearly circular orbits, but eccentricities might increase later as the result of gravitational interactions between planets. (Note that we cannot measure the inclinations of the orbits of these planets, so we do not know whether they are confined to a plane as are the planets in our own system.)

Most of the planets are apparently alone; that is, only one member of a system has been found. If there are multiple planets, the others are too small or too far out to have been discovered. However, even with the limitation of our present data, about 10 percent of these systems contain two or even three planets.

One of the most interesting results from the first hundred external planets is illustrated in Figure 13.17. Here we have plotted the planets according to their mass. As expected from the search techniques, most are more massive than Jupiter (since these are the easiest to find). But only two have more than 11 Jupiter masses, and none of them is in the range above 15 Jupiter masses where the distinction between planets and very low mass stars becomes blurred. This result shows us that there really is a difference be-

tween planets and stars. Planets are not just small or failed stars. They form by a different process, presumably from a spinning circumstellar disk as we have described.

Note also in Figure 13.17 that the numbers of detected planets increases toward smaller sizes, even though small planets are much harder to detect. Although we cannot detect terrestrial planets, this result suggests that there should be many more terrestrial mass planets than those of jovian mass.

13.5.2 Confrontation Between Theory and Data

The planetary systems discovered so far do not resemble our own solar system. In particular, the presence of hot Jupiters was totally unexpected. Consequently, it has been necessary to revisit the "standard models" for the formation of planetary systems. Science sometimes works in this way, with new data contradicting our expectations. The press often talks about a scientist making experiments to "confirm" a theory. Indeed, it is comforting when new data support a hypothesis or theory and increase our confidence in a previous result. But the most exciting and productive moments in science often come when the new data contradict existing theories, forcing scientists to rethink their positions and develop new and deeper insights into the way nature works.

Nothing about the new planetary systems contradicts the basic idea that planets form from aggregations of material within circumstellar disks. To test this hypothesis we need to find multiplanet systems and determine whether all the planets orbit the star in the same direction and in the plane of the star's rotation. The new data don't help us much in this respect, since we are missing all the smaller planets and we can't measure orbital inclinations in any case.

The biggest problem is posed by the hot Jupiters. As far as we know, there is no way to form a giant planet without the condensation of water ice, and water ice is not stable close to a star. It therefore seems likely that all the giant planets, hot or "normal," formed at a distance of several AU from the star. Some of these giant planets must have subsequently migrated inward to explain the cases where we see giant planets orbiting in short-period orbits, which is true for between 5 and 10 percent of the solar-type stars investigated.

There are several suggestions for ways this migration might have occurred. Most involve interaction between the giant planets and the remnant circumstellar disk that might have taken place when the system was very young, before the disk dissipated. In such cases, there are processes that can transfer angular momentum from the planet to the disk, pulling the planet inward. The next question is how to stop the planet from plunging into the star. There are some ideas here, but no really convincing hypothesis. Maybe a number of inward-migrating giant planets were lost (eaten by the star) and we see only the planets that had not been consumed when the disk went away and the migration stopped.

One thing seems certain: If a planetary system initially formed with a configuration like our own, having the giant planets on the outside and terrestrial planets on the inside, then such a migration process would destroy the inner terrestrial planets. We will return to some of these ideas in the final chapter, where we consider the search for habitable planets and life beyond our own solar system.

13.6 PLANETARY EVOLUTION

While we await more discoveries of other planetary systems, let us look again at the early history of our own solar system, after the dissipation of the disk. The era of giant impacts was probably confined to the first 100 million years of solar system history, ending by about 4.4 billion years ago. Shortly thereafter, the planets cooled and began to assume their present aspects. Up until about 4 billion years ago they continued to acquire volatiles, and their surfaces were heavily cratered from the remaining debris that hit them. However, as external influences declined, all of the terrestrial planets, as well as the satellites of the outer planets, began to follow their own evolutionary courses. The nature of this evolution depended on the composition of each object, its mass, and its distance from the Sun.

13.6.1 Geological Activity

We have seen a wide range in the level of geological activity on the terrestrial planets and icy satellites. Internal sources of such activity (as opposed to pummeling from above) require energy, in the form of primordial heat left over from the formation of a planet or from the decay of radioactive elements in the interior. The larger the planet or satellite, the more likely it is to retain its internal heat and the more slowly it cools; this is the "baked potato effect" mentioned in Chapter 6. Therefore we are more likely to see evidence of continuing geological activity on the surface of larger worlds (Figure 13.18). Jupiter's satellite Io is an interesting exception to this rule; we saw that it has an unusual source of heat—the gravitational flexing of its interior by the tidal pull of Jupiter.

The Moon, the smallest of the terrestrial worlds, was internally active until about 3.3 billion years ago, when its major volcanism ceased. Since that time its mantle has cooled and become solid, and today even internal seismic activity has declined to almost zero. The Moon is a geologically dead world. Although we know much less about Mercury, it seems likely that this planet, too, ceased most volcanic activity about the same time the Moon did.

Mars represents an intermediate case, and it has been much more active than the Moon. The southern hemisphere

FIGURE 13.18

Stages in the Geological History of a Terrestrial Planet In this graph, time increases downward along the left axis and the stages are labeled. Each planet is shown roughly in its present stage. The smaller the planet, the more quickly it passes through these stages.

Geological history of the terrestrial planets

Accretion, heating, differentiation

Formation of solid crust, heavy cratering

Widespread mare-like volcanism

Reduced volcanism, possible plate tectonics

Mantle solidification, end of tectonic activity

Cool interior, no activity

Moon   Mercury   Mars   Venus   Earth

Olympus Mons

Mt. Everest

Mauna Loa

Maxwell Mountains

Sea level

■ **FIGURE 13.19**

The Highest Mountains on Mars, Venus, and Earth   Mountains can rise taller on Mars because the surface gravity is less and there are no moving plates. The vertical scale is exaggerated by a factor of three to make comparison easier.

crust had formed by 4 billion years ago, and the northern hemisphere volcanic plains seem to be contemporary with the lunar maria. However, the Tharsis bulge formed somewhat later, and activity in the large Tharsis volcanoes has apparently continued intermittently to the present.

Earth and Venus are the largest and most active terrestrial planets. Our planet experiences global plate tectonics driven by convection in its mantle. As a result, our surface is continually reworked, and most of the Earth's surface material is less than 200 million years old. Venus has generally similar levels of volcanic activity, but unlike the Earth, it has not experienced plate tectonics. Most of its surface appears to be roughly 500 million years old. We did see that the surface of our sister planet is being modified by a kind of "blob tectonics"—where hot material from below puckers and bursts through the surface, leading to coronae, pancake volcanoes, and other such features. A better understanding of the geological differences between Venus and Earth is a high priority for planetary geologists.

The geological evolution of the icy satellites has been distinct from that of the terrestrial planets. The energy sources were similar, but the materials are different. On these outer worlds, we see evidence of low-temperature volcanism, with the silicate lava of the inner planets being supplemented by sulfur and sulfur compounds on Io, and replaced by water and other ices on other outer-planet satellites. We also see the effects of tidal heating, especially among the satellites of Jupiter.

13.6.2   Elevation Differences

The mountains on the terrestrial planets have very different origins. On the Moon and Mercury, the major mountains are ejecta thrown up by the large basin-forming impacts that took place billions of years ago. Most large mountains on Mars are volcanoes, produced by repeated eruptions of lava from the same vents. There are similar (but smaller) volcanoes on Earth and Venus. However, the highest mountains on Earth and Venus are the result of compression and uplift of the surface. On the Earth, this crustal compression results from collisions of one continental plate with another.

It is interesting to compare the maximum heights of the volcanoes on Earth, Venus, and Mars (Figure 13.19). On Venus and Earth, the maximum elevation differences between these mountains and their surroundings are about 10 km. Olympus Mons, in contrast, towers 26 km above its surroundings, and nearly 30 km above the lowest elevation areas on Mars (Figure 13.20).

One reason Olympus Mons is so much larger than its terrestrial counterparts is that the crustal plates on Earth never stop long enough to let a really large volcano grow. Instead, the moving plate creates a long row of volcanoes like the Hawaiian Islands. On Mars (and perhaps Venus) the crust remains stationary with respect to the underlying hot spot, and so the volcano can continue to grow for hundreds of millions of years.

A second difference relates to the force of gravity on the three planets. The surface gravity on Venus is nearly the same as that on Earth, but on Mars it is only about one-third as great. In order for a mountain to survive, its internal strength must be great enough to support its weight against the force of gravity. Volcanic rocks have known strengths, and we can calculate that on Earth 10 km is about the limit. For instance, when new lava is added to the top of Mauna Loa in Hawaii, the mountain slumps downward under its own weight. The same height limit applies on Venus, where the force of gravity is the same as ours. On Mars, however, with its lesser surface gravity, much greater elevation differences can be supported, which helps explain why Olympus Mons is more than twice as high as the tallest mountains of Venus or Earth.

The same kind of calculation that determines the limiting height of a mountain can be used to ascertain the largest body that can have an irregular shape. Gravity, if it can, pulls all objects into the most "efficient" shape (where all the outside points are equally distant from the center). All of the planets and larger satellites are nearly spherical, due to the force of their own gravity having pulled them into a sphere. But the smaller the object, the greater is the departure from spherical shape that the strength of its rocks can support. For silicate bodies, the limiting diameter is about 400 km; larger objects will always be approximately spherical, while smaller ones can have almost any shape (as we see in photographs of asteroids, such as Figure 13.21).

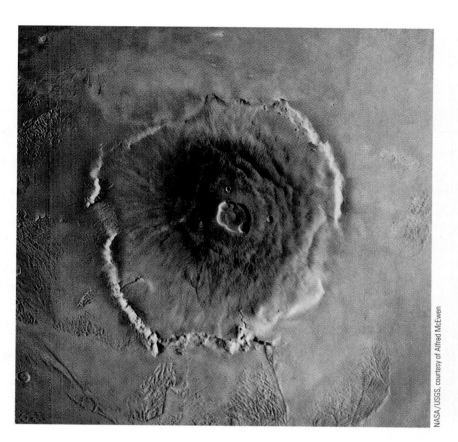

NASA/USGS, courtesy of Alfred McEwen

■ **FIGURE 13.20**

Olympus Mons The largest martian volcano is seen here from above on this spectacular composite image created from many Viking orbiter photographs. The volcano is nearly 500 km wide at its base and 26 km high.

JPL/NASA

■ **FIGURE 13.21**

Irregular Asteroid Small objects such as asteroid Ida (shown here in multiple views taken by the Galileo spacecraft camera as it flew past) are generally irregular or elongated; they do not have a strong enough gravity to pull them into into a spherical shape.

13.6.3 Atmospheres

The atmospheres of the planets were formed by a combination of gas escaping from their interiors and the impacts of volatile-rich debris from the outer solar system. Each of the terrestrial planets must have originally had similar atmospheres, but Mercury was too small and too hot to retain its gas. The Moon probably never had an atmosphere, since the material composing it was depleted in volatiles.

The predominant volatile gas on the terrestrial planets is now carbon dioxide (CO_2), but initially there were probably also hydrogen-containing gases. In this more chemically reduced environment, there should have been large amounts of carbon monoxide (CO) and perhaps also traces of ammonia (NH_3) and methane (CH_4). Ultraviolet light from the Sun destroyed the reducing gases in the inner solar system, however, and most of the hydrogen escaped, leaving behind the oxidized atmospheres we see today on Earth, Venus, and Mars.

The fate of water was different for each of these three planets, depending on its size and distance from the Sun. Early in its history, Mars apparently had a thick atmosphere with abundant liquid water, but it could not retain those conditions. As we saw in Chapter 9, the CO_2 necessary for a substantial greenhouse effect was lost, the temperature dropped, and eventually the water froze. On Venus the reverse process took place, with a runaway greenhouse effect leading to the permanent loss of water. Only the Earth managed to maintain the delicate balance that permits liquid water to persist.

With the water gone, Venus and Mars each ended up with an atmosphere of about 96 percent carbon dioxide and a few percent nitrogen. On Earth, the presence first of water and then of life led to a very different kind of atmosphere. The CO_2 was removed to be deposited in marine sediment, while proliferation of life that could do photosynthesis eventually led to the release of enough oxygen to overcome natural chemical reactions that eliminate this gas from the atmosphere. As a result, we on Earth find ourselves with a great deficiency of CO_2 and with the only planetary atmosphere that contains free oxygen.

In the outer solar system, Titan is the only satellite with a substantial atmosphere. Apparently this object contained sufficient volatiles—such as ammonia, methane, and nitrogen—to form an atmosphere. It was also large and cold enough to retain these gases. Thus, as we saw in Chapter 11, today Titan's atmosphere consists primarily of nitrogen. Compared with those on the inner planets, temperatures on Titan are too low for either carbon dioxide or water to be in vapor form. With these two common volatiles frozen solid, it is perhaps not too surprising that nitrogen has ended up as the primary atmospheric constituent.

We see that nature, starting with one set of chemical constituents, can fashion a wide range of final atmospheres appropriate to the conditions and history of each world. The atmosphere we have on Earth is the result of many eons of evolution and adaptation. And, as we saw, it can be changed by the actions of the life-forms that inhabit the planet.

13.7 CONCLUSION

Although we have learned a great deal about the other planets during the past few decades of spacecraft exploration, much remains unknown. Now that we have adopted the comparative approach to study these different worlds and the processes that form them and their environments, we see how limited our knowledge really is. More recently, we have had some surprises in store when we found planets around a number of nearby stars. Even such generalizations that giant planets form only in the outer parts of a planetary system may need reexamination in light of some of these new discoveries.

For now, we end by noting that we do not even understand our own planet well enough to assess the consequences of the changes humans are inadvertently inflicting on the atmosphere and oceans. The exploration of the solar system is one of the greatest human adventures, and it will continue, if only so that we can learn to appreciate and protect our own planet.

SURFING THE WEB

 Meteor Showers Site:
medicine.wustl.edu/~kronkg/index.html
Amateur astronomer Gary Kronk, author of a book on meteor showers, has set up a site with all the information anyone could want on major and minor showers, background and historical explanations, meteor legends, fireballs, observing hints, and links.

Sky On Line Meteor Page:
skyandtelescope.com/observing/objects/meteors
Another excellent site for serious or casual meteor observers, with lots of up-to-date information on upcoming showers, hints for viewing, and links to organizations of meteor watchers around the world.

Mars Meteorites: www.jpl.nasa.gov/snc/index.html
Good resource with information and links about meteorites from Mars.

 Nine Planets Meteors, Meteorites, and Impacts Page:
seds.lpl.arizona.edu/nineplanets/nineplanets/meteorites.html
A quick overview with a long list of links.

For images of dust disks around other stars, check the gallery of Hubble Space Telescope images at: hubblesite.org/newscenter/archive

 Extra-Solar Planets Encyclopedia:
www.obspm.fr/encycl/encycl.html
Jean Schneider of the Paris Observatory keeps a detailed catalog of confirmed and unconfirmed new planets, with basic data, useful background information, and references.

 Searching for Extrasolar Planets: exoplanets.org
A site from Geoff Marcy's team about the search for planets with charts, summaries, papers, popular articles, links, and an excellent graphic summary.

 Lists of Planet Search Programs Worldwide:

- Darwin Site: ast.star.rl.ac.uk/darwin/searches.html
- Extrasolar Encyclopedia Site:
  www.obspm.fr/encycl/searches.html
  Excellent annotated lists of links to all the research programs that are working on or are proposed for working on the search for planets outside the solar system, including many we have not had room to discuss in the chapter.

SUMMARY

13.1 When a fragment of interplanetary dust strikes the Earth's atmosphere, it burns up to create a **meteor.** Streams of dust particles traveling through space together produce **meteor showers,** in which we see meteors diverging from a spot in the sky called the *radiant* of the shower. Many meteor showers recur each year and are associated with particular comets.

13.2 Meteorites are the debris from space (mostly asteroid fragments) that survive to reach the surface of the Earth. Meteorites are called *finds* or *falls* according to how they are found; the most productive source today is the Antarctic ice cap. Meteorites are classified as **irons, stony-irons,** or **stones** according to their composition. Most stones are primitive objects, dated to the origin of the solar system 4.5 billion years ago. The most primitive are the **carbonaceous meteorites,** such as Murchison, Allende, and Tagish Lake. These can contain a number of organic molecules.

13.3 A viable theory of solar system formation must take into account motion constraints, chemical constraints, and age constraints. Meteorites, comets, and asteroids are survivors of the *solar nebula* out of which the solar system formed. This nebula was the result of the collapse of an interstellar cloud of gas and dust, which contracted (conserving its angular momentum) to form the *protosun* surrounded by a thin, spinning disk of dust and hot vapor. Condensation in the disk led to the formation of *planetesimals*, which became the building blocks of the planets. **Accretion** of infalling materials heated the planets, leading to their differentiation. The giant planets were also able to at-

tract and hold gas from the solar nebula. After a few million years of violent impacts, most of the debris was swept up or ejected, leaving only the asteroids and cometary remnants surviving to the present. All of these violent events had terminated by about 4.4 billion years ago.

13.4 At present we can search for planets around nearby stars by looking for changes either in the star's position on the sky over time or in its radial velocity (as seen in the Doppler shift of the lines in its spectrum). Radial velocity searches in the past 5 years have been stunningly successful. We now know of more than 100 such planets, most of them of Jupiter mass or greater, and many in orbital configurations very different from those in our own solar system.

13.5 The first external planets challenge our ideas about planetary system formation. Roughly 15 percent of stars like the Sun have jovian planets in relatively short-period orbits, and at least 5 percent have "hot Jupiters" inside 1 AU from their stars. Many of the new planets are also in eccentric orbits. These systems are fundamentally different from our own, and they suggest that giant planets can migrate inward from their place formation where it is cold enough for ice to form. Astronomers eagerly await the development of more sensitive tools so that we can detect smaller planets and perhaps find other systems that look more like our own; current data hint that there should be many more small (terrestrial-type) planets than giants.

13.6 After their common beginning, each of the planets evolved on its own path. The different possible outcomes are illustrated by comparison of the terrestrial planets

(Earth, Venus, Mars Mercury, and the Moon). All are rocky, differentiated objects. The level of geological activity is proportional to mass: greatest for Earth and Venus, less for Mars, and absent for the Moon and Mercury. Mountains can result from impacts, volcanism, or uplift. Whatever their origin, higher mountains can be supported on smaller planets where the surface gravity is less. All the terrestrial planets may have acquired atmospheric volatiles from comet impacts. The Moon and Mercury lost their atmospheres; most volatiles on Mars are frozen due to its greater distance from the Sun; and Venus retained CO_2 but lost H_2O when it developed a massive greenhouse effect. Only Earth still has liquid H_2O on its surface and hence can support life.

INTER-ACTIVITY

A Ever since the true (cosmic) origin of meteorites was understood, people have tried to make money selling them to museums and planetaria. More recently, a growing number of private collectors have been interested in purchasing meteorite fragments, and a network of dealers (some more reputable than others) has sprung up to meet this need. What does your group think of all this? Who should own a meteorite? The person on whose land it falls, the person who finds it, or the local, state, or federal government where it falls? Why? Should there be any limit to what people charge for meteorites? Or should all meteorites be the common property of humanity? (If you can, try to do research on what the law is now in this area.)

B Your group has been formed to advise a very rich person who wants to buy some meteorites but is afraid of being cheated and being sold some Earth rocks. What are some ways you would advise your client to make sure that the meteorites she buys are authentic?

C Your group is a committee set up to give advice to NASA about how to design satellites and telescopes in space so that the danger of meteor impacts is minimized. Remember that the heavier a satellite is, the harder (more expensive) it is to launch. What would you include in your recommendations?

D Discuss what you would do if you suddenly found that a small meteorite had crashed in or near your home. Whom would you call first, second, third? What would you do with the sample? (And would any damage to your home be covered by your insurance?)

E A friend of your group really wants to see a meteor shower. The group becomes a committee to assist her in fulfilling this desire. What time of year would be best? What equipment would you recommend she get? What advice would you give her?

F The search for planets around other stars is an example of a research problem where selection effects must be taken into account in order to determine whether the conclusions are valid. "Selection effect" is a fancy way of saying that the observations are in some way biased or not truly representative of the actual situation. Have your group discuss how selection effects are relevant in answering the following questions:

1. No terrestrial (Earth-like) planets have yet been discovered through measurements of radial velocities. Therefore, a friend of yours asserts that there must be no habitable planets in the universe other than the Earth. Is your friend right?

2. So far no brown dwarfs (objects with masses in the range, say, of 20–100 times the mass of Jupiter) have been found with orbits less than about 5 AU around other stars. A friend of yours says that therefore brown dwarfs are not likely to be found in close orbits around other stars. Is your friend right?

G Your group is a subcommittee of scientists examining the issue of whether any of the "hot Jupiters" (giant planets closer to their stars than Mercury is to the Sun) could have life on or near them. Can you come up with places on or near such planets where life could develop or forms of life that might survive there?

H Even before Earth-like planets have been found, some people consider the discovery of any planets (even hot Jupiters) around other stars one of the most important events in the history of astronomical research. Some astronomers have been surprised that the public is not more excited about the planet discoveries. One reason that has been suggested for this lack of public surprise and excitement is that science-fiction stories have long prepared us for there being planets around other stars. (The Starship Enterprise on the *Star Trek*™ series found some in just about every weekly episode!) What does your group think? Did you know about the discovery of planets around other stars before taking this course? Do you consider it exciting? Were you surprised to hear about it? Are science-fiction movies and books good or bad tools for astronomy education in general, do you think?

I What if future space instruments reveal an Earth-like planet with oxygen and methane in its atmosphere orbiting another star. Suppose the star is 50 LY away. What does your group suggest astronomers do next? How much effort and money would you recommend be put into finding out more about this planet, and why?

1. A friend of yours who has not taken astronomy sees a meteor shower. The next day she confides in you that she was concerned that the stars in the Big Dipper (her favorite star pattern) might be the next ones to go. How would you put her mind at ease?

2. In what ways are meteorites different from meteors? What is the probable origin of each?

3. What do we mean by primitive material? How can we tell whether a meteorite is primitive?

4. Describe the solar nebula, and outline the sequence of events within the nebula that gave rise to the planetesimals.

5. Why do the giant planets and their satellites have compositions different from those of the terrestrial planets?

6. Why is it so hard to see planets around other stars and so easy to see them around our own?

7. What techniques have been used to search for planets around other stars?

8. How were the first planets around stars like the Sun discovered?

9. Explain what the current ideas are about how "hot Jupiters" formed around other stars.

10. What do observations of other planetary systems tell us about the origin of our solar system?

11. Explain the role of impacts in planetary evolution, including both giant impacts and more modest ones.

12. Why are some planets more geologically active than others?

13. Summarize the origins and evolutions of the atmospheres of Venus, Earth, and Mars.

THOUGHT QUESTIONS

14. What sorts of methods do scientists use to distinguish a meteorite from terrestrial material?

15. Explain why we do not believe meteorites come from comets, or meteors from asteroids.

16. Why do iron meteorites represent a much higher percentage of finds than of falls?

17. Why is it more useful to classify meteorites according to whether they are primitive or differentiated, rather than whether they are stones, irons, or stony-irons?

18. Which meteorites are the most useful for defining the age of the solar system? Why?

19. Suppose a new primitive meteorite is discovered (sometime after it falls in a field of soybeans) and analysis reveals that it contains a trace of amino acids, all of which show the same rotational symmetry (unlike the Murchison meteorite). What might you conclude from this finding?

20. How do we know when the solar system formed? Usually we say that the solar system is about 4.5 billion years old. To what does this age correspond?

21. Why were giant planets close to their stars the first ones to be discovered? Why has the same technique not been used yet to discover giant planets at the distance of Saturn?

22. Imagine what would happen to our planetary system if Jupiter began to migrate inward to the orbit of Mercury. What would be the fate of the Earth and other terrestrial planets?

23. We have seen how Mars can support greater elevation differences than can Earth or Venus. According to the same arguments, the Moon should have higher mountains than any of the other terrestrial planets, yet we know it does not. What is wrong with applying the same line of reasoning to the mountains on the Moon?

FIGURING FOR YOURSELF

We can use the concept of angular momentum to trace the evolution of the collapsing solar nebula. The angular momentum of an object is proportional to the square of its size times its period of rotation (D^2P). If angular momentum is conserved, then any change in size must be compensated for by a proportional change in period, in order to keep D^2P a constant. Suppose that the solar nebula began with a diameter of 10,000 AU and a rotation period of 1 million years. Let's calculate its rotation period when it has shrunk to the size of Pluto's orbit. If the subscript 1 denotes the initial state before the collapse and 2 denotes the final state at Pluto's orbit, then $P_2/P_1 = (D_2/D_1)^2 = (80/10,000)^2 = 0.000064$. The new rotation period is 64 years, which is already faster than the Keplerian orbital velocity at Pluto's distance.

24. What would be the rotation period when the solar nebula had shrunk to the size of Jupiter's orbit? Of the Earth's orbit?

25. Consider the differentiated meteorites. We think the irons are from the cores, the stony-irons from the interfaces between mantles and cores, and the stones are from the mantles of their differentiated parent bodies. If these parent bodies were like the Earth, what fraction of the meteorites would you expect to consist of irons, stony-irons, and stones? Is this consistent with the observed numbers of each?

26. Estimate the maximum height of the mountains on a hypothetical planet similar to the Earth but with twice the surface gravity of our planet.

27. When astronomers found the first giant planets with orbits of only a few days, they did not know whether these planets were gaseous like Jupiter or rocky like Mercury. The observations of HD 209458 settled this question because observations of the transit of the planet made it possible to determine the radius of the planet. Use the data given in the text to estimate the density of this planet and then use that information to explain why it must be a gas giant.

SUGGESTIONS FOR FURTHER READING

On Meteors
Bagnall, P. "Watching Halley's Debris" in *Astronomy*, May 1992, p. 78. Describes the two meteor showers caused by Comet Halley.
Bone, N. *Meteors*. 1993, Sky Publishing. A good manual and information source.
Kronk, G. "Meteor Showers" in *Mercury*, Nov./Dec. 1988, p. 162.
Litmann, M. *The Heavens on Fire*. 1998, Cambridge U. Press. Best introduction to meteor showers in history and science.

On Meteorites
Chaikin, A. "A Stone's Throw from the Planets" in *Sky & Telescope*, Feb. 1983, p. 122. Discusses meteorites from the Moon and Mars.
Dodd, R. *Thunderstones and Shooting Stars: The Meaning of Meteorites*. 1986, Harvard U. Press. A clear guide to the science of meteorites.
Gallant, R. "Saga of the Lump: The Pallas Meteorite" in *Sky & Telescope*, Jan. 1999, p. 50.
Hutchinson, R. *The Search for Our Beginnings*. 1983, Oxford U. Press. Focuses on the role of meteorites in helping us understand the solar system.
McSween, H. *Meteorites and Their Parent Planets*. 1987, Cambridge U. Press. A good general introduction by a geologist.
Norton, O. *Rocks from Space,* 2nd ed. 1998, Mountain Press Publishing. A guide for amateurs, with information on science, folklore, and collecting.
Schaefer, B. "Meteor[ite]s That Changed the World" in *Sky & Telescope,* Dec. 1998, p. 68. On famous cases, including Tunkuska, Chicxulub, and the Black Stone of the Ka'aba.
Spratt, C. and Stephens, S. "Against All Odds: Meteorites That Have Struck [the Earth]" in *Mercury*, Mar./Apr. 1992, p. 50.

On Other Worlds
Boss, A. *Looking for Earths: The Race to Find New Solar Systems.* 1998, Wiley.
Croswell, K. *Planet Quest.* 1997, Harcourt Brace.
Doyle, L. "Searching for Shadows of Other Earths" in *Scientific American*, Sept. 2000, p. 58.
Fischer, D. "Prowling for Planets" in *Mercury* (the magazine of the Astronomical Society of the Pacific), July/Aug. 2000, p. 13. Review by the first woman in history to discover a planet.
Goldsmith, D. *Worlds Unnumbered: The Search for Extraterrestrial Planets.* 1998, Dutton.
Lewis, J. *Worlds Without End: The Exploration of Planets Known and Unknown.* 1998, Perseus Books.
Marcy, G. and Butler, R. "The Diversity of Planetary Systems" in *Sky & Telescope,* Mar. 1998, p. 30. A progress report on planet discoveries by the leading team. (See also their "Hunting Planets Beyond" in *Astronomy*, Mar. 2000, p. 43.)
McInnis, D. "Wanted: Life-Bearing Planets" in *Astronomy*, Apr. 1998, p. 38. On future instruments that could find planets more like the Earth.
Schilling, G. "The Race to Epsilon Eridani" in *Sky & Telescope*, June 2001, p. 34. On the star and its possible planet.
Stephens, S. "Planet Hunters" in *Astronomy,* July 1998, p. 58. Profile of Geoff Marcy and Paul Butler, who have found the largest number of extrasolar planets.

On the Evolution of the Solar System
See the readings recommended in Chapter 6, plus
Jayawardhana, R. "Spying on Planetary Nurseries" in *Astronomy*, Nov. 1998, p. 62. On dust disks around other stars that may be the birthplaces of other planetary systems.

Kasting, J. "How Climate Evolved on the Terrestrial Planets" in *Scientific American,* Feb. 1988, p. 90.

Rubin, A. "Microscopic Astronomy" in *Sky & Telescope,* July 1995, p. 24. Discusses the study of solar system history from analysis of samples.

Wood, J. "Forging the Planets: The Origin of Our Solar System" in *Sky & Telescope,* Jan. 1999, p. 36. Good overview.

Yulsman, T. "From Pebbles to Planets" in *Astronomy,* Feb. 1998, p. 56. Comparing the origin of our system to others.

Space Weather and Its Impact on Earth The first image shows a storm on the Sun, which blasts matter into space. This matter, which is composed of charged particles, mainly the nuclei of hydrogen atoms, speeds out into the solar system. If headed in the right direction, the particles will slam into the Earth. In the middle frame we see the northern hemisphere of the Earth, and the glow shows where charged particles are entering the Earth's atmosphere. The final image, taken by Jan Curtis, shows the visible evidence of space weather—an aurora in the sky over Alaska.

14

The Sun: A Garden-Variety Star

The adventure of the Sun is the great natural drama by which we live, and not to have joy in it and awe of it, not to share in it, is to close a dull door on nature's sustaining and poetic spirit.

Henry Beston, from "Midwinter" in *The Outermost House* (1928)

THINKING AHEAD

Can we trust the Sun? This may seem like a silly question—after all, the Sun has been rising in the east and setting in the west every day for billions of years. But the Sun has *not* put out exactly the same amount of heat during all that time; it has been both hotter and cooler than it is today. Since life on Earth ultimately depends on the Sun's heat and light, the way the Sun's energy output changes, and on what timescales, are questions of personal importance to those of us who share the Sun's cosmic neighborhood.

The Sun is, by many measures, a rather ordinary star—not unusually hot or cold, old or young, large or small. Indeed, we are lucky that the Sun is typical. By studying it, we learn much that helps us understand the stars in general. Just as studies of the Earth help us understand observations of the more distant planets, so too the Sun serves as a guide to astronomers in interpreting the messages contained in the light we receive from distant stars.

In this chapter we describe what the Sun looks like, how it changes with time, and how those changes affect the Earth. Some of the basic characteristics of the Sun are listed in Table 14.1, although many of the terms in that table may be unfamiliar to you until you've read further. Figure 14.1 shows what the Sun would look like if we could see all parts of it from the center to its outer atmosphere; the terms in the figure will also become familiar to you as you read on.

TABLE 14.1 *Solar Data*

Datum	How Found	Value
Mean distance	Radar reflection from planets	1 AU (149,597,892 km)
Maximum distance from Earth		1.521×10^8 km
Minimum distance from Earth		1.471×10^8 km
Mass	Orbit of Earth	333,400 Earth masses (1.99×10^{30} kg)
Mean angular diameter	Direct measure	31′59″.3
Diameter of photosphere	Angular size and distance	$109.3 \times$ Earth diameter (1.39×10^6 km)
Mean density	Mass/volume	1.41 g/cm^3 (1400 kg/m^3)
Gravitational acceleration at photosphere (surface gravity)	GM/R^2	$27.9 \times$ Earth surface gravity = 273 m/s^2
Solar constant	Instrument sensitive to radiation at all wavelengths	1370 W/m^2
Luminosity	Solar constant $\times$ area of spherical surface 1 AU in radius	3.8×10^{26} W
Spectral class	Spectrum	G2 V
Effective temperature	Derived from luminosity and radius of Sun	5800 K
Rotation period at equator	Sunspots and Doppler shift in spectra taken at the edge of the Sun	24 d 16 h
Inclination of equator to ecliptic	Motions of sunspots	7°10′.5

14.1 THE VISIBLE SUN

The Sun, like all stars, is an enormous ball of extremely hot gas, shining under its own power. And we do mean enormous. The Sun has enough volume (takes up enough space) to hold about 1.3 million Earths. Most of the Sun's regions are shielded from our view by its outside layers, just as the Earth shows only its surface and atmosphere to our direct observation. The Sun has no solid surface, but it does have an extensive atmosphere—the only part of the Sun we can actually see.

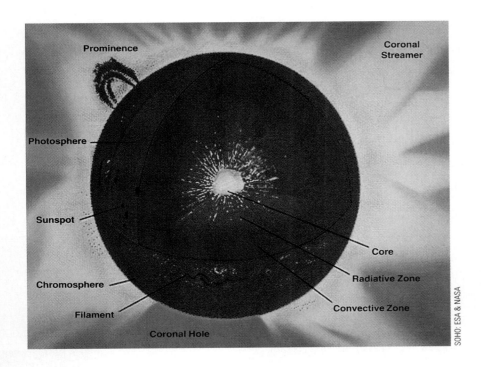

■ FIGURE 14.1

The Parts of the Sun This drawing shows the different parts of the Sun, from the hot core where the energy is generated through regions where energy is transported outward, first by radiation and then by convection and then out through the solar atmosphere. The parts of the atmosphere are also labeled—the photosphere, chromosphere, and corona. The artist shows some typical features in the atmosphere, such as coronal holes and prominences.

14.1.1 The Composition of the Sun

Let's begin by asking what the Sun's atmosphere is made of. As explained in Chapter 4, we can use its *absorption line spectrum* to determine what elements are present. It turns out that the Sun contains the same elements as the Earth, but *not* in the same proportions. About 73 percent of the Sun's mass is hydrogen, and another 25 percent is helium. All the other chemical elements (including those we know and love in our own bodies, such as carbon, oxygen, and nitrogen) make up only 2 percent of our star. The ten most abundant gases in the Sun's visible surface layer are listed in Table 14.2. Examine that table and notice how different the composition of the Sun's outer layer is from the Earth's crust, where we live. (In our planet's crust, the three most abundant elements are oxygen, silicon, and aluminum.) But the makeup of the Sun is quite typical of the stars in general.

The fact that the Sun and the stars all have similar compositions and are made up of mostly hydrogen and helium was first shown in a brilliant thesis in 1925 by Cecilia Payne-Gaposchkin, the first woman to get a PhD in astronomy in the United States (Figure 14.2). However, the idea that the simplest light gases, hydrogen and helium, were the most abundant elements in stars was so unexpected and so shocking that she assumed her analysis of the data must be wrong. At the time, she wrote, "The enormous abundance derived for these elements in the stellar atmosphere is almost certainly not real." Even scientists sometimes find it hard to accept new ideas that do not agree with what everyone "knows" to be right!

Before Payne's work, everyone assumed that the composition of the Sun and stars would be much like that of the Earth. It was three years after her thesis that other studies proved beyond doubt that the enormous abundance of hydrogen and helium in the Sun is indeed real. (And, the composition of the stars is much more typical of the makeup of

■ **FIGURE 14.2**

Cecilia Payne-Gaposchkin Her 1925 PhD thesis laid the foundations for understanding the composition of the Sun and the stars. Yet she was not given a formal appointment at Harvard, where she worked, until 1938 and was not appointed a professor until 1956.

the universe than the odd concentration of heavier elements that characterizes our planet.)

Most of the elements found in the Sun are in the form of atoms, but several types of molecules, including water vapor and carbon monoxide, have been identified in the light from the Sun's cooler regions, such as the sunspots (see Section 14.2). The atoms and molecules contained in the Sun are all in the form of gases: The Sun is so hot that no matter can survive as a liquid or a solid.

In fact, the Sun is so hot that many of the atoms in it are *ionized*—stripped of one or more of their electrons. This separation between electrons and protons means that the Sun is an electrically charged environment—quite different from the neutral one in which you are reading this book. As we saw in Chapter 4, when charged particles flow, magnetic fields are generated. As we shall see, the strong, complicated magnetism of the Sun plays a vital role in how it appears to us.

14.1.2 The Solar Photosphere

The Earth's air is generally transparent. But on a smoggy day it can become opaque, which prevents us from seeing through it past a certain point. Something similar happens in the Sun. Its outer atmosphere is transparent, allowing us to look a short distance through it. But when we try to look through the atmosphere deeper into the Sun, our view is blocked. The **photosphere** is the layer where the Sun becomes opaque and marks the boundary past which we cannot see (Figure 14.3).

The energy that emerges from the photosphere was originally generated deep inside the Sun. This energy is in the form of photons, which make their way slowly toward

Element	Percentage by Number of Atoms	Percentage by Mass
TABLE 14.2 *The Abundance of Elements in the Sun*		
Hydrogen	92.0	73.4
Helium	7.8	25.0
Carbon	0.02	0.20
Nitrogen	0.008	0.09
Oxygen	0.06	0.8
Neon	0.01	0.16
Magnesium	0.003	0.06
Silicon	0.004	0.09
Sulfur	0.002	0.05
Iron	0.003	0.14

Approx. size of Earth ➞ ·

SOHO: NASA & ESA

■ FIGURE 14.3

Solar Photosphere plus Sunspots This photograph shows the photosphere—the visible surface of the Sun. Also shown is an enlarged image of one of the groups of sunspots; the size of the Earth is shown for comparison. Sunspots appear darker because they are cooler than their surroundings. The typical temperature at the center of a large sunspot is about 4000 K, while the photosphere has a temperature of about 5800 K.

the solar surface, taking about a million years to reach the photosphere. Inside the Sun, a photon of energy emitted by one atom travels only a short distance before it is "grabbed" (i.e., absorbed) by another. Outside the Sun, we can observe only those photons that are emitted in the solar photosphere, where the density of atoms is low and the photons can finally escape from the Sun without being captured.

As an analogy, imagine you are attending a big campus rally, and you have found a prime spot near the center of the action. Your friend arrives late and calls you on your cell phone to ask you to join her at the edge of the crowd. You decide that friendship is worth more than a prime spot, and so you work your way out through the dense crowd to meet her. You can move only a short distance before bumping into someone, changing direction, and trying again, making your way slowly to the outside edge of the crowd. Your friend can't see you until you get very close to the edge because of all the bodies in the way. So too photons making their way through the Sun are constantly bumping into atoms, changing direction, working their way slowly outward, and becoming visible only when they reach the atmosphere of the Sun, where the density of atoms is too low to block their outward progress.

Astronomers have found that the solar atmosphere changes from almost perfectly transparent to almost completely opaque in a distance of just over 400 kilometers (km); it is this thin region that we call the *photosphere,* a word that comes from the Greek for "light sphere." When astronomers speak of the "diameter" of the Sun, they mean the size of the region surrounded by the photosphere.

The photosphere looks sharp only from a distance. If you were falling into the Sun, you would not feel any surface but would just sense a gradual increase in the density of the gas surrounding you. It is much the same as falling through a cloud while skydiving. From far away, the cloud looks as if

it has a sharp surface, but you do not feel a surface as you fall into it. (One big difference between these two scenarios, however, is temperature. The Sun is so hot that you would be vaporized long before you reached the photosphere. Skydiving in the Earth's atmosphere is much safer!)

We might note that the atmosphere of the Sun is not a very dense layer compared to the air in the room where you are reading this book. At a typical point in the photosphere, the pressure is less than 10 percent of the Earth's pressure at sea level, and the density is about one ten-thousandth of the Earth's atmospheric density at sea level.

Observations with telescopes show that the photosphere has a mottled appearance resembling grains of rice spilled on a dark tablecloth. This structure of the photosphere is now generally called **granulation** (see Figure 14.8, p. 325). Granules, which are typically 700 to 1000 km in diameter, appear as bright areas surrounded by narrow, darker regions. The lifetime of an individual granule is only 5 to 10 minutes.

The motions of the granules can be studied by examining the Doppler shifts in the spectra of gases just above them (see Section 4.6). The bright granules are columns of hotter gases rising at speeds of 2 to 3 km/s from below the photosphere. As this rising gas reaches the photosphere, it spreads out and sinks down again into the darker regions between the granules. The centers of the granules are hotter than the intergranular regions by 50 to 100 degrees Kelvin (K).

The granules are the tops of *convection* currents of gas rising through the photosphere. Convection is the rising of a hot liquid or gas, transferring energy from the hotter layers underneath to the cooler layers above. As the gas cools, it falls back down, and the area looks darker from above. (You can see the same sort of hot currents rising in a pot of boiling water as you make more coffee to stay awake while reading your astronomy book!)

14.1.3 The Chromosphere

The Sun's outer gases extend far beyond the photosphere (Figure 14.4). Because they are transparent to most visible radiation and emit only a small amount of light, these outer layers are difficult to observe. The region of the Sun's atmosphere that lies immediately above the photosphere is called the **chromosphere.** Until this century, the chromosphere was visible only when the photosphere was concealed by the Moon during a total solar eclipse (see Section 3.7). In the 17th century, several observers described what appeared to them as a narrow red "streak" or "fringe" around the edge of the Moon during a brief instant after the Sun's photosphere had been covered. The name *chromosphere,* from the Greek for "colored sphere," was given to this red streak.

Observations made during eclipses show that the chromosphere is about 2000 to 3000 km thick, and its spectrum consists of bright emission lines, indicating that this layer is composed of hot, transparent gases emitting light at discrete wavelengths. The reddish color of the chromosphere arises from one of the strongest emission lines in the visible part of its spectrum—the bright red line caused by hydrogen, the element that, as we have already seen, dominates the composition of the Sun.

In 1868, observations of the chromospheric spectrum revealed a yellow emission line that did not correspond to any previously known element on Earth. Scientists quickly realized they had found a new element and named it *he-*lium (after *helios,* the Greek word for "Sun"). It took until 1895 for helium to be discovered on our planet. Today, students are probably most familiar with it as the light gas used to inflate balloons, although it turns out to be the second most abundant element in the universe.

The temperature of the chromosphere is about 10,000 K. This means that the chromosphere is hotter than the photosphere, which should seem surprising. In all the situations we are familiar with, temperatures fall as one moves away from the source of heat, and the chromosphere is farther from the center of the Sun than the photosphere is. We will describe one possible explanation for its high temperature after we describe even hotter regions of the solar atmosphere.

14.1.4 The Transition Region

The increase in temperature does not stop with the chromosphere. Above it is a region in the solar atmosphere where the temperature changes from 10,000 K, typical of the chromosphere, to nearly a million degrees. The hottest part of the solar atmosphere, which has a temperature of a million degrees or more, is called the **corona.** Appropriately, the part of the Sun where the rapid temperature rise occurs is called the **transition region.** It is probably only a few tens of kilometers thick. Figure 14.5 summarizes how the temperature of the solar atmosphere changes from the photosphere outward.

■ **FIGURE 14.4**

The Sun's Atmosphere Composite image showing the three components of the solar atmosphere: the photosphere or surface of the Sun taken in ordinary light; the chromosphere, imaged in the light of the strong red spectral line of hydrogen (H-alpha); and the corona as seen with x rays.

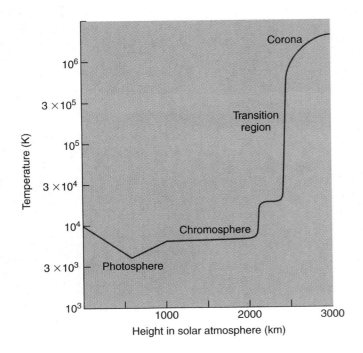

■ FIGURE 14.5

Temperatures in the Solar Atmosphere On this graph, temperature is shown increasing upward, and height above the photosphere is shown increasing to the right. Note the very rapid increase in temperature over a very short distance in the transition region between the chromosphere and the corona.

■ FIGURE 14.6

The Corona During an Eclipse This image of the Sun was taken during the solar eclipse on March 18, 1988. Since the light from the brilliant surface (photosphere) of the Sun is blocked by the Moon, it is possible to see the tenuous outer atmosphere of the Sun, which is called the corona.

Figure 14.5 makes the Sun seem rather like an onion, with smooth spherical shells, each one with a different temperature. For a long time, astronomers did indeed think of the Sun this way. However, we now know that while this idea of layers—photosphere, chromosphere, transition region, corona—describes the big picture fairly well, the Sun's atmosphere is really more complicated, with hot and cool regions intermixed. For example, clouds of carbon monoxide gas with temperatures colder than 4000 K have now been found at the same height above the photosphere as the much hotter gas of the chromosphere.

This complexity should not seem too surprising. You may have learned that temperatures in the Earth's atmosphere change as high- and low-pressure areas circulate and as the jet stream moves about. In the same way, the atmosphere of the Sun is constantly changing, with matter flowing up and down and changing temperature as it does so. Fortunately, the weather extremes on the Earth are small relative to the violent events on the Sun. Even hurricanes pale in comparison to "solar weather," as we shall see later in this chapter.

14.1.5 The Corona

The outermost part of the Sun's atmosphere is called the corona. Like the chromosphere, the corona was first observed during total eclipses (Figure 14.6). Unlike the chro-

mosphere, the corona has been known for many centuries: It was referred to by the Roman historian Plutarch and was discussed in some detail by Kepler. The corona extends millions of kilometers above the photosphere and emits half as much light as the full moon. The reason we don't see this light until an eclipse occurs is the overpowering brilliance of the photosphere. Just as bright city lights make it difficult to see faint starlight, so too the intense light from the photosphere hides the faint light from the corona. While the best time to see the corona is during a total solar eclipse, its brighter parts can now be photographed with special instruments even at other times, and it can be observed easily from orbiting spacecraft.

Studies of its spectrum show the corona to be very low in density. At the bottom of the corona there are only about 10^9 atoms per cubic centimeter, compared with about 10^{16} atoms per cubic centimeter in the upper photosphere and 10^{19} molecules per cubic centimeter at sea level in the Earth's atmosphere. The corona thins out very rapidly at greater heights, where it corresponds to a high vacuum by laboratory standards.

The corona is very hot. We know this because the spectral lines produced in it come from highly ionized atoms of iron, nickel, argon, calcium, and other elements. For example, astronomers observe lines of iron whose atoms have lost 16 electrons in the corona. Such a high degree of ionization requires a temperature of millions of degrees

■ FIGURE 14.7

Coronal Loops This image shows gas in the corona at a temperature of about 1 million K. High arches of gas reach to a height of approximately 120,000 km. The distinctive arch shape results from strong magnetic fields in the region that force the hot ionized gas to follow their loops.

Kelvin. Because of its high temperature, the corona is also very bright at x-ray wavelengths (see Figure 14.10b, p. 326).

Why is the corona so hot? Observation and theory together have identified magnetic energy as the culprit. The Sun, as we mentioned earlier, is a giant magnet, with a complicated magnetic field that can control the motion of the thin, electrically charged gas in the outer regions of the Sun. The surface of the Sun is covered with magnetic loops, much like the loops of a carpet (Figure 14.7), and the charged atoms flow along these loops.

If a loop in which the magnetic field has one polarity is jostled by turbulence into close proximity to a loop of the opposite polarity, the two loops can "reconnect," much as two bar magnets held side by side with their north poles in opposite directions will snap together. The reconnection of the solar loops releases energy, which can accelerate and heat material that rises into the corona. Magnetic fields also play the key role in heating the chromosphere.

14.1.6 The Solar Wind

One of the most remarkable discoveries about the Sun's atmosphere is that it produces a stream of charged particles (mainly protons and electrons) that we call the **solar wind.** These particles flow outward from the Sun into the solar system at a speed of about 400 km/s (about 900,000 mi/h)! The solar wind exists because the gases in the corona are so hot and moving so rapidly that they cannot be held back by solar gravity. (This wind was actually discovered by its effects on the charged tails of comets; in a sense, we can see the comet tails blow in the solar breeze the way wind socks or curtains in open windows flutter on Earth.)

Although the solar wind material is very, very rarified, the Sun has an enormous surface area. Astronomers estimate that the Sun is losing about 10 million tons of material each year. While this amount of lost mass seems large by Earth standards, it is completely insignificant for the Sun.

In visible light photographs, the solar corona appears fairly uniform and smooth. X-ray pictures, however, show that the corona has loops, plumes, and both bright and dark regions (see Figure 14.10b). Large dark regions of the corona that are relatively cool and quiet are called **coronal holes.** In these regions, magnetic field lines stretch far out into space away from the Sun, rather than looping back to the surface. The solar wind comes predominantly from coronal holes, where gas can stream away from the Sun into space unhindered by magnetic fields. Hot coronal gas, on the other hand, is present mainly where magnetic fields have trapped and concentrated it.

At the surface of the Earth, we are protected to some degree from the solar wind by our atmosphere and the Earth's magnetic field. However, the magnetic field lines come into the Earth at the north and south magnetic poles. Here charged particles accelerated by the solar wind can follow the field down into our atmosphere. As the particles strike molecules of air, they cause them to glow, producing beautiful curtains of light called the **auroras** (or the northern and southern lights). See the opening figure of this chapter for what an auroral display looks like.

14.2 THE CHANGING SUN

Before the invention of the telescope, the Sun was thought to be an unchanging and perfect sphere. We now know, however, that the Sun is in a perpetual state of change; its surface is a seething, bubbling cauldron of hot gas. Areas that are darker and cooler than the rest of the surface come and go. Vast plumes of gas erupt into the chromosphere and corona. Occasionally, there are even giant explosions on the Sun that send enormous streamers of charged particles and energy hurtling toward the Earth. When they

VOYAGERS IN ASTRONOMY

Art Walker: Doing Astronomy in Space

Art Walker

Courtesy of Stanford University

Since the Sun's corona is a rich source of high-energy photons, observations at x-ray wavelengths are a good way to learn more about our star. Unfortunately, x rays do not penetrate the Earth's atmosphere, so such observations must be done from rockets and satellites. Art Walker (1936–2001) was one of the pioneers in designing and flying instruments in space to learn more about the Sun.

Walker received a PhD in physics from the University of Illinois, supported in his graduate training by the Air Force Reserve officer program. After receiving his degree, he began active service at the Air Force Weapons Laboratory, which was beginning a space research program in the physics of the Sun. (See Making Connections: *Space Weather* to find out why the military is interested in the Sun.) Using a specially designed spectrometer aboard an Air Force satellite, Walker was able to take one of the first x-ray spectra of the Sun and identify several new emission lines in those spectra. Such measurements allowed him and his co-workers to probe the temperature, composition, and structure of the

corona, and get a much better sense of what the Sun's atmosphere is like.

After that, Walker worked with a novel design for x-ray telescopes that makes them similar to some of the telescopes astronomers use on the ground. These instruments allow scientists to take very high-resolution x-ray images of the active Sun; they are also relatively easy to build and can be flown aboard sounding rockets whose paths are spectacular arcs trav-

eling 100 or more miles above the Earth's surface.

In an interview, Walker emphasized one of the key challenges of doing astronomy from space: "In the laboratory, if you build an experiment and it doesn't quite work right, you can always tinker with it until you get the instrumentation to operate. In the case of space observations, you have to build the instrument anticipating all the things that might go wrong and eliminating each possibility of failure. Once the instrument is launched, it is out of your control, and you just have to hope that you have anticipated everything."

After working for the Aerospace Corporation, Walker became a professor at Stanford University, where he also served as Dean of the Graduate College. He worked hard to achieve greater inclusion of women and ethnic minorities in the sciences. Perhaps his best-known student was Sally Ride, who received her PhD in astrophysics and went on to become America's first woman in space. Walker also served on the Presidential Commission that investigated the accident that destroyed the shuttle *Challenger*.

---

arrive, these can cause power outages and other serious effects on our planet.

14.2.1 Sunspots

The first evidence that the Sun varies came from studies of **sunspots,** which are large, dark features seen on the surface of the Sun (Figure 14.8). Occasionally, these spots are large enough to be visible to the unaided eye, and we have records going back over a thousand years of observers who noticed them when haze or mist reduced the Sun's intensity. (We emphasize what your mother surely told you: Looking at the Sun for even a brief time can cause permanent eye damage. This is the one area of astronomy where we don't encourage you to do your own observing without getting careful instructions or filters from your instructor!)

Today we know that sunspots are darker than the photosphere in which they are embedded because the gases in sunspots are as much as 1500 K cooler than the surround-

ing gases. Sunspots are nevertheless hotter than the surfaces of many stars. If they could be removed from the Sun, they would shine brightly. They appear dark only in contrast with the hotter, brighter photosphere around them.

Individual sunspots have lifetimes that range from a few hours to a few months. If a spot lasts and develops, it usually consists of two parts: an inner darker core, the *umbra,* and a surrounding less dark region, the *penumbra.* Many spots become much larger than the Earth, and a few have reached diameters of 50,000 km. Frequently, spots occur in groups of 2 to 20 or more. The largest groups are very complex and may have over 100 spots. Like storms on the Earth, sunspots are not fixed in position, but they drift slowly compared with the Sun's rotation.

By recording the apparent motions of the sunspots as the turning Sun carried them across its disk (Figure 14.9), Galileo in 1612 demonstrated that the Sun rotates on its axis with a rotation period of approximately one month. Modern measurements show that the Sun's rotation period

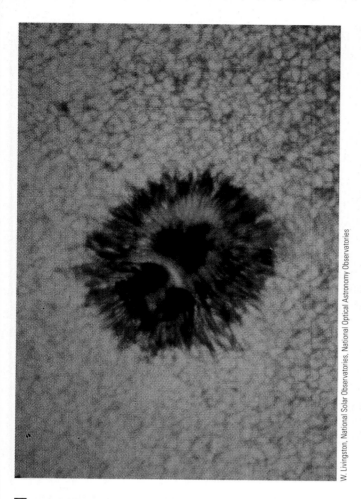

■ FIGURE 14.8

A Sunspot This image of a sunspot, a cooler and thus darker region on the Sun, was taken with the McMath-Pierce Solar Telescope. The spot has a dark central region (called the umbra) surrounded by a less dark region (the penumbra). Although sunspots appear dark when seen next to the hotter gases of the photosphere, an average sunspot, cut out of the solar surface and left standing in the night sky, would be about as bright as the full moon. Note that you can also see the granulation on the Sun's surface.

is about 25 days at the equator, 28 days at latitude 40°, and 36 days at latitude 80°. Our star turns in a west-to-east direction, like the orbital motions of the planets. Note that the Sun, being composed of gas, does not have to rotate rigidly, the way a solid body does.

14.2.2 The Sunspot Cycle

Between 1826 and 1850, Heinrich Schwabe, a German pharmacist and amateur astronomer, kept daily records of the number of sunspots. What he was really looking for was a planet inside the orbit of Mercury, which he hoped to find by observing its dark silhouette as it passed between the Sun and the Earth. He failed to find the hoped-for planet, but his diligence paid off with an even more important discovery—the **sunspot cycle.** He found that the number of sunspots varied systematically in cycles of about ten years.

What Schwabe observed was that, although individual spots are short-lived, the total number visible on the Sun at any one time is likely to be very much greater at certain times—the periods of sunspot maximum—than at other times—the periods of sunspot minimum. We now know that sunspot maxima occur at an *average* interval of 11 years, but the intervals between successive maxima have ranged from as short as 8 years to as long as 16 years. During sunspot maxima, more than 100 spots can often be seen at once, but even so usually less than half a percent of the surface is covered by spots (Figure 14.10). During sunspot minima, sometimes no spots are visible. The Sun's activity reached its most recent maximum in 2001.

14.2.3 Magnetism and the Solar Cycle

It is the Sun's changing magnetic field that drives the sunspot cycle. The solar magnetic field is measured using a property of atoms called the *Zeeman effect.* Recall from Chapter 4 that an atom has many energy levels and that spectral lines are formed when electrons shift from one

■ FIGURE 14.9

Sunspots Rotate Across Sun's Surface This sequence of photographs of the Sun's surface tracks a large group of sunspots as the Sun turns. The series of exposures follows the rotation of the sunspots across the visible hemisphere of the Sun. The top sequence shows the Sun in ordinary light; the bottom sequence shows emission from the chromosphere.

MARCH 7 MARCH 8 MARCH 9 MARCH 10 MARCH 13 MARCH 14 MARCH 15 MARCH 16 MARCH 17, 1989

(a)

Karel Schrijver and NASA

(b)

■ **FIGURE 14.10 The Solar Cycle**

National Optical Astronomy Observatories

■ FIGURE 14.11

The Zeeman Effect These photographs show how magnetic fields in sunspots are measured by means of the Zeeman effect. The vertical black line in the picture at right indicates the position of the spectrograph slit through which light passed in order to obtain the spectrum in the picture at left. Note that the second strongest spectral line in the left-hand picture is split into three components, which indicates that a strong magnetic field is present.

level to another. If each energy level is precisely defined, then the difference between them is also quite precise. As an electron changes levels, the result is a sharp, narrow spectral line (either an absorption or emission line, depending on whether the electron's energy increases or decreases in the transition).

In the presence of a strong magnetic field, however, each energy level is separated into several levels very close to one another. The separation of the levels is proportional to the strength of the field. As a result, spectral lines formed in the presence of a field are not single lines, but a series of very closely spaced lines corresponding to the subdivisions of the atomic energy levels. This splitting of lines in the presence of a magnetic field is termed the Zeeman effect.

Measurements of the Zeeman effect in the spectra of the light from sunspot regions show them to have strong

■ FIGURE 14.10

The Solar Cycle These dramatic sequences show the activity cycle of the Sun. (a) The image shows ten maps of the magnetic field on the surface of the Sun, spanning a period of seven and a half years. The two polarities (N and S) of the magnetic field are shown against a blue disk as dark blue to black and as light blue to white, respectively. The earliest image, taken on January 8, 1992, is at the lower left and was taken just after a maximum in solar activity. The arc of images shows how the magnetic fields change during the decline of one cycle and the increase in activity associated with the next cycle. (Each image, from left to right around the arc, was taken one half to one year later than the preceding one.) The last image was taken on July 25, 1999, just as the Sun was approaching the most recent solar maximum. Note a few striking patterns in the magnetic maps: The direction from white to black polarity on the southern solar hemisphere is opposite from that on the northern hemisphere. Moreover, the direction from white to black polarity in each hemisphere alternates from one cycle to the next. The complete magnetic cycle of the Sun therefore averages around 22 years. The magnetic field heats the outer solar atmosphere to several million degrees wherever it is strong. The hot gas, forming the solar corona, emits x rays. (b) This series of images was taken by the YOHKOH satellite and shows striking changes in the intensity of coronal x-ray emission during the same period.

magnetic fields (Figure 14.11). Whenever sunspots are observed in pairs, or in groups containing two principal spots, one of the spots usually has the magnetic polarity of a north-seeking magnetic pole and the other has the opposite polarity. Moreover, during a given cycle, the leading spots of pairs (or leading principal spots of groups) in the northern hemisphere all tend to have the same polarity, while those in the southern hemisphere all tend to have the opposite polarity.

During the next sunspot cycle, however, the polarity of the leading spots is reversed in each hemisphere. For example, if during one cycle the leading spots in the northern hemisphere all had the polarity of a north-seeking pole, then the leading spots in the southern hemisphere would have the polarity of a south-seeking pole. During the next cycle, the leading spots in the northern hemisphere would have south-seeking polarity, while those in the southern hemisphere would have north-seeking polarity. Therefore, the sunspot cycle does not repeat itself in regard to magnetic polarity until two maxima have passed. The solar activity cycle, fundamentally a magnetic cycle, thus averages 22 years in length, not 11.

Why does the Sun's magnetic field change in strength and polarity in a nearly regular way? Calculations from detailed models of how the Sun works show that rotation and convection just below the solar surface can distort the magnetic fields. This causes them to grow and then decay, regenerating with opposite polarity approximately every 11 years. The calculations also show that as the fields grow stronger near solar maximum, they flow from the interior of the Sun toward its surface in the form of loops. When a large loop emerges from the solar surface, much like the loop formed by a snake as it slithers along, it creates regions of sunspot activity.

The "ends" of the loop, where they penetrate the solar surface, have different polarities. This idea of magnetic loops offers a natural explanation of why the leading and trailing sunspots in an active region have opposite polarity. The leading sunspot coincides with one end of the loop and the trailing spot with the other end. Smaller loops form the "magnetic carpet" we described earlier and appear to be responsible for heating the corona.

Magnetic fields also hold the key to explaining why sunspots are cooler and darker than the regions without strong magnetic fields. The forces produced by the magnetic field resist the motions of the bubbling columns of rising hot gases. Since these columns carry most of the heat from inside the Sun to the surface by means of convection, less heating occurs where there are strong magnetic fields. As a result, these regions are seen as darker, cooler sunspots.

High-Altitude Observatory

■ **FIGURE 14.12**

Plages on the Sun This image of the Sun was taken with a filter that transmits only the light of the spectral line produced by singly ionized calcium. The bright cloud-like regions are the plages.

14.3 ACTIVITY ABOVE THE PHOTOSPHERE

Sunspots are not the only features that vary during a solar cycle. There are extraordinary changes in the chromosphere and corona as well. To see what happens in the chromosphere, we must observe the emission lines from elements such as hydrogen and calcium. The hot corona, on the other hand, can be studied by observations of x rays.

14.3.1 Plages and Prominences

Emission lines of hydrogen and calcium are produced in the hot gases of the chromosphere. Astronomers routinely photograph the Sun through filters that transmit light only at the wavelengths that correspond to these emission lines. Pictures taken through these special filters show bright

"clouds" in the chromosphere around sunspots; these bright regions are known as **plages** (Figure 14.12). The plages are regions within the chromosphere that have higher temperature and density than their surroundings. The plages actually contain all of the elements in the Sun, not just hydrogen and calcium. It just happens that the spectral lines of hydrogen and calcium produced by these clouds are bright and easy to observe.

Moving higher into the Sun's atmosphere, we come to the spectacular phenomena called **prominences** (Figure 14.13), which usually originate near sunspots. Eclipse observers often see prominences as red, flame-like protuberances rising above the eclipsed Sun and reaching high into the corona. Some, the quiescent prominences, are graceful loops that can remain nearly stable for many hours or even days. The relatively rare eruptive

NASA/SOHO

■ **FIGURE 14.13**

A Collage of Four Different Prominences A prominence is a huge cloud of relatively cool (about 60,000 K in this case), fairly dense gas suspended in the much hotter corona. These pictures are color coded so that white corresponds to the hottest temperatures and dark red to cooler ones. The four images were taken, moving clockwise from the upper left, on May 15, 2001; March 28, 2000; January 18, 2000; and February 2, 2001.

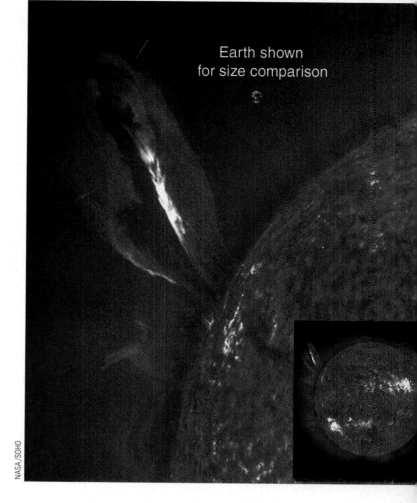

Earth shown
for size comparison

NASA/SOHO

■ FIGURE 14.14

An Eruptive Prominence This image of an eruptive prominence was taken in the light of singly ionized helium in the extreme ultraviolet part of the spectrum. The prominence, which occurred on July 24, 1999, is a particularly large one. An image of the Earth is shown at the same scale for comparison.

prominences appear to send matter upward into the corona at high speeds, and the most active surge prominences may move as fast as 1300 km/s (almost 3 million mi/h). Some eruptive prominences have reached heights of more than 1 million km above the photosphere; the Earth would be completely lost inside one of these awesome displays (Figure 14.14).

14.3.2 Flares

The most violent event on the surface of the Sun is a rapid eruption called a **solar flare.** A typical flare lasts for 5 to 10 minutes and releases a total amount of energy equivalent to that of perhaps a million hydrogen bombs. The largest flares last for several hours and emit enough energy to power the entire United States at its current rate of electrical consumption for 100,000 years. Near sunspot maximum, small flares occur several times per day, and major ones may occur every few weeks.

Flares are often observed in the red light of hydrogen, but the visible emission is only a tiny fraction of the energy released when a solar flare explodes. At the moment of the explosion, the matter associated with the flare is heated to

temperatures as high as 10 million K. At such high temperatures, a flood of x-ray and ultraviolet radiation is emitted.

Flares seem to occur when magnetic fields pointing in opposite directions release energy by interacting with and destroying each other—much as a stretched rubber band releases energy when it breaks. We have discussed this idea before, when we talked about heating the corona. What is different about flares is that their magnetic interactions cover a large volume in the solar corona and release a tremendous amount of electromagnetic radiation. In some cases, immense quantities of coronal material—mainly protons and electrons—may also be ejected at high speeds (500–1000 km/s) into interplanetary space. Such **coronal mass ejections** (CMEs) can affect the Earth in several ways. (See Making Connections: *Space Weather*.)

14.3.3 Active Regions

Sunspots, flares, and bright regions in the chromosphere and corona tend to occur together on the Sun. That is, they all tend to have similar longitudes and latitudes, but they are located at different heights in the atmosphere. Because they all occur together, they all vary with the sunspot cycle.

Space Weather

As if scientists did not have enough trouble trying to predict weather on Earth, they are now facing the challenge of predicting the effects of solar storms on Earth. This new field of research is called *space weather*. With thousands of satellites in orbit, astronauts taking up long term residence in the International Space Station, millions of people using cell phones and wireless communication, and nearly everyone relying on the availability of dependable electrical power, governments are now making major investments in trying to learn how to predict when solar storms will occur and how strongly they will affect the Earth.

Three solar phenomena—coronal holes, solar flares, and coronal mass ejections (CMEs)—account for most of the space weather. Coronal holes allow the solar wind to flow freely away from the Sun, unhindered by solar magnetic fields. When the solar wind reaches the Earth, it causes the Earth's magnetosphere (the region around the Earth that is filled with charged particles) to contract and then expand after the solar wind passes by. These changes can cause (usually mild) electromagnetic disturbances on Earth.

More serious are solar flares, which shower the upper atmosphere of the Earth with x rays, energetic particles, and intense ultraviolet radiation. The x rays and ultraviolet radiation can ionize atoms in the upper atmosphere, and the freed electrons can build up a charge on the surface of a spacecraft. When this static charge discharges, it can damage the electronics in the spacecraft—just as you can receive a shock when you walk across a carpet in your stocking feet in a dry climate and then touch a light switch or some metal object.

Most disruptive are the coronal mass ejections. A CME is an erupting bubble of tens of millions of tons of gas blown away from the Sun into space. When this bubble reaches the Earth a few days after leaving the Sun, it heats the ionosphere, which expands and reaches farther into space. As a consequence, friction between the atmosphere and spacecraft increases, dragging satellites to lower altitudes. At the time of a particularly strong flare in March 1989, the system responsible for tracking some 19,000 objects orbiting the Earth temporarily lost track of 11,000 of them because their orbits were changed by the expansion of the Earth's atmosphere. During solar maximum, a number of satellites are brought to such a low altitude that they are destroyed by friction with the atmosphere. Both the Hubble Space Telescope and the International Space Station require reboosts to higher altitude so that they can remain in orbit.

When a CME reaches the Earth, it distorts the Earth's magnetic field. Since a changing magnetic field induces electrical currents, the CME accelerates electrons, sometimes to very high speeds. These "killer electrons" can penetrate deep into satellites, sometimes destroying their electronics and permanently disabling operation. This has happened with some communications satellites.

Disturbances in the Earth's magnetic field can cause disruptions in communications, especially cell phone and wireless systems. In fact, disruptions can be expected to occur several times a year during solar maximum. Changes in the Earth's magnetic field due to CMEs can also cause surges in power lines large enough to burn out transformers and cause major power outages. For example, in 1989 parts of Montreal and Quebec Province in Canada were without power for up to 9 hours as a result of a major solar storm. Electrical outages due to CMEs are more likely to occur in North America than in Europe because North America is closer to the Earth's magnetic pole where the currents induced by CMEs are strongest.

Solar storms also expose astronauts, passengers in high-flying airplanes, and even people on the surface of the Earth to streams of high energy particles. Astronauts, for example, are limited in the total amount of radiation that they can be exposed to during their careers. A single ill-timed solar outburst could end an astronaut's career. This problem becomes increasingly serious as astronauts spend more time in space. For example, the typical daily dose of radiation aboard the Russian Mir was equivalent to about eight chest x rays. One of the major challenges in planning the human exploration of Mars is devising a way to protect astronauts from high-energy solar radiation.

Warnings of solar storms would help to minimize their disruptive effects. Power networks could be run at less than their full capacity so that they could absorb the effects of power surges. Communications networks could be prepared for malfunctions and have backup plans in place. Spacewalks could be timed to avoid major solar outbursts. Scientists are now trying to find ways to predict where and when flares and CMEs will occur and whether they will be big, fast events or small, slow ones with little consequence for Earth. The strategy is to relate changes in the appearance of small regions on the Sun and changes in local solar magnetic fields to subsequent eruptions. However, right now this predictive capability is poor, and so the only real warning we have is from actually seeing CMEs and flares occur. Since a CME travels outward at about 300 km/s, an observation of an eruption provides several days warning at the distance of the Earth. However, the severity of the impact on the Earth depends on how the magnetic field associated with the CME is oriented relative to that of the Earth. The orientation can be measured only when the

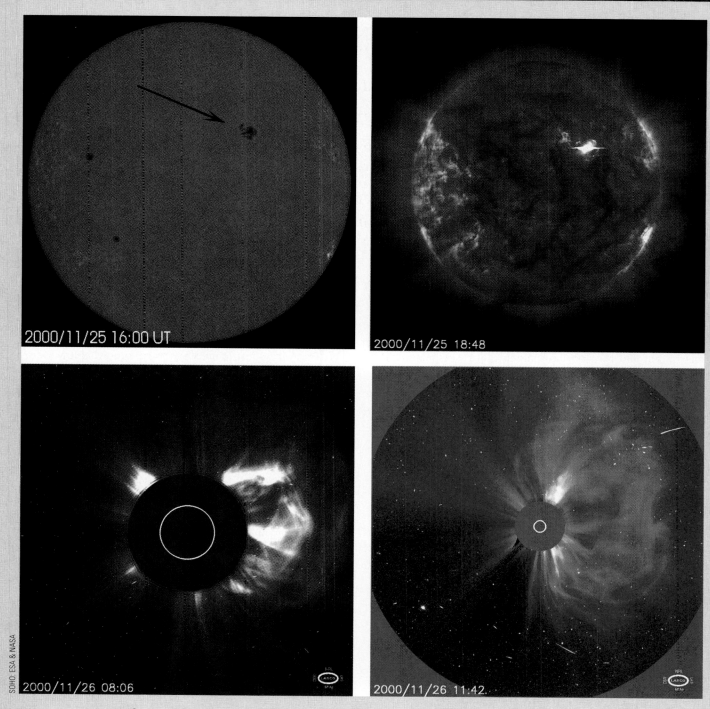

A Flare and a Coronal Mass Ejection This sequence of four images shows the evolution over time of a giant eruption on the Sun. The event began at the location of a sunspot group, and a flare is seen in far ultraviolet light in the upper right image. Fourteen hours later, a CME is seen blasting out into space. Three hours later, this CME has expanded to form a giant cloud of particles escaping from the Sun and beginning the journey out into the solar system. The white circle in the lower two pictures shows the diameter of the solar photosphere. The larger dark area shows where the light from the Sun has been blocked out by a specially designed instrument in order to make it possible to see the faint emission from the corona.

Space Weather (continued)

CME flows past a satellite, which is located only about an hour upstream from the Earth.

Space weather predictions are available on the Web (www.noaa.gov/solar.html). Outlooks are given a week ahead, bulletins are issued when there is an event that is likely to be of interest to the public, and warnings and alerts are posted when an event is imminent or already under way.

Fortunately, we can expect calmer space weather for the next few years, since the most recent solar maximum occurred in 2001 and we expect a solar minimum in 2005–2006. Between now and then, more satellites will be launched that will allow us to determine whether CMEs are headed toward Earth and how big they are. Models are being developed that will then allow scientists to use information about the CME to predict its likely impact on the Earth. The hope is that by the time of the next maximum, solar weather forecasting will have some of the predictive capability that meteorologists have achieved for terrestrial weather at the Earth's surface. However, the most difficult events to predict are the largest and most damaging storms—hurricanes on Earth and extreme, rare events on the Sun. Thus it is inevitable that the Sun will continue to surprise us.

For example, flares are more likely to occur near sunspot maximum, and the corona is much more conspicuous at that time (see Figure 14.10). A place on the Sun where these phenomena are seen is called an **active region** (Figure 14.15). Active regions are always associated with strong magnetic fields.

14.4 IS THE SUN A VARIABLE STAR?

The Sun rises faithfully every day at a time that can be calculated precisely. Each day it deposits energy on the Earth, warming it and sustaining life. But over the last decade, scientists have accumulated evidence that the Sun is not truly constant but varies over the centuries by a small amount—probably less than 1 percent. Still, that variation is enough to have profound effects on the Earth and its climate.

14.4.1 Variations in the Number of Sunspots

You already read that the number of sunspots varies, with the time between sunspot maxima being about 11 years. However, the number of sunspots at maximum is not always the same. Considerable evidence shows that between the years 1645 and 1715, the number of sunspots, even at sunspot maximum, was much lower than it is now. This interval of extremely low activity was first noted by Gustav Spörer in 1887 and then by E. W. Maunder in 1890; it is now called the **Maunder Minimum.** The variation in the number of sunspots over the past three centuries is shown in Figure 14.16. Besides the Maunder Minimum in the 17th century, sunspot numbers were somewhat lower during the first part of the 19th century than they are now; this period is called the Little Maunder Minimum.

When the number of sunspots is high, the Sun is active in various other ways as well, and this activity affects the Earth directly. For example, there are more auroral displays when the sunspot number is high. Auroras are caused when energetic charged particles from the Sun interact with the Earth's magnetosphere, and the Sun is more likely to spew out particles when it is active and the sunspot number is high. Historical accounts indicate that auroral activity was abnormally low throughout the several decades of the Maunder Minimum.

The Maunder Minimum was a time of exceptionally low temperatures in Europe—so low that this period is described as a Little Ice Age. The river Thames in London froze at least 11 times during the 17th century, ice appeared in the oceans off the coasts of southeast England, and low summer temperatures led to short growing seasons and poor harvests (Figure 14.17).

Changes in climate can have profound impacts on human history. For example, explorers from Norway first colonized Iceland and then reached Greenland by 986. From there, they were able to make repeated visits to the northeastern coasts of North America, including Newfoundland, between about 1000 and 1350. (The ships of the time did not allow the Norse explorers to travel all the way to North America directly, but only from Greenland, which served as a station for further exploration.)

Most of Greenland is covered by ice, and the Greenland station was never self-sufficient; rather it depended on imports of food and other goods from Norway for its survival. When a little ice age began in the 13th century,

Vertical text along right side of image:

■ **FIGURE 14.15**

Active Regions as Seen at Different Levels in the Solar Atmosphere   If we
observe the Sun at different wavelengths, we can see different temperature regions and
different levels in the solar atmosphere. The central image is a three-color composite of the
solar corona, observed in extreme ultraviolet light with the Transition Region and Coronal
Explorer (TRACE) satellite. The colors in this image represent different temperatures: 1 million K gas is represented by green, 1.5 million K gas by blue, and 2 million K gas by red.
The surrounding images are, clockwise starting from the top: a magnetic map taken by the
SOHO satellite, with opposite magnetic polarities shown in black and white; an image of
the solar photosphere taken in ordinary white light; two images of the chromosphere taken
in the ultraviolet by the TRACE satellite; and four images of the corona taken at the three
wavelengths used to construct the central image. You can see that coronal x-ray emission,
chromospheric emission, strong magnetic fields, and sunspots tend to occur in the same
locations—the active regions. The composite was prepared by Joe Covington.

FIGURE 14.16

The Numbers of Sunspots As They Change with Time This diagram shows how the number of sunspots has changed with time since counts of the numbers of spots began to be recorded on a consistent scale. Note the low number of spots during the early years of the 19th century—the Little Maunder Minimum.

FIGURE 14.17

Europe's Little Ice Age During the "Little Ice Age" in Europe, bodies of water froze every winter. This painting by Robert van den Hoecke from 1649 is entitled "Skating in the Town Moat of Brussels."

voyaging became very difficult, and support of the Greenland colony was no longer possible. The last known contact with it was made by a ship from Iceland blown off course in 1410. When European ships again began to visit Greenland in 1577, the entire colony there had disappeared.

The estimated dates for these patterns of migration follow what we know about solar activity. Solar activity was unusually high between 1100 and 1250, which includes the time the first European visits were made to North America. Activity was low from 1280 to 1340 and there was a little ice age, which was about the time regular contact with North America and between Greenland and Europe stopped.

14.4.2 The Effects of the Sun on the Earth's Climate

The historical record therefore suggests that climate and solar activity are correlated. But is this an accidental correlation? Or do changes in the Sun *cause* changes in the Earth's climate? There are two reasons that it is urgent to answer this question. First, active debate is under way about whether greenhouse gases that we humans are adding to the Earth's atmosphere are affecting the climate. If solar variability also causes changes in climate, then our estimate of the impact of human activities would be affected. Second, we can test models that predict how the atmosphere will respond to changing temperatures. We have data concerning both solar variability and climate changes that span several centuries, and we can see whether atmospheric models predict accurately the relationship between the energy output of the Sun and the climate here on Earth. If the models can explain the Earth's response to solar variability, then we will have greater confidence in the predictions of the consequences of increases in greenhouse gases.

Two steps are required to show that changes in solar luminosity affect the Earth's climate: (1) A long record of correlation between weather and the energy emitted by the Sun must be established. (2) The mechanism must be found whereby small changes in the output of solar energy are translated into significant changes in climate. Let's deal with each of these steps in turn.

14.4.3 Solar Activity and the Earth's Climate: The Correlation

It is only with the advent of the space age that we have been able to measure the luminosity of the Sun accurately enough to see whether its total energy output changes with activity level. To our surprise, it turns out that the Sun is a tenth of 1 percent or so brighter during solar maximum. We might have expected the Sun to be fainter when it is covered with a large number of dark sunspots. The extra radiation at sunspot maximum probably comes from bright regions, like plages, which are more abundant at that time.

Incidentally, we may be fortunate to live at a time when the Sun is exceptionally stable. If we look at activity cycles in stars that are similar to the Sun, we find that the Sun's current variations (just 0.1 percent of the total energy emitted) are unusually small. The luminosity of most stars varies by 0.3 percent and some by as much as 1 percent. Therefore, it seems likely that the Sun too might sometimes vary by larger amounts than it does currently, with consequently even greater effects on climate.

Since we have no historical measurements of solar luminosity (the changes were much too small to be measured until recently), we will follow the recent results from satellites and assume that the level of solar activity is an indicator of the energy emitted by the Sun. It turns out that nature provides us with an estimate of the level of solar activity going back for thousands of years.

The best quantitative evidence of long-term variations in the level of solar activity comes from studies of the radioactive isotope carbon-14. The Earth is constantly bombarded by cosmic rays—high-energy charged particles that include protons and nuclei of heavier elements. The rate at which cosmic rays from sources outside the solar system reach the upper atmosphere depends on the level of solar activity. When the Sun is active, its charged particles, streaming out into the solar system, carry the Sun's strong magnetic field with them. This magnetic field shields the Earth from incoming cosmic rays. At times of low activity, when the Sun's magnetic field is weak, cosmic rays reach the Earth in larger numbers.

When the energetic cosmic-ray particles strike the upper atmosphere, they produce several different radioactive isotopes (see Section 4.4). One such isotope is carbon-14, which is produced when nitrogen is struck by high-energy cosmic rays. The rate of carbon-14 production is higher when the activity of the Sun is lower and the solar magnetic field does not shield the Earth from bombardment by cosmic rays.

Some of the radioactive carbon atoms then form carbon dioxide molecules, which are ultimately incorporated into trees through photosynthesis. By measuring the amount of radioactive carbon in tree rings, we can estimate the historical levels of solar activity. Correlations with visual estimates of sunspot numbers over the past 300 years indicate that the carbon-14 estimates of solar activity are indeed valid. Because it takes an average of about 10 years for a carbon dioxide molecule to be absorbed from the atmosphere or ocean into plants, this technique cannot provide data on the 11-year solar cycle. It can be used, however, to look for long-term changes in the level of solar activity.

Estimates of the amount of carbon-14 in tree rings show that variations in solar activity levels have occurred throughout the last several thousand years, and at different times the Sun has been both more and less active than it is now. The measurements confirm that the amount of carbon-14 was unusually high, with solar activity correspondingly low, during both the Maunder Minimum and the Little

Observing the Sun

Looking directly at the Sun is very dangerous! Even a brief exposure can burn your retina and cause severe eye damage. Looking at the Sun directly through an unfiltered telescope is even worse because the telescope concentrates the Sun's radiation and hence damages your eye even more quickly. But there are indirect ways of observing the Sun that are both safe and instructive.

Not long after Galileo began training his first telescope on the sky, his assistant and disciple Benedetto Castelli came up with a way of observing sunspots, whose nature was a big controversy in those days. Castelli projected the image of the Sun made by the telescope onto a sheet of paper, where the sunspots could be sketched safely. Using this technique, Galileo was able to show that sunspots were not small, dark planets blocking sunlight, as some had suggested, but dark areas on the Sun itself moving around as the Sun rotated. You can try this experiment for yourself.

Set up a small telescope on a steady mount so that it points at the Sun. Remember, you have to do this by trial and error; never look through the telescope at the Sun to see how well you're doing! Instead, watch the telescope's shadow; it will be smallest when the telescope is pointing directly at the Sun. Then you'll be able to see an image of the Sun on a sheet of white cardboard or other light surface. Sketch the sunspots if there are any. Repeat the observation over the course of a week or two, and, orienting your sketches the same way, trace how the spots have moved over the face of the Sun.

Looking through a telescope at the Sun is dangerous, but you can always view the Sun safely with a small telescope by projecting its image onto a sheet of white cardboard.

Putting a cardboard collar around the telescope makes the image easier to see by casting a shadow on it and thus making the sphere of the Sun stand out. If you observe the Sun for a while, don't be surprised if its image moves across and off the paper over time. This is caused by the turning of the Earth, which also makes the Sun appear to rise and set. And don't be upset if you don't see any sunspots, especially if you are observing during the solar activity minimum.

Maunder Minimum. Activity was also low from 1410 to 1530 and from 1280 to 1340. Between about 1100 and 1250, the level of solar activity may have been even higher than it is now.

Nature also provides us with a way to estimate the average temperature of the Earth over a time period of thousands of years. Ice in the far north, near the coasts of Canada, Greenland, and Iceland, picks up microscopic rocky debris as it moves over land. This ice then floats southward into the North Atlantic, eventually melts, and drops the debris onto the ocean floor. In cold years, the ice floats farther south before melting, occasionally reaching as far south as Ireland. Studies of the material picked up from the ocean floor show that the amount of debris jumped every 1500 years or so as the ice moved farther out into the (temporarily) colder Atlantic.

In each case, peaks in the amount of debris coincided with peaks in the amount of carbon-14 measured. This result implies that times of low temperature correspond to times of low solar activity. The measurements show that the climate has warmed and cooled nine times in the last 12,000 years, each time in step with changes in the level of solar activity as determined from carbon-14. The Sun really does affect the Earth's climate.

14.4.4 Solar Variability and the Earth's Climate: The Mechanism

The evidence then is that the Earth is on average cooler when solar activity is low. But why?

The changes in the Sun's luminosity over the past several hundred years are too small to change the overall temperature of the Earth dramatically. It is also true that the temperature changes are not the same everywhere during an event like the Little Ice Age, which affected Europe more strongly than some other parts of the world. People who compute models of the Earth's atmosphere have been trying to identify mechanisms that amplify the small changes in solar luminosity. One recent model suggests that wind may be the culprit. Measurements show that the ultraviolet energy emitted by the Sun varies by 10 to 100 times more than the energy put out in the visible-light region of the spectrum. At times of high solar activity, more ultraviolet radiation is emitted and as a consequence more ozone is formed in the Earth's atmosphere. Ozone can absorb more sunlight and increase the temperature of the stratosphere. This in turn can affect the patterns of the winds aloft, modify the paths of storms, and change the temperature in lower layers of the Earth's atmosphere by differing amounts at different latitudes.

This atmospheric model shows that at times of low solar activity, the westerly winds in the northern hemisphere will be reduced. In winter, the oceans are relatively warmer than the land because water retains heat more efficiently. If winds are reduced, then less warm air will be blown from the Pacific over North America and from the Atlantic over Europe, thereby resulting in colder winters. These models predict that while global temperatures averaged over the whole Earth during the Maunder Minimum were only 0.3–0.4°C cooler, Europe and North America were 1.8–2.7°C cooler, which is what is needed to account for the Little Ice Age in the northern hemisphere.

Other factors also alter the Earth's climate. For example, the eruption of the Pinatubo volcano in June 1991 spewed out 20 million tons of sulfur dioxide (SO_2), which then spread around the globe. The presence of these particles in the atmosphere cooled the northern hemisphere by about 0.5°C on average for the next two years. This event retarded the effects of global warming, but also allowed scientists to calibrate the effects of volcanoes on climate. The 0.6°C warming of the Earth that has occurred over the past century cannot be accounted for by changes in the Sun and volcanic eruptions alone. In order to account for this warming, it is necessary to include greenhouse gases from human activities in the model.

Many of these ideas are still new and must be strengthened by further observations and modeling. Still, such studies clearly give us reason to continue to study the behavior of our star and its influences on Earth.

SURFING THE WEB

General Information Sites About the Sun:

- *The Nine Planets Site:*
  seds.lpl.arizona.edu/nineplanets/nineplanets/sol.html

- *Views of the Solar System Site:*
  www.solarviews.com/eng/sun.htm

- *Stanford SOLAR Center:*
  solar-center.stanford.edu/

- *NASA's Sun-Earth Connection:*
  sunearth.ssl.berkeley.edu

Sites for Space Missions Examining the Sun:

- *Ulysses:*
  ulysses.jpl.nasa.gov

- *SOHO:*
  sohowww.nascom.nasa.gov/

- *TRACE:*
  vestige.lmsal.com/TRACE/

- *Yohkoh:*
  www.lmsal.com/SXT/homepage.html

- *HESSI:*
  hesperia.gsfc.nasa.gov/hessi/

Other Relevant Sites

- *Images of the Sun Today:*
  umbra.nascom.nasa.gov/images/latest.html

- *Active Sun Monitor:*
  www.bbso.njit.edu/arm/latest

- *Space Weather Site:*
  www.spaceweather.com

- *Space Environment Center:*
  www.sec.noaa.gov

- *The Theory of Solar Flares:*
  hesperia.gsfc.nasa.gov/sftheory/index.htm

- *Climate Time Line Tool (overview of the history of our planet's climate):* www.ngcd.noaa.gov/paleo/ctl/

SUMMARY

14.1 The Sun, our star, is surrounded by a number of layers that make up the solar atmosphere. In order of increasing distance from the center of the Sun, they are the **photosphere,** with a temperature that ranges from 4500 K to about 6800 K; the **chromosphere,** with a typical temperature of 10^4 K; the **transition region,** a zone that may be only a few kilometers thick, where the temperature increases rapidly from 10^4 K to 10^6 K; and the **corona,** with temperatures of a few million degrees Kelvin. The Sun's surface is mottled with upwelling currents seen as hot, bright **granules. Solar wind** particles stream out into the solar system through **coronal holes.** When such particles reach the vicinity of the Earth, they produce **auroras,** which are strongest near the Earth's magnetic poles. Hydrogen and helium together make up 98 percent of the mass of the Sun, whose composition is much more characteristic of the universe at large than is the composition of the Earth.

14.2 Sunspots are dark regions where the temperature is up to 1500 K cooler than the surrounding photosphere. Their motion across the Sun's disk allows us to calculate how fast the Sun turns on its axis. The Sun rotates more rapidly at its equator, where the rotation period is about 25 days, than near the poles, where the period is slightly longer than 36 days. The number of visible sunspots varies according to a **sunspot cycle** that averages 11 years in length. Spots frequently occur in pairs. During a given 11-year cycle, all leading spots in the northern hemisphere have the same magnetic polarity, while all leading spots in the southern hemisphere have the opposite polarity. In the subsequent 11-year cycle, the polarity reverses. For this reason, the magnetic activity cycle of the Sun is often said to last for 22 years.

14.3 Sunspots, **prominences,** and bright regions, including **plages, solar flares,** and **coronal mass ejections,** all tend to occur in **active regions**—that is, in places on the Sun with the same latitude and longitude but at different heights in the atmosphere. Solar flares and coronal mass ejections can cause auroras, disrupt communications, damage satellites, and cause power outages.

14.4 Over long periods of time (100 years or more), there are changes in the level of solar activity and the number of sunspots seen at solar maximum. For example, the number of sunspots was unusually low from 1645 to 1715, a period now called the **Maunder Minimum.** Data now show that there is a correlation between the temperature on Earth and the energy output of the Sun. Models suggest that heating of the upper atmosphere by ultraviolet radiation affects wind patterns, which in turn affect weather near the Earth's surface. However, the warming of the Earth seen over the last century cannot be accounted for by changes in the Sun alone, even when combined with the effects of volcanic eruptions. The recent temperature rise is accounted for only if greenhouse gases produced by human activities are included in the models.

INTER-ACTIVITY

A Have your group make a list of all the ways the Sun affects your life on Earth. How long a list can you come up with? (Be sure you consider the everyday effects as well as the unusual effects due to high solar activity.)

B Long before the nature of the Sun was fully understood, astronomer (and planet discoverer) William Herschel (1738–1822) proposed that the hot Sun may have a cool interior and may be inhabited. Have

338 CHAPTER 14 THE SUN: A GARDEN-VARIETY STAR

your group discuss this proposal and come up with modern arguments against it.

 C We discussed how the migration of Europeans to North America was apparently affected by climate change. If the Earth were to become significantly hotter, either because of changes in the Sun or because of greenhouse warming, one effect would be an increase in the rate of melting of the polar ice caps. How would this affect modern civilization?

D Suppose we experience another Maunder Minimum on Earth, with a drop in the average temperatures. Have your group discuss how this would affect civilization and international politics. Make a list of the most serious effects you can think of.

E Watching sunspots move across the disk of the Sun is one way to show that our star rotates on its axis. Can your group come up with other ways to show the Sun's rotation?

REVIEW QUESTIONS

1. Describe the main differences between the composition of the Earth and that of the Sun.
2. Make a sketch of the Sun's atmosphere showing the locations of the photosphere, chromosphere, and corona. What is the approximate temperature of each of these regions?
3. Why do sunspots look dark?
4. What is the Zeeman effect, and what does it tell us about the Sun?

5. Describe three different types of solar activity.
6. How does activity on the Sun affect the Earth?
7. Which aspects of the Sun's activity cycle have a period of about 11 years? Which vary during intervals of about 22 years?
8. Summarize the evidence indicating that over several decades or more there have been variations in the level of solar activity.

THOUGHT QUESTIONS

9. Table 14.1 indicates that the density of the Sun is 1.41 g/cm³. What kinds of materials have similar densities? One such material is ice. How do you know that the Sun is not made of ice?

10. If the rotation period of the Sun is determined by observing the apparent motions of sunspots, must any correction be made for the orbital motion of the Earth? If so, explain what the correction is and how it arises. If not, explain why the Earth's orbital revolution does not affect the observations. Making some sketches may help answer this question.

11. Suppose an (extremely hypothetical) elongated sunspot forms that extends from a latitude of 30° to a latitude of 40° along a fixed line of longitude on the Sun. How will the appearance of that sunspot change as the Sun rotates?

12. Suppose you live in northern Canada and an extremely strong flare is reported on the Sun. What precautions might you take? What could compensate you for your troubles?

13. Give some reasons why it is difficult to determine whether or not small changes in the amount of energy radiated by the Sun have an effect on the Earth's climate.

FIGURING FOR YOURSELF

A basic equation you use almost automatically to figure out trips in your car is also useful in figuring out when events on the Sun will impact the Earth. The equation is:

$$\text{distance} = \text{velocity} \times \text{time} \quad or \quad D = v \times t$$

Dividing both sides by v, we get

$$t = \frac{D}{v}$$

If you are driving at 90 km/h and you drive 180 km, the trip will take 180/90 = 2 hours.

When you use this equation, make sure you use the same units for velocity (say, km/s) as you do for distance (km) and time (seconds).

14. Suppose you observe a major solar flare while astronauts are orbiting the Earth in the shuttle. Use the data in Section 14.3 to calculate how long it will be before the charged particles ejected from the Sun during the flare reach the shuttle.

15. Suppose an eruptive prominence rises at a speed of 150 km/s. If it does not change speed, how far from the

photosphere will it extend after 3 hours? How does this distance compare with the diameter of the Earth?

16. From the information in the figure that accompanies the Making Connections box in this chapter, estimate the speed with which the particles in the CME in the final two frames are moving away from the Sun.

17. From the Doppler shifts of the spectral lines in the light coming from the east and west edges of the Sun, astronomers find that the radial velocities of the two edges differ by about 4 km/s. Find the approximate period of rotation of the Sun. (It may help to sketch the Sun and show the velocities at the east and west edges in order to work out this problem.)

The equation for density is:

$$\text{density} = \frac{\text{mass}}{\text{volume}}$$

The volume of a sphere is given by

$$V = \left(\frac{4}{3}\right)\pi R^3$$

where R is the radius of the sphere.
Again, be sure that you use consistent units.

18. Use the data in Table 14.1 to calculate the density of the Sun. How does its density compare with the densities of other objects given in this text (e.g., planets, minerals,

ice)? Does the density alone allow you to say anything about the composition of the Sun? If so, what conclusions would you draw?

19. The edge of the Sun doesn't have to be absolutely sharp in order to look that way to us. It just has to go from being transparent to being completely opaque in a distance that is smaller than your eye can resolve. Remember from Chapter 5 on telescopes that the ability to resolve detail depends on the size of the telescope. The pupil of your eye is very small relative to the size of a telescope and therefore is very limited in the amount of detail you can see. In fact, your eye cannot see details that are smaller than 1/30 of the diameter of the Sun (about 1 arcminute). Nearly all the light from the Sun emerges from a layer that is only about 400 km thick. What fraction is this of the diameter of the Sun? How does this compare with the ability of the human eye to resolve detail? Suppose we could see light emerging directly from a layer that was 300,000 km thick. Would the Sun appear to have a sharp edge?

20. Table 14.2 shows what percentage of the atoms in the Sun consists of hydrogen and what percentage of the total mass of the Sun is composed of hydrogen. Can you explain why the two numbers are different? Show that the statement that 92 percent of the atoms are hydrogen is consistent with the statement that 73 percent of the mass is made up of hydrogen. (*Hint:* Make the simplifying assumption, which is nearly correct, that the Sun is made up entirely of hydrogen and helium.)

SUGGESTIONS FOR FURTHER READING

Akasofu, S. "The Shape of the Solar Corona" in *Sky & Telescope*, Nov. 1994, p. 24.

Baliunas, S. and Soon, W. "The Sun-Climate Connection" in *Sky & Telescope*, Dec. 1996, p. 38.

Eddy, J. "The Case of the Missing Sunspots" in *Scientific American*, May 1977, p. 80.

Emslie, A. "Explosions in the Solar Atmosphere" in *Astronomy*, Nov. 1987, p. 18. Discusses solar flares.

Frank, A. "Blowin' in the Solar Wind" in *Astronomy*, Oct. 1998, p. 60. On results from the SOHO spacecraft.

Freeman, J. *Storms in Space*. 2001, Cambridge U. Press. On solar-terrestrial relationships.

Friedman, H. *Sun and Earth*. 1986, W. H. Freeman. Contains good sections about the Sun's effects on the Earth.

Golub, L. "Heating the Sun's Million-Degree Corona" in *Astronomy*, May 1993, p. 27.

Golub, L. and Pasachoff, J. M. *Nearest Star: The Surprising Science of Our Sun*. 2001, Harvard U. Press. An introduction to the Sun, with especially good chapters on solar eclipses and space missions.

Hufbauer, K. *Exploring the Sun: Solar Science Since Galileo*. 1991, Johns Hopkins U. Press. A good history.

Jaroff, L. "Fury on the Sun" in *Time*, July 3, 1989, p. 46. Nice introduction to solar activity and the Sun's interior.

Kippenhahn, R. *Discovering the Secrets of the Sun*. 1994, John Wiley. Excellent modern introduction for the beginner.

Lang, K. "SOHO Reveals the Secrets of the Sun" in *Scientific American*, Mar. 1997.

Nichols, R. "Solar Max: 1980–1989" in *Sky & Telescope*, Dec. 1989, p. 601.

Schaefer, B. "Sunspots That Changed the World" in *Sky & Telescope*, Apr. 1997, p. 34. Historical events connected with sunspots and solar activity.

Schrijver, C. and Title, A. "Today's Science of the Sun" in *Sky & Telescope*, Feb. 2001, p. 34 and Mar. 2001, p. 34. Excellent review of recent results about the solar atmosphere.

Verschuur, G. "The Day the Sun Cut Loose" in *Astronomy*, Aug. 1989, p. 48. Examines events surrounding a huge flare.

Zirker, J. *Journey from the Center of the Sun*. 2001, Princeton U. Press. Covers both the solar interior and atmosphere.

Astrobiology: The Road to Life in the Universe   In this fanciful montage produced by a NASA artist, we see one roadmap for discovering life in the universe. Learning more about the origins and domains of life on Earth, we can explore Mars, which may have had life billions of years ago when conditions were warmer. Next is Jupiter's satellite Europa, under whose icy surface a liquid ocean may harbor some kind of life. The road winds onward to other stars, which may well have planets around them that are hospitable to some kind of life.

We shall not cease from
    exploration,
And the end of all our
    exploring,
Will be to arrive where
    we started,
And know the place for
    the first time.

**T. S. Eliot, *Little
Gidding* (from *The
Four Quartets* in
*The Collected Poems
of T. S. Eliot*, 1934,
1936, Harcourt Brace
& World)**

THINKING AHEAD

Suppose that one day we receive a message from a civilization around another star. In a series of pictures transmitted across space, they tell us a bit about their world and themselves. Do we reply? And if we do, will all humanity speak with a single voice, or will we send a bewildering variety of answers—from different countries, religions, and ethnic groups? Who speaks for Earth?

Our voyages have taken us through planetary systems—both here and around other stars. As we have learned more about the universe, we have naturally wondered whether there might be other forms of life out there—perhaps even other creatures who also take astronomy courses and think about worlds beyond their own.

In this final chapter, we want to consider how life began on Earth, to ask whether the same processes could have led to life on other worlds, and then to suggest some strategies for finding out about life elsewhere. These are all topics that are central to the science of life in the universe, variously called exobiology, bioastronomy, or astrobiology. (We'll use the last term in our book.)

Sometimes the media confuse the various topics that we discuss in this chapter. Remember that the search for life on other planets is not the same as the search for *intelligent* life, which (if it exists) is surely much more rare. Even on Earth today, most life is microbial, and humans have existed on our planet for only 0.1 percent of the age of the planet. Also, it is important to distinguish between the scientific search for life, including intelligent life, and "belief in UFOs." As we will discuss below, most scientists think it is likely that the universe is teeming with life, but none of them find the evidence for UFOs and extraterrestrial visitors convincing.

15.1 THE COSMIC CONTEXT FOR LIFE

Several lines of evidence indicate that the universe was born in a "big bang" about 14 billion years ago. After the initial hot, dense fireball of creation cooled sufficiently for atoms to exist, all matter consisted of hydrogen and helium (with a very small amount of lithium). As the universe aged, processes within stars created the other elements, including those that make up the Earth (such as iron, silicon, magnesium and oxygen) and those of prime biological interest (such as carbon, oxygen, and nitrogen). These and other elements combined in space to produce a wide variety of compounds, including complex organic chemicals that form the basis of life on Earth. While we do not understand the details of the origin of life, it is clear that these events took place within the context of the chemical evolution of the universe.

15.1.1 Where Were the Atoms in Your Body Billions of Years Ago?

Let's take a look at the history of the cosmos by surveying the history of the atoms in your body, starting long before they became a part of you. Note that in this section we refer to some information that was not covered in this book but comes from other areas of astronomical research.

The universe began as hydrogen and helium, and the "first-generation" stars contained little else. The hydrogen atoms in the water in your body formed at this early time; they are the oldest atoms that are part of you. But it is not possible to make an organism as complex and interesting as you with only these elements. More complex atoms had to be "cooked" in the only places in the universe hot enough to do the job—the centers of stars. (Fusing together simpler nuclei to make more complex ones is what stars "do for a living" and also the way they produce the energy we see coming from them.) The most massive stars not only produced the greatest variety of new nuclei, but then had the courtesy to explode, scattering the newly minted atoms into space (Figure 15.1).

Astronomers know from observations of distant parts of the universe that stars must have formed, lived out their lives, and exploded within the first billion years or so, because spectra of distant (and therefore ancient) galaxies already show the presence of some of the heavier elements. Over the years, thanks to such explosions (and mass loss in all kinds of stars), the gas between the stars became increasingly enriched with heavier elements. In the cooler outer layers of old stars, atoms frequently combined into solid particles that we call interstellar dust. The next generations of stars and planets then formed from reservoirs of enriched gas and dust. They thus contained atoms of carbon, nitrogen, silicon, iron, and the rest of the familiar elements. One of the most remarkable discoveries of modern astronomy is that life on Earth is mostly composed of just those elements that stars find easiest to make.

About 5 billion years ago, a cloud of gas and dust in this cosmic neighborhood began to collapse under its own weight. Out of this cloud formed the Sun and its planets, together with all the smaller bodies that also orbit the Sun (Figure 15.2). The third planet from the Sun, as it cooled, developed an atmosphere that served to moderate temperature extremes and allow the formation of large quantities of liquid water on its surface. The chemicals available on

■ FIGURE 15.1

A Star That Exploded The Crab Nebula is the remnant of an exploding star, first seen in 1054. Now almost 11 LY across, this object still glows with tremendous energy in many bands of the electromagnetic spectrum. It is powered inside by the compressed whirling remnant (corpse) of the original star, whose beams of energy still stir and excite the atoms thrown out by the explosion. The left-hand image, taken with a ground-based telescope, shows the full nebula. The right-hand image, taken with the Hubble Space Telescope, is a close-up of the Crab's central region. The stellar corpse can be seen as the left member of a pair of stars near the center of the frame. HST was able to observe wisps of material streaming away from it at half the speed of light! Such exploding stars recycle new elements made during the star's life into the general supply of raw material from which new generations of stars and planets then form.

J. Hester, P. Scowen, and NASA

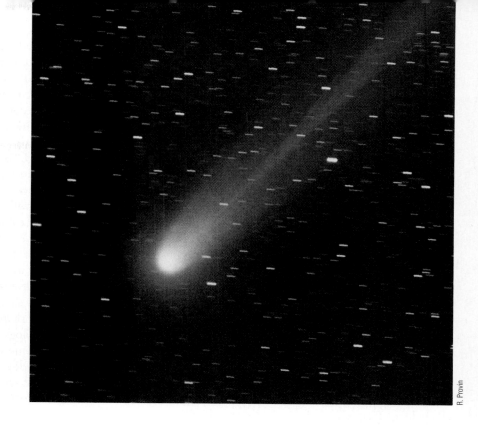

■ **FIGURE 15.2**

A Comet Comet Hyakutake captured in 1996 by amateur astrophotographer Robert Provin of California State University, Northridge. This 12-min exposure was taken on a clear night in the Mojave Desert with a camera and telephoto lens mounted on a telescope tripod. As the moving comet was held steady in the camera, the stars appeared to streak.

the cooling Earth were probably further enriched by the addition of complex molecules frozen in the nuclei of comets that collided with our planet.

The chemical variety and moderate conditions on Earth eventually led to the formation of self-reproducing molecules and the beginnings of life. Over the billions of years of Earth history, life slowly evolved and became more complex. The course of evolution was punctuated by occasional planetwide changes caused by collisions with those smaller bodies that did not make it into the Sun or one of its accompanying worlds. Mammals may owe their domination of the Earth's surface to just such a collision 65 million years ago.

Through many twisting turns, the course of evolution on Earth produced a creature with self-consciousness, able to ask questions about its own origins and place in the cosmos (Figure 15.3). Like most of the Earth, this creature is composed of atoms that were forged in earlier generations of stars—in this case, assembled rather cleverly into brains, kidneys, fingers, and faces. We might say that through the thoughts of human beings, the matter in the universe can become aware of itself.

Think about those atoms in your body for a minute. They are merely on loan to you from the lending library of atoms that make up our local corner of the universe. Atoms of many kinds circulate through your body and then leave it with the breath you exhale and the food you excrete. Even the atoms that take up more permanent residence in your tissues will not be part of you much longer than you are alive. Ultimately, you will return your atoms to the vast reservoir of the Earth, where they will be incorporated into other structures and even other living things in the millennia to come.

This picture of *cosmic evolution*, of our descent from the stars, has been obtained through the efforts of scientists in many fields over many decades. Some of its details are still tentative and incomplete, but we feel reasonably confident in its broad outlines. It is remarkable how much we have been able to learn in the short time we have had the instruments to probe the physical nature of the universe.

■ **FIGURE 15.3**

A Young Human Human beings have the intellect to wonder about their planet and what lies beyond it. Through them, the universe becomes aware of itself.

15.1.2 The Copernican Principle

Our study of astronomy has taught us that we have always been wrong in the past whenever we have claimed that the Earth is somehow unique. Copernicus and Galileo showed us that the Earth is not the center of the solar system but merely one of a number of objects orbiting the Sun. Our study of the stars has demonstrated that the Sun itself is a rather undistinguished star, just like so many billions of others. There seems nothing special about our position in the Milky Way Galaxy either, and nothing surprising about our Galaxy's position in the larger scheme of the universe.

The recent discovery of planets around other stars confirms our idea that the formation of planets is probably a natural consequence of the formation of many kinds of stars. While our current techniques of finding planets allow us to identify only giant planets, there is no reason to believe that other systems could not contain smaller planets like the Earth as well.

Philosophers of science sometimes call the idea that there is nothing special about our place in the universe the *Copernican principle*. Although it may be tempting to consider ourselves the central and unique focus of all creation, no evidence for such a belief is found in any of the observations discussed in this book.

Most scientists, therefore, would be surprised if life were limited to our planet and had started nowhere else. There are billions of stars in our Galaxy old enough for life to have developed on a planet around them, and there are billions of other galaxies as well. Astronomers and biologists have long conjectured that a series of events similar to those on the early Earth probably led to living organisms on many planets around other stars. And, where conditions are right, such life might even have evolved to become what we would call intelligent—that is, aware of and interested in its own cosmic history. (In this sense, we conclude—with tongue firmly in cheek—that taking an astronomy class is the supreme example of intelligent behavior in the universe!)

Such arguments from the Copernican principle, however interesting they may be for philosophers, are nonetheless insufficient for scientists. Science demands data. We would like to find actual evidence for the existence of life (and perhaps even of intelligent life) elsewhere. Despite the sensationalistic claims of unidentified flying objects (UFOs) and alien abductions in the tabloid media, no such evidence has yet been found. But because many scientists feel that such a discovery would be a defining moment in the history of the human species, we are beginning to search seriously for evidence of life beyond the Earth.

15.1.3 So Where Are They?

If the Copernican principle is applied to life, then biology may be rather common among planets. Taken to its logical limit, the Copernican principle also suggests that intelligent life might be common. Intelligence like ours has some very special properties, including an ability to progress very rapidly through the application of technology. Indeed, some philosophers today think that we will soon construct artificial intelligence that is superior to that of humans, in which case future evolution may be dominated by "creatures" composed of silicon, manipulating increasingly sophisticated systems of data and software. It seems entirely possible that such super-intelligent machines might decide to move beyond their own planetary systems and explore the Galaxy, or even beyond. Or organic life around other (older) stars may have started a billion years earlier than we did on Earth and may thus have had a lot more time to send out probes.

Faced with such a prospect, physicist Enrico Fermi asked the question half a century ago that is now called the *Fermi paradox:* Where are they? If life and intelligence are common and have such tremendous capacity for growth, why is there not a network of galactic civilizations whose presence extends even into a backwater solar system like ours?

Several solutions have been suggested to the Fermi paradox. Perhaps life is common but intelligence (or at least technological civilization) is rare. Perhaps such a network will come about in the future but has not yet had the time to develop. Maybe there are invisible streams of data flowing past us all the time that we are not advanced enough or sensitive enough to detect. Or maybe advanced species make it a practice not to interfere with immature, developing consciousness such as our own. We do not yet know whether any advanced life is out there and, if it is, why we are not aware of it. Still, you might want to think about these issues as you read the rest of this chapter.

15.2 ASTROBIOLOGY

In the last decade of the 20th century, many discoveries in both astronomy and biology stimulated the development of **astrobiology.** Astrobiologists study the origin, distribution, and ultimate fate of life in the universe. The field brings together astronomers, planetary scientists, biochemists, environmentalists, geologists, geneticists, and microbiologists to work on the same problems from their own perspectives.

Among the issues that astrobiologists are grappling with are the conditions under which life arose on the Earth and the reasons for the extraordinary adaptability of life on our planet. They are also involved in planning the continuing search for evidence of life on Mars and Jupiter's moon Europa and in trying to understand where habitable zones might be found around other stars. Let's look at some of these issues in more detail.

15.2.1 The Building Blocks of Life

While no unambiguous evidence for life has yet been found anywhere beyond the Earth, life's chemical building blocks have been detected in a wide range of extraterrestrial environments. Meteorites, rocks that fall to Earth from space, have yielded many amino acids (the molecular building blocks of proteins) whose chemical structures mark them as having an extraterrestrial origin. When we examine the

Photo by David Malin, © Anglo-Australian Observatory

■ FIGURE 15.4

A Cloud of Gas and Dust This cloud of gas and dust in the constellation of Scorpius is the sort of region where complex molecules are found. It is also the sort of cloud where new stars form from the reservoir of gas and dust in the cloud. Radiation from a group of hot stars (off the picture to the bottom left) called the Scorpius OB Association is "eating into" the cloud, sweeping it into an elongated shape and causing the reddish glow seen at its tip.

gas and dust around comets, we also find a number of **organic molecules**—those carbon-based compounds that on Earth are associated with the chemistry of life.

One of the most interesting results of modern radio astronomy has been the discovery of organic molecules in giant clouds of gas and dust between stars. More than 100 different molecules have been identified in these reservoirs of cosmic raw material, including formaldehyde ("embalming fluid"), alcohol, and others we know as important stepping stones in the development of life on Earth. Using radio telescopes and radio spectrometers, astronomers can measure the abundances of various chemicals in these clouds. We find organic molecules most readily in regions where the interstellar dust is most abundant, and it turns out these are precisely the regions where star formation (and probably planet formation) happens most easily (Figure 15.4).

Starting in the early 1950s, scientists have tried to duplicate in their laboratories the chemical pathways that led to life on our planet. In a series of experiments pioneered by Stanley Miller and Harold Urey at the University of Chicago, biochemists have simulated conditions on the early Earth and have been able to produce many of the fundamental building blocks of life, including those that go into forming proteins and nucleic acids (Figure 15.5).

Although that's encouraging, there are problems with the experiments. The most interesting chemistry from a biological perspective takes place with hydrogen-rich or *reducing* gases, such as ammonia and methane. However, the early atmosphere of the Earth was probably dominated by carbon dioxide (as Venus's atmosphere still is today) and never contained abundant reducing gases. Thus, while we know it is relatively easy to make complex carbon compounds of biological interest, we are not sure that the con-

APPARATUS USED BY MILLER FOR THE SYNTHESIS OF AMINO ACIDS BY ELECTRIC DISCHARGE

CH_4
NH_3
H_2O
H_2
Gases

Electrodes

Spark discharge

Water out
Condenser
Water in

Water containing organic compounds

Boiling water   Trap

F. Drake and the Astronomical Society of the Pacific

■ FIGURE 15.5

Simulating the Early Earth The Miller-Urey experiment, performed in 1953, studies chemical reactions that might have taken place on the early Earth. An "atmosphere" consisting of methane, ammonia, water vapor, and hydrogen was subjected to electrical sparks to simulate lightning. Water at the bottom of the apparatus provided an "ocean" into which materials synthesized in the atmosphere could fall. When its contents were analyzed, the "ocean" was found to contain a variety of amino acids, the building blocks of proteins. Today we understand that the atmosphere of the early Earth probably had more carbon dioxide and much less hydrogen than Miller and Urey used, and this reduces the number of organic molecules formed in such experiments.

Space Telescope Science Institute/AURA/NASA; © Bonnie Sue Photography

■ **FIGURE 15.7**

Yellowstone Hot Spring with Colorful Microbial Life
Grand Prismatic Spring in Yellowstone National Park supports a wide range of microbial life under conditions that we would consider extreme. The rings of color mark the habitats of microbes with different temperature tolerances. The water, which wells up near the center at temperatures close to boiling (100°C), cools as it spreads from the center and overflows the lip of the pool. You can see people on the boardwalk winding around at the left.

meaning to like or love, as in "anglophile," someone who likes the British). At the low end, life can often survive at temperatures even below 0°C in the presence of some kind of natural "antifreeze," although metabolism slows down and often stops. Microbes that were frozen and dormant for tens of thousands of years in the Antarctic ice have been revived in our laboratories.

At high temperatures, thermophiles have developed mechanisms to make the chemical repairs that are needed as carbon-based compounds begin to come apart. Unlike the dormant state at low temperatures, adaptation to high temperatures requires active chemical intervention. Many microbes flourish in the Yellowstone hot springs at temperatures up to 100°C. The record for a thermophile is 113°C at deep-sea vents, where the pressure of the water above the vent is so great that 113°C is still below the boiling point of water (Figure 15.8).

Peter Ryan/Scripps/Science Photo Library/Photo Researchers, Inc.

■ **FIGURE 15.8**

"Black Smoker" on the Ocean Floor   Starting in 1977, marine geologists and biologists have been discovering that hot, mineral-rich water can emerge from vents on the ocean floor. As you can see here, when the hot vent water mixes with the much colder ocean water, the dissolved minerals can fall out of solution and make a kind of "black smoke." The regions around these hydrothermal vents are full of life; more than 350 new species have been discovered close to them. In the pitch darkness of the ocean floor, these forms of life derive their energy not from sunlight, but from the chemistry of the vents.

Three related environmental extremes involve moisture, salt, and acid. Many microbes can survive extremely dry conditions in much the way they do low temperatures—by going into a dormant state and waiting for better conditions to return. Some microbes are so tolerant of high salinity that they can survive even in the waters of the Dead Sea. The range of acidity in which life has been found goes from pH less than 0 to greater than 9. The numbers in pH, which stands for potential of hydrogen, measure whether a substance is acidic (0–7), neutral (exactly 7), or alkaline (7–14). One example of an acid environment is the Rio Tinto (Red River) of southern Spain, which originates in a region of extensive mineral deposits and maintains a steady pH of 2.5 all the way to its mouth (Figure 15.9). A rich microbe community inhabits this river, and indeed some of the microbes help to maintain the constant pH of 2.5 because that is the environment they like.

One of the most surprising cases of tolerance to extremes is exhibited by the bacterium called *Deinococcus radiodurans,* which is found (among other places) in the cooling water of nuclear reactors. (It is called an atomophile!) With its highly developed chemical repair mechanisms, *radiodurans* can survive ultraviolet or particle radiation up to 6000 rads per hour, a thousand times more than you would be able to tolerate. It is also resistant to many unpleasant industrial chemicals, and it is commonly found in toxic waste dumps. No one knows what evolutionary pathways enabled *radiodurans* to develop this amazing tolerance for otherwise lethal conditions.

So we see that life has learned to survive on Earth in a remarkable range of environments on the surface, in the water and air, and even deep underground. Nearly every conceivable ecological niche seems to be filled, although in many cases the rate of metabolism and other biological processes is very low. But there are some exceptions. No organism has learned to extract the water it needs from ice: The ice sheets of Greenland are not green! There are also no organisms that carry out their life cycle entirely in the air. Life (as we know it) requires liquid water and something substantial such as the land or ocean to make a home.

15.3 SEARCHING FOR LIFE BEYOND EARTH

Astronomers and planetary scientists are on the forefront of the search for life in the universe. In this section we discuss two kinds of searches. First is the direct exploration of planets within our own solar system, especially Mars and Europa. Second is the even more difficult task of searching for evidence of life—*biomarkers*—on planets circling other stars. Finally, in Section 15.4, we will examine SETI, the search for extraterrestrial *intelligence.* As you will see, the approaches taken in these three cases are very different, even though the goal of each is the same—to determine whether life on Earth is unique in the universe.

■ **FIGURE 15.9**
Rio Tinto, the Acid River of Spain   The waters of Rio Tinto are acidic, with a pH of 2.5. The reddish color comes from dissolved minerals. Yet the river supports microbes that are adapted to the acidic conditions and actually help maintain the river at a constant pH of 2.5 from its source to its mouth.

David Morrison

15.3.1 Life on Mars

In 1996, an enormous amount of public and media attention was focused on a suggestion by an interdisciplinary team of scientists that they had found evidence of life on Mars. They had done detailed laboratory analysis of an ancient martian rock sample recovered in Antarctica, a 4-billion-year-old meteorite called ALH 84001. Their work showed that this bit of martian crust (knocked into space by a powerful impact; Figure 15.10) had experienced wet conditions about 3.5 billion years ago. At that time, Mars's atmosphere was thicker and water could flow—leaving traces of carbonate minerals and some organic compounds embedded in the potato-sized piece of rock that eventually fell to Earth. The team further hypothesized that tiny structures seen in the rock at extremely high magnification were the fossilized remains of ancient microbial life.

If there are indeed fossils in ancient martian rocks, this would be one of the most spectacular discoveries in the history of science. As we write in 2003, it appears to most scientists that this meteorite was contaminated with terrestrial chemicals during the millennia it lay trapped in the Antarctic ice and that the "fossils" are too small to represent anything once living. But even if the rock's specific features are not organic in origin, the work on ALH 84001 verified that wet, apparently Earth-like conditions once existed on Mars. And the energetic debate about martian meteorites has stimulated thinking about how we might detect fossil life on other planets.

Much of the exploration of Mars by spacecraft has focused on either the environmental conditions there or the search for evidence of life. As we saw in Chapter 9, there is abundant geological evidence that Mars once had a warmer climate and liquid water. The 1976 Viking landers carried instruments designed specifically to detect microbial life in the soil, but the results were negative. Although Mars has the most Earth-like environment of any planet in the solar system, the surface is too dry, too cold, and too affected by solar ultraviolet radiation to meet the requirements for habitability.

However, there is no reason to think that life could not have begun on Mars about 4 billion years ago, since Earth and Mars apparently had similar surface conditions then. This brings us back to ALH 84001 and the arguments about fossil microbes. Astrobiologists today are learning how to identify fossil microbes in ancient Earth rocks and preparing to apply these lessons to Mars. There are plans for future missions that will include the return of samples selected from sedimentary rocks at sites (such as ancient lake beds or hot springs) that once held water (Figure 15.11). The most powerful searches for martian life (past or present) will thus be carried out in our laboratories here on Earth.

If there is evidence from these rocks or elsewhere that Mars once had life, then scientists will accelerate their search for possible survivors. They will look for life-forms that may have evolved to deal with the deteriorating climate of Mars, perhaps by finding some oasis or refuge that is warmer and wetter than most of the martian surface. NASA's theme in searching for life is "follow the water." The most likely source of liquid water on Mars today is deep below the surface, where extensive *aquifers* (layers containing groundwater) may exist. Perhaps someday astronauts on Mars will drill deep wells down to this layer of liquid water and finally encounter living alien life.

One interesting twist on the search for life is emphasized by the presence of ALH 84001 and other Mars rocks on Earth. Mars and Earth are close enough together that they have exchanged material in this way throughout their history (although most of the traffic has been from Mars to Earth, as a consequence of the lower surface gravity on Mars). It is entirely possible that some of these rocks may have contained viable microorganisms, not just fossils. Thus Mars might have seeded Earth, or the two planets could have exchanged biological material. It is thus conceivable that if we eventually find living things on Mars they will be genetically similar to terrestrial life, for the good reason that they are our distant cousins.

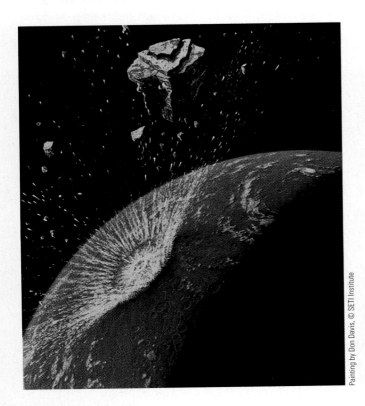

Painting by Don Davis, © SETI Institute

■ **FIGURE 15.10**

Martian Impact This painting shows how a large rock impacting Mars might have ejected some smaller chunks from the red planet. After millions of years of orbiting the Sun, some fragments intersected the Earth and wound up in the Antarctic, where scientists later found them.

15.3.2 Life in the Outer Solar System

Another location of great interest to astrobiologists is Jupiter's satellite Europa (see Chapter 11). Galileo spacecraft data on this Moon-sized world indicate the presence of a global ocean of liquid water beneath a thick ice crust.

USGS/NASA/JPL

■ **FIGURE 15.11**

Potential Mars Rover Landing Site The most promising sites in the search for fossil microbes on Mars are places where water once pooled, such as ancient ponds or lakes. One of the places the Mars Exploration Rover mission plans to land is at Gusev Crater, an impact basin about 160 km in diameter, seen at the top of this image. A 600-km-long channel called Ma'adim Vallis drains into it, and there is speculation, based on the appearance of the crater floor, that significant water may have accumulated in Gusev in the past.

As the only location in the solar system beyond Earth where there is known to be a large body of liquid water, Europa beckons us to further exploration (Figure 15.12).

Life requires an energy source, and sunlight does not penetrate below the thick icy crust of Europa. Thus any life-forms there are unlikely to have evolved photosynthesis. What other energy sources might be present on Europa? To remain liquid, the global ocean must be warmed by heat generated by tides and now escaping from the interior of Europa. Hot (or at least warm) springs might be active there, analogous to those we have discovered in the deep oceans of the Earth. Europa might therefore support life that derives all its energy from the mineral-laden water in such springs.

Some people have suggested that Europa is the most likely place beyond the Earth to find life in the solar system. On the other hand, we might question whether life could originate in a dark ocean heated only by hot springs. Since we do not know exactly how life formed on Earth, it is impossible to evaluate this possibility for Europa. One thing is clear, however: If there is life in the europan

JPL/NASA

■ **FIGURE 15.12**

Icy Surface of Europa The Galileo spacecraft demonstrated that Jupiter's satellite Europa, which is about the same size as our Moon, has a global ocean of liquid water roughly 100 km deep. Unfortunately this ocean lies beneath a thick crust of ice, and no sunlight penetrates into the cold dark waters beneath. This picture illustrates the cracked surface of Europa (top) on the same scale as the San Francisco peninsula (bottom). The round, flat, smooth area at left on the Europa image is probably an area where some fluid coming up from below buried the cracks and ridges that crisscross the satellite's surface.

oceans, it is likely to be unrelated to terrestrial life. There is no exchange of rocks between Earth and Europa to provide the possibility of cross-contamination, as is the case for Mars. Thus Europa holds out the tantalizing prospect of a clear *second genesis*—an independent origin of life. If so, we can hardly guess what that life might be like. Will it be carbon based? Will it utilize protein chemistry? Will it possess a genetic material something like DNA or RNA? Or will it be more exotic than we can now imagine?

15.3.3 Habitable Planets Orbiting Other Stars

Although our robot probes are still confined to the worlds of our own solar system, we certainly do not want to limit our search for life to our own neighborhood. How can we gather evidence about the possibility of life among the stars?

As we have discussed in Chapter 13, one of the most important recent developments in astronomy has been the ability to detect planets circling distant solar-type stars—with more than 100 such planets discovered by the end of 2002. So far, we are able to detect only giant planets (like Jupiter and Saturn) and these are unlikely as the home of life. Still, their existence holds out the prospect of smaller worlds (either Earth-like planets or satellites orbiting the giants) that might support liquid water and the other conditions necessary for biology.

In evaluating the prospect for life in distant planetary systems, astrobiologists have developed the idea of a **habitable zone**—a region around a star where suitable conditions might exist for life. Since the focus is on the presence of liquid water, the usual definition of a habitable zone is the distance where water will be liquid on the surface of a terrestrial-type planet (that is, a planet with roughly the mass of the Earth).

Obviously the Earth is in the habitable zone for the solar system, although there are periods in the past when most of the oceans froze over (we call these times "snowball Earth"). Note, however, that the Earth is warmer than the freezing temperature of water only because the greenhouse effect in our atmosphere raises the average temperature by about 25°C over the value it would have if there were no greenhouse gases in the atmosphere. Thus, we must consider the nature of any atmosphere as well as the distance from the star in evaluating the range of habitability.

The luminosity of stars like the Sun changes over their main-sequence lifetime, and this further complicates the picture. Calculations indicate that the power output of the Sun has increased by at least 30 percent over the past 4 billion years.[1] Yet geological evidence suggests that liquid water has been present on the Earth's surface for that entire time, and indeed the coldest periods (the snowball Earth times) took place within the past billion years, not earlier when the Sun was fainter. Apparently the Earth had a larger greenhouse effect early on, and the declining greenhouse effect has roughly compensated for the increasing luminosity of the Sun. Such changes must be factored into our estimates of where a star's habitable zone might lie at any given time.

Within the solar system, Venus has evolved through a runaway greenhouse effect into an oven where life is impossible, but it was once inside the habitable zone. Mars today is too cold and dry for surface life, but in the past it had a thicker atmosphere and apparently supported surface water (although perhaps its lakes and seas were ice-covered). Today Mars seems to be outside the habitable zone, but if the Earth (with its greater ability to retain an atmo-

sphere) were in the orbit of Mars, it might still be relatively warm. In other words, the current inhospitable nature of Mars is as much a consequence of its small mass as its distance from the Sun.

This is all very complicated, and scientists still differ in what they consider to be a habitable zone. Most would agree that for the solar system, habitability means having a mass at least as great as that of Mars (11 percent that of the Earth) and orbiting between about 0.9 and 1.2 AU. Stars that have lower luminosity than the Sun will have a smaller habitable zone that is closer in to the star. Stars with somewhat higher luminosities will have larger zones, but their ultraviolet output will be greater and may pose a hazard for life. Stars that have really large masses and luminosities won't last long enough for life to develop around them, so we won't worry about them.

15.3.4 Biomarkers

Our prime candidate worlds in the search for life beyond the solar system are terrestrial-type planets within the habitable zones of their stars. In addition, we require that the stars themselves be reasonably constant in brightness (no variable stars allowed) and that the orbits of the planets have low eccentricity (look more like circles than ellipses), so that their surface temperatures are stable. We are unable to detect such planets with our current technology, but within a decade or so, space missions should allow us to determine how common such habitable planets are and to identify nearby candidate systems for further study.

The fact that a planet is within the habitable zone and is therefore considered able to support life does not ensure, of course, that it is actually inhabited. Indeed, one of the most important questions in astrobiology is just that: Will life arise naturally when the environmental conditions are correct? It is thus very important to consider how we might recognize the signature of life on a distant planet.

Even with the largest space-based telescope we can contemplate, we will never be able to obtain an image of a distant planet, as we do the worlds in our own solar system, let alone visit with robot spacecraft. Astronomers therefore need a global **biomarker**—something distinctive to separate a live world from a dead one. To be detectable, these biomarkers should involve changes in atmospheric or surface chemistry that can *only* be the result of life.

If we observed Earth from a great distance and took sensitive visible-light and infrared spectra, we might just see such biomarkers. The most easily detectable evidence is the presence of abundant free oxygen in the atmosphere, which produces distinctive features in near infrared spectra (Figure 15.13). Because oxygen is a highly reactive element, any planet with lots of it must have a way of producing fresh supplies regularly. On Earth, oxygen is the by-product of photosynthesis; if life on Earth should cease, the oxygen in the atmosphere would disappear within a few thousand years. The oxygen, therefore, is a product of life, which in effect uses sunlight to manufacture atmospheric

---

[1] The Sun is very gradually brightening because it is converting more and more of its core hydrogen into helium through fusion. When four hydrogen atoms are turned into one helium atom, the core pressure is reduced (fewer particles to do the pressing). This means the outer layers of the Sun can squeeze harder on the core. When squeezed, the core gets hotter and more hydrogen atoms fuse. This very gradually raises the energy output of our star.

FIGURE 15.13

Spectrum of Planet Earth This shows what the spectrum of our planet in infrared radiation would look like from another world. Note the absorption feature from ozone (a form of oxygen).

oxygen. A similar atmospheric gas is methane, produced by microbes in swamps (it used to be called swamp gas) and in the digestive tracts of some mammals. Without the presence of life, methane would be quickly oxidized and disappear from the atmosphere. In fact, the most distinctive biomarker for Earth is the simultaneous presence of these two mutually inconsistent gases: oxygen and methane.

The only problem with using oxygen and methane as biomarkers is that for more than half the history of the Earth, there was no free oxygen. Even after the biological invention of photosynthesis, it took a long time before oxygen achieved measurable concentrations. Thus, for 2 billion years Earth was a living planet but without the oxygen/methane biomarker in its atmosphere. Today scientists are beginning to look in more detail at the way early life functioned on the Earth and to try to identify other biomarkers that would be useful in the study of a faint and distant world (Figure 15.14). Alternatively, we can look for one quite dramatic biomarker of an intelligent technological life-form: the broadcast of radio signals.

15.4 THE SEARCH FOR EXTRATERRESTRIAL INTELLIGENCE

For the first time in human history, we know that our planetary system is not unique. Given all the developments discussed in this chapter, it may also be that life has developed on many planets around other stars. If that life is microbial, or even if it includes metazoans (large animals), such as those that have populated the Earth for the most recent several hundred million years, we do not yet have the technology to discover it. But suppose in a few cases intelligent, technical civilizations have arisen, as has happened on Earth in the most recent blink of cosmic time. If there is such an intelligent species out there, how could we make contact with it?

This problem is similar to making contact with people who live in a remote part of the Earth. If students in the United States want to converse with students in Australia,

FIGURE 15.14

Microbial Mats Microbes that can do photosynthesis have played the dominant role in producing the Earth's oxygen for 3 billion years. This photo shows a cross-section of a mat of such organisms living today in the salt flats of Baja California. Hundreds of different species of microbes "work together" in a miniature ecosystem that is less than a centimeter thick. A coin is shown to give you a sense of scale.

David DesMarais, NASA Ames

for example, they have two choices. Either one group gets on an airplane and travels to meet the other, or they communicate via some message medium (today probably by telephone, fax, or e-mail). Given how expensive airline tickets are, most students would probably select the message route.

In the same way, if we want to get in touch with intelligent life around other stars, we can travel or we can try to exchange messages. Because of the great distances involved, interstellar space travel is very slow and very expensive. The fastest spacecraft the human species has built so far would take almost 80,000 years to get to the nearest star. While we could certainly design a faster craft, the more quickly we require it to travel, the greater the energy cost involved. To reach neighboring stars in less than a human life span, we would have to travel close to the speed of light. In that case, however, the expense would become truly astronomical.

15.4.1 Interstellar Travel

Bernard Oliver, then Vice President of the Hewlett-Packard Corporation and an engineer with an abiding interest in life elsewhere, made a revealing calculation about the costs of rapid space travel. Since we do not know what sort of technology we (or other civilizations) might someday develop, Oliver considered a trip to the nearest star in a spaceship with a "perfect engine"—one that would convert its fuel into energy with 100 percent efficiency. (No future technology can possibly do better than this. In reality, nature is unlikely to yield efficiency even close to the perfect value.) Even with a perfect engine, the energy cost of a single round-trip journey at 70 percent the speed of light turns out to be equivalent to about 500,000 years worth of total U.S. electrical energy consumption!

In case you are wondering why this figure is so high, you must remember that the voyagers could not depend on finding "gas stations" open at their destination; therefore, they would have to carry the fuel for the return legs of the journey with them, and getting that huge mass of fuel up to 70 percent the speed of light would be very expensive. The important thing about Oliver's calculation is that it does not depend on present-day technology (since it assumes a perfect engine) but only on the known laws of science. What it shows is that no matter who does the traveling, it is very expensive to go fast enough to get to the stars within the course of a single human life.

This is one reason astronomers are so skeptical about claims that UFOs are spaceships from extraterrestrial civilizations. Given the distance and expense involved, it seems unlikely that the dozens of UFOs (and even UFO abductions) claimed each year could all be visitors from other stars so fascinated by Earth civilization that they are willing to expend fantastically large amounts of energy or time to reach us. Nor does it seem credible that these visitors have made this long and expensive journey and then systematically avoided contacting our governments or political and intellectual leaders.

In fact, a sober evaluation of UFO reports often converts them to IFOs (identified flying objects) or NFOs (not-at-all flying objects). While some are hoaxes, others are natural phenomena, such as ball lightning, fireballs, bright planets, or even flocks of birds with reflective bellies. Still others are human craft, such as private planes with some lights missing, or secret military aircraft. It is also interesting that the group of people who most avidly look at the night sky, the amateur astronomers, have never reported UFO sightings. Further, not a single UFO has ever left behind any physical evidence that can be tested in a laboratory and shown to be of nonterrestrial origin.[2]

15.4.2 Messages on Spacecraft

While space travel by living creatures seems very difficult, robot probes can travel over long distances and over long periods of time without needing a good cup of coffee. Four spacecraft built by the human species—two Pioneers and two Voyagers—having finished their program of planetary exploration, are now leaving the solar system. At their coasting speeds, they will take hundreds of thousands or millions of years to get anywhere close to another star. They were the first products of human technology to go beyond our home system, however, so we wanted to put messages on board to show where they came from.

Each Pioneer carries a plaque with a pictorial message engraved on a gold-anodized aluminum plate (Figure 15.15). The Voyagers, launched in 1977, have audio and video records attached (Figure 15.16), which allowed the inclusion of more than 100 photographs and a selection of music from around the world. (Included among the excerpts from Bach, Beethoven, folk music, tribal chants, and others is one piece of rock—"Johnny B. Goode" by Chuck Berry.) Given the enormity of the space between stars in our section of the Galaxy, it is very unlikely that these messages will ever be received by anyone. They are more like a note in a bottle thrown into the sea by a shipwrecked sailor, with no realistic expectation of its being found soon, but a slim hope that perhaps someday, somehow, someone will know of the sender's fate.

15.4.3 Communicating with the Stars

If direct visits to stars are unlikely, we must turn to the alternative for making contact—exchanging messages. Here the news is a lot better. We already know (and have learned to use) a messenger—electromagnetic radiation—that moves through space at the fastest speed in the universe.

---

[2] If you are interested in pursuing the topic of what UFOs are and aren't, we recommend the following books: P. Klass, *UFO Abductions: A Dangerous Game* (1988, Prometheus Books); C. Peebles, *Watch the Skies: A Chronicle of the Flying Saucer Myth* (1994, Smithsonian Institution Press); and R. Shaeffer, *UFO Sightings: The Evidence* (1998, Prometheus Books). Also see: www.astrosociety.org/education/resources/pseudobib.html.

FIGURE 15.15

Interstellar Message This is the image engraved on the plaques aboard the Pioneer 10 and 11 spacecraft. The human figures are drawn in proportion to the spacecraft, which is shown behind them. The Sun and planets in the solar system can be seen at the bottom, with the trajectory that the spacecraft followed. The lines and markings in the left center show the positions and pulse periods for a number of pulsars, which might help locate the spacecraft's origins in space and time.

NASA

(a)

(b)

NASA

FIGURE 15.16

The Voyager Record (a) Encoded onto a gold-coated copper disk, the Voyager record contains 118 photographs, 90 minutes of music from around the world, greetings in almost 60 languages, and other audio material. It is a summary of the sights and sounds of Earth. (b) One of the images encoded onto the record. Originally, the team devising the record wanted to send a photograph from a medical book of a nude man and a pregnant woman. However, NASA was concerned about offending some people on Earth, and so artist Jon Lomberg drew a silhouette version, which allowed a fetus inside the woman to be shown as well.

Traveling at the speed of light, radiation reaches the nearest star in only four years and does so at a tiny fraction of the cost of sending material objects. These advantages are so clear and obvious that we assume they will occur to any other species of intelligent beings who develop technology.

However, we have access to a wide spectrum of electromagnetic radiation, ranging from the longest-wavelength radio waves to the shortest-wavelength gamma rays. Which would be the best for interstellar communication? It would not be smart to select a wavelength that is easily absorbed by interstellar gas and dust, or one that is unlikely to penetrate the atmosphere of a planet like ours. Nor would we want to pick a wave that has lots of competition for attention in our neighborhood.

One final criterion makes the selection easier: We want the radiation to be inexpensive to produce in large quantities. When we consider all these requirements, radio waves turn out to be the best answer. Being the lowest-frequency (and lowest-energy) band of the spectrum, they are not very expensive to produce (permitting us to use them extensively for communications on the Earth). They are not significantly absorbed by interstellar dust and gas. With some exceptions, they easily pass through the Earth's atmosphere and through the atmospheres of the other planets we are acquainted with.

15.4.4 The Cosmic Haystack

For these reasons, many astronomers have decided that the radio band is probably the best place in the spectrum for communication among intelligent civilizations. In particular, we are focusing on the part of the radio band called microwaves, where both the universe and the atmosphere of planets turn out to produce the least amount of static. Having made such a decision, however, we still have many questions and a daunting task ahead of us. Shall we *send* a message, or try to *receive* one? Obviously, if every civilization decides to receive only, then no one will be sending, and everyone will be disappointed. On the other hand, it may be appropriate for us to *begin* by listening, since we are likely to be among the most primitive civilizations in the Galaxy who are interested in exchanging messages.

We do not make this statement to insult the human species (which, with certain exceptions, we are rather fond of). Instead, we base it on the fact that humans have had the ability to receive (or send) a radio message across interstellar distances for only a few decades. Compared to the ages of the stars and the Galaxy, this is a mere instant. If there are civilizations out there that are ahead of us in development by even a short time (in the cosmic sense), they are likely to have a head start of many, many years. If there are civilizations behind us, chances are they are sufficiently far behind that they have not yet developed radio communications.

In other words, we, who have just started, may well be the "youngest" species in the Galaxy with this capability (see the Figuring for Yourself section at the end of the chapter). Just as the youngest members of a community are often told to be quiet and listen to their elders for a while before they say something foolish, so we may want to begin our exercise in extraterrestrial communication by listening.

Even restricting our activities to listening, however, leaves us with an array of challenging questions. For example, you know from your own experience with radio transmissions that a typical signal comes in on only one channel (which means it is carried by one small frequency band of radio waves). The owners of your favorite station are confident—because there aren't that many channels on the radio dial—that you will find them despite this (although a few actually transmit over two different frequencies, one on the AM band and one on the FM band). Many of us, when we first arrive in a new city, scan up and down the radio band until we find the stations we like, but if we have only an AM radio in our car, and the stations that play our favorite music are all FM, then we are out of luck.

In the same way, it would be very expensive (and perhaps even ill mannered) for an extraterrestrial civilization to broadcast on a huge number of channels. Most likely, they select one or a few channels for their particular message. (Communicating on a narrow band of channels also

TABLE 15.1 *The Cosmic Haystack Problem: Some Questions About an Extraterrestrial Message*

- From which direction (which star) is the message coming?
- On what channels (or frequencies) is the message being broadcast?
- How wide in frequency is the channel?
- How strong is the signal (can our radio telescopes detect it)?
- Is the signal continuous, or does it shut off at times (as, for example, a lighthouse beam does when it turns away from us)?
- Does the signal drift (change) in frequency because of the changing relative motion of the source and the receiver?
- How is the message encoded in the signal (how do we decipher it)?
- Can we even recognize a message from a completely alien species? Might it take a form we don't at all expect?

helps distinguish an artificial message from the radio static that comes from natural cosmic processes.) But the radio band of the electromagnetic spectrum contains an astronomically large number of possible channels. How can we know in advance which one they have selected and how they have coded their message into the radio signal?

There is another problem that may make it hard for you to find your favorite type of radio station. If your radio has a poor antenna, then it may not pick up the signal from a weak station some distance away. You may not learn about the existence of that station (and others like it) until you buy better equipment. The same will be true for interstellar transmissions. If an extraterrestrial civilization's signal is just too weak for our present-day radio telescopes, they may be broadcasting their little alien hearts out, but we will miss it completely.

Table 15.1 summarizes these and other factors that scientists must grapple with when they try to tune into radio messages from distant civilizations. Because their success depends on either guessing right about so many factors or searching through all the possibilities for each factor, some scientists have compared their quest to looking for a needle in a haystack. Thus, they like to say that the list of factors in Table 15.1 defines the *cosmic haystack problem*.

15.4.5 Radio Searches

Although the cosmic haystack problem seems daunting, many other research problems in astronomy also require a large investment of time, equipment, and patient effort. And, as several astronomers have pointed out, if we don't search, we're sure not to find anything. Therefore, several groups of radio astronomers have undertaken searches for extraterrestrial messages during the last four decades.

The very first search was conducted by astronomer Frank Drake in 1960, using the 85-ft antenna at the National Radio Astronomy Observatory (Figure 15.17). Called Project Ozma, after the queen of the exotic Land of Oz in

National Radio Astronomy Observatory

■ **FIGURE 15.17**

Project Ozma A 25th-anniversary photo of some members of the Project Ozma team standing in front of the 85-ft radio telescope with which the 1960 search for extraterrestrial messages was performed. Frank Drake is in the back row, second from the right.

the children's stories of L. Frank Baum, Drake's experiment involved looking at about 7200 channels and two nearby stars over a period of 200 hours. Although he found nothing, Drake demonstrated the feasibility of such a search and set the stage for the more sophisticated projects that followed. (It is interesting to note that what took 200 hours in 1960 can be done with today's automated systems in about a thousandth of a second.)

More than 60 additional radio searches have been carried out by scientists around the world, each exploring a minuscule region of the cosmic haystack. Although a number of initially interesting signals emerged, all either have been traced to local interference or have lasted such a short time that the signals could not be verified. Scientists are continuing several of the searches, always trying to improve their equipment and beat the odds against finding that elusive needle of a message.

In 1992, NASA began the most comprehensive SETI observations ever undertaken, only to have Congress cut the funding for their project after less than a year. Using private donations, the nonprofit SETI Institute has undertaken to continue the search. They now call it Project Phoenix, since the program has risen from the ashes of its funding crisis.

Using modern electronics and computers, the Phoenix system can tune to tens of millions of channels simultaneously (Figure 15.18). Its software searches for promising signals (such as continuous waves or pulses) and alerts the

experimenters if an interesting signal persists on any one channel or moves between channels because of Doppler shift. The intent of the project (which resumed in 1995) is to search 2 billion channels for each of about 1000 nearby stars. To make it onto their list, a star must be roughly similar to the Sun and at least 3 billion years old. That means it could have had enough time for intelligent life to develop on any Earth-like planets around it. The majority of the initial candidate stars have now been observed. Other SETI programs search only a selected range of channels, but sweep the sky in the hope of encountering a message. So far, no signal has been found; the search will continue using other large radio telescopes and even better equipment.

Receivers are constantly improving, and the sensitivity of SETI programs advances rapidly. One of the next steps will be to build much larger radio telescopes that are optimized for SETI. The SETI Institute is constructing the Allen Telescope Array, which uses many relatively small and inexpensive radio dishes linked together to provide the equivalent of 1 hectare of collecting area (Figure 15.19). A hectare is an area of 10,000 square meters, or about 2.5 acres. There is hope of expanding this concept to a 100-hectare or 1-square-kilometer area, which would be a powerful telescope for radio astronomy as well as SETI.

What kind of signals do we hope to pick up? We on Earth are inadvertently sending out a flood of radio signals, dominated by high-power television and military radar systems. This is a kind of leakage signal, similar to the wasted

(a)

(b)

■ **FIGURE 15.18**

Project Phoenix (a) The 64-m radio telescope at Parkes, Australia, was used in 1995 to search for radio signals from possible extraterrestrial civilizations around 200 stars. (b) Phoenix Project scientists Jill Tarter and Peter Backus are shown at the telescope controls during their observing run.

Jill Tarter: Trying to Make Contact

1997 was quite a year for Jill Cornell Tarter, one of the world's leading scientists in the SETI field. The SETI Institute announced that she was the recipient of its first endowed chair (the equivalent of an endowed research professorship) named in honor of Bernard Oliver. The National Science Foundation approved a proposal by a group of scientists and educators she headed to develop an innovative hands-on high school curriculum based on the ideas of cosmic evolution (the topics of this chapter). And, at roughly the same time, she was besieged with requests for media interviews as news reports identified her as the model for Ellie Arroway, the protagonist of *Contact*, Carl Sagan's best-selling novel about SETI. The book had been made into a high-budget science-fiction film starring Jodie Foster, who had actually talked with Tarter before taking the role.

Tarter is quick to point out that "Carl Sagan wrote a book about a woman who does what I do, not about me." Still, as the only woman in such a senior position in the small field of SETI, she was the center of a great deal of public attention. (However, colleagues and reporters pointed out that this was nothing compared to what would happen if her search for radio signals from other civilizations recorded a success.)

Being the only woman in a group is not a new situation for Tarter, who

Jill Tarter

Photo by Seth Shostak/SETI Institute

was often the only woman in her advanced science and math classes. She had been encouraged by her father in both her interest in science and her "tinkering," and as an undergraduate at Cornell University, she majored in engineering physics. That training now comes in very handy as she puts together and maintains the complex devices that automatically scan for signals from other civilizations.

Switching to astrophysics for her graduate studies, Tarter wrote a PhD thesis that, among other topics, considered the formation of failed stars— those whose mass was not sufficient

for the nuclear reactions that power the stars to begin inside them. Tarter coined the term *brown dwarf* for these small, dim objects, and it has remained the name astronomers use. Later, her postgraduate work involved thinking about "dark matter," so-far unseen material that seems (from its gravitational pull) to be a very substantial component of galaxies and groups of galaxies. Now that she is working full time searching for extraterrestrial civilizations, Tarter likes to joke that her entire career seems to have involved searching for the most elusive things in the universe.

It was while she was still in graduate school that Stuart Bowyer, one of her professors at Berkeley, gave her a study on the possibilities of SETI (chaired, as it turns out, by Bernard Oliver). Bowyer asked her if she wanted to be involved in a small experiment to siphon off a bit of radiation from a radio telescope as astronomers used it year in and year out and see whether there was any hint of an intelligently coded radio message buried in the radio noise. Her engineering and computer programming skills became essential to the project, and soon she was hooked on the search for life elsewhere. Years later, as Tarter plans for the Allen Telescope Array, she may one day (like Ellie Arroway in the film) be the first person on Earth to know that humanity is not alone in the cosmos.

light energy that is beamed upward by poorly designed street lights and advertising signs. Could we detect a similar leakage of radio signals from another civilization? The answer is just barely, but only for the nearest stars. For the most part, therefore, current radio SETI searches are looking for beacons, not trying to eavesdrop on someone else's communication system. Our prospects for success depend on how often civilizations arise, how long they last, and how patient they are about broadcasting their locations to the

cosmos (see the Figuring for Yourself section at the end of the chapter).

15.4.6 SETI Outside the Radio Realm

For the reasons discussed above, most SETI programs search for signals at radio wavelengths. It seems reasonable to us that if another civilization were broadcasting a beacon for purposes of interstellar communication, it would use

■ **FIGURE 15.19**

Allen Telescope Array The most ambitious expansion of SETI today, the Allen Telescope Array will eventually consist of 350 small antennas linked together. This photo shows the first three antennas, which began test operations in California in December 2002.

microwaves. But maybe the civilization isn't broadcasting on purpose. In that case, is there any other way we could pick up other evidence for its existence?

Recently technology has allowed astronomers to expand the search into the domain of visible light. You might think it would be hopeless to try to detect a flash of visible light from a planet, given the brilliance of the star it orbits. This is why we cannot measure the reflected light of planets around other stars. The feeble light of the planet is simply swamped by the "big light" in the neighborhood. So another civilization would need a mighty strong beacon to compete with its star.

However, in recent years, human engineers have learned how to make flashes of light brighter than the Sun. The trick is to "turn on" the light for a very brief time, so that the costs are manageable. But ultra-bright, ultra-short laser pulses (operating for periods of a billionth of a second) can pack a lot of energy and can be coded to carry a message. We also have the technology to detect such short pulses—not with human senses, but with special detectors that can be "tuned" to hunt automatically for such short bursts of light from nearby stars.

Why would any civilization try to outshine its own star in this way? It turns out that the cost of sending an ultra-short laser pulse in the direction of a few promising stars can be less than the cost of sweeping a continuous radio message across the whole sky. Or perhaps they too have a special fondness for light messages because one of their senses evolved using light. Several programs are now experimenting with "optical SETI," which can be done with only a modest telescope (Figure 15.20). (The term *optical* here means using visible light.)

If we let our imaginations expand, we might think of other possibilities. What if a truly advanced civilization should decide to (or need to) renovate its planetary system to maximize the area for life? It could do so by breaking

■ **FIGURE 15.20**

Optical SETI at the University of California Undergraduate physics major Shelley Wright designed and built one of the first optical SETI systems, designed to detect extremely short-duration flashes of visible light. She is shown here with the Lick Observatory 1.1-m telescope where the observations are being made.

apart some planets or moons and building a sphere that completely encloses the star and intercepts all its light. The star would then become invisible, but the huge artificial sphere might glow very brightly at infrared wavelengths as all the starlight was eventually converted to heat and reradiated into space. In fact, we might imagine engineering on an even larger scale, capturing the energy from many stars. When astronomers find new and mysterious objects in the universe, there is always the temptation to ask ourselves whether these could be the constructions of distant, powerful minds with knowledge and skills almost beyond our imagining. It is a wonderful idea, but so far no evidence of such a civilization has been discovered.

15.4.7 What If We Succeed

No one can predict when or whether SETI searches will be successful. It may well be that civilizations technologically far in advance of our own use other forms of communication that we are not yet aware of. After all, 150 years ago we did not have an inkling of the possibilities of radio communication, while today it is difficult to imagine our civilization without it. On the other hand, we would never dream of giving a preschooler a book like the one you are reading. Young children learning to read are given simple books until they have mastered the basics. We hope that advanced civilizations remember their own youth and send out messages that even youngsters like us can find and interpret.

What will happen if we do find a radio signal that is unambiguously the product of an extraterrestrial intelligence? The existence of the signal itself will be of tremendous philosophical importance, demonstrating that we are not alone in the cosmos. But unless we are able to interpret the message, there may not be much practical value in the discovery. If we can eventually work out a method of mutual communication, however, an interesting question will arise: Who will speak for planet Earth?

Suppose we know that a star 35 LY away has a technological civilization around it and that they have given us a kind of "code" by which we can make ourselves understood to each other. (An easy way to begin interacting might be to send pictures.) Who on Earth decides what to send? Does the whole planet try to agree on one set of messages, or can any individual or group send a separate communication? There may be many countries, religious groups, cultural or-

ganizations, corporations, and individuals who can afford a radio antenna and are interested in getting their message out. Among ourselves, we rarely speak with one voice; should we try to do so in addressing the universe? Confronting such questions may be a good test of whether there is intelligent life on Earth.

Whether or not we ultimately turn out to be the only intelligent species in our part of the Galaxy, our exploration of the cosmos will surely continue. An important part of that exploration will still be the search for biomarkers from inhabited planets that have not produced technological creatures that send out radio signals. After all, creatures like butterflies and dolphins may never build radio antennas, but we are happy to share our planet with them and would be delighted to find their counterparts on other worlds.

A humble acknowledgment of how much we have left to learn about the universe is one of the fundamental hallmarks of science. This should not, however, prevent us from feeling exhilarated about how much we have already managed to discover, and curious about what else we might find out.

Our progress report on the ideas of astronomy ends here, but we hope that your interest in the universe does not. We hope you will keep up with developments in astronomy through the media, or by going to an occasional public lecture by a local scientist. Who, after all, can even guess all the amazing things that future research projects will reveal about both the universe and our connection with it?

SURFING THE WEB

The Astrobiology Web: www.astrobiology.com
This private site, maintained by a space enthusiast, collects a vast amount of information about many aspects of astrobiology.

NASA Astrobiology Site: astrobiology.arc.nasa.gov
NASA's Ames Research Center is taking a leading role in astrobiology and its site can point you to NASA's plans for missions as well as useful background information.

SETI Institute Home Page: www.seti.org
This is the main scientific organization searching for signals from extraterrestrial civilizations (which also works to understand the formation and prevalence of life beyond the Earth).

The Planetary Society's Search:
planetary.org/html/UPDATES/seti/index.html
This membership and advocacy organization helps pay for several SETI projects.

Drake Equation Calculators:
www.seti.org/seti/seti_science/drake_calculator.html
www.activemind.com/Mysterious/Topics/SETI/drake_equation.html

SETI@Home: setiathome.ssl.berkeley.edu
This screen saver allows anyone with a home computer to analyze a tiny bit of SETI data to see if there are any messages buried in it.

SUMMARY

15.1 The Earth and its life-forms are a product of cosmic evolution. The universe began with only the very simplest elements. More complex nuclei are produced via nuclear fusion at the hot centers of stars; some stars explode at the

end of their lives and thus recycle the newly formed elements into space. Our solar system formed about 5 billion years ago from a cloud of gas and dust enriched by several generations of heavier element production in stars. The

Copernican principle, which suggests that there is nothing special about our place in the universe, implies that if life could develop on Earth, it should be able to develop in other places as well. The Fermi paradox asks why more advanced life-forms have not visited us if life is common.

15.2 The study of life in the universe, including its origin on Earth, is called **astrobiology.** Carbon-based or **organic molecules** are abundant in space, but the actual origin of life—of cells that can extract energy from their environment and reproduce themselves—remains mysterious. Proliferation of life on Earth is possible because of **photosynthesis,** the ability to extract chemical energy from sunlight. Life has evolved to occupy many diverse ecological niches, with **extremophiles** living in conditions of heat, cold, dryness, salinity, acidity, or high radiation that seem extreme to us. The basic requirements of liquid water, organic compounds, and an energy source point us toward searching for life on Earth-like or terrestrial planets.

15.3 Astronomers and astrobiologists are searching for life within our solar system but are also beginning to turn their attention toward planets around other stars. Mars is the most promising site, and the hypothesis of fossil life in the Mars rock ALH 84001 has stimulated additional interest. The strategy of "follow the water" also leads us to Europa, Jupiter's moon that has an ocean of liquid water. The same requirement of liquid water is part of the concept of a **habitable zone** around other stars—regions where life may

be able to develop. When future telescopes are able to detect the light or infrared from Earth-like planets within the habitable zones of other stars, we will search for **biomarkers**—global changes (such as oxygen in the atmosphere) that provide evidence of life.

15.4 Astronomers are also engaged in the search for extraterrestrial *intelligent* life (**SETI**). Because the stars are so far away, traveling to them is either very slow or extremely expensive (in terms of energy required). Despite the many UFO reports and tremendous media publicity, there is no evidence that any of these are related to extraterrestrial spacecraft. The spacecraft we have sent beyond the planets (two Pioneers and two Voyagers) each contain a plaque or recording with information about humanity. Scientists have determined that the best way to communicate with any intelligent civilizations out there is by using electromagnetic waves, and radio waves seem best suited to the task. Since 1960, more than 60 projects have searched for radio messages from civilizations around other stars. So far, they have only begun to comb the many different possible stars, frequencies, signal types, and other factors that make up what we call the *cosmic haystack problem.* Another approach, called optical SETI, searches for extremely strong, but very brief pulses of light. If we do find a signal someday, deciding whether to answer and what to answer may be two of the greatest challenges humanity will face.

INTER-ACTIVITY

 A If one of the rocks from Mars does turn out to have unambiguous signs of ancient life that formed on Mars, what does your group think would be the implications of such a discovery for science and for our view of life elsewhere? Would such a discovery have any long-term effects on your own thinking?

 B Suppose we receive a radio message from a civilization around another star (which shows clear evidence of intelligence, even if we can't quite decipher all the details of what they send). What does your group think the implications of this discovery would be? How would your own thinking or philosophy be affected by such a discovery?

C A radio message has been received from a civilization around a star 40 LY away, which contains (in pictures) quite a bit of information about the beings that sent the message. The president of the United States has appointed your group a high-level commission to advise whether humanity should answer the message (which was not particularly directed at us, but comes from a beacon that, like a lighthouse, sweeps out a

circle in space). How would you advise the president? Does your group agree, or do you have a majority or minority view to present?

D If there is no evidence that the so-called UFOs are extraterrestrial visitors, why does your group think that television shows, newspapers, and movies spend so much time and effort publicizing the point of view that UFOs are craft from other worlds? Who stands to gain by exaggerating stories of unknown lights in the sky or simply fabricating stories that alien visitors are already here?

E Do you think scientists should just ignore all the media publicity about UFOs, or should they try to respond? Why?

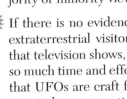 **F** Suppose your group had been the team planning the sights and sounds of Earth for the Voyager spacecraft audio/video record. What would you have put on the record to represent our planet to another civilization?

 G Let's suppose Earth civilization has decided to broadcast a radio (or television) message announcing our existence to other possible civilizations among the stars. Your group is part of a large task force of scientists, communications specialists, and people from the humanities charged with deciding the form and content of our message. What would you recommend?

 H If we do send messages into space to announce our presence, would it be useful to start with a reading of

one of Shakespeare's plays coded as a very powerful FM broadcast? Why or why not?

I Think of examples of contact with aliens you have seen in movies and on TV. Discuss with your group how realistic these have been, given what you have learned in this class. Why do you think Hollywood does so many shows and films that are not scientifically valid?

REVIEW QUESTIONS

1. What is the Copernican principle? Make a list of scientific discoveries that confirm it.
2. Where in the solar system (and beyond) have scientists found evidence of organic molecules?
3. Give a short history of the atoms that are now in your little finger, going back to the beginning of the universe.
4. What is a biomarker? Give some possible examples of biomarkers we might look for beyond the solar system?
5. Why are Mars and Europa the top targets for the study of astrobiology? Explain.
6. Why is traveling between the stars difficult?

7. What are the advantages to using radio waves for communication between civilizations that live around different stars? List as many as you can.
8. What is the "cosmic haystack problem"? List as many of its components as you can think of.
9. In what ways is *Project Phoenix* an improvement over the first search ever conducted for radio signals from extraterrestrial civilizations?
10. What is optical SETI? Why is starlight not an overwhelming problem for such searches?

THOUGHT QUESTIONS

11. Would a human have been possible during the first generation of stars that formed right after the big bang? Why or why not?

12. How would you demonstrate that a rock found in Antarctica is from Mars and not from Earth?

13. If we do find life on Mars, what might be some ways to check whether it formed separately from Earth life, or whether exchanges of material between the two planets meant that the two forms of life have a common origin?

14. What kind of evidence do you think would convince astronomers that an extraterrestrial spacecraft has landed on the Earth? (*Hint:* See question 12.)

15. What are some reasons that more advanced civilizations might want to send out some messages to other star systems?

16. What are some answers to the Fermi paradox? Can you think of some that are not discussed in this chapter?

FIGURING FOR YOURSELF: THE DRAKE EQUATION

The first scientific meeting devoted to SETI (the search for extraterrestrial intelligence) was held in 1961 at the U.S. National Radio Astronomy Observatory, where Frank Drake had just completed his pioneering Project Ozma observations. To provide some structure to the wide-ranging discussions about life in the universe, Drake wrote an equation on the blackboard that took the difficult question of estimating the number of civilizations in the Galaxy and broke it down into a series of

smaller more manageable questions. Ever since then, both astronomers and students have used this Drake equation as a means of approaching the most challenging question: Are we alone? Since this is at present an unanswerable question, astronomer Jill Tarter has called the Drake equation a "way of organizing our ignorance."

The form of the Drake equation is very simple. To estimate the number of communicating civilizations that currently exist in the Galaxy (we will define these terms more carefully in a moment), we multiply the rate of formation of such civilizations (number per year) by their average lifetime (in years). In symbols,

$$N = R(\text{total}) \times L$$

This problem is similar to estimating the number of frogs in a fresh pond. If 50 new frogs are born in the pond each year (the rate of formation) and each frog lives an average of 8 years, then, when things settle down after a decade or so, the number of frogs in the pond at any given time should be 50×8, or 400.

To make his formula easier to use (and more interesting!), however, Drake separated the rate of formation $R(\text{total})$ into a series of probabilities:

$$R(\text{total}) = R(\text{star}) \times f_p \times f_e \times f_l \times f_i \times f_c$$

$R(\text{star})$ is the rate of formation of stars like the Sun in our Galaxy, which is about 10 stars per year. Each of the other terms is a fraction or probability (less than or equal to 1.0), and the product of all these probabilities is itself the total probability that each star will have an intelligent, technological, communicating civilization that we might want to talk to. We have:

f_p = the fraction of these stars with planets (thanks to recent discoveries, this is now known to be between 0.1 and 1.0)

f_e = the fraction of the planetary systems that include Earth-like or habitable planets (maybe all of them if they are like the solar system, but maybe a smaller fraction if hot Jupiters spiral inward in some systems)

f_l = the fraction of habitable planets that actually support life (from the Copernican principle, you might say it's near 1.0, but we just don't know)

f_i = the fraction of inhabited planets that develop advanced intelligence (it happened on Earth, but only after more than 4 billion years of evolution; it's hard to say how likely it is without having other examples)

f_c = the fraction of these intelligent civilizations that develop science and the technology to build radio telescopes and transmitters for communication (something that has happened on Earth only in the last few decades and could probably happen only on worlds where the inhabitants can mine metals)

Each of these factors can be discussed and perhaps evaluated, but we must guess at many of the values. In particular, we don't know how to calculate the probability of something that happened once on Earth but has not been observed elsewhere—and these include the development of life, of intelligent life, and of technological life (the last three factors in the equation).

Even if we don't know the answers, you will find that it is instructive to make some guesses and calculate the resulting number N. Let's start with the optimism

implicit in the Copernican principle and set the last three terms equal to 1.0. If R is 10, f_p is 0.1, and f_e is 1.0, the equation becomes

$$N = R(\text{total}) \times L = L$$

Now we see the importance of the term L, the lifetime of a communicating civilization. We have had this capability for only a few decades. Suppose we assume that this stage in our history lasts only one century; then, with our optimistic assumptions about the other factors, $L = 100$ years and $N = 100$ such civilizations in the entire Galaxy. In that case there are so few other civilizations like ours that we are unlikely to detect any signals in a SETI search. But suppose the average lifetime is a million years; in that case there are a million such civilizations in the Galaxy and several are likely to be close enough for communication. (You may be amused to know that Frank Drake drives a car with license plate "NEQLSL.")

The most important conclusion from this calculation is that, even if we are extremely optimistic about the probabilities, the only way we can expect success from SETI is if other civilizations are much older (and hence probably much more advanced) than ours. We must be the youngest kid on the block (in our corner of the Galaxy). Most other civilizations are likely to be much older and perhaps much smarter (or at least more experienced) than we. It's a sobering thought. If we had made less optimistic assumptions, this conclusion would be even stronger.

17. Using the information from this and previous chapters, pick values that make sense to you for the probabilities in the Drake equation. What is the resulting value of N, the number of communicating civilizations?

18. Select the most optimistic and pessimistic values that you think are reasonable (and this choice requires some judgments, since we really don't know the values of these numbers) and calculate the corresponding range in values of N.

19. Suppose astronomers discover a radio message from a civilization whose planet orbits a star 35 LY away. Their message encourages us to send a radio answer, which we decide to do. Suppose our governing bodies take two years to decide whether and how to answer. When our answer arrives there, their governing bodies take one year to frame an answer to us. How long after we get their first message can we hope to get their reply to ours? (Once communication gets going, should we continue to wait for a reply before we send the next message?)

20. Think of our Milky Way Galaxy as a flat disk of diameter 100,000 LY. Suppose we are one of 1000 civilizations, randomly distributed through the disk, interested in communicating via radio waves. How far away would the nearest such civilization be from us (on average)?

SUGGESTIONS FOR FURTHER READING

Croswell, K. "Interstellar Trekking" in *Astronomy*, June 1998, p. 46. Science perspective on how to travel to the stars.

Darling, D. *Life Everywhere: The Maverick Science of Astrobiology.* 2001, Basic Books. Short overview of the science of astrobiology.

Davies, P. *The Fifth Miracle: The Search for the Origin and Meaning of Life.* 1999, Simon & Schuster. Scientific and philosophical issues associated with the origin of life and the search for life elsewhere.

deDuve, C. *Vital Dust: The Origin and Evolution of Life on Earth.* 1995, Basic Books. A Nobel laureate recounts our modern understanding of how life emerged on our planet.

Gross, M. *Life on the Edge.* 1998, Plenum. A British biologist discusses how life can thrive under remarkably extreme conditions, and how we can search for such life elsewhere.

Jakosky, B. *The Search for Life on Other Planets.* 1998, Cambridge University Press. Popular account of the prospects for life in our solar system and beyond by one of the pioneers of astrobiology.

Krauss, L. *Atom: An Odyssey from the Big Bang to Life on Earth . . . and Beyond.* 2001, Little Brown. Traces the cosmic history of an oxygen atom.

Lemonick, M. *Other Worlds: The Search for Life in the Universe.* 1998, Simon & Schuster. *Time* magazine senior science correspondent summarizes the new ideas and discoveries.

LePage, A. and MacRobert, A. "SETI Searches Today" in *Sky & Telescope*, Dec. 1998, p. 44. On the scientific searches for signals from other civilizations.

Nadis, S. "Using Lasers to Detect ET" in *Astronomy*, Sept. 2002, p. 44. On optical SETI, with profiles of the main players.

Pendleton, Y. and Farmer, J. "Life: A Cosmic Imperative?" in *Sky & Telescope*, July 1997, p. 42. On the rise of life and environments that make it possible or impossible.

Shilling, G. "The Chance of Finding Aliens: Re-evaluating the Drake Equation" in *Sky & Telescope*, Dec. 1998, p. 36.

Shostak, S. "The Future of SETI" in *Sky & Telescope*, Apr. 2001, p. 42. New searches, new strategies; nice long review.

Shostak, S. "When ET Calls Us" in *Astronomy*, Sept. 1997, p. 36. What do we do when we receive a message?

Shostak, S. *Sharing the Universe: Perspectives on Extraterrestrial Life.* 1998, Berkeley Hills. Wide-ranging and often humorous discussions by one of the most popular speakers and writers on SETI.

Tarter, J. and Chyba, C. "Is There Life Elsewhere in the Universe?" in *Scientific American*, Dec. 1999, p. 118. On future search for life in the solar system and beyond.

Treiman, A. "Microbes in a Martian Meteorite: The Evidence Has Grown More Cloudy" in *Sky & Telescope*, Apr. 1999, p. 52.

Webb, S. *Where Is Everybody?* 2002, Copernicus/Praxis. On 50 solutions to the Fermi paradox.

In just a few short years, the World Wide Web has become a planet-wide library, full of information, images, opinion, and argument. Astronomers, always eager to use technological tools like computers, jumped on the Web bandwagon early, and thousands of sites are dedicated to astronomical ideas.

This great wealth of astronomy pages also presents a problem: Unlike a textbook that is written by (and checked by) experts in their fields, Web sites can be posted by anyone. Thus many Web sites contain misinformation, exaggeration, misunderstanding, and downright nonsense about the universe. Some of these sites are offered by well-meaning but inexperienced enthusiasts, while others are by charlatans or hoaxers out to make a quick buck or promote a hidden cause.

Throughout *Voyages* we have recommended reliable Web sites for each chapter and topic as we went along. In this appendix, we list some more general sites that are especially good at guiding astronomy beginners to helpful information, the latest images, or reliable organizations. Most of the sites we list include links to other sites that their Webmasters have found useful.

Note that Web addresses and sites can change faster than any mere mortal can follow. For the latest information on Web sites we recommend, check the *Voyages* Web site at:
info.brookscole.com/voyages

1. Magazines That Cover Astronomy

Astronomy **Magazine** (www.astronomy.com) has the largest circulation of any magazine devoted to the universe and is designed especially for astronomy hobbyists and armchair astronomers. The site features many sections on the hobby of astronomy and on things you can do to become involved.

Sky & Telescope **Magazine** (skyandtelescope.com) is an older and somewhat higher-level magazine for astronomy hobbyists. The site is especially rich in observational guides, software and equipment reviews, and lists of astronomy clubs around the country.

Scientific American **Magazine** (www.sciam.com) offers one astronomy article about every second issue. These articles, a number of which are reproduced on their Web site, are at a slightly higher level than *Voyages,* but—often written by the astronomers who have done the work being described—are authoritative and up-to-date.

Mercury **Magazine** (www.astrosociety.org/pubs/mercury/mercury.html) is a popular-level magazine published by the Astronomical Society of the Pacific that features articles and columns for astronomers, educators, and astronomy enthusiasts. The site often deals with more philosophical issues and the influence of astronomy on other areas of human thought.

Other Magazines That Cover Astronomy

- *Science News:* www.sciencenews.org
- *Griffith Observer:* www.griffithobs.org/Observer.html
- *Astronomy Now:* www.astronomynow.com

2. Organizations That Deal with Astronomy for the Public

The Astronomical Society of the Pacific (www.astrosociety.org) is the largest and oldest national astronomy organization in the United States. The membership of the society, founded in 1889, includes astronomers, educators, and astronomy enthusiasts. The site features a nonprofit catalog of interesting astronomy items (posters, videos, software, etc.) and also a large quantity of information related to astronomy education.

The Planetary Society (www.planetary.org), founded by the late Carl Sagan, works to encourage planetary exploration and the search for life elsewhere. Its Web site has many pages of information in these two areas.

The Astronomical League (www.astroleague.org) is the umbrella organization of astronomy clubs in the United States. The site has good lists of astronomy clubs around the country, activities and hints for astronomy hobbyists, and information on regional and national programs.

The Royal Astronomical Society of Canada (www.rasc.ca) is a membership organization that unites professional and amateur astronomers in Canada and has 26 centers around that country with local activities. Specific calendars for each location can be found on the site.

NASA (www.nasa.gov) has an astronomical amount of information on its many Web sites; the trick is to find what you need. Their site has improved in recent years, with better navigation and search functions. Most projects and NASA centers have their own Web sites and we have recommended many of these throughout the chapters.

3. Sites for Astronomy and Space Science News

- www.skyandtelescope.com
- www.astronomy.com
- space.com
- www.universe.com
- www.spacedaily.com
- science.nasa.gov

Each of these sites is run by people with a science background and tries to feature (and explain) new developments as they occur.

4. Selected Sources for the Latest Astronomy Images

Anglo-Australian Observatory Image Collection:
www.aao.gov.au/images.html
A marvelous library of images (focusing on nebulae and galaxies) taken using large Australian telescopes. Many are by David Malin, who is acknowledged to be one of the finest astronomical photographers of our time. The images include captions and ordering information.

Hubble Space Telescope Images:
hubblesite.org/newscenter/archive
All the magnificent Hubble Space Telescope images are here, with captions; many have detailed background information and new research results included. You can check the latest images, browse the ones the staff considers the Hubble's "greatest hits," or search for objects of interest to you. Best of all, the images can all be downloaded in various formats completely free.

National Optical Astronomy Observatories Image Gallery: www.noao.edu/image_gallery/
NOAO includes a number of major telescopes in the United States and the Southern Hemisphere. Some of the best images from NOAO instruments are collected and organized at this site.

Planetary Photojournal: photojournal.jpl.nasa.gov
This site, run by the Jet Propulsion Laboratories and the U.S. Geological Survey Branch of Astrogeology in Flagstaff, is one of the most useful resources on the Web. It features thousands of the best NASA images from planetary exploration, with detailed captions and excellent indexing, and more are being added all the time. You can dial up images by world, feature name, date, or catalog number, and download images in a number of popular formats.

Astronomy Picture of the Day:
antwrp.gsfc.nasa.gov/apod/astropix.html
Astronomers Robert Nemiroff and Jerry Bonell feature one relatively new celestial image each day with a brief nontechnical caption. Over the years, some of the best astronomical images have been featured here and an index is available on site.

European Southern Observatory:
www.eso.org/outreach/gallery/astro
This growing album contains images from the large telescopes in the Southern Hemisphere run by a consortium of European countries. With the advent of the Very Large Telescope, there is an increasing number of important new images and results on this site.

Infrared Image Gallery:
www.ipac.caltech.edu/Outreach/Gallery
This site collects infrared images from space missions and telescopes on the ground and also offers an intriguing set of comparisons showing images at visible and invisible wavelengths.

Other Ways to Find Astronomy and Space Images

- *Google image search:*
www.google.com/advanced_image_search?hl=en
- *NASA Image Exchange:* nix.nasa.gov

- *Great Images at NASA (mostly historical):* grin.hq.nasa.gov/
- *Index to NASA Image Sites:* www.nasa.gov/gallery/photo/index.html

5. Debunking Pseudo-science

Astronomical Pseudo-science: A Skeptic's Resource List:
www.astrosociety.org/education/resources/pseudobib.html
An annotated list of written and Web resources for dealing with astrology, UFOs, the Face on Mars, ancient astronauts, and many similar fringe topics.

Debunking Pseudo-science (in general): www.csicop.org
For students or instructors who want assistance in dealing with topics that mix up science with new-age philosophy, paranormal beliefs, conspiracy theories, and utter silliness, the premier organization is the awkwardly named, but very effective, Committee for the Scientific Investigation of Claims of the Paranormal (CSICOP). Composed of scientists, educators, magicians, legal experts, doctors, and others who are tired of the uncritical acceptance of pseudo-science by the media and the public, CSICOP seeks to provide the skeptical, rational side of many controversial topics, such as UFOs, astrology, psychic power, the Bermuda Triangle, crop circles, and so on. The site is a treasure trove of information, anecdotes, experiments, and links.

6. Sources and Reviews of Astronomical Software

Sky & Telescope's Software Pages:
skyandtelescope.com/resources/software
Lists and explains a wide range of astronomical freeware and shareware, programs you can download and copy.

John Mosley's Software Directory:
www.griffithobs.org/software.html
The editor of the magazine for planetarium educators lists a wide range of astronomical software by publisher, plus references to published reviews.

Dan Bruton's Software List:
www.physics.sfasu.edu/astro/software.html
Includes listings and links to over 200 pieces of astronomical software.

Bill Arnett's Software List:
www.seds.org/billa/astrosoftware.html
A long listing of both commercial and noncommercial packages with links.

7. Miscellaneous Sites

Science Fiction Stories with Good Astronomy and Physics:
www.astrosociety.org/education/resources/scifi.html
A list of novels and stories organized by topic in astronomy, with a brief description of each.

Astronomical Observing Tips: skyandtelescope.com/howto/
This series of excellent articles from *Sky & Telescope* magazine introduces beginners to observational astronomy. It includes good advice on getting started with just your eyes, binoculars, small telescopes, star wheels, photography, and much more.

History of Astronomy Site: www.astro.uni-bonn.de/~pbrosche/astoria.html
Professor Wolfgang Dick maintains this comprehensive series of links to sites about all aspects of astronomical history.

Among its best features is the ability to select the name of a historical astronomer and find links that provide instant biographical information.

AstroWeb: Astronomy/Astrophysics on the Internet:
www.cv.nrao.edu/fits/www/astronomy.html
This site, maintained by a consortium of astronomy institutions, is a collection of thousands of astronomy-related links, organized by subject. Some links are to highly technical sites, some to pages that are just right for beginners. You can check whether an astronomer has a personal Web page, look up what astronomy departments or observatories have put on the Web, and delve further into many specialized areas.

Space Calendar: wwwjpl.nasa.gov/calendar
Ron Baalke at JPL keeps this amazing listing of space- and astronomy-related events, including upcoming launches, significant anniversaries, and things to see in the sky. Most entries in the calendar have a Web link where you can find more information.

With the growth of the World Wide Web, even print magazines and all organizations dedicated to working with the public have set up Web sites of basic information. Many of their Web sites are given in Appendix 1. Still, you may want to contact some of the magazines and organizations by mail to subscribe or ask for specific materials. Here we list a few of the best-known public information providers in astronomy.

1. Popular-level Astronomy Magazines

Astronomy, Kalmbach Publishing, P.O. Box 1612, Waukesha, WI 53187. The astronomy publication with the largest circulation in the world; it is very colorful and basic.

Griffith Observer, Griffith Observatory, 2800 E. Observatory Rd., Los Angeles, CA 90027. A small magazine specializing in historical topics, edited with humor and verve.

Mercury, Astronomical Society of the Pacific, 390 Ashton Ave., San Francisco, CA 94112. The magazine of the largest general membership astronomy organization in the United States, with many interesting features, particularly about the influence of astronomy on other fields, astronomy teaching, and the stories behind the new discoveries.

Planetary Report, Planetary Society, 65 N. Catalina Ave., Pasadena, CA 91106. The magazine of an organization that promotes the exploration of the solar system and the search for life elsewhere.

Stardate, McDonald Observatory, RLM 15.308, University of Texas, Austin, TX 78712. Accompanies the popular radio program on astronomy.

Sky & Telescope, P.O. Box 9111, Belmont, MA 02178. Found in most libraries, this is the popular astronomy magazine "of record," with many excellent articles on astronomy and amateur astronomy.

Sky News, Box 10, Yarker, Ontario K0K 3N0 Canada; www.skynewsmagazine.com. This is Canada's popular-level astronomy magazine, with astronomy news and features for hobbyists.

The following magazines (found in most libraries) also contain reports on astronomical developments: *Discover, National Geographic, Science News,* and *Scientific American.*

2. Organizations for Astronomy Enthusiasts

American Association of Variable Star Observers, 25 Birch St., Cambridge, MA 02138 (617-354-0484); www.aavso.org. An organization of amateur astronomers devoted to making serious observations of stars whose brightness changes.

Association of Lunar and Planetary Observers, c/o Donald Parker, 12911 Lerida St., Coral Gables, FL 33156; www.lpl.arizona.edu/alpo. An organization of amateur astronomers who monitor and report on objects in our solar system.

Astronomical League, 9201 Ward Pkwy., Kansas City, MO 64114; www. astroleague.org. Umbrella organization of amateur astronomy clubs in the United States, with many activities. If you write to this all-volunteer organization, be sure to enclose a stamped, self-addressed envelope.

Astronomical Society of the Pacific, 390 Ashton Ave., San Francisco, CA 94112 (415-337-1100); www.astrosociety.org. This international organization brings together scientists, teachers, and people with a general interest in astronomy; its name is merely a reminder of its origins on the West Coast of the United States in 1889. Write for the catalog of interesting astronomy materials for all levels.

Committee for the Scientific Investigation of Claims of the Paranormal (CSICOP), P.O. Box 703, Buffalo, NY 14226 (716-636-1425); www.csicop.org. An organization of scientists, educators, magicians, and other skeptics that seeks to inform the public about the rational perspective on such pseudo-sciences as astrology, UFOs, and psychic power. The committee publishes *The Skeptical Inquirer* magazine, full of great debunking articles, and holds meetings and workshops around the world.

International Dark Sky Association, 3225 N. First Ave., Tucson, AZ 85719; www.darksky.org. Small non-profit organization devoted to fighting light pollution and educating politicians, lighting engineers, and the public about the importance of not spilling light where it will interfere with astronomical observations.

The Planetary Society, 65 N. Catalina Ave., Pasadena, CA 91106 (818-793-1675); www.planetary.org. Large national membership organization was founded by Carl Sagan and others. It lobbies for more planetary exploration and SETI, and publishes a colorful magazine.

The Royal Astronomical Society of Canada, 136 Dupont St., Toronto, Ontario M5R 1V2 (416-924-7973); www.rasc.ca. The main organization of amateur astronomers in Canada, with local centers in each province. Write for information on contacting the center near you.

Note: The American Astronomical Society is the professional organization for astronomers in the United States. The journals and activities are mostly for research scientists, but the Web site does have an excellent introduction for high school students to becoming an astronomer. See: http://www.aas.org/education/career.html

Glossary

Note: The number in parentheses is the chapter where the term is first defined or explained (I for the Interlude). Terms not used in the book but often used by instructors are also included.

absolute brightness (absolute magnitude)   A measure of the luminosity of an object like a star. *See* luminosity.

absolute zero   A temperature of −273°C (or 0 K), where all molecular motion stops. (4)

absorption spectrum   Dark lines superimposed on a continuous spectrum. (4)

accelerate   To change velocity; to speed up, slow down, or change direction. (1)

accretion   Gradual accumulation of mass, as by a planet forming from colliding particles in the solar nebula or gas falling into a black hole. (13)

accretion disk   A disk of matter spiraling in toward a massive object; the disk shape is the result of conservation of angular momentum.

active galactic nucleus   A galaxy is said to have an active nucleus if unusually violent events are taking place in its center, emitting large quantities of electromagnetic radiation. Seyfert galaxies and quasars are examples of galaxies with active nuclei.

active region   An area on the Sun where magnetic fields are concentrated; sunspots, prominences, and flares all tend to occur in active regions. (14)

adaptive optics   Systems used with telescopes that change to compensate for distortions in an image introduced by the atmosphere, thus producing sharper images. (5)

albedo   The fraction of sunlight reaching it that a planet or asteroid reflects; reflectivity.

alpha particle   The nucleus of a helium atom, consisting of two protons and two neutrons. (4)

amplitude   The range in variability, as in the light from a variable star.

angular diameter   Angle subtended by the diameter of an object.

angular momentum   A measure of the momentum associated with motion about an axis or fixed point. (2)

antapex (solar)   Direction away from which the Sun is moving with respect to the local standard of rest.

anthropic principle   The idea that physical laws must be the way they are because otherwise we could not be here to measure them.

Antarctic Circle   Parallel of latitude 66°30′S; at this latitude the noon altitude of the Sun is 0° on the date of the summer solstice. (3)

antimatter   Matter consisting of antiparticles: antiprotons (protons with negative rather than positive charge), positrons (positively charged electrons), and antineutrons.

aperture   The diameter of an opening, or of the primary lens or mirror of a telescope. (5)

apex (solar)   The direction toward which the Sun is moving with respect to the local standard of rest.

aphelion   Point in its orbit where a planet (or other orbiting object) is farthest from the Sun. (2)

apogee   Point in its orbit where an Earth satellite is farthest from the Earth. (2)

apparent brightness   A measure of the observed light received from a star or other object at the Earth—that is, how bright an object appears in the sky, as contrasted with its luminosity. (4)

apparent solar time   Time as measured by the position of the Sun in the sky (the time that would be indicated by a sundial). (3)

Arctic Circle   Parallel of latitude 66°30′N; at this latitude the noon altitude of the Sun is 0° on the date of the winter solstice. (3)

array (interferometer)   A group of several telescopes that are connected together to make observations at high angular resolution. (5)

association   A loose group of young stars whose spectral types, motions, and positions in the sky indicate that they probably had a common origin.

asterism   An especially noticeable star pattern in the sky, such as the Big Dipper. (1)

asteroid   A stony or metallic object orbiting the Sun that is smaller than a major planet but that shows no evidence of an atmosphere or of other types of activity associated with comets. (6)

asteroid belt   The region of the solar system between the orbits of Mars and Jupiter in which most asteroids are located. The main belt, where the orbits are generally the most stable, extends from 2.2 to 3.3 AU from the Sun. (12)

astrobiology   The multidisciplinary study of life in the universe: its origin, evolution, distribution, and fate; similar terms are *exobiology* and *bioastronomy*. (15)

astrology   The pseudoscience that deals with the supposed influences on human destiny of the configurations and locations in the sky of the Sun, Moon, and planets; a primitive belief system that had its origin in ancient Babylonia. (1)

astronomical unit (AU)   Originally meant to be the semi-major axis of the orbit of the Earth; now defined as the semi-major axis of the orbit of a hypothetical body with the mass and period that Gauss assumed for the Earth. The semimajor axis of the orbit of the Earth is actually 1.000000230 AU. (2)

atom The smallest particle of an element that retains the properties that characterize that element.

atomic mass unit *Chemical:* one-sixteenth of the mean mass of an oxygen atom. *Physical:* one-twelfth of the mass of an atom of the most common isotope of carbon. The atomic mass unit is approximately the mass of a hydrogen atom, 1.67×10^{-27} kg.

atomic number The number of protons in each atom of a particular element.

atomic weight The mean mass of an atom of a particular element in atomic mass units.

aurora Light radiated by atoms and ions in the ionosphere, mostly in the magnetic polar regions. (7)

autumnal equinox The intersection of the ecliptic and celestial equator where the Sun crosses the equator from north to south. A time when every place on Earth has 12 hours of daylight and 12 hours of night. (3)

axis An imaginary line about which a body rotates. (1)

azimuth The angle along the celestial horizon, measured eastward from the north point to the intersection of the horizon with the vertical circle passing through an object. (1)

Balmer lines Emission or absorption lines in the spectrum of hydrogen that arise from transitions between the second (or first excited) and higher energy states of the hydrogen atom. (4)

bands (in spectra) Emission or absorption lines, usually in the spectra of chemical compounds, so numerous and closely spaced that they coalesce into broad emission or absorption bands. (4)

bar A force of 100,000 newtons acting on a surface area of 1 square meter. The average pressure of the Earth's atmosphere at sea level is 1.013 bars. (7)

barred spiral galaxy Spiral galaxy in which the spiral arms begin from the ends of a "bar" running through the nucleus rather than from the nucleus itself.

barycenter The center of mass of two mutually revolving bodies.

basalt Igneous rock produced by the cooling of lava. Basalts make up most of Earth's oceanic crust and are also found on other planets that have experienced extensive volcanic activity. (7)

Big Bang theory A theory of cosmology in which the expansion of the universe began with a primeval explosion.

binary stars Two stars that revolve about each other.

binding energy The energy required to separate completely the constituent parts of an atomic nucleus.

bioastronomy *See* astrobiology.

biomarker Evidence of the presence of life, especially a global indication of life on a planet that could be detected remotely (such as an unusual atmospheric composition). (15)

blackbody A hypothetical perfect radiator, which absorbs and reemits all radiation incident upon it. (4)

black dwarf The final state of evolution for a low-mass star, in which all of its energy sources are exhausted and it no longer emits significant radiation.

black hole A collapsed massive star (or other collapsed body) whose velocity of escape is equal to or greater than the speed of light; thus no radiation can escape from it.

Bohr atom A particular model of an atom, invented by Niels Bohr, in which the electrons are described as revolving about the nucleus in circular orbits. (4)

brown dwarf An object intermediate in size between a planet and a star. The approximate mass range is from about 1/100 of the mass of the Sun up to the lower mass limit for self-sustaining nuclear reactions, which is 0.072 the mass of the Sun.

caldera Volcanic depression or crater, usually produced by the collapse of a volcano's summit. (9)

carbon-nitrogen-oxygen (CNO) cycle A series of nuclear reactions in the interiors of stars involving carbon as a catalyst, by which hydrogen is transformed to helium.

carbonaceous meteorite A primitive meteorite made primarily of silicates but including chemically bound water, free carbon, and complex organic compounds. Also called carbonaceous chondrites. (13)

carbonate Chemical compound that contains CO_2, such as calcium carbonate ($CaCO_2$); carbonates can "store" carbon dioxide in solid form.

Cassegrain focus An optical arrangement in a reflecting telescope in which light is reflected by a second mirror to a point behind the primary mirror. (5)

CBR *See* cosmic background radiation.

CCD *See* charge-coupled device.

celestial equator A great circle on the celestial sphere 90° from the celestial poles; where the celestial sphere intersects the plane of the Earth's equator. (1)

celestial meridian An imaginary line on the celestial sphere passing through the north and south points on the horizon and through the zenith.

celestial poles Points about which the celestial sphere appears to rotate; intersections of the celestial sphere with the Earth's polar axis. (1)

celestial sphere Apparent sphere of the sky; a sphere of large radius centered on the observer. Directions of objects in the sky can be denoted by their position on the celestial sphere. (1)

center of gravity Center of mass.

center of mass The average position of the various mass elements of a body or system, weighted according to their distances from that center of mass; that point in an isolated system that moves with constant velocity according to Newton's first law of motion.

cepheid variable A star that belongs to a class of yellow supergiant pulsating stars. These stars vary periodically in brightness, and the relationship between their periods and luminosities is useful in deriving distances to them.

Chandrasekhar limit The upper limit to the mass of a white dwarf (equals 1.4 times the mass of the Sun).

charge-coupled device (CCD) An array of high-sensitivity electronic detectors of electromagnetic radiation, used at the focus of a telescope (or camera lens) to record an image or spectrum. (5)

chemical condensation sequence The calculated order of chemical compounds and minerals that would form at different temperatures in a cooling gas of cosmic composition; used to infer the composition of grains that formed in the solar nebula at different distances from the protosun.

chondrite A primitive stony meteorite

chondrule Small spheres (roughly the size of a pea or smaller) of once-melted rock that are found in most chondrite meteorites.

chromosphere That part of the solar atmosphere that lies immediately above the photospheric layers. (14)

circular satellite velocity   The critical speed that a revolving body must have in order to follow a circular orbit. (2)

circumpolar zone   Those portions of the celestial sphere near the celestial poles that are either always above or always below the horizon. (1)

circumstellar disk   Thin disk of gas and dust surrounding a very young star or protostar out of which a planetary system may form; similar to the solar nebula that gave birth to our solar system (6)

closed universe   A model of the universe in which the curvature of space is such that straight lines eventually curve back upon themselves; in this model, the universe expands from a big bang, stops, and then contracts to a big crunch.

cluster of galaxies   A system of galaxies containing several to thousands of member galaxies.

color index   Difference between the magnitudes of a star or other object measured in light of two different spectral regions—for example, blue minus visual (B − V) magnitudes.

coma (of comet)   The cloud of evaporated gas and dust around a comet nucleus. (12)

comet   A small body of icy and dusty matter that revolves about the Sun. When a comet comes near the Sun, some of its material vaporizes, forming a large head of tenuous gas and often a tail. (6)

compound   A substance composed of two or more chemical elements.

conduction   The transfer of energy by the direct passing of energy or electrons from atom to atom.

conservation of angular momentum   The law that the total amount of angular momentum in a system remains the same (in the absence of any force not directed toward or away from the point or axis about which the angular momentum is referred). (2)

constellation   One of 88 sectors into which astronomers divide the celestial sphere; many constellations are named after a prominent group of stars within them that represents a person, animal, or legendary creature from ancient mythology.

continental drift   A gradual movement of the continents over the surface of the Earth due to plate tectonics. (7)

continuous spectrum   A spectrum of light composed of radiation of a continuous range of wavelengths or colors rather than only certain discrete wavelengths. (4)

convection   The transfer of energy by moving currents in a fluid. (7)

core (of a planet)   The central part of a planet, consisting of higher-density material. (7)

corona (of Sun)   Outer atmosphere of the Sun. (3)

corona (on Venus)   Distinctive large tectonic feature on Venus, apparently caused by a rising plume of hot magma in the mantle of the planet (9)

coronal hole   A region in the Sun's outer atmosphere that appears darker because there is less hot gas there. (14)

cosmic background radiation (CBR)   The microwave radiation coming from all directions that is the redshifted glow of the big bang.

cosmic rays   Atomic nuclei (mostly protons) that are observed to strike the Earth's atmosphere with exceedingly high energies.

cosmological constant   A term in the equations of general relativity that represents a repulsive force in the universe. *See also* dark energy.

cosmological principle   The assumption that, on the large scale, the universe at any given time is the same everywhere—isotropic and homogeneous.

cosmology   The study of the organization and evolution of the universe.

crater   A circular depression (from the Greek word for "bowl"), generally of impact origin. (8)

crescent moon   One of the phases of the Moon when it appears less than half full. (3)

critical density   In cosmology, the density that provides enough gravity to bring the expansion of the universe just to a stop after infinite time.

crust   The outer layer of a terrestrial planet. (6)

cryovolcanism   Geological processes on cold worlds (such as the outer-planet satellites) that mimic volcanic activity but involve fluids such as water rather than silicate magma. (11)

dark energy   The energy that is causing the expansion of the universe to accelerate. Its existence is inferred from observations of distant supernovae.

dark matter   Nonluminous mass, whose presence can be inferred only because of its gravitational influence on luminous matter. The composition of the dark matter is not known.

dark nebula   A cloud of interstellar dust that obscures the light of more distant stars and appears as a small opaque region in the sky.

declination   Angular distance north or south of the celestial equator. (3)

degenerate gas   A gas in which the allowable states for the electrons have been filled; it behaves according to different laws from those that apply to "perfect" gases and resists further compression.

density   The ratio of the mass of an object to its volume. (2)

deuterium   A "heavy" form of hydrogen, in which the nucleus of each atom consists of one proton and one neutron.

differential galactic rotation   The rotation of the Galaxy, not as a solid wheel but so that parts adjacent to one another do not always stay close together.

differentiated meteorite   Meteorite derived from a differentiated parent body and hence not primitive.

differentiation (geological)   Gravitational separation of materials of different density into layers in the interior of a planet or satellite. (6)

disk (of Galaxy)   The central plane or "wheel" of our Galaxy, where most of the luminous mass is concentrated.

dispersion   Separation of the different wavelengths of white light through being refracted by different amounts. (4)

Doppler effect   Apparent change in wavelength or frequency of the radiation from a source due to its relative motion away from or toward the observer. (4)

dust tail (of comet)   Dust loosened by the sublimation of ice in a comet that is then pushed by photons from the Sun into a curved stream. (12)

Earth-approaching asteroid   An asteroid with an orbit that either crosses the Earth's orbit or will at some time cross the Earth's orbit as it evolves under the influence of the planets' gravity. *See also* near-Earth object. (12)

eccentricity (of ellipse)   Ratio of the distance between the foci to the major axis. (2)

eclipse   The cutting off of all or part of the light of one body by another; in planetary science, the passing of one body into the shadow of another. (3)

eclipsing binary star   A binary star in which the plane of revolution of the two stars is nearly edge-on to our line of sight, so that the light of one star is periodically diminished by the other passing in front of it.

ecliptic   The apparent annual path of the Sun on the celestial sphere. (1)

effective temperature   *See* temperature (effective).

ejecta   Material excavated from an impact crater, such as the blanket of material surrounding lunar craters and crater rays. (8)

electromagnetic force   One of the four fundamental forces or interactions of nature; the force that acts between charges and binds atoms and molecules together.

electromagnetic radiation   Radiation consisting of waves propagated through regularly varying electric and magnetic fields; includes radio, infrared, visible light, ultraviolet, x rays, and gamma rays. (4)

electromagnetic spectrum   The whole array or family of electromagnetic waves, from radio to gamma rays. (4)

electron   A negatively charged subatomic particle that normally moves around the nucleus of an atom. (4)

element   A substance that cannot be decomposed by chemical means into simpler substances.

elementary particle   One of the basic particles of matter. The most familiar elementary particles are the proton, neutron, and electron.

ellipse   A curve for which the sum of the distances from any point on the ellipse to two points inside (called the foci) is always the same. (2)

elliptical galaxy   A galaxy whose shape is an ellipse and that contains no conspicuous interstellar material.

ellipticity   The ratio (in an ellipse) of the major axis minus the minor axis to the major axis.

emission line   A discrete bright line in the spectrum. (4)

emission nebula   A gaseous nebula that derives its visible light from the fluorescence of atoms using ultraviolet light from a star in or near the nebula.

emission spectrum   A spectrum consisting of emission lines. (4)

energy flux   The amount of energy passing through a unit area (for example, 1 square meter) per second. The units of flux are $J\,m^{-2}\,s^{-1}$. (4)

energy level   A particular level, or amount, of energy possessed by an atom or ion above the energy it possesses in its least energetic state; also used to refer to the states of energy an electron can have in an atom. (4)

epicycle   A circular orbit of a body in the Ptolemaic system, the center of which revolves about another circle (the deferent). (1)

equator   A great circle on the Earth, 90° from (or equidistant from) each pole. (1)

equinox   One of the intersections of the ecliptic and celestial equator; one of the two times during the year when the lengths of the day and the night are the same. (3)

equivalence principle   Principle that a gravitational force and a suitable acceleration are indistinguishable within a sufficiently local environment.

escape velocity   The velocity a body must achieve to break away from the gravity of another body and never return to it. (2)

eucrite meteorite   One of a class of basaltic meteorites believed to have originated on the asteroid Vesta.

event horizon   The surface through which a collapsing star passes when its velocity of escape is equal to the speed of light—that is, when the star becomes a black hole.

excitation   The process of giving an atom or an ion an amount of energy greater than it has in its least-energy state. (4)

exclusion principle   *See* Pauli exclusion principle.

exobiology   *See* astrobiology

extinction   Reduction of the light from a celestial body produced by the Earth's atmosphere, or by interstellar absorption. *See also* mass extinction.

extragalactic   Beyond our own Milky Way Galaxy.

extrasolar planet   A planet orbiting a star other than the Sun.

extremophile   An organism (usually a microbe) that tolerates or even thrives under conditions that we and most of the life around us would consider hostile, such as very high or low temperature or acidity. (15)

eyepiece   A magnifying lens used to view the image produced by the objective of a telescope. (5)

fall (of meteorites)   Meteorites seen in the sky and recovered on the ground.

fault   In geology, a crack or break in the crust of a planet along which slippage or movement can take place, accompanied by seismic activity. (7)

Fermi paradox   The question why, if there are many advanced civilizations in the Galaxy, none of them are visiting us or have left indications of their existence.

field   A mathematical description of the effect of forces, such as gravity, that act on distant objects. For example, a given mass produces a gravitational field in the space surrounding it, which produces a gravitational force on objects within that space. (4)

find (of meteorites)   A meteorite that has been recovered but was not seen to fall.

fireball   A spectacular meteor seen for longer than an instant in the sky.

fission   The breakup of a heavy atomic nucleus into two or more lighter ones.

flare   A sudden and temporary outburst of light from an extended region of the Sun's surface. (14)

fluorescence   The absorption of light of one wavelength and the reemission of it at another wavelength; especially the conversion of ultraviolet into visible light.

flux   The rate at which energy or matter crosses a unit area of a surface; so, energy per unit time per unit area. (4)

focal length   The distance from a lens or mirror to the point where light converged by it comes to a focus. (5)

focus (of ellipse)   One of two fixed points inside an ellipse from which the sum of the distances to any point on the ellipse is a constant. (2)

focus (of telescope)   Point where the rays of light converged by a mirror or lens meet. (5)

forbidden lines   Spectral lines that are not usually observed under laboratory conditions because they result from atomic transitions that are highly improbable on Earth.

force   That which can change the momentum of a body; numerically, the rate at which the body's momentum changes. That which causes masses to accelerate. (2)

Fraunhofer line   An absorption line in the spectrum of the Sun or of a star. (4)

Fraunhofer spectrum   The array of absorption lines in the spectrum of the Sun or a star. (4)

frequency   Number of vibrations per unit time; number of waves that cross a given point per unit time (in radiation). (4)

fusion   The building up of heavier atomic nuclei from lighter ones.

galactic cannibalism The process by which a larger galaxy strips material from or completely swallows a smaller one.

galactic cluster An "open" cluster of stars located in the spiral arms or disk of the Galaxy.

galaxy A large assemblage of stars; a typical galaxy contains millions to hundreds of billions of stars.

Galaxy The galaxy to which the Sun and our neighboring stars belong; the band of light in the sky we call the Milky Way is light from many remote stars in the disk of the Galaxy.

gamma rays Photons (of electromagnetic radiation) of energy higher than those of x rays; the most energetic form of electromagnetic radiation. (4)

gamma-ray burst A brief pulse of gamma rays; primarily originating in distant galaxies. (I)

general relativity theory Einstein's theory relating acceleration, gravity, and the structure (geometry) of space and time.

geocentric Centered on the Earth. (1)

giant (star) A star of large luminosity and radius.

giant molecular cloud Large, cold interstellar clouds with diameters of dozens of light years and typical masses of 10^5 solar masses; found in the spiral arms of galaxies, these clouds are where massive stars form.

giant planet *See* jovian planet.

globular cluster One of about 150 large spherical star clusters that form a system of clusters centered on the center of the Galaxy.

grand unified theories (GUTS) Physical theories that attempt to describe the four interactions (forces) of nature as different manifestations of a single force.

granite The type of igneous silicate rock that makes up most of the continental crust of the Earth. (7)

granulation The rice-grain-like structure of the solar photosphere; granulation is produced by upwelling currents of gas that are slightly hotter, and therefore brighter, than the surrounding regions, which are flowing downward into the Sun. (14)

gravity The mutual attraction of material bodies or particles. (2)

gravitational energy Energy that can be released by the gravitational collapse, or partial collapse, of a system—that is, by particles that fall in toward the center of gravity.

gravitational lens A massive object, such as a galaxy, whose gravity deflects the light (as predicted by Einstein's general theory of relativity) of an object behind it, thus showing us distorted or multiple images of the distant source.

gravitational redshift The redshift of electromagnetic radiation caused by a gravitational field. The slowing of clocks in a gravitational field.

gravitational waves Disturbances in the curvature of spacetime caused by changes in how matter is distributed; gravitational waves propagate at (or near) the speed of light.

great circle Circle on the surface of a sphere that is the curve of intersection of the sphere with a plane passing through its center. (3)

greenhouse effect The blanketing (absorption) of infrared radiation near the surface of a planet—for example, by carbon dioxide in its atmosphere. (7)

ground state The lowest energy state of an atom. (4)

H or H₀ *See* Hubble constant.

HI region Region of neutral hydrogen in interstellar space.

HII region Region of ionized hydrogen in interstellar space.

habitable zone Region around a star in which liquid water could exist on the surface of terrestrial-size planets, hence the most probable place to look for life. (15)

half-life The time required for half of the radioactive atoms in a sample to disintegrate. (6)

halo (of galaxy) The outermost extent of our Galaxy or another, containing a sparse distribution of stars and globular clusters in a more or less spherical distribution. More recently, astronomers also speak of a halo of dark matter around galaxies.

heavy elements In astronomy, usually those elements of higher atomic number than helium.

helio- Prefix referring to the Sun.

heliocentric Centered on the Sun. (1)

helium flash The nearly explosive ignition of helium in the triple-alpha process in the dense core of a red giant star.

Herbig-Haro (HH) object Luminous knots of gas in an area of star formation, which are set to glow by jets of material from a protostar.

hertz A unit of frequency: one cycle per second. Named for Heinrich Hertz, who first produced radio radiation. (4)

Hertzsprung-Russell (H-R) diagram A plot of luminosity against surface temperature (or spectral type) for a group of stars.

highlands The lighter, heavily cratered regions of the Moon, which are generally several kilometers higher than the maria; also similar heavily cratered regions on Mars. (8)

homogeneous Having a consistent and even distribution of matter that is the same everywhere.

horizon (astronomical) A great circle on the celestial sphere 90° from the zenith; more popularly, the circle around us where the dome of the sky meets the Earth. (1)

horoscope A chart used by astrologers, showing the positions along the zodiac and in the sky of the Sun, Moon, and planets at some given instant and as seen from a particular place on Earth—usually corresponding to the time and place of a person's birth. (1)

Hubble constant Constant of proportionality between the velocities of remote galaxies and their distances.

Hubble law (or the law of the redshifts) The rule that the radial velocities of remote galaxies are proportional to their distances from us.

hydrostatic equilibrium A balance between the weights of various layers, as in a star or the Earth's atmosphere, and the pressures that support them.

hypothesis A tentative supposition or scientific model advanced to explain certain facts or phenomena and subject to further tests and verification.

igneous rock Any rock produced by cooling from a molten state. (7)

inclination (of an orbit) The angle between the orbital plane of a revolving body and some fundamental plane—usually the plane of the celestial equator or of the ecliptic.

inertia The property of matter that requires a force to act on it to change its state of motion; the tendency of objects to continue doing what they are doing in the absence of outside forces. (2)

inertial system A system of coordinates that is not itself accelerated but that either is at rest or is moving with constant velocity.

inflationary universe A theory of cosmology in which the universe is assumed to have undergone a phase of very rapid expansion during the first 10^{-30} s. After this period of rapid expansion, the standard big bang and inflationary models are identical.

infrared cirrus Patches of interstellar dust that emit infrared radiation and look like cirrus clouds on infrared images of the sky.

infrared radiation Electromagnetic radiation of wavelength longer than the longest (red) wavelengths that can be perceived by the eye, but shorter than radio wavelengths. (4)

interference A phenomenon of waves that mix together such that their crests and troughs can alternately reinforce and cancel one another. (5)

interferometer An instrument that combines electromagnetic radiation (or gravitational waves) from one or more telescopes to obtain a resolution equivalent to what would be obtained with a single telescope with a diameter equal to the baseline separating the individual separate telescopes. (5)

international date line An arbitrary line on the surface of the Earth near longitude 180° across which the date changes by one day. (3)

interstellar dust Tiny solid grains in interstellar space, thought to consist of a core of rocklike material (silicates) or graphite surrounded by a mantle of ices. Water, methane, and ammonia are probably the most abundant ices.

interstellar extinction The attenuation or absorption of light by dust in the interstellar medium.

interstellar medium or interstellar matter Gas and dust between the stars in a galaxy.

inverse-square law (for light) The amount of energy (light) flowing through a given area in a given time (flux) decreases in proportion to the square of the distance from the source of energy or light. (4)

ion An atom that has become electrically charged by the addition or loss of one or more electrons. (4)

ion tail (of comet) A stream of ionized particles evaporated from a comet and then swept away from the Sun by the solar wind. (12)

ionization The process by which an atom gains or loses electrons. (4)

ionosphere The upper region of the Earth's atmosphere in which many of the atoms are ionized. (7)

iron meteorite A meteorite composed primarily of iron and nickel. (13)

irregular galaxy A galaxy without rotational symmetry; neither a spiral nor an elliptical galaxy.

irregular satellite A planetary satellite with an orbit that is retrograde, or of high inclination or eccentricity.

isotope Any of two or more forms of the same element whose atoms all have the same number of protons but different numbers of neutrons. (4)

isotropic The same in all directions.

joule The metric unit of energy; the work done by a force of 1 newton (N) acting through a distance of 1 m.

jovian planet or giant planet Any of the planets Jupiter, Saturn, Uranus, and Neptune in our solar system, or planets of roughly that mass and composition in other planetary systems. (6)

KBO *See* Kuiper belt object.

kinetic energy Energy associated with motion.

Kuiper belt A region of space beyond Neptune that is dynamically stable (like the asteroid belt); the source region for most short-period comets. (12)

Kuiper belt object (KBO) An object in the Kuiper belt beyond Neptune; sometimes a comet that is derived from the Kuiper belt. (12)

laser An acronym for light amplification by stimulated emission of radiation; a device for amplifying a light signal at a particular wavelength into a coherent beam.

latitude A north-south coordinate on the surface of the Earth; the angular distance north or south of the equator measured along a meridian passing through a place. (3)

law of areas Kepler's second law: The radius vector from the Sun to any planet sweeps out equal areas in the planet's orbital plane in equal intervals of time. (2)

law of the redshifts *See* Hubble law.

leap year A calendar year with 366 days, inserted approximately every four years to make the average length of the calendar year as nearly equal as possible to the tropical year. (3)

light or visible light Electromagnetic radiation that is visible to the eye. (4)

light curve A graph that displays the time variation of the light from a variable or eclipsing binary star or, more generally, from any other object whose radiation output changes with time.

light year The distance light travels in a vacuum in one year; $1 \text{ LY} = 9.46 \times 10^{12}$ km, or about 6×10^{12} mi.

line broadening The phenomenon by which spectral lines are not precisely sharp but have finite widths.

Local Group The small cluster of galaxies to which our Galaxy belongs.

local standard of rest A coordinate system that shares the average motion of the Sun and its neighboring stars about the galactic center.

long-period comet A comet with orbital period longer than about a century; mostly derived from the Oort comet cloud. (12)

longitude An east-west coordinate on the Earth's surface; the angular distance, measured east or west along the equator, from the Greenwich meridian to the meridian passing through a place. (3)

luminosity The rate at which a star or other object emits electromagnetic energy into space; the total power output of an object.

luminosity class A classification of a star according to its luminosity within a given spectral class. Our Sun, a G2 V star, has luminosity class V.

luminous matter Material that gives off light or other electromagnetic radiation (as opposed to dark matter).

lunar eclipse An eclipse of the Moon, in which the Moon moves into the shadow of the Earth. (3)

Lyman lines A series of absorption or emission lines in the spectrum of hydrogen that arise from transitions to and from the lowest-energy states of the hydrogen atom. (4)

Magellanic Clouds Two neighboring galaxies visible to the naked eye from southern latitudes.

magma High-temperature molten state of rock, usually of silicate mineral composition and with dissolved gases and other volatiles. (7)

magnetic field The region of space near a magnetized body within which magnetic forces can be detected.

magnetic pole One of two points on a magnet (or a planet) at which the greatest density of lines of force emerge. A compass needle aligns itself along the local lines of force on the

Earth and points more or less toward the magnetic poles of the Earth.

magnetosphere   The region around a planet in which its intrinsic magnetic field dominates the interplanetary field carried by the solar wind; hence, the region within which charged particles can be trapped by the planetary magnetic field. (7)

magnitude   A system of measuring the amount of light flux received from a star or other luminous object. The higher the magnitude, the less radiation we receive from the object. (1)

main sequence   A sequence of stars on the Hertzsprung–Russell diagram, containing the majority of stars, that runs diagonally from the upper left to the lower right.

major axis (of ellipse)   The maximum diameter of an ellipse. (2)

mantle (of Earth)   The greatest part of the Earth's interior, lying between the crust and the core. (7)

mare (pl. maria)   Latin for "sea"; name applied to the dark, relatively smooth features that cover 17 percent of the Moon's surface. (8)

mass   A measure of the total amount of material in a body; defined either by the inertial properties of the body or by its gravitational influence on other bodies (2).

mass extinction   The sudden disappearance in the fossil record of a large number of species of life, to be replaced by new species in subsequent layers. Mass extinctions are indications of catastrophic changes in the environment, such as might be produced by a large impact on the Earth. (7)

mass-to-light ratio   The ratio of the total mass of a galaxy to its total luminosity, usually expressed in units of solar mass and solar luminosity. The mass-to-light ratio gives a rough indication of the types of stars contained within a galaxy and whether or not substantial quantities of dark matter are present.

mass-luminosity relation   An empirical relation between the masses and luminosities of many (principally main-sequence) stars.

Maunder Minimum   The interval from 1645 to 1715 when solar activity was very low. (14)

mean solar day   Average length of the apparent solar day. (3)

mean solar time   Time based on the rotation of the Earth; mean solar time passes at a constant rate, unlike apparent solar time. (3)

merger (of galaxies)   When galaxies (of roughly comparable size) collide and form one combined structure.

meridian (celestial)   The great circle on the celestial sphere that passes through an observer's zenith and the north (or south) celestial pole. (3)

meridian (terrestrial)   The great circle on the surface of the Earth that passes through a particular place and the North and South Poles of the Earth. (3)

Messier catalog   A catalog of nonstellar objects compiled by Charles Messier in 1787 (includes nebulae, star clusters, and galaxies).

metals   In general, any element or compound whose electron structure makes it a good conductor of electricity. Among astronomers, the term is often used to mean all elements beyond hydrogen and helium.

metamorphic rock   Any rock produced by physical and chemical alteration (without melting) under high temperature and pressure. (7)

meteor   A flash of light observed when a small piece of solid matter enters the Earth's atmosphere and burns up; popularly called a "shooting star." (6)

meteor shower   Many meteors appearing to radiate from one point in the sky; produced when the Earth passes through a cometary dust stream. (13)

meteorite   A portion of a meteoroid that survives passage through the atmosphere and strikes the ground. (6)

meteorite shower   The breakup of an object entering the Earth's atmosphere at high speed to produce multiple meteorites falling over an area of many square kilometers.

meteoroid   A particle or chunk of typically rocky or metallic material in space before any encounter with the Earth.

micron   Old term for micrometer (10^{-6} m).

microwave   Electromagnetic radiation at radio wavelengths, somewhat longer than the longest infrared waves. (4)

Milky Way   The band of light encircling the sky, which is due to the many stars and diffuse nebulae lying near the plane of the Galaxy.

minerals   The solid compounds (often primarily silicon and oxygen) that form rocks.

model atmosphere or photosphere   The result of a theoretical calculation of the run of temperature, pressure, density, and so on through the outer layers of the Sun or a star.

molecule   A combination of two or more atoms bound together; the smallest particle of a chemical compound or substance that exhibits the chemical properties of that substance.

momentum   A measure of the inertia or state of motion of a body; the momentum of a body is the product of its mass and velocity. In the absence of a force, momentum is conserved. (2)

near-Earth object (NEO)   A comet or asteroid whose path intersects the orbit of the Earth. (12)

nebula   Cloud of interstellar gas or dust; the term is typically used for clouds that are seen to glow with visible light or infrared.

neutrino   A fundamental particle that has little or no rest mass and no charge but that does have spin and energy. Neutrinos rarely interact with ordinary matter.

neutron   A subatomic particle with no charge and with mass approximately equal to that of the proton.

neutron star   A star of extremely high density composed almost entirely of neutrons. (12)

nonthermal radiation   *See* synchrotron radiation.

north celestial pole   *See* celestial poles.

nova   A star that experiences a sudden outburst of radiant energy, temporarily increasing its luminosity by hundreds to thousands of times.

nuclear   Referring to the nucleus of the atom.

nuclear bulge   Central part of our or another galaxy.

nuclear transformation   The change of one atomic nucleus into another, as in nuclear fusion.

nucleosynthesis   The building up of heavy elements from lighter ones by nuclear fusion.

nucleus (of atom)   The heavy part of an atom, composed mostly of protons and neutrons, and about which the electrons revolve. (4)

nucleus (of comet)   The solid chunk of ice and dust in the head of a comet. (12)

nucleus (of galaxy)   Concentration of matter at the center of a galaxy.

occultation   The passage of an object of large angular size in front of a smaller object, such as the Moon in front of a distant star or the rings of Saturn in front of the Voyager spacecraft. *Compare with* transit. (11)

Oort comet cloud The large spherical region around the Sun from which most "new" comets come; a reservoir of objects with aphelia at about 50,000 AU. (12)

opacity Absorbing power; capacity to impede the passage of light.

open cluster A comparatively loose cluster of stars, containing from a few dozen to a few thousand members, located in the spiral arms or disk of the Galaxy; sometimes referred to as a galactic cluster.

open universe A model of the universe in which gravity is not strong enough to bring the universe to a halt; it expands forever. In this model the geometry of spacetime is such that if you go in a straight line, not only can you never return to where you started, but even more space opens up than you would expect from Euclidean geometry.

optical In astronomy, relating to the visible-light band of the electromagnetic spectrum. Optical observations are those made with visible light. (4)

optical double star Two stars at different distances that are seen nearly lined up in projection so that they appear close together, but that are not really gravitationally associated.

orbit The path of an object that is in revolution about another object or point. (2)

organic compound A compound containing carbon, especially a complex carbon compound; not necessarily produced by life.

organic molecule A molecule that contains carbon, especially complex hydrocarbons. *See also* organic compound. (15)

oscillation A periodic motion; in the case of the Sun, a periodic or quasi-periodic expansion and contraction of the whole Sun or some portion of it.

ozone A heavy molecule of oxygen that contains three atoms rather than the more normal two; designated O_3. (7)

parabola A conic section of eccentricity 1.0; the curve of the intersection between a circular cone and a plane parallel to a straight line in the surface of the cone.

parallax An apparent displacement of a nearby star that results from the motion of the Earth around the Sun; numerically, the angle subtended by 1 AU at the distance of a particular star.

parsec A unit of distance in astronomy, equal to 3.26 light years. At a distance of 1 parsec, a star has a parallax of one arcsecond.

Pauli exclusion principle Quantum mechanical principle by which no two electrons (or similar particles) can have the same position and momentum.

peculiar velocity The velocity of a star with respect to the local standard of rest—that is, its space motion, corrected for the motion of the Sun with respect to our neighboring stars.

penumbra The outer, not completely dark part of a shadow; the region from which the source of light is not completely hidden. (3)

perfect radiator or blackbody A body that absorbs and subsequently reemits all radiation incident upon it. (4)

periastron The place in the orbit of a star in a binary star system where it is closest to its companion star.

perigee The place in the orbit of an Earth satellite where it is closest to the center of the Earth. (2)

perihelion The place in the orbit of an object revolving about the Sun where it is closest to the Sun's center. (2)

period-luminosity relation An empirical relation between the periods and luminosities of certain variable stars.

perturbation The small disturbing effect on the motion of a body produced by a third body or other external agent. (2)

phase A particular point in the time of a cycle (for example, the time after maximum for a variable star) or any distinct time period in a sequence of events (as in the phases of the Moon). (3)

photochemistry Chemical changes caused by electromagnetic radiation. (10)

photometry The measurement of light (or other electromagnetic wave) intensities. (5)

photon A discrete unit of electromagnetic energy. (4)

photosphere The region of the solar (or a stellar) atmosphere from which continuous radiation escapes into space. (14)

photosynthesis Complex sequence of chemical reactions through which some living things can use sunlight to manufacture products that store energy (such as carbohydrates), releasing oxygen as one by-product. (7)

pixel An individual picture element in a detector—for example, a particular silicon diode in a CCD.

plage A bright region of the solar surface observed in the light of some spectral line. (14)

Planck's constant The constant of proportionality relating the energy of a photon to its frequency. (4)

planet Any of the nine largest objects revolving about the Sun, or any similar objects that orbit other stars. Unlike stars, planets do not (for the most part) give off their own light but only reflect the light of their parent star. (1)

planetarium An optical device for projecting on a screen or domed ceiling the stars and planets and their apparent motions in the sky. (1)

planetary nebula A shell of gas ejected by and expanding away from an extremely hot low-mass star that is nearing the end of its life.

planetesimals Objects, from tens to hundreds of kilometers in diameter, that formed in the solar nebula as an intermediate step between tiny grains and the larger planetary objects we see today. The comets and some asteroids may be leftover planetesimals. (6)

plasma A hot ionized gas.

plate tectonics The motion of segments or plates of the outer layer of the Earth over the underlying mantle. (7)

populations I and II Two classes of stars (and systems of stars) classified according to their spectral characteristics, chemical compositions, radial velocities, ages, and locations in the Galaxy.

positron An electron with a positive rather than negative charge; an antielectron.

potential energy Stored energy that can be converted into other forms; especially gravitational energy.

power The rate at which work is done (energy used per second)—that is, the rate at which energy is transferred or absorbed. The metric unit of power is the watt, and 1 watt is equivalent to 1 joule/s.

precession (of Earth) A slow, conical motion of the Earth's axis of rotation caused principally by the gravitational pull of the Moon and Sun on the Earth's equatorial bulge. (1)

precession of the equinoxes Slow westward motion of the equinoxes along the ecliptic that results from precession. (1)

pressure Force per unit area; expressed in units of atmospheres or pascals.

prime focus The point in a telescope where the objective focuses the light. (5)

prime meridian The terrestrial meridian passing through the site of the old Royal Greenwich Observatory; longitude 0°. (3)

primitive In planetary science and meteoritics, an object or rock that is little changed, chemically, since its formation. Also used to refer to the chemical composition of an atmosphere that has not undergone extensive chemical evolution. (7)

primitive meteorite A meteorite that has not been greatly altered chemically since its condensation from the solar nebula; called in meteoritics a chondrite (either ordinary chondrite or carbonaceous chondrite).

primitive rock Any rock that has not experienced great heat or pressure and therefore remains representative of the original condensates from the solar nebula. (7)

principle of equivalence Principle that a gravitational force and a suitable acceleration are indistinguishable within a sufficiently local environment.

prism A wedge-shaped piece of glass that is used to disperse white light into a spectrum. (5)

prominence A phenomenon in the solar corona that commonly appears like a flame above the edge of the Sun. (14)

proper motion The angular change per year in the direction of a star as seen from the Sun.

proton A heavy subatomic particle that carries a positive charge; one of the two principal constituents of the atomic nucleus. (4)

proton–proton cycle A series of thermonuclear reactions by which nuclei of hydrogen are built up into nuclei of helium.

protoplanet or -star or -galaxy A very young planet (or a star or galaxy) still in the process of formation.

pulsar A variable radio source of small angular size that emits very rapid radio pulses in very regular periods that range from fractions of a second to several seconds.

pulsating variable A variable star that pulsates in size and luminosity.

quantum mechanics The branch of physics that deals with the structure of atoms and their interactions with one another and with radiation. (4)

quasar An object of very high redshift that looks like a star but is extragalactic and highly luminous; an active galactic nucleus where the galaxy appears faint from Earth.

radar The technique of transmitting radio waves to an object and then detecting the radiation that the object reflects back to the transmitter; used to measure the distance to, and motion of, a target object or to form images of it. (5)

radial velocity The component of relative velocity that lies in the line of sight; motion toward or away from the observer. (4)

radial-velocity curve A plot of the variation of radial velocity with time for a binary or variable star.

radiant (of meteor shower) The point in the sky from which the meteors belonging to a shower seem to radiate. (13)

radiation A mode of energy transport whereby energy is transmitted through a vacuum; also the transmitted energy itself. (4)

radiation pressure The transfer of momentum carried by electromagnetic radiation to a body that the radiation impinges upon.

radio galaxy A galaxy that emits greater amounts of radio radiation than average.

radio telescope A telescope designed to make observations in radio wavelengths. (5)

radioactive dating The technique of determining the ages of rocks or other specimens by the decay of certain radioactive elements contained therein. (6)

radioactivity (radioactive decay) The process by which certain kinds of atomic nuclei naturally decompose, with the spontaneous emission of subatomic particles and gamma rays. (6)

red giant A large, cool star of high luminosity; a star occupying the upper right portion of the Hertzsprung–Russell diagram.

reddening (interstellar) The reddening of starlight passing through interstellar dust because dust scatters blue light more effectively than red.

redshift A shift to longer wavelengths of light, typically a Doppler shift caused by the motion of the source or a shift due to the expansion of space.

reducing In chemistry, referring to conditions in which hydrogen dominates over oxygen, so that most other elements form compounds with hydrogen. In very reducing conditions, free hydrogen (H_2) is present and free oxygen (O_2) cannot exist. (6)

reflecting telescope A telescope in which the principal light collector is a concave mirror. (5)

reflection nebula A relatively dense dust cloud in interstellar space that is illuminated by reflected starlight.

refracting telescope A telescope in which the principal light collector is a lens or system of lenses. (5)

refraction The bending of light rays passing from one transparent medium (or a vacuum) to another. (3)

relativistic particle (or electron) A particle (electron) moving at nearly the speed of light.

relativity A theory formulated by Einstein that describes the relationships between measurements of physical phenomena by two different observers who are in relative motion at constant velocity (the special theory of relativity) or that describes how a gravitational field can be replaced by a curvature of spacetime (the general theory of relativity).

resolution The detail in an image; specifically the smallest angular (or linear) features that can be distinguished. (5)

resonance An orbital condition in which one object is subject to periodic gravitational perturbations by another; most commonly arising when two objects orbiting a third have periods of revolution that are simple multiples or fractions of each other. (11)

retrograde (rotation or revolution) Backward with respect to the common direction of motion in the solar system; clockwise as viewed from the north, and going from east to west rather than from west to east. (1)

retrograde motion An apparent westward motion of a planet on the celestial sphere or with respect to the stars. (1)

revolution The motion of one body around another.

rift zone In geology, a place where the crust is being torn apart by internal forces, generally associated with the injection of new material from the mantle and with the slow separation of tectonic plates. (7)

right ascension A coordinate for measuring the east–west positions of celestial bodies; the angle measured eastward along the celestial equator from the vernal equinox to the hour circle passing through a body. (3)

Roche limit *See* tidal stability limit.

rotation Turning of a body about an axis running through it.

RR Lyrae variable One of a class of giant pulsating stars with periods shorter than one day.

runaway greenhouse effect An irreversible process whereby the heating of a planet leads to an increase in its atmospheric greenhouse effect and thus to further heating, thereby significantly altering the composition of its atmosphere and the temperature of its surface. (9)

satellite An object that revolves about a planet.

Schwarzschild radius *See* event horizon.

scientific method The procedure scientists follow to understand the natural world: (1) the observation of phenomena or the results of experiments; (2) the formulation of hypotheses that describe these phenomena and that are consistent with the body of knowledge available; (3) the testing of these hypotheses by noting whether or not they adequately predict and describe new phenomena or the results of new experiments; and (4) the modification or rejection of hypotheses that are not confirmed by observations or experiment.

sedimentary rock Any rock formed by the deposition and cementing of fine grains of material. (7)

seeing The unsteadiness of the Earth's atmosphere, which blurs telescopic images. Good seeing means the atmosphere is steady. (5)

seismic waves Vibrations that travel through the interior of the Earth or any other object; on Earth these are generally caused by earthquakes. (7)

seismology (solar) The study of small changes in the radial velocity of the Sun as a whole or of small regions on the surface of the Sun. Analyses of these velocity changes can be used to infer the internal structure of the Sun.

seismology (terrestrial) The study of earthquakes, the conditions that produce them, and the internal structure of the Earth as deduced from analyses of seismic waves. (7)

semimajor axis Half the major axis of a conic section, such as an ellipse. (2)

SETI The search for extraterrestrial intelligence, usually applied to searches for radio signals from other civilizations. (15)

Seyfert galaxy A galaxy with an active galactic nucleus; one whose nucleus shows bright emission lines; first described by C. Seyfert.

shepherd satellite Informal term for a satellite that is thought to maintain the structure of a planetary ring through its close gravitational influence. (11)

short-period comet A comet with orbital period shorter than about a century; most are derived from the Kuiper belt and are gravitationally linked to Jupiter (called Jupiter-family comets). (12)

sidereal day The Earth's rotation period as defined by the positions of the stars in the sky; the time between successive passages of the same star through the meridian. (3)

sidereal period The period of revolution of one body about another measured with respect to the stars.

sidereal time Time on Earth measured with respect to the stars rather than the Sun; the local hour angle of the vernal equinox. (3)

sidereal year Period of the Earth's revolution about the Sun with respect to the stars. (3)

sign (of zodiac) Astrological term for any of 12 equal sections along the ecliptic, each of length 30°. Because of precession, these signs today are no longer lined up with the constellations from which they received their names. (1)

singularity A theoretical point of zero volume and infinite density to which any object that becomes a black hole must collapse, according to the general theory of relativity.

solar activity Phenomena that cause changes in the appearance or energy output of the solar atmosphere, such as sunspots, plages, and flares. (14)

solar antapex Direction away from which the Sun is moving with respect to the local standard of rest.

solar apex The direction toward which the Sun is moving with respect to the local standard of rest.

solar day The Earth's rotation period as defined by the position of the Sun in the sky; the time between successive passages of the Sun through the meridian. (3)

solar eclipse An eclipse of the Sun by the Moon, caused by the passage of the Moon in front of the Sun. Solar eclipses can occur only at the time of the new moon. (3)

solar motion Motion of the Sun, or the velocity of the Sun, with respect to the local standard of rest.

solar nebula The cloud of gas and dust from which the solar system formed. *See also* circumstellar disk. (6)

solar seismology The study of pulsations or oscillations of the Sun in order to determine the characteristics of the solar interior.

solar system The system of the Sun and the planets, their satellites, asteroids, comets, KBOs, and other objects revolving around the Sun. (6)

solar time A time based on the Sun; usually the hour angle of the Sun plus 12 h. (3)

solar wind A flow of hot charged particles leaving the Sun. (14)

solstice Either of two points on the celestial sphere where the Sun reaches its maximum distances north and south of the celestial equator; time of the year when the daylight is the longest or the shortest. (3)

south celestial pole *See* celestial poles.

space velocity or space motion The velocity of a star with respect to the Sun.

spacetime A system of one time and three spatial coordinates, with respect to which the time and place of an event can be specified.

spectral class (or type) The classification of stars according to their temperatures using the characteristics of their spectra; the types are O, B, A, F, G, K, M, with L and T added recently for cooler stars that recent surveys have revealed.

spectral line Radiation at a particular wavelength of light produced by the emission or absorption of energy by an atom. (4)

spectral sequence The sequence of spectral classes of stars arranged in order of decreasing temperatures of stars of those classes; O B A F G K M L T.

spectrometer An instrument for obtaining a spectrum; in astronomy, usually attached to a telescope to record the spectrum of a star, galaxy, or other astronomical object. (5)

spectroscopic binary star A binary star in which the components are not resolved but whose binary nature is indicated by periodic variations in radial velocity, indicating orbital motion.

spectroscopic parallax A parallax (or distance) of a star that is derived by comparing the apparent magnitude of the star with its absolute magnitude as deduced from its spectral characteristics.

spectroscopy   The study of spectra.

spectrum   The array of colors or wavelengths obtained when light (or other radiation) from a source is dispersed, as in passing it through a prism or grating. (4)

speed   The rate at which an object moves without regard to its direction of motion; the numerical or absolute value of velocity. (2)

spicule   A jet of rising material in the solar chromosphere. (14)

spiral arms   Arms (or long denser regions) of interstellar material and young stars that wind out in a plane from the central nucleus of a spiral galaxy.

spiral density wave   A mechanism for the generation of spiral structure in galaxies; a density wave interacts with interstellar matter and triggers the formation of stars. Spiral density waves are also seen in the rings of Saturn. (11)

spiral galaxy   A flattened, rotating galaxy with pinwheel-like arms of interstellar material and young stars winding out from its nucleus.

spring tide   The highest tidal range of the month, produced when the Moon is near either the full or the new phase. (3)

standard bulb   An astronomical object of known luminosity; such an object can be used to determine distances; also sometimes called a standard candle.

star   A sphere of gas shining under its own power; massive enough to support the fusion of hydrogen to helium in the interior.

star cluster   An assemblage of stars held together by their mutual gravity.

Stefan-Boltzmann law   A formula from which the rate at which a blackbody radiates energy can be computed; the total rate of energy emission from a unit area of a blackbody is proportional to the fourth power of its absolute temperature. (4)

stellar evolution   The changes that take place in the characteristics of stars as they age.

stellar model   The result of a theoretical calculation of the physical conditions in the different layers of a star's interior.

stellar parallax   *See* parallax.

stellar wind   The outflow of gas, sometimes at speeds as high as hundreds of kilometers per second, from a star.

stony-iron meteorite   A type of differentiated meteorite that is a blend of nickel-iron and silicate materials. (13)

stony meteorite   A meteorite composed mostly of stony material, either primitive or differentiated. (13)

stratosphere   The layer of the Earth's atmosphere above the troposphere (where most weather takes place) and below the ionosphere. (7)

strong nuclear force or strong interaction   The force that binds together the parts of the atomic nucleus.

subduction zone   In terrestrial geology, a region where one crustal plate is forced under another, generally associated with earthquakes, volcanic activity, and the formation of deep ocean trenches. (7)

summer solstice   The point on the celestial sphere where the Sun reaches its farthest distance north of the celestial equator; the day with the longest amount of daylight. (3)

Sun   The star about which the Earth and other planets revolve.

sunspot   A temporary cool region in the solar photosphere that appears dark by contrast against the surrounding hotter photosphere. (14)

sunspot cycle   The semiregular 11-year period with which the frequency of sunspots fluctuates. (14)

supercluster   A large region of space (more than 100 million LY across) where groups and clusters of galaxies are more concentrated; a cluster of clusters of galaxies.

supergiant   A star of very high luminosity and relatively low temperature.

supernova   An explosion that marks the final stage of evolution of a star. A Type I supernova occurs when a white dwarf accretes enough matter to exceed the Chandrasekhar limit, collapses, and explodes. A Type II supernova marks the final collapse of a massive star.

surface gravity   The weight of a unit mass at the surface of a body.

synchrotron radiation   The radiation emitted by charged particles being accelerated in magnetic fields and moving at speeds near that of light. (10)

tail (of a comet)   *See* dust tail *or* ion tail.

tangential (transverse) velocity   The component of a star's space velocity that lies in the plane of the sky.

tectonic   Geological features that result from stresses and pressures in the crust of a planet. Tectonic forces can lead to earthquakes and motion of the crust. (7)

temperature   A measure of how fast the particles in a body are moving or vibrating in place; a measure of the average heat energy in a body.

temperature (Celsius)   Temperature measured on scale where water freezes at 0° and boils at 100°.

temperature (color)   The temperature of a star as estimated from the intensity of the stellar radiation at two or more colors or wavelengths.

temperature (effective)   The temperature of a blackbody that would radiate the same total amount of energy that a particular object, such as a star, does.

temperature (excitation)   The temperature of a star as estimated from the relative strengths of lines in its spectrum that originate from atoms in different stages of excitation.

temperature (Fahrenheit)   Temperature measured on a scale where water freezes at 32° and boils at 212°.

temperature (ionization)   The temperature of a star as estimated from the relative strengths of lines in its spectrum that originate from atoms in different stages of ionization.

temperature (Kelvin)   Absolute temperature measured in Celsius degrees, with the zero point at absolute zero ($-273°C$).

temperature (radiation)   The temperature of a blackbody that radiates the same amount of energy in a given spectral region as does a particular body.

terrestrial planet   Any of the planets Mercury, Venus, Earth, or Mars; sometimes the Moon is included in the list. Also any planet with mass roughly in the interval from 10 Earths to 1/10 Earth, which orbits another star. (6)

theory   A set of hypotheses and laws that have been well demonstrated to apply to a wide range of phenomena associated with a particular subject.

thermal energy   Energy associated with the motions of the molecules or atoms in a substance.

thermal equilibrium   A balance between the input and outflow of heat in a system.

thermal radiation   The electromagnetic radiation emitted by any object or gas that is not at absolute zero.

thermonuclear energy   Energy associated with thermonuclear reactions or that can be released through thermonuclear reactions.

thermonuclear reaction   A nuclear reaction or transformation that results from encounters between particles that are given high velocities (by heating them).

tidal force A differential gravitational force acting on two sides of an object that tends to deform that object. (3)

tidal stability limit The distance—approximately 2.5 planetary radii from the center—within which differential gravitational forces (or tides) are stronger than the mutual gravitational attraction between two adjacent orbiting objects. Within this limit, fragments are not likely to accrete or assemble themselves into a larger object. Also called the Roche limit. (11)

tide Deformation of a body by the differential gravitational force exerted on it by another body; in the Earth, the deformation of the ocean surface by the differential gravitational forces exerted by the Moon and Sun. (3)

transit Passage of an object of small angular size in front of another of larger size, as in the transit of a planet in front of a star. *Compare with* occultation. (11)

transition region The region in the Sun's atmosphere where the temperature rises very rapidly from the relatively low temperatures that characterize the chromosphere to the high temperatures of the corona. (14)

triple-alpha process A sequence of two nuclear reactions by which three helium nuclei are built up into one carbon nucleus.

Trojan asteroid An asteroid with the same orbital period as Jupiter (12 years) that is part of a cluster either 60 degrees in front of or 60 degrees behind Jupiter. (12)

Tropic of Cancer The parallel (circle) of latitude 23.5° N. (3)

Tropic of Capricorn The parallel (circle) of latitude 23.5° S. (3)

tropical year Period of revolution of the Earth about the Sun with respect to the vernal equinox. (3)

troposphere Lowest level of the Earth's atmosphere, where most weather takes place. (7)

turbulence Random motions of gas masses, as in the atmosphere of a star.

21-cm line A line in the spectrum of neutral hydrogen at the radio wavelength of 21 cm.

ultraviolet radiation Electromagnetic radiation of wavelengths shorter than the shortest visible wavelengths; radiation of wavelengths in the approximate range 10 to 400 nm. (4)

umbra The central, completely dark part of a shadow. (3)

universe The totality of all matter, radiation, and space; everything accessible to our observations.

variable star A star that varies in luminosity.

velocity The speed and the direction a body is moving—for example, 44 km/s toward the north galactic pole. (2)

velocity of escape The speed with which an object must move in order to enter a parabolic orbit about another body (such as the Earth) and hence move permanently away from the vicinity of that body. (2)

vernal equinox The point on the celestial sphere where the Sun crosses the celestial equator passing from south to north;

a time in the course of the year when the day and night are roughly equal. (3)

very long baseline interferometry A technique of radio astronomy whereby signals from telescopes thousands of kilometers apart combine to obtain very high resolution by letting waves from different sites interfere with each other. (5)

visual binary star A binary star in which the two components are telescopically resolved.

void A region between clusters and superclusters of galaxies that appears relatively empty of galaxies.

volatile materials Materials that are gaseous at fairly low temperatures. This is a relative term, usually applied to the gases in planetary atmospheres and to common ices (H_2O, CO_2, and so on), but it is also sometimes used for elements such as cadmium, zinc, lead, and rubidium that form gases at temperatures up to 1000 K. (These are called volatile elements, as opposed to refractory elements.) (7)

volume A measure of the total space occupied by a body. (2)

watt A unit of power (energy per unit time; a joule per second).

wavelength The spacing of the crests or troughs in a wave (4).

weak nuclear force or weak interaction The nuclear force involved in radioactive decay. The weak force is characterized by the slow rate of certain nuclear reactions, such as the decay of the neutron, which occurs with a half-life of 11 min.

weight A measure of the force due to gravitational attraction.

white dwarf A low-mass star that has exhausted most or all of its nuclear fuel and has collapsed to a very small size; such a star is near its final stage of life.

Wien's law Formula that relates the temperature of a blackbody to the wavelength at which it emits the greatest intensity of radiation. (4)

winter solstice Point on the celestial sphere where the Sun reaches its farthest distance south of the celestial equator; the time of the year with the shortest amount of daylight. (3)

x rays Photons of wavelengths intermediate between those of ultraviolet radiation and gamma rays. (4)

year The period of revolution of the Earth around the Sun. (1)

Zeeman effect A splitting or broadening of spectral lines due to magnetic fields. (14)

zenith The point on the celestial sphere opposite to the direction of gravity; the point directly above the observer. (1)

zero-age main sequence Main sequence on the H-R diagram for a system of stars that have completed their contraction from interstellar matter and are now deriving all their energy from nuclear reactions, but whose chemical composition has not yet been altered by nuclear reactions.

zodiac A belt around the sky about 18° wide centered on the ecliptic. (1)

zone of avoidance A region near the Milky Way where obscuration by interstellar dust is so heavy that few or no exterior galaxies can be seen.

Powers-of-Ten Notation

In astronomy (and other sciences), it is often necessary to deal with very large or very small numbers. In fact, when numbers become truly large in everyday life, such as the U.S. national debt, we call them astronomical. Among the ideas astronomers must routinely deal with is that the Earth is 150,000,000,000 m from the Sun, and the mass of the hydrogen atom is 0.00000000000000000000000000167 kg. No one in his or her right mind would want to continue writing so many zeros!

Instead, scientists have agreed on a kind of shorthand notation, which not only is easier to write but (as we shall see) also makes multiplication and division of large and small numbers much less difficult. If you have never used this powers-of-ten notation or scientific notation, it may take a bit of time to get used to it, but you will soon find it much easier than keeping all those zeros.

Writing Large Numbers

The convention in this notation is that we generally have only one number to the left of the decimal point. If a number is not in this format, it must be changed. The number 6 is already in the right format because for integers, we understand there to be a decimal point to the right of them. So 6 is really 6., and there is indeed only one number to the left of the decimal point. But the number 465 (which is 465.) has three numbers to the left of the decimal point and is thus ripe for conversion.

To change 465 to proper form, we must make it 4.65 and then keep track of the change we have made. (Think of the number as a weekly salary and suddenly it makes a lot of difference whether we have $465 or $4.65.) We keep track of the number of places we moved the decimal point by expressing it as a power of ten. So 465 becomes 4.65×10^2, or 4.65 multiplied by ten to the second power. The small raised 2 is called an exponent, and it tells us how many times we moved the decimal point to the left.

Note that 10^2 also designates 10 squared or 10×10, which equals 100. And 4.65×100 is just 465, the number we started with. Another way to look at scientific notation is that we separate out the messy numbers out front, and leave the smooth units of ten for the exponent to denote. So a number like 1,372,568 becomes 1.372568 times a million (10^6), or 1.372568 times 10 multiplied by itself 6 times. We had to move the decimal point six places to the left (from its place after the 8) to get the number into the form where there is only one digit to the left of the decimal point.

The reason we call this powers-of-ten notation is that our counting system is based on increases of ten; each place in our numbering system is ten times greater than the place to the right of it. As you have probably learned, this got started because human beings have ten fingers and we started counting with them.

It is interesting to speculate that if we ever meet intelligent life-forms with only eight fingers, their counting system would probably be a powers-of-eight notation!

So, in the example we started with, the number of meters from the Earth to the Sun is 1.5×10^{11}. Elsewhere in the book, we mention that a string a light year long would fit around the Earth's equator 236 million or 236,000,000 times. In scientific notation, this becomes 2.36×10^8. Now if you like expressing things in millions, as the annual reports of successful companies do, you might like to write this number as 236×10^6. However, the usual convention is to have only one number to the left of the decimal point.

Writing Small Numbers

Now take a number like 0.00347, which is also not in the standard (agreed-to) form for scientific notation. To put it into that format, we must make the first part of it 3.47 by moving the decimal point three places to the right. Note that this motion to the right is the opposite of the motion to the left that we discussed above. To keep track, we call this change negative and put a minus sign in the exponent. Thus 0.00347 becomes 3.47×10^{-3}.

In the example we gave at the beginning, the mass of the hydrogen atom would then be written as 1.67×10^{-27} kg. In this system, 1 is written as 10^0, a tenth as 10^{-1}, a hundredth as 10^{-2}, and so forth. Note that any number, no matter how large or how small, can be expressed in scientific notation.

Multiplication and Division

The powers-of-ten notation is not only compact and convenient, it also simplifies arithmetic. To multiply two numbers expressed as powers of ten, you need only multiply the numbers out front and then add the exponents. If there are no numbers out front, as in $100 \times 100,000$, then you just add the exponents (in our notation, $10^2 \times 10^5 = 10^7$). When there are numbers out front, you do have to multiply them, but they are much easier to deal with than numbers with many zeros in them. Here are two examples:

$$3 \times 10^5 \times 2 \times 10^9 = 6 \times 10^{14}$$

$$0.04 \times 6,000,000 = 4 \times 10^{-2} \times 6 \times 10^6$$

$$= 24 \times 10^4 = 2.4 \times 10^5$$

Note in the second example that when we added the exponents, we treated negative exponents as we do in regular arithmetic (-2 plus 6 equals 4). Also, notice that the first result we got had a 24 in it, which was not in the acceptable form, having two places to the left of the decimal point, and we therefore changed it to 2.4.

To divide, you divide the numbers out front and subtract the exponents. Here are three examples:

$$1,000,000 \div 1000 = 10^6 \div 10^3 = 10^{6-3} = 10^3$$

$$9 \times 10^{12} \div 2 \times 10^3 = 4.5 \times 10^9$$

$$2.8 \times 10^2 \div 6.2 \times 10^5 = 0.452 \times 10^{-3} = 4.52 \times 10^{-4}$$

If this is the first time you have met scientific notation, we urge you to practice using it (you might start by solving the exercises below). Like any new language, the notation looks complicated at first, but gets easier as you practice it.

Exercises

1. On April 8, 1996, the Galileo spacecraft was 775 million kilometers from Earth. Convert this number to scientific notation. How many astronomical units is this? (An astronomical unit is the distance from the Earth to the Sun; see above, but remember to keep your units consistent!)

2. During the first six years of its operation, the Hubble Space Telescope circled the Earth 37,000 times, for a total of 1,280,000,000 km. Use scientific notation to find the number of kilometers in one orbit.

3. In a college cafeteria, a soybean-vegetable burger is offered as an alternative to regular hamburgers. If 489,875 burgers were eaten during the course of a school year, and 997 of them were veggie-burgers, what fraction of the burgers does this represent?

4. In a June 1990 Gallup poll, 27 percent of adult Americans thought that alien beings have actually landed on Earth. The number of adults in the United States in 1990 was (according to the census) about 186,000,000. Use scientific notation to determine how many adults believe aliens have visited the Earth.

5. In 1995, 1.7 million degrees were awarded by American colleges and universities. Among these were 41,000 PhD degrees. What fraction of the degrees were PhDs? Express this number as a percent. (Now find jobs for all those PhDs!)

6. A star 60 light years away has been found to have a large planet orbiting it. Your uncle wants to know the distance to this planet in old-fashioned miles. If light travels 186,000 miles per second, and there are 60 seconds in a minute, 60 minutes in an hour, 24 hours in a day, and 365 days in a year, how many miles away is that star?

In the American system of measurement (originally developed in England), the fundamental units of length, weight, and time are the foot, pound, and second, respectively. There are also larger and smaller units, which include the ton (2240 lb), the mile (5280 ft), the rod (16½ ft), the yard (3 ft), the inch (¹⁄₁₂ ft), the ounce (¹⁄₁₆ lb), and so on. Such units, whose origins in decisions by British royalty have been forgotten by most people, are inconvenient for conversion or doing calculations.

In science, therefore, it is more usual to use the *metric system,* which has been adopted in virtually all countries except the United States. Its great advantage is that every unit increases by a factor of ten, instead of the strange factors in the American system. The fundamental units of the metric system are

length: 1 meter (m)

mass: 1 kilogram (kg)

time: 1 second (s)

A meter was originally intended to be 1 ten-millionth of the distance from the equator to the North Pole along the surface of the Earth. It is about 1.1 yd. A kilogram is the mass that on Earth results in a weight of about 2.2 lb. The second is the same in both metric and American units.

The most commonly used quantities of length and mass of the metric system are listed next.

Length

1 kilometer (km) = 1000 meters = 0.6214 mile

1 meter (m) = 0.001 km = 1.094 yards = 39.37 inches

1 centimeter (cm) = 0.01 meter = 0.3937 inch

1 millimeter (mm) = 0.001 meter = 0.1 cm

1 micrometer (μm) = 0.000001 meter = 0.0001 cm

1 nanometer (nm) = 10^{-9} meter = 10^{-7} cm

To convert from the American system, use these helpful factors:

1 mile = 1.61 km

1 inch = 2.54 cm

Mass

Although we don't make the distinction very carefully in everyday life on Earth, strictly speaking the kilogram is a unit of mass (measuring how many atoms a body has) and the pound is a unit of weight (measuring how strongly the Earth's gravity pulls on a body).

1 metric ton = 10^6 grams = 1000 kg (and it produces a weight of 2.205×10^3 lb on Earth)

1 kg = 1000 grams (and it produces a weight of 2.2046 lb on Earth)

1 gram (g) = 0.0353 oz (and the equivalent weight is 0.002205 lb)

1 milligram (mg) = 0.001 g

A weight of 1 lb is equivalent on Earth to a mass of 0.4536 kg, while a weight of 1 oz is produced by a mass of 28.35 g.

Temperature

Three temperature scales are in general use:

1. Fahrenheit (F); water freezes at 32°F and boils at 212°F.

2. Celsius or centigrade° (C); water freezes at 0°C and boils at 100°C.

3. Kelvin or absolute (K); water freezes at 273 K and boils at 373 K.

All molecular motion ceases at about −459°F = −273°C = 0 K, a temperature called *absolute zero.* Kelvin temperature is measured from this lowest possible temperature. It is the temperature scale most often used in astronomy. Kelvins have the same value as centigrade or Celsius degrees, since the difference between the freezing and boiling points of water is 100 degrees in each. (Note that we just say "Kelvins," not Kelvin degrees.)

On the Fahrenheit scale, the difference between the freezing and boiling points of water is 180 degrees. Thus, to convert Celsius degrees or Kelvins to Fahrenheit degrees, it is necessary to multiply by 180/100 = 9/5. To convert from Fahrenheit degrees to Celsius degrees or Kelvins, it is necessary to multiply by 100/180 = 5/9. The full conversion formulas are:

$$K = °C + 273$$
$$°C = 0.555 \times (°F - 32)$$
$$°F = (1.8 \times °C) + 32$$

° *Celsius* is the correct term; it has a more modern standardization but differs from the old centigrade scale by less than 0.1°.

Some Useful Constants for Astronomy

Physical Constants

speed of light $(c) = 2.9979 \times 10^8$ m/s

gravitational constant $(G) = 6.672 \times 10^{-11}$ N m^2/kg^2

Planck's constant $(h) = 6.626 \times 10^{-34}$ joule · s

mass of a hydrogen atom $(m_H) = 1.673 \times 10^{-27}$ kg

mass of an electron $(m_e) = 9.109 \times 10^{-31}$ kg

Rydberg constant $(R) = 1.0974 \times 10^7$ per m

Stefan-Boltzmann constant $(\sigma) = 5.670 \times 10^{-8}$ joule/(s · m^2 deg^4)
    [deg stands for degrees Celsius or Kelvins]

constant in Wien's law $(\lambda_{max}T) = 2.898 \times 10^{-3}$ m deg

electron volt (energy) (eV) $= 1.602 \times 10^{-19}$ joules

energy equivalent of 1 ton TNT $= 4.3 \times 10^9$ joules

Astronomical Constants

astronomical unit (AU) $= 1.496 \times 10^{11}$ m

light year (LY) $= 9.461 \times 10^{15}$ m

parsec (pc) $= 3.086 \times 10^{16}$ m $= 3.262$ LY

sidereal year (yr) $= 3.158 \times 10^7$ s

mass of Earth $(M_E) = 5.977 \times 10^{24}$ kg

equatorial radius of Earth $(R_E) = 6.378 \times 10^6$ m

obliquity of ecliptic $= 23° 27'$

surface gravity of Earth $(g) = 9.807$ m/s^2

escape velocity of Earth $(v_E) = 1.119 \times 10^4$ m/s

mass of Sun $(M_{Sun}) = 1.989 \times 10^{30}$ kg

equatorial radius of Sun $(R_{Sun}) = 6.960 \times 10^8$ m

luminosity of Sun $(L_{Sun}) = 3.83 \times 10^{26}$ watts

solar constant (flux of energy received at Earth)
    $S = 1.37 \times 10^3$ watts/m^2

Hubble constant $(H_0) =$ approximately 20 km/s per million LY
    or approximately 70 km/s per megaparsec

Physical and Orbital Data for the Planets

Physical Data for the Planets

Planet	Diameter (km)	Diameter (Earth = 1)	Mass (Earth = 1)	Mean Density (g/cm³)	Rotation Period (days)	Inclination of Equator to Orbit (°)	Surface Gravity (Earth = 1)	Velocity of Escape (km/s)
Mercury	4,878	0.38	0.055	5.43	58.6	0.0	0.38	4.3
Venus	12,104	0.95	0.82	5.24	−243.0	177.4	0.91	10.4
Earth	12,756	1.00	1.00	5.52	0.997	23.4	1.00	11.2
Mars	6,794	0.53	0.107	3.9	1.026	25.2	0.38	5.0
Jupiter	142,800	11.2	317.8	1.3	0.41	3.1	2.53	60
Saturn	120,540	9.41	94.3	0.7	0.43	26.7	1.07	36
Uranus	51,200	4.01	14.6	1.3	−0.72	97.9	0.92	21
Neptune	49,500	3.88	17.2	1.7	0.67	29	1.18	24
Pluto	2,300	0.18	0.0025	2.0	−6.387	118	0.09	1

Orbital Data for the Planets

Planet	Semimajor Axis AU	Semimajor Axis 10⁶ km	Sidereal Period Tropical Years	Sidereal Period Days	Mean Orbital Speed (km/s)	Orbital Eccentricity	Inclination of Orbit to Ecliptic (°)
Mercury	0.3871	57.9	0.24085	88	47.9	0.206	7.004
Venus	0.7233	108.2	0.61521	225	35.0	0.007	3.394
Earth	1.0000	149.6	1.000039	365	29.8	0.017	0.0
Mars	1.5237	227.9	1.88089	687	24.1	0.093	1.850
(Ceres)	2.7671	414	4.603		17.9	0.077	10.6
Jupiter	5.2028	778	11.86		13.1	0.048	1.308
Saturn	9.538	1427	29.46		9.6	0.056	2.488
Uranus	19.191	2871	84.07		6.8	0.046	0.774
Neptune	30.061	4497	164.82		5.4	0.010	1.774
Pluto	39.529	5913	248.6		4.7	0.248	17.15

Selected Satellites of the Planets

Note: As this book goes to press, more than 100 satellites are now known in the solar system and more are being discovered on a regular basis. Many of these are small bodies, most likely captured asteroids. It's no longer practical to list every known satellite in such an appendix, and we therefore restrict ourselves to the more significant bodies that orbit each planet. See the Web sites listed for Chapters 6 and 11 for information on all the known satellites.

Planet	Satellite Name	Discovery	Semimajor Axis (km × 1000)	Period (days)	Diameter (km)	Mass (10^{20} kg)	Density (g/cm³)
Earth	Moon	—	384	27.32	3476	735	3.3
Mars	Phobos	Hall (1877)	9.4	0.32	23	1×10^{-4}	2.0
	Deimos	Hall (1877)	23.5	1.26	13	2×10^{-5}	1.7
Jupiter	Metis	Voyager (1979)	128	0.29	20	—	—
	Adrastea	Voyager (1979)	129	0.30	40	—	—
	Amalthea	Barnard (1892)	181	0.50	200	—	—
	Thebe	Voyager (1979)	222	0.67	90	—	—
	Io	Galileo (1610)	422	1.77	3630	894	3.6
	Europa	Galileo (1610)	671	3.55	3138	480	3.0
	Ganymede	Galileo (1610)	1,070	7.16	5262	1482	1.9
	Callisto	Galileo (1610)	1,883	16.69	4800	1077	1.9
	Leda	Kowal (1974)	11,090	239	15	—	—
	Himalia	Perrine (1904)	11,480	251	180	—	—
	Lysithea	Nicholson (1938)	11,720	259	40	—	—
	Elara	Perrine (1905)	11,740	260	80	—	—
	Ananke	Nicholson (1951)	21,200	631 (R)	30	—	—
	Carme	Nicholson (1938)	22,600	692 (R)	40	—	—
	Pasiphae	Melotte (1908)	23,500	735 (R)	40	—	—
	Sinope	Nicholson (1914)	23,700	758 (R)	40	—	—
Saturn	Unnamed	Voyager (1985)	118.2	0.48	15?	3×10^{-5}	—
	Pan	Voyager (1985)	133.6	0.58	20	3×10^{-5}	—
	Atlas	Voyager (1980)	137.7	0.60	40	—	—
	Prometheus	Voyager (1980)	139.4	0.61	80	—	—
	Pandora	Voyager (1980)	141.7	0.63	100	—	—
	Janus	Dollfus (1966)	151.4	0.69	190	—	—
	Epimetheus	Fountain, Larson (1980)	151.4	0.69	120	—	—

(Table continues)

Planet	Satellite Name	Discovery	Semimajor Axis (km × 1000)	Period (days)	Diameter (km)	Mass (10^{20} kg)	Density (g/cm^3)
Saturn	Mimas	Herschel (1789)	186	0.94	394	0.4	1.2
	Enceladus	Herschel (1789)	238	1.37	502	0.8	1.2
	Tethys	Cassini (1684)	295	1.89	1048	7.5	1.3
	Telesto	Reitsema et al. (1980)	295	1.89	25	—	—
	Calypso	Pascu et al. (1980)	295	1.89	25	—	—
	Dione	Cassini (1684)	377	2.74	1120	11	1.4
	Helene	Lecacheux, Laques (1980)	377	2.74	30	—	—
	Rhea	Cassini (1672)	527	4.52	1530	25	1.3
	Titan	Huygens (1655)	1,222	15.95	5150	1346	1.9
	Hyperion	Bond, Lassell (1848)	1,481	21.3	270	—	—
	Iapetus	Cassini (1671)	3,561	79.3	1435	19	1.2
	Phoebe	Pickering (1898)	12,950	550 (R)	220	—	—
Uranus	Cordelia	Voyager (1986)	49.8	0.34	40?	—	—
	Ophelia	Voyager (1986)	53.8	0.38	50?	—	—
	Bianca	Voyager (1986)	59.2	0.44	50?	—	—
	Cressida	Voyager (1986)	61.8	0.46	60?	—	—
	Desdemona	Voyager (1986)	62.7	0.48	60?	—	—
	Juliet	Voyager (1986)	64.4	0.50	80?	—	—
	Portia	Voyager (1986)	66.1	0.51	80?	—	—
	Rosalind	Voyager (1986)	69.9	0.56	60?	—	—
	Belinda	Voyager (1986)	75.3	0.63	60?	—	—
	Unnamed	Voyager (1986)	73.5	0.63	40?	—	—
	Puck	Voyager (1985)	86.0	0.76	170	—	—
	Miranda	Kuiper (1948)	130	1.41	485	0.8	1.3
	Ariel	Lassell (1851)	191	2.52	1160	13	1.6
	Umbriel	Lassell (1851)	266	4.14	1190	13	1.4
	Titania	Herschel (1787)	436	8.71	1610	35	1.6
	Oberon	Herschel (1787)	583	13.5	1550	29	1.5
	Caliban	Gladman et. al. (1997)	7,000	562	60?	—	—
	Sycorax	Nicholson et. al. (1997)	12,000	1,261	120?	—	—
Neptune	Naiad	Voyager (1989)	48	0.30	50	—	—
	Thalassa	Voyager (1989)	50	0.31	90	—	—
	Despina	Voyager (1989)	53	0.33	150	—	—
	Galatea	Voyager (1989)	62	0.40	150	—	—
	Larissa	Voyager (1989)	74	0.55	200	—	—
	Proteus	Voyager (1989)	118	1.12	400	—	—
	Triton	Lassell (1846)	355	5.88 (R)	2720	220	2.1
	Nereid	Kuiper (1949)	5,511	360	340	—	—
Pluto	Charon	Christy (1978)	19.7	6.39	1200	—	—

(R) = Retrograde orbit

Upcoming (Total) Eclipses

1 Total Eclipses of the Sun

Date	Duration of Totality (min)	Where Visible
2003 Nov. 23	2.0	Antarctica
2005 April 8	0.7	South Pacific
2006 March 29	4.1	Africa, western Asia, Russia
2008 Aug. 1	2.4	Arctic Ocean, Siberia, China
2009 July 22	6.6	India, China, South Pacific
2010 July 11	5.3	South Pacific
2012 Nov. 13	4.0	Northern Australia, South Pacific
2013 Nov. 3	1.7	Atlantic, Central Africa
2015 March 20	4.1	North Atlantic, Arctic
2016 March 9	4.5	Indonesia, Pacific
2017 Aug. 21	2.7	Pacific, United States, Atlantic
2019 July 2	4.5	South Pacific, South America
2020 Dec. 14	2.2	South Pacific, South America, South Atlantic
2021 Dec. 4	1.9	Antarctica
2023 April 20	4.5	South Pacific, Mexico, eastern United States
2026 Aug. 12	2.3	Arctic, Greenland, North Atlantic, Spain
2027 Aug. 2	6.4	North Africa, Arabia, Indian Ocean
2028 July 22	5.1	Indian Ocean, Australia, New Zealand
2030 Nov. 25	3.7	South Africa, Indian Ocean, Australia

2 Total Eclipses of the Moon

2003 Nov. 9
2004 May 4
2004 Oct. 28
2007 March 3
2007 Aug. 28
2008 Feb. 21
2010 Dec. 21
2011 June 15
2011 Dec. 10
2014 Apr. 15

Note: Lunar eclipses are much more "democratic" than solar ones. They are visible over the entire hemisphere of the Earth where the Moon can be seen (see Chapter 3).

| Name (catalog number) | Distance (LY) | Spectral Type | Location[1] | | Luminosity (Sun = 1) |
			RA	Dec	
Sun	—	G2V	—	—	1.0
Proxima Centauri	4.2	M5V	14 30	−62 41	6×10^{-6}
Alpha Centauri A	4.4	G2V	14 40	−60 50	1.5
Alpha Centauri B	4.4	K0V	14 40	−60 50	0.5
Barnard's Star (Gliese 699)	6.0	M4V	17 58	+04 42	4×10^{-4}
Wolf 359 (Gliese 406)	7.8	M6V	10 56	+07 03	2×10^{-5}
Lalande 21185 (HD 95735)	8.3	M2V	11 03	+35 58	5×10^{-3}
Sirius A	8.6	A1V	06 45	−16 43	24
Sirius B	8.6	w.d.[2]	06 45	−16 43	3×10^{-3}
Luyten 726-8 A (Gliese 65A)	8.7	M5V	01 39	−17 57	6×10^{-5}
Luyten 726-8 B (UV Ceti)	8.7	M6V	01 39	−17 58	4×10^{-5}
Ross 154 (Gliese 729)	9.7	M4V	18 50	−23 50	5×10^{-4}
Ross 248 (Gliese 905)	10.3	M6V	23 42	+44 11	1×10^{-4}
Epsilon Eridani (Gliese 144)	10.5	K2V	03 33	−09 27	0.3
Lacaille 9352 (Gliese 887)	10.7	M1V	23 06	−35 51	1×10^{-2}
Ross 128 (Gliese 447)	10.9	M4V	11 48	+00 48	3×10^{-4}
Luyten 789-6 A (Gliese 866A)	11.3	M5V[3]	22 39	−15 18	1×10^{-4}
Luyten 789-6 B	11.3	—	22 39	−15 18	—
Luyten 789-6 C	11.3	—	22 39	−15 18	—
Procyon A	11.4	F51V	07 39	+05 13	7.7
Procyon B	11.4	w.d.[2]	07 39	+05 13	6×10^{-4}
61 Cygni A (Gliese 820A)	11.4	K5V	21 07	+38 45	8×10^{-2}
61 Cygni B	11.4	K7V	21 07	+38 45	4×10^{-2}
Gliese 725 A	11.5	M3V	18 43	+59 38	3×10^{-3}
Gliese 725 B	11.5	M4V	18 43	+59 38	2×10^{-3}
Gliese 15 A	11.6	M1V	00 18	+44 01	$6 \times 10-3$
Gliese 15 B	11.6	M3V	00 18	+44 01	4×10^{-4}
Epsilon Indi (Gliese 845)	11.8	K5V	22 03	−56 47	0.14
GJ 1111 (DX Cancri)	11.8	M7V	08 30	+26 47	1×10^{-5}
Tau Ceti (Gliese 71)	11.9	G8V	01 44	−15 56	0.45

(Table continues)

[1] Location (right ascension and declination) given for Epoch 2000.0
[2] White dwarf star
[3] The stars in this system are so close to each other that it is not possible to measure their spectral types and luminosities separately.

| Name (catalog number) | Distance (LY) | Spectral Type | Location [1] | | Luminosity (Sun = 1) |
			RA	Dec	
GJ 1061	11.9	M5V	03 36	−44 31	8×10^{-5}
Luyten 725-32 (YZ Ceti)	12.1	M5V	01 12	−16 60	3×10^{-4}
Gliese 273 (Luyten's Star)	12.4	M4V	07 27	+05 14	1×10^{-3}
Gliese 191 (Kapteyn's Star)	12.8	M1V	05 12	−45 01	4×10^{-3}
Gliese 825 (AX Microscopium)	12.9	M0V	21 17	−38 52	3×10^{-2}

[1] Location (right ascension and declination) given for Epoch 2000.0

With many thanks to Dr. Todd Henry and the RECONS team for updated information. For the latest data on nearby star systems, see their Web page at http://joy.chara.gsu.edu/RECONS/TOP100.htm

11 The Brightest Stars

Note: These are the stars that appear the brightest visually, as seen from our vantage point on Earth. They are not necessarily the stars that are intrinsically the brightest.

Name[1]	Luminosity (Sun = 1)	Distance[2] (LY)	Spectral Type	Proper Motion (arcsec/yr)	Right Ascension (Epoch 2000.0) (h)	(m)	Declination (Epoch 2000.0) (deg)	(min)
Sirius (α CMa)	24	8.6	A1V	1.34	06	45.1	−16	43
Canopus (α Car)	7.3×10^3	228	F0II	0.02	06	24.0	−52	42
Alpha Centauri	2	4	G2V	3.68	14	39.6	−60	50
Arcturus (α Boo)	187	37	K1.5II	2.28	14	15.7	+19	11
Vega (α Lyr)	50	25	A0V	0.35	18	36.9	+38	47
Capella (α Aur)	145	42	G8III	0.43	05	16.7	+46	00
Rigel (β Ori)	6×10^4	772	B8Ia	0.00	05	14.5	−08	12
Procyon (α Cmi)	7	11	F5IV-V	1.26	07	39.3	+05	13
Betelgeuse (α Ori)	7×10^5	427	M1Iab	0.03	05	55.2	+07	24
Achernar (α Eri)	2800	144	B3V	0.10	01	37.7	−57	14
Beta Centauri	6.5×10^4	525	B1III	0.04	14	03.8	−60	22
Altair (α Acll)	10	17	A7V	0.66	19	50.8	+08	52
Aldebaran (α Tau)	450	65	K5III	0.20	04	35.9	+16	31
Spica (α Vir)	1.2×10^4	262	B1III	0.05	13	25.2	−11	10
Antares (α Sco)	8.5×10^5	604	M1.5Ib	0.03	16	29.4	−26	26
Pollux (β Gem)	40	34	K0III	0.62	07	45.3	+28	02
Fomalhaut (α PsA)	16	25	A3V	0.37	22	57.6	−29	37
Deneb (α Cyg)	2.4×10^5	3228	A2Ia	0.00	20	41.4	+45	17
Beta Crucis	1.6×10^4	352	B0.5IV	0.05	12	47.7	−59	41
Regulus (α Leo)	230	77	B7V	0.25	10	08.3	+11	58

[1]The brightest stars typically have names from antiquity (although most fainter stars are not given names but merely catalog designations). Next to each star's ancient name, we have put its name in the system originated by Bayer. The abbreviations of the constellations are given in Appendix 14.
[2]The distances of the more remote stars are estimated from their spectral types and apparent brightnesses and are only approximate. The luminosities for those stars are approximate to the same degree.

The Brightest Members of the Local Group of Galaxies

Galaxy	Type[1]	Right Ascension[2] (h)	(m)	Declination (degrees)	Distance[3] (1000 LY)	Luminosity (L_{Sun})	Apparent Magnitude	Diameter (1000 LY)	Mass[4] ($10^6 M_{Sun}$)
Milky Way	SBb	17	46	−29	—	1.3×10^{10}	—	100	10^6
Andromeda (M31, NGC224)	Sb	00	43	+41	2900	2.6×10^{10}	3.4	160	4×10^5
M33 (NGC598)	Sc	01	34	31	3000	3.4×10^9	5.7	60	25,000
Large Magellanic Cloud	Irr	05	20	−69	179	2.1×10^9	0.1	35	20,000
Small Magellanic Cloud	Irr	00	52	−73	210	3.7×10^8	2.3	18	6,000
IC10	Irr	00	20	+59	4200	9.3×10^7	10.3	9	
NGC205	E5pec	00	41	+42	2900	2.3×10^8	8.5	17	10,000
M32 (NGC221)	E2	00	43	+41	2900	3.4×10^8	8.1	10	3,000
NGC6822	Irr	19	45	−15	1700	5.3×10^7	9		
WLM	Irr	00	02	−15	4200	5.6×10^7	10.9	15	
NGC185	E3pec	00	39	+48	2500	9.3×10^7	9.2	11	
IC1613	Irr	01	05	+02	2900	1.2×10^8	9.2	17	
NGC147	E5	00	33	+48	2400	6.4×10^7	9.5	11	
Leo A	Irr	09	59	+31	7000				
Pegasus	Irr	23	29	+15	6000				
Fornax	E2	02	40	−34	530	1.2×10^7	8.1	2	
DD0210	Irr	20	47	−13	2000	7.7×10^5	13.9	1	
Sagittarius Dwarf[5]	DwE7	18	55	−31	80				
Sagittarius	Irr	19	30	−18	2000				
Sculptor	E3	01	00	−34	300	4.1×10^5	10.5		
Andromeda I	E3	00	46	+38	2900	3.4×10^6	13.2		
Andromeda III	E2	00	35	+36	2900	2.3×10^6	13.5		
Andromeda II	E0	01	16	+33	2900	3.7×10^6	13		
Pisces (LGS3)	Irr	01	04	+22	3000	4.4×10^5	15.4	2	
Leo I	E3	10	09	+12	880	7.0×10^6	9.8		
Leo II	E0	11	14	+22	800				
Ursa Minor	E4	15	09	+67	240	1.8×10^5	10.9	3	
Draco	E0	17	20	+58	280	2.6×10^5	10.9	4	

(Table continues)

Galaxy	Type[1]	Right Ascension[2] (h)	Right Ascension[2] (m)	Decli-nation (degrees)	Distance[3] (1000 LY)	Lumi-nosity (L_{Sun})	Apparent Magni-tude	Diameter (1000 LY)	Mass[4] ($10^6 M_{Sun}$)
Carina	E3	06	42	−51	300				
Andromeda V	DwE	01	10	+48	2900	1.2×10^6			
Phoenix	Irr	01	51	−44	1600	6.4×10^5	—		
Sextans	DwE3	10	13	−02	300	4.9×10^5			
Tucana	DwE5	22	42	−64	3000	5.3×10^5			
Andromeda VI	DwE	23	52	+25	2900	—			

[1] S means spiral, SB means barred spiral, E means elliptical, Irr means irregular, Dw means dwarf, pec means peculiar; the numbers represent sub-groups into which Hubble and others divided these broad categories.
[2] Coordinates are given for Epoch 2000.0.
[3] Many of the distances (and therefore the diameters calculated from them) are only approximate.
[4] Masses have been measured for only a few of the brightest and/or nearest galaxies. Determining masses for irregular galaxies requires the measurement of spectra of a large number of faint stars, and this work has not yet been done for most of the stars in the Local Group.
[5] This close neighbor galaxy is so extended on the sky that giving a magnitude would not make sense.

13 The Chemical Elements

Element	Symbol	Atomic Number	Atomic Weight° (chemical scale)	Number of Atoms per 10^{12} Hydrogen Atoms
Hydrogen	H	1	1.0080	1×10^{12}
Helium	He	2	4.003	8×10^{10}
Lithium	Li	3	6.940	2×10^{3}
Beryllium	Be	4	9.013	3×10^{1}
Boron	B	5	10.82	9×10^{2}
Carbon	C	6	12.011	4.5×10^{8}
Nitrogen	N	7	14.008	9.2×10^{7}
Oxygen	O	8	16.00	7.4×10^{8}
Fluorine	F	9	19.00	3.1×10^{4}
Neon	Ne	10	20.183	1.3×10^{8}
Sodium	Na	11	22.991	2.1×10^{6}
Magnesium	Mg	12	24.32	4.0×10^{7}
Aluminum	Al	13	26.98	3.1×10^{6}
Silicon	Si	14	28.09	3.7×10^{7}
Phosphorus	P	15	30.975	3.8×10^{5}
Sulfur	S	16	32.066	1.9×10^{7}
Chlorine	Cl	17	35.457	1.9×10^{5}
Argon	Ar(A)	18	39.944	3.8×10^{6}
Potassium	K	19	39.100	1.4×10^{5}
Calcium	Ca	20	40.08	2.2×10^{6}
Scandium	Sc	21	44.96	1.3×10^{3}
Titanium	Ti	22	47.90	8.9×10^{4}
Vanadium	V	23	50.95	1.0×10^{4}
Chromium	Cr	24	52.01	5.1×10^{5}
Manganese	Mn	25	54.94	3.5×10^{5}
Iron	Fe	26	55.85	3.2×10^{7}
Cobalt	Co	27	58.94	8.3×10^{4}
Nickel	Ni	28	58.71	1.9×10^{6}
Copper	Cu	29	63.54	1.9×10^{4}
Zinc	Zn	30	65.38	4.7×10^{4}
Gallium	Ga	31	69.72	1.4×10^{3}
Germanium	Ge	32	72.60	4.4×10^{3}
Arsenic	As	33	74.91	2.5×10^{2}
Selenium	Se	34	78.96	2.3×10^{3}

° Where average atomic weights have not been well determined, the atomic mass numbers of the most stable isotopes are given in parentheses.

(Table continues)

Element	Symbol	Atomic Number	Atomic Weight° (chemical scale)	Number of Atoms per 10^{12} Hydrogen Atoms
Bromine	Br	35	79.916	4.4×10^2
Krypton	Kr	36	83.80	1.7×10^3
Rubidium	Rb	37	85.48	2.6×10^2
Strontium	Sr	38	87.63	8.8×10^2
Yttrium	Y	39	88.92	2.5×10^2
Zirconium	Zr	40	91.22	4.0×10^2
Niobium (Columbium)	Nb(Cb)	41	92.91	2.6×10^1
Molybdenum	Mo	42	95.95	9.3×10^1
Technetium	Tc(Ma)	43	(99)	—
Ruthenium	Ru	44	101.1	68
Rhodium	Rh	45	102.91	13
Palladium	Pd	46	106.4	51
Silver	Ag	47	107.880	20
Cadmium	Cd	48	112.41	63
Indium	In	49	114.82	7
Tin	Sn	50	118.70	1.4×10^2
Antimony	Sb	51	121.76	13
Tellurium	Te	52	127.61	1.8×10^2
Iodine	I (J)	53	126.91	33
Xenon	Xe(X)	54	131.30	1.6×10^2
Cesium	Cs	55	132.91	14
Barium	Ba	56	137.36	1.6×10^2
Lanthanum	La	57	138.92	17
Cerium	Ce	58	140.13	43
Praseodymium	Pr	59	140.92	6
Neodymium	Nd	60	144.27	31
Promethium	Pm	61	(147)	—
Samarium	Sm(Sa)	62	150.35	10
Europium	Eu	63	152.00	4
Gadolinium	Gd	64	157.26	13
Terbium	Tb	65	158.93	2
Dysprosium	Dy(Ds)	66	162.51	15
Holmium	Ho	67	164.94	3
Erbium	Er	68	167.27	9
Thulium	Tm(Tu)	69	168.94	2
Ytterbium	Yb	70	173.04	8
Lutecium	Lu(Cp)	71	174.99	2
Hafnium	Hf	72	178.50	6
Tantalum	Ta	73	180.95	1
Tungsten	W	74	183.86	5
Rhenium	Re	75	186.22	2
Osmium	Os	76	190.2	27
Iridium	Ir	77	192.2	24
Platinum	Pt	78	195.09	56
Gold	Au	79	197.00	6
Mercury	Hg	80	200.61	19
Thallium	Tl	81	204.39	8

° Where average atomic weights have not been well determined, the atomic mass numbers of the most stable isotopes are given in parentheses.

Element	Symbol	Atomic Number	Atomic Weight* (chemical scale)	Number of Atoms per 10^{12} Hydrogen Atoms
Lead	Pb	82	207.21	1.2×10^2
Bismuth	Bi	83	209.00	5
Polonium	Po	84	(209)	—
Astatine	At	85	(210)	—
Radon	Rn	86	(222)	—
Francium	Fr(Fa)	87	(223)	—
Radium	Ra	88	226.05	—
Actinium	Ac	89	(227)	—
Thorium	Th	90	232.12	1
Protactinium	Pa	91	(231)	—
Uranium	U(Ur)	92	238.07	1
Neptunium	Np	93	(237)	—
Plutonium	Pu	94	(244)	—
Americium	Am	95	(243)	—
Curium	Cm	96	(248)	—
Berkelium	Bk	97	(247)	—
Californium	Cf	98	(251)	—
Einsteinium	E	99	(254)	—
Fermium	Fm	100	(253)	—
Mendeleevium	Mv	101	(256)	—
Nobelium	No	102	(253)	—
Lawrencium	Lr	103	(262)	—
Rutherfordium	Rf	104	(261)	—
Dubnium	Db	105	(262)	—
Seaborgium	Sg	106	(263)	—
Bohrium	Bh	107	(262)	—
Hassium	Hs	108	(264)	—
Meitnerium	Mt	109	(266)	—
Ununnilium	Uun	110	(269)	—
Unununium	Uuu	111	(272)	—
Ununbium	Uub	112	(277)	—
Ununquadium	Uuq	114	(285)	—
Ununhexium	Uuh	116	(289)	—
Ununoctium	Uuo	118	(293)	—

* Where average atomic weights have not been well determined, the atomic mass numbers of the most stable isotopes are given in parentheses.

The Constellations

Constellation (Latin name)	Genitive Case Ending[1]	English Name or Description	Abbre-viation	Approximate Position	
				α	δ
				h	°
Andromeda	Andromedae	Princess of Ethiopia	And	1	+40
Antila	Antilae	Air pump	Ant	10	−35
Apus	Apodis	Bird of paradise	Aps	16	−75
Aquarius	Aquarii	Water bearer	Aqr	23	−15
Aquila	Aquilae	Eagle	Aql	20	+5
Ara	Arae	Altar	Ara	17	−55
Aries	Arietis	Ram	Ari	3	+20
Auriga	Aurigae	Charioteer	Aur	6	+40
Boötes	Boötis	Herdsman	Boo	15	+30
Caelum	Caeli	Engraving tool	Cae	5	−40
Camelopardalis	Camelopardalis	Giraffe	Cam	6	+70
Cancer	Cancri	Crab	Cnc	9	+20
Canes Venatici	Canum Venaticorum	Hunting dogs	CVn	13	+40
Canis Major	Canis Majoris	Big dog	CMa	7	−20
Canis Minor	Canis Minoris	Little dog	CMi	8	+5
Capricornus	Capricorni	Sea goat	Cap	21	−20
Carina°	Carinae	Keel of the Argonauts' ship	Car	9	−60
Cassiopeia	Cassiopeiae	Queen of Ethiopia	Cas	1	+60
Centaurus	Centauri	Centaur	Cen	13	−50
Cepheus	Cephei	King of Ethiopia	Cep	22	+70
Cetus	Ceti	Sea monster (whale)	Cet	2	−10
Chamaeleon	Chamaeleontis	Chameleon	Cha	11	−80
Circinus	Circini	Compasses	Cir	15	−60
Columba	Columbae	Dove	Col	6	−35
Coma Berenices	Comae Berenices	Berenice's hair	Com	13	+20
Corona Australis	Coronae Australis	Southern crown	CrA	19	−40
Corona Borealis	Coronae Borealis	Northern crown	CrB	16	+30
Corvus	Corvi	Crow	Crv	12	−20

[1] When a constellation name is by itself in a sentence, we use the form in the first column (for example, the comet last night was seen in Crux). But when the constellation is part of the name of an object, we use the form in the second column (for example, the brightest star in that constellation is Alpha Crucis).

° The four constellations Carina, Puppis, Pyxis, and Vela originally formed the single constellation Argo Navis.

Constellation (Latin name)	Genitive Case Ending	English Name or Description	Abbre-viation	Approximate Position α h	δ °
Crater	Crateris	Cup	Crt	11	−15
Crux	Crucis	Cross (southern)	Cru	12	−60
Cygnus	Cygni	Swan	Cyg	21	+40
Delphinus	Delphini	Porpoise	Del	21	+10
Dorado	Doradus	Swordfish	Dor	5	−65
Draco	Draconis	Dragon	Dra	17	+65
Equuleus	Equulei	Little horse	Equ	21	+10
Eridanus	Eridani	River	Eri	3	−20
Fornax	Fornacis	Furnace	For	3	−30
Gemini	Geminorum	Twins	Gem	7	+20
Grus	Gruis	Crane	Gru	22	−45
Hercules	Herculis	Hercules, son of Zeus	Her	17	+30
Horologium	Horologii	Clock	Hor	3	−60
Hydra	Hydrae	Sea serpent	Hya	10	−20
Hydrus	Hydri	Water snake	Hyi	2	−75
Indus	Indi	Indian	Ind	21	−55
Lacerta	Lacertae	Lizard	Lac	22	+45
Leo	Leonis	Lion	Leo	11	+15
Leo Minor	Leonis Minoris	Little lion	LMi	10	+35
Lepus	Leporis	Hare	Lep	6	−20
Libra	Librae	Balance	Lib	15	−15
Lupus	Lupi	Wolf	Lup	15	−45
Lynx	Lyncis	Lynx	Lyn	8	+45
Lyra	Lyrae	Lyre or harp	Lyr	19	+40
Mensa	Mensae	Table Mountain	Men	5	−80
Microscopium	Microscopii	Microscope	Mic	21	−35
Monoceros	Monocerotis	Unicorn	Mon	7	−5
Musca	Muscae	Fly	Mus	12	−70
Norma	Normae	Carpenter's level	Nor	16	−50
Octans	Octantis	Octant	Oct	22	−85
Ophiuchus	Ophiuchi	Holder of serpent	Oph	17	0
Orion	Orionis	Orion, the hunter	Ori	5	+5
Pavo	Pavonis	Peacock	Pav	20	−65
Pegasus	Pegasi	Pegasus, the winged horse	Peg	22	+20
Perseus	Persei	Perseus, hero who saved Andromeda	Per	3	+45
Phoenix	Phoenicis	Phoenix	Phe	1	−50
Pictor	Pictoris	Easel	Pic	6	−55
Pisces	Piscium	Fishes	Psc	1	+15
Piscis Austrinus	Piscis Austrini	Southern fish	PsA	22	−30
Puppis°	Puppis	Stern of the Argonauts' ship	Pup	8	−40

[1] When a constellation name is by itself in a sentence, we use the form in the first column (for example, the comet last night was seen in Crux). But when the constellation is part of the name of an object, we use the form in the second column (for example, the brightest star in that constellation is Alpha Crucis).

° The four constellations Carina, Puppis, Pyxis, and Vela originally formed the single constellation Argo Navis.

(Table continues)

Constellation (Latin name)	Genitive Case Ending	English Name or Description	Abbreviation	α h	δ °
Pyxis° (= Malus)	Pyxidus	Compass of the Argonauts' ship	Pyx	9	−30
Reticulum	Reticuli	Net	Ret	4	−60
Sagitta	Sagittae	Arrow	Sge	20	+10
Sagittarius	Sagittarii	Archer	Sgr	19	−25
Scorpius	Scorpii	Scorpion	Sco	17	−40
Sculptor	Sculptoris	Sculptor's tools	Scl	0	−30
Scutum	Scuti	Shield	Sct	19	−10
Serpens	Serpentis	Serpent	Ser	17	0
Sextans	Sextantis	Sextant	Sex	10	0
Taurus	Tauri	Bull	Tau	4	+15
Telescopium	Telescopii	Telescope	Tel	19	−50
Triangulum	Trianguli	Triangle	Tri	2	+30
Triangulum Australe	Trianguli Australis	Southern triangle	TrA	16	−65
Tucana	Tucanae	Toucan	Tuc	0	−65
Ursa Major	Ursae Majoris	Big bear	UMa	11	+50
Ursa Minor	Ursae Minoris	Little bear	UMi	15	+70
Vela°	Velorum	Sail of the Argonauts' ship	Vel	9	−50
Virgo	Virginis	Virgin	Vir	13	0
Volans	Volantis	Flying fish	Vol	8	−70
Vulpecula	Vulpeculae	Fox	Vul	20	+25

[1]When a constellation name is by itself in a sentence, we use the form in the first column (for example, the comet last night was seen in Crux). But when the constellation is part of the name of an object, we use the form in the second column (for example, the brightest star in that constellation is Alpha Crucis).

° The four constellations Carina, Puppis, Pyxis, and Vela originally formed the single constellation Argo Navis.

The Messier Catalog of Nebulae and Star Clusters

M	NGC or (IC)	Right Ascension (2000)		Declination (2000)		Apparent Visual Magnitude	Description
		h	m	(deg)	(min)		
1	1952	5	34.5	+22	01	8.4	"Crab" nebula in Taurus; remains of SN1054
2	7089	21	33.5	−0	50	6.4	Globular cluster in Aquarius
3	5272	13	42.2	+28	23	6.3	Globular cluster in Canes Venatici
4	6121	16	23.6	−26	32	5.9	Globular cluster in Scorpius
5	5904	15	18.6	+2	05	5.8	Globular cluster in Serpens
6	6405	17	40.1	−32	13	4.2	Open cluster in Scorpius
7	6475	17	53.9	−34	49	3.3	Open cluster in Scorpius
8	6523	18	03.8	−24	23	5.1	"Lagoon" nebula in Sagittarius
9	6333	17	19.2	−18	31	7.9	Globular cluster in Ophiuchus
10	6254	16	57.1	−4	06	6.7	Globular cluster in Ophiuchus
11	6705	18	51.1	−6	16	5.8	Open cluster in Scutum
12	6218	16	47.2	−1	57	6.6	Globular cluster in Ophiuchus
13	6205	16	41.7	+36	28	5.9	Globular cluster in Hercules
14	6402	17	37.6	−3	15	7.6	Globular cluster in Ophiuchus
15	7078	21	30.0	+12	10	6.4	Globular cluster in Pegasus
16	6611	18	18.8	−13	58	6.6	Open cluster with nebulosity in Serpens
17	6618	18	20.8	−16	11	7.5	"Swan" or "Omega" nebula in Sagittarius
18	6613	18	19.9	−17	08	6.9	Open cluster in Sagittarius
19	6273	17	02.6	−26	16	6.9	Globular cluster in Ophiuchus
20	6514	18	02.3	−23	02	8.5	"Trifid" nebula in Sagittarius
21	6531	18	04.6	−22	30	5.9	Open cluster in Sagittarius
22	6656	18	36.4	−23	54	5.1	Globular cluster in Sagittarius
23	6494	17	56.8	−19	01	5.5	Open cluster in Sagittarius
24	6603	18	16.9	−18	29	4.6	Star field with open cluster in Sagittarius
25	(4725)	18	31.6	−19	15	4.6	Open cluster in Sagittarius
26	6694	18	45.2	−9	24	8.0	Open cluster in Scutum
27	6853	19	59.6	+22	43	8.1	"Dumbbell" planetary nebula in Vulpecula
28	6626	18	24.5	−24	52	6.9	Globular cluster in Sagittarius
29	6913	20	23.9	+38	32	7.0	Open cluster in Cygnus
30	7099	21	40.4	−23	11	7.5	Globular cluster in Capricornus
31	224	0	42.7	+41	16	3.5	Andromeda galaxy

(Table continues)

M	NGC or (IC)	Right Ascension (2000)		Declination (2000)		Apparent Visual Magnitude	Description
		h	m	(deg)	(min)		
32	221	0	42.7	+40	52	8.2	Elliptical galaxy; companion to M31
33	598	1	33.9	+30	39	5.7	Spiral galaxy in Triangulum
34	1039	2	42.0	+42	47	5.2	Open cluster in Perseus
35	2168	6	08.9	+24	20	5.1	Open cluster in Gemini
36	1960	5	36.1	+34	08	6.5	Open cluster in Auriga
37	2099	5	52.4	+32	33	5.6	Open cluster in Auriga
38	1912	5	28.7	+35	50	6.4	Open cluster in Auriga
39	7092	21	32.2	+48	26	4.6	Open cluster in Cygnus
40	—	12	22.4	+58	05	8.0	Close double star in Ursa Major
41	2287	6	47.0	−20	44	4.5	Loose open cluster in Canis Major
42	1976	5	35.4	−5	27	4.0	Orion nebula
43	1982	5	35.6	−5	16	9.0	Northeast portion of Orion nebula
44	2632	8	40.1	+19	59	3.1	Praesepe; open cluster in Cancer
45	—	3	47.0	+24	07	1.2	The Pleiades; open cluster in Taurus
46	2437	7	41.8	−14	49	6.1	Open cluster in Puppis
47	2422	7	36.6	−14	30	4.4	Loose group of stars in Puppis
48	2548	8	13.8	−5	48	5.8	"Cluster of very small stars"
49	4472	12	29.8	+8	00	8.5	Elliptical galaxy in Virgo
50	2323	7	03.2	−8	20	6.0	Loose open cluster in Monoceros
51	5194	13	29.9	+47	12	8.4	"Whirlpool" spiral galaxy in Canes Venatici
52	7654	23	24.2	+61	35	6.9	Loose open cluster in Cassiopeia
53	5024	13	12.9	+18	10	7.7	Globular cluster in Coma Berenices
54	6715	18	55.1	−30	29	7.7	Globular cluster in Sagittarius
55	6809	19	40.0	−30	58	7.0	Globular cluster in Sagittarius
56	6779	19	16.6	+30	11	8.3	Globular cluster in Lyra
57	6720	18	53.6	+33	02	9.0	"Ring" nebula; planetary nebula in Lyra
58	4579	12	37.7	+11	49	9.8	Spiral galaxy in Virgo
59	4621	12	42.0	+11	39	9.8	Spiral galaxy in Virgo
60	4649	12	43.7	+11	33	8.8	Elliptical galaxy in Virgo
61	4303	12	21.9	+4	28	9.7	Spiral galaxy in Virgo
62	6266	17	01.2	−30	07	6.6	Globular cluster in Scorpius
63	5055	13	15.8	+42	02	8.6	Spiral galaxy in Canes Venatici
64	4826	12	56.7	+21	41	8.5	Spiral galaxy in Coma Berenices
65	3623	11	18.9	+13	05	9.3	Spiral galaxy in Leo
66	3627	11	20.2	+12	59	9.0	Spiral galaxy in Leo; companion to M65
67	2682	8	50.4	+11	49	6.9	Open cluster in Cancer
68	4590	12	39.5	−26	45	8.2	Globular cluster in Hydra
69	6637	18	31.4	−32	21	7.7	Globular cluster in Sagittarius
70	6681	18	43.2	−32	18	8.1	Globular cluster in Sagittarius
71	6838	19	53.8	+18	47	8.0	Globular cluster in Sagittarius
72	6981	20	53.5	−12	32	9.3	Globular cluster in Aquarius
73	6994	20	59.0	−12	38	9.1	Faint open cluster in Aquarius
74	628	1	36.7	+15	47	9.3	Spiral galaxy in Pisces
75	6864	20	06.1	−21	55	8.6	Globular cluster in Sagittarius

M	NGC or (IC)	Right Ascension (2000)		Declination (2000)		Apparent Visual Magnitude	Description
		h	m	(deg)	(min)		
76	650	1	42.4	+51	34	11.4	Planetary nebula in Perseus
77	1068	2	42.7	0	01	8.9	Spiral galaxy in Cetus
78	2068	5	46.7	0	03	8.3	Small emission nebula in Orion
79	1904	5	24.5	−24	33	7.8	Globular cluster in Lepus
80	6093	16	17.0	−22	59	7.3	Globular cluster in Scorpius
81	3031	9	54.2	+69	04	7.0	Spiral galaxy in Ursa Major
82	3034	9	55.8	+69	41	8.4	Irregular galaxy in Ursa Major
83	5236	13	37.0	-29	52	7.6	Spiral galaxy in Hydra
84	4374	12	25.1	+12	53	9.3	Elliptical galaxy in Virgo
85	4382	12	25.4	+18	11	9.3	Elliptical galaxy in Coma Berenices
86	4406	12	26.2	+12	57	9.2	Elliptical galaxy in Virgo
87	4486	12	30.8	+12	24	8.7	Elliptical galaxy inVirgo
88	4501	12	32.0	+14	25	9.5	Spiral galaxy in Coma Berenices
89	4552	12	35.7	+12	33	9.8	Elliptical galaxy in Virgo
90	4569	12	36.8	+13	10	9.5	Spiral galaxy in Virgo
91	4548	12	35.4	+14	30	10.2	Spiral galaxy in Coma Berenices [1]
92	6341	17	17.1	+43	08	6.4	Globular cluster in Hercules
93	2447	7	44.6	-23	52	6.5	Open cluster in Puppis
94	4736	12	50.9	+41	07	8.2	Spiral galaxy in CanesVenatici
95	3351	10	44.0	+11	42	9.7	Barred spiral galaxy in Leo
96	3368	10	46.8	+11	49	9.2	Spiral galaxy in Leo
97	3587	11	14.8	+55	01	11.1	"Owl" nebula; planetary nebula in Ursa Major
98	4192	12	13.8	+14	54	10.1	Spiral galaxy in Coma Berenices
99	4254	12	18.8	+14	25	9.9	Spiral galaxy in Coma Berenices
100	4321	12	22.9	+15	49	9.4	Spiral galaxy in Coma Berenices
101	5457	14	03.2	+54	21	7.9	Spiral galaxy in Ursa Major
102	5866(?)	15	06.5	+55	46	10.5	Spiral galaxy (identification in doubt) [1]
103	581	1	33.2	+60	42	7.4	Open cluster in Cassiopeia
104°	4594	12	40.0	−11	37	8.3	Spiral galaxy in Virgo
105°	3379	10	47.8	+12	35	9.3	Elliptical galaxy in Leo
106°	4258	12	19.0	+47	18	8.4	Spiral galaxy in Canes Venatici
107°	6171	16	32.5	−13	03	8.2	Globular cluster in Ophiuchus
108°	3556	11	11.5	+55	40	10.0	Spiral galaxy in Ursa Major
109°	3992	11	57.6	+53	23	9.8	Spiral galaxy in Ursa Major
110°	205	0	40.4	+41	41	8.1	Elliptical galaxy; companion to M31

[1] According to some versions of Messier's list, M91 is actually a repeat of M58 and M102 is the same as M101, but here we give the galaxies that modern lists usually cite.

° Not in Messier's original (1781) list; added later by others.

Credits

This page constitutes an extension of the copyright page. We have made every effort to trace the ownership of all copyrighted material and to secure permission from copyright holders. In the event of any question arising as to the use of any material, we will be pleased to make the necessary corrections in future printings. Thanks are due to the following authors, publishers, and agents for permission to use the material indicated.

Prologue. Opener: JPL/NASA. **P.1:** USGS. **P.2:** JPL/NASA. **P.3:** C.R. O'Dell and NASA. **P.4:** NASA. **P.5:** ESA. **P.8:** JPL/NASA. **P.11:** Eastern Southern Observatory. **P.12:** Roger Angel, Steward Observatory/University of Arizona. **P.13:** Anglo-Australian Telescope Board/David Malin Images. **P.14:** Photo taken with the U.K. Schmidt Telescope; © Anglo-Australian Telescope Board/David Malin Images. **P.15:** Tony Hallas. **P.16:** Photo taken with the U.K. Schmidt Telescope; © Anglo-Australian Telescope Board/David Malin Images.

Chapter 1. CO1: U. of Toronto. **Fig.1.3:** "National Optical Astronomy Observatories". **Fig.1.7:** "J.M. Pasachoff and the Chapin Library". **Fig.1.8:** Kent Wood. **Fig.1.14:** Yerkes Observatory. **Fig.1.15:** "Crawford Collection, Royal Observatory, Edinburgh". **Fig.1.17:** Yerkes Observatory. **Fig.1.18:** "Instituto e Museo di Storia della Scienza de Florenza".

Chapter 2. CO2: NASA. **Fig.2.1:** Granger Collection, New York. **Fig.2.2:** AIP/Niels Bohr Library. **Fig.2.6:** AIP/Niels Bohr Library. **Fig.2.7:** NASA. **Fig.2.9:** NASA. **Fig.2.11b:** "Crawford Collection, Royal Observatory, Edinburgh". **Fig.2.12:** US Space Command, NORAD. **Fig.2.13:** NASA/ARC. **Fig.2.14a:** Yerkes Observatory. **Fig.2.14b:** Corbis/Bettmann.

Chapter 3. CO3: NASA. **Fig.3.2:** © 1993 Hal Berol/Visuals Unlimited. **Fig.3.3:** © Bob Emott, Photographer. **Fig.3.11:** David Morrison. **Fig.3.12:** David Morrison. **Fig.3.17a:** Courtesy Nova Scotia Tourism. **Fig.17b:** Courtesy Nova Scotia Tourism. **Fig.3.21:** © 1991 Stephen J. Edberg. **Voyagers:** © Royal Society.

Chapter 4. CO4: Martha Haynes and Riccardo Giovanelli, Cornell University. **Fig.4.1:** American Institute of Physics, Niels Bohr Library. **Fig.4.2:** © 1993 Comstock, Inc. **Fig.4.6:** Max Planck Institute for Extraterrestrial Physics. **Fig.4.10:** National Solar Observatory/NOAO.

Chapter 5. CO5: Johnson Space Ctr/NASA. **Fig.5.1a:** Infrared Processing and Analysis Center/JPL. **Fig.5.1b:** Max Planck Institut fur Extraterrestrische Physik. **Fig.5.1c:** JPL/NASA. **Fig.5.2a:** David Morrison. **Fig.5.2b:** David Morrison. **Fig.5.7:** Gemini Observatory. **Fig.5.8a:** California Institute of Technology. **Fig.5.8b:** Gemini Observatory. **Fig.5.9:** California Association for Research in Astronomy. **Fig.5.10:** ESO. **Fig.5.11:** © William Keck Observatory. **Fig.5.13:** Bell Laboratories. **Fig.5.14:** NRAO/AUI (Nat'l Radio Astronomy Observatory). **Fig.5.15:** NRAO. **Fig.5.16:** NRAO. **Fig.5.17:** National Astronomy and Ionosphere Center. **Fig.5.18:** NASA/CXC/SAO. **Fig.5.19:** ESO. **Voyagers:** G.E. Hale: Caltech. The Yerkes telescope: Yerkes Observatory.

Chapter 6. CO6: JPL/NASA. **Fig.6.1:** NASA/JSC. **Fig.6.3:** NASA. **Fig.6.4:** NASA. **Fig.6.5:** NASA/JPL. **Fig.6.6:** NASA/JPL. **Fig.6.7:** E. Kakorchka, U. of Arizona; and NASA. **Fig.6.8:** JPL/NASA. **Fig.6.9:**

STScl/NASA. **Fig.6.10:** NASA. **Fig.6.12:** Art by planetary astronomer William Hartmann. **Fig.6.13a:** C.R. O'Dell/Rice University/NASA. **Fig.6.13b:** C.R. O'Dell/Rice University/NASA. **Fig.6.13c:** C.R. O'Dell/Rice University/NASA. **Voyagers:** unknown.

Chapter 7. CO7: G. Briggs/USGS. **Fig.7.1:** NASA. **Fig.7.3:** J. Frawley and NASA. **Fig.7.5:** David Morrison. **Fig.7.8:** David Porter/Science Photo Library/Photo Researchers, Inc. **Fig.7.9:** David Morrison. **Fig.7.11:** NASA. **Fig.7.12:** NASA/ARC, courtesy of David DesMarais. **Fig.7.16:** Novosty. **Fig.7.17:** Meteor Crater, Northern Arizona. **Fig.7.18:** Painting by Don Davis. **Voyagers:** Historical Pictures Service, Chicago.

Chapter 8. CO8: NASA. **Fig.8.1:** NASA. **Fig.8.2:** NASA. **Fig.8.3:** NASA. **Fig.8.4:** NASA. **Fig.8.5:** USGS. **Fig.8.6:** NASA. **Fig.8.7:** NASA. **Fig.8.8:** NASA. **Fig.8.9:** NASA. **Fig.8.10:** NASA. **Fig.8.11:** NASA. **Fig.8.14:** NASA. **Fig.8.16a-d:** Courtesy of A.G.W. Cameron, Center for Astrophysics. **Fig.8.17:** University of Arizona, Courtesy of Robert Strom. **Fig.8.19:** Mosaic courtesy of the Astrogeology Team, USGS, Flagstaff. **Fig.8.20:** NASA/JPL. **Fig.8.21:** NASA/JPL.

Chapter 9. CO9: NASA/USGS, courtesy of Alfred McEwen. **Fig.9.1:** NASA/ARC. **Fig.9.2:** Steve Larson, University of Arizona. **Fig.9.3:** Lowell Observatory. **Fig.9.4:** NASA/JPL. **Fig.9.5a:** NASA/JPL. **Fig.9.5b:** NASA/JPL. **Fig.9.6:** NASA/JPL. **Fig.9.7:** NASA/JPL. **Fig.9.8:** NASA/JPL. **Fig.9.9:** USSR Academy of Science and Carle Pieters, Brown University. **Fig.9.11:** Space Telescope Science Institute/NASA. **Fig.9.12:** NASA/JPL/Caltech. **Fig.9.13:** NASA/MOLA. Team led by David Smith at Goddard Space Flight Center. **Fig.9.14:** "Kees Veenenbos/veenen@tiscali.nl." **Fig.9.15:** NASA/USGS. **Fig.9.16a:** NASA/JPL. **Fig.9.16b:** NASA/JPL. **Fig.9.16c:** NASA/JPL. **Fig.9.17:** NASA/JPL. **Fig.9.18:** Malin Space Science Systems/NASA. **Fig.9.19:** NASA/JPL/Malin Space Science Systems. **Fig.9.2:** NASA/JSC. **Fig.9.21:** NASA/JPL/Malin Space Science Systems. **Fig.9.22:** NASA/from Mars Digital Image Map, processing by Brian Fessler/LPI. **Fig.9.23:** NASA/Malin Space Science Systems. **Fig.9.24:** NASA/JPL. **Fig.9.25a:** Malin Space Science Systems/NASA. **Fig.9.25b:** NASA/JPL/Malin Space Science Systems. **Fig.9.26:** Malin Space Science Systems/NASA. **Voyagers:** Lowell Observatory. **Making Connections (1):** JPL/NASA. **Making Connections (2):** NASA/Malin Space Science Systems.

Chapter 10. CO10: NASA/JPL. **Fig.10.1:** NASA/JPL/University of Arizona. **Fig.10.2:** NASA/JPL. **Fig.10.3:** NASA/ARC. **Fig.10.4:** NASA/JPL. **Fig.10.5:** Erich Karkoschka, Univ. of Arizona and NASA. **Fig.10.8:** Imke dePater/NRAO. **Fig.10.9:** NASA/JPL. **Fig.10.10:** Space Telescope Science Institute/NASA. **Fig.10.11:** NASA/JPL. **Fig.10.13:** NASA/JPL. **Fig.10.14, Right two:** Andrew Ingersoll and Don Davis, courtesy of Sky Publishing Corp. **Fig.10.14, Left two:** NASA/JPL. **Fig.10.15:** NASA/JPL. **Fig.10.16a&b:** Caltech/NASA. **Fig.10.16c:** Cornell University/Caltech/NASA. **Voyagers:** Photo by D. Hakanson.

Chapter 11. CO11: DLR/NASA. **Fig.11.1:** NASA/JPL. **Fig.11.2:** NASA/JPL. **Fig.11.3:** Arizona State Univ./JPL/NASA. **Fig.11.4:** Galileo Imaging Team/JPL/NASA. **Fig.11.5:** NASA/JPL. **Fig.11.6:** NASA/JPL. **Fig.11.7:** NASA/JPL. **Fig.11.8:** NASA/JPL. **Fig.11.9:** NASA/JPL. **Fig.11.10:** NASA/JPL. **Fig.11.11:** DLR (German Aerospace Center)/NASA. **Fig.11.12:** NASA/USGS. **Fig.11.13:** NASA/JPL. **Fig.11.14:**

University of Arizona/NASA. **Fig.11.15:** University of Arizona/NASA. **Fig.11.16:** University of Arizona/NASA. **Fig.11.17:** NASA/JPL. **Fig.11.19:** NASA/Image processing at USGS. **Fig.11.20:** Lowell Observatory. **Fig.11.21:** A. Stern, M. Buie, NASA and ESA. **Fig.11.23:** NASA/JPL. **Fig.11.24:** NASA/JPL. **Fig.11.25:** NASA/JPL. **Fig.11.26:** NASA/JPL. **Fig.11.27:** NASA/JPL. **Voyagers:** Lowell Observatory.

Chapter 12. CO12: Tony Hallas, AstroPhoto. **Fig.12.1:** Yerkes Observatory. **Fig.12.2:** Edward Bowell, Lowell Observatory. **Fig.12.4:** NASA. **Fig.12.5:** "NASA/JPL." **Fig.12.6:** NASA/JPL. **Fig.12.7:** Malin Space Systems/JPL/NASA. **Fig.12.9:** NASA/JPL, courtesy of Steven Ostro. **Fig.12.10:** NASA/JHUAPL. **Fig.12.11:** NASA/JHUAPL. **Fig.12.12:** NASA/JHUAPL. **Fig.12.13:** Anglo-Australian Observatory/David Malin Images. **Fig.12.15:** Donald Brownlee, University of Washington. **Fig.12.16a:** Max Planck Institut fur Aeronomie, and Ball Aerospace Corporation, courtesy of Harold Reitsema. **Fig.12.16b:** NASA/USGS. **Fig.12.19:** Caltech/Palomar Observatory. **Fig.12.20:** Eastern Southern Observatory. **Fig.12.21a:** University of Hawaii. **Fig.12.21b:** NASA, Harold Weaver (Johns Hopkins University), and the HST Comet Linear Investigation. **Fig.12.22:** NASA/STScI. **Fig.12.23:** NASA/STScI. **Voyagers:** Yerkes Observatory, University of Chicago.

Chapter 13. CO13: Painting by Don Dixon. **Fig.13.1:** P. Parviainen, Polar Image. **Fig.13.2:** Photo © 1986 John Sanford. **Fig.13.4:** unknown. **Fig.13.5:** J.O. Wheeler; courtesy Department of Library Services, American Museum of Natural History. **Fig.13.6:** NASA. **Fig.13.7:** Chip Clark. **Fig.13.9:** Adapted from a diagram published by John Lewis, University of Arizona. **Fig.13.10:** M. McCaughrean, C.R. O'Dell, and NASA. **Fig.13.11:** Geoffrey Bryden/JPL/NASA. **Fig.13.12:** Diagram courtesy of G. Marcy, San Francisco State University. **Fig.13.13:** M. Mayor and D. Queloz. **Fig.13.14:** Painting by Lynette Cook. **Fig.13.15:** San Francisco State University. **Fig.13.16:** Adapted from the California &

Carnegie Planet Search Web site: exoplanet.org (Marcy, Butler, et al.). **Fig.13.17:** G. Marcy, P. Butler, D. Fischer, S. Vogt. **Fig.13.20:** NASA/USGS, courtesy of Alfred McEwen. **Fig.13.21:** JPL/NASA. **Making Connections:** © 1982, 1992 Dan Haar/The Hartford Courant.

Chapter 14. CO14: SOHO/NASA. **Fig.14.2:** Harvard University Archives. **Fig.14.3:** SOHO/NASA. **Fig.14.4:** David Alexander; NASA/Yohkoh. **Fig.14.6:** High Altitude Observatory/NCAR. **Fig.14.7:** "TRACE/Stanford-Lockheed Institute for Space Research." **Fig.14.8:** "W. Livingston, National Solar Observatories/NOAO." **Fig.14.9:** National Solar Observatory/NOAO. **Fig.14.10a:** LMATC/NSO/NASA. **Fig.14.10b:** LMATC/NSO/NASA. **Fig.14.11:** NOAO. **Fig.14.12:** Data courtesy of J. Harvey, National Solar Observatory (Tucson/Kitt Peak, AZ). **Fig.14.13:** SOHO/NASA. **Fig.14.14:** Solar & Heliospheric Observatory/NASA. **Fig.14.15:** LMATC/NSO/NASA. **Fig.14.17:** "Kunsthistorisches Museum, Vienna; Photo by Erich Lessing. Art Resource." **Voyagers:** Courtesy of Stanford University. **Making Connections:** SOHO/NASA.

Chapter 15. CO15: NASA; Artist: Roger Arno. **Fig.15.1:** J. Hester, P. Scowen, and NASA. **Fig.15.2:** R. Provin. **Fig.15.3:** Photo by A. Fraknoi. **Fig.15.4:** Photo by David Malin, © Anglo-Australian Observatory. **Fig.15.5:** F. Drake and the Astronomical Society of the Pacific. **Fig.15.6:** William Schopf, UCLA. **Fig.15.7:** Space Telescope Science Institute/AURA/NASA. **Fig.15.8:** Peter Ryan/Scripps/Science Photo Library/Photo Researchers, Inc. **Fig.15.9:** David Morrison. **Fig.15.10:** Painting by Don Davis, © SETI Institute. **Fig.15.11:** USGS/JPL. **Fig.15.12:** Galileo Team/Landsat/JPL. **Fig.15.14:** David DesMarais, NASA Ames. **Fig.15.15:** NASA. **Fig.15.16:** NASA. **Fig.15.17:** National Radio Astronomy Observatory. **Fig.15.18:** Photos by Seth Shostak, SETI Institute. **Fig.15.19:** SETI Institute. **Fig.15.20:** SETI Institute. **Voyagers:** Photo by Seth Shostak/SETI Insititute.

Index

Note: *Italicized* page numbers indicate figures, illustrations, or tables.

atomic mass unit, 373
atomic number, 373
atomic weight, 373
atoms, 15, 97–100
    Bohr's model of, 99–100, *101*
    definition of, 373
    energy levels of, 100
    excited state of, 101
    ground state of, *98*, 100–101
    ionization of, 102–103
    nucleus of, 97–99
    Rutherford's experiment with, 97, *98*
    spectral formation and, 100
    structure of, 97–100
auroras, 156, 323, 332, 373
Australia Telescope, *124*
Australia Tracking Station, *124*
autumnal equinox, 66, 373
axis, 373
    of Earth, 21, 64
    major and semimajor, 45
azimuth, 373

Babylonians, 31
Backus, Peter, *360*
Bacon, Francis, 158
Bagenal, Fran, 262
Balmer series, 101, 373
bands, spectral, 373
bar, 161, 373
barometer, 161
barred spiral galaxies, 373
barycenter, 373
basalt, 154, 373
Baum, L. Frank, 360
BepiColombo spacecraft, 190
BeppoSAX satellite, *128*
Beston, Henry, 317
Big Bang, 14, 373
Big Dipper, 21, 25
binary stars, 373
binding energy, 373
binoculars, 113, *113*
bioastronomy. *See* astrobiology
biomarkers, 351, 354–355, 373
biopolymers, 349
Biot, Jean Baptiste, 293
blackbodies, 92*n*, 93, 373
black dwarfs, 373
black holes, 373
Blanco Telescope, *115*
blink comparator, 257
blueshift, 104
Bohr, Niels, 99
    atomic model of, 99–100, *101*, 373
    Web site on work of, 105
Bolshoi Teleskop, *115*
Bonell, Jerry, 17
Bossidy, John, 197
Bowyer, Stuart, 361
Bragg, Sir William, 85
Brecht, Bertolt, 1
brightest stars, 394
bright lines, 96
Bronze Age, 69
brown dwarfs, 10, 361, 373
Butler, Paul, 302, *303*
Byrd, Robert C., *123*

C asteroids, 269, *270*
Caesar, Julius, 71
calculus, 50
caldera, 201, *206*, 373
calendars, 69–71
    early forms of, 69–71

Gregorian, 71
    historic challenge of, 69
    traditional functions of, 69
    Web sites about, 80
Callisto, 142, 244, 245–247, *245*, *246*, *250*
Caloris basin, *188*, 189
Canada-France-Hawaii telescope, *115*
canyons on Mars, *194*, 207, *207*
carbon, radioactive, 335
carbon-14 isotope, 335, 337
carbonaceous meteorites, 296, 373
carbonate, 373
carbon-based molecules, 349
carbon dioxide
    on Earth, 164, 165–166, *166*
    on Mars, 210
    on Venus, 203–204
carbon-nitrogen-oxygen (CNO) cycle, 373
Carson, Johnny, 148
Cassegrain focus, *113*, 373
Cassini, Gian Domenico, 259
Cassini Division, 259, 261
Cassini spacecraft explorations
    of Saturn, 225, 226, *226*
    Web site on, 239
Castelli, Benedetto, 336
Catholic Church, 35, 36, 37
celestial equator, 21–22, *24*, 373
celestial meridian, 373
celestial poles, 21–22, *22*, *23*, 373
celestial sphere, 20–21, *21*, 373
celestial tilt, *24*
Centaur asteroids, 281
center of gravity, 373
center of mass, 373
cepheid variable stars, 373
Ceres, 268, 269
CFCs (chlorofluorocarbons), 162
Challis, James, 56
Chandrasekhar limit, 373
Chandra x-ray telescope, 127–128, *127*, *128*
channels on Mars, 212, *212*, *213*
charge-coupled devices (CCDs), 119, 373
Charon, 76, 256, *257*, *258*
chemical condensation sequence, 373
chemical elements, 397–399
Chicxulub crater, 169, *169*
Chinese calendar, 70–71
Chiron, 272
chondrite, 373
chondrule, 373
chromosphere, 321, *321*, 373
circular satellite velocity, 53, 374
circumpolar zone, 374
circumstellar disks, 146, 300, *300*, *301*, 374
classification
    of asteroids, 269, *270*
    of meteorites, 294–295, *295*, *296*
Clementine spacecraft, *178*
climate
    on Earth, 162–163, 332, *334*, 335, 337
    on Mars, 214–215
    See also weather
closed universe, 374
Clough, Arthur Hugh, 52
clusters
    of galaxies, 13, *14*, 374
    of stars, 12
cold fusion, 35
color
    false-color images, *141*
    spectral shifts in, 104
color index, 374
coma (of comet), 277, 374
Comamara Chaos, *248*

Comet Borrelly, 278, *278*, 279
Comet Hale-Bopp, *266*, 276
Comet Halley, 275–276, 277, *277*, 278, *278*,
    279, *279*, 281, 282
Comet Hyakutake, 276, *345*
Comet LINEAR, 282
Comet Mrkos, *280*
Comet Shoemaker-Levy 9, 143, 168, 282,
    283–284, *283*, *284*
Comet Swift-Tuttle, 292
comets, 139, 142, 267–268, 275–284
    appearance of, 275
    atmosphere of, 277, 279
    breaking apart of, 282–283, *282*
    collisions with Jupiter, 143, *143*, 283–284,
    283, *284*
    definition of, 374
    dust captured from, 277–278, *278*
    fate of, 281–284
    impact hazards of, 169–170
    Jupiter-family, 276, 281
    Kuiper belt and, 279–280, *280*, 281
    list of well-known, 277
    naming of, 141
    nucleus of, 277–278, *278*
    Oort comet cloud and, 280–281
    orbits of, 53, *53*, 275–276
    origin and evolution of, 280–281
    parts of, *278*
    searching for, 282–283
    solar system formation and, 299
    spacecraft images of, 278, *278*
    tail of, 275, 279, *279*
    Web sites about, 170, 284, 285
Committee for the Scientific Investigation of
    Claims of the Paranormal (CSICOP),
    215, 371
comparative planetology, 136
compounds, 374
Compton Gamma-Ray Observatory, *128*
conduction, 374
conic sections, 45, *45*
conservation of momentum, 47
constants, 387
constellations, 25, 400–402
    astrological signs and, 31–32
    definition of, 13, 25, 374
    ecliptic and, *24*
Contact (Sagan), 148, 361
continental drift, 158, 374
continuous spectrum, 92, *94*, 96, 102, *102*, 374
convection, 157, 200, 320, 374
convection currents, 235, 320
Copernican principle, 346
Copernicus, Nicolaus, 33–35, *34*
Cordelia, 261
core, 155, 374
corona, 374
    solar, 78, *79*, 322–323, *322*, *323*
    volcanic, 201, *201*
coronal holes, 323, 330, 374
coronal mass ejections (CMEs), 329, 330–
    332, *331*
Cosmic Background Explorer (COBE), *128*
cosmic background radiation (CBR), 374
Cosmic Connection, The (Sagan), 135, 148
cosmic evolution, 345
cosmic haystack problem, 358–359, *359*
cosmic rays, 374
cosmic time, 15–16, *16*
cosmological constant, 374
cosmological principle, 374
cosmology, 26, 374
Cosmos (Sagan), 148
Crab Nebula, *344*

THE NIGHT SKY IN FEBRUARY

Latitude of chart is 34°N, but it is practical throughout the continental United States.

To use: Hold chart vertically and turn it so the direction you are facing shows at the bottom.

Chart time (Local Standard):
10 p.m. First of month
9 p.m. Middle of month
8 p.m. Last of month

Star Chart from *GRIFFITH OBSERVER*, Griffith Observatory, Los Angeles

THE NIGHT SKY IN MARCH

Latitude of chart is 34°N, but it is practical throughout the continental United States.

To use: Hold chart vertically and turn it so the direction you are facing shows at the bottom.

Chart time (Local Standard):
10 p.m. First of month
9 p.m. Middle of month
8 p.m. Last of month

Star Chart from *GRIFFITH OBSERVER*, Griffith Observatory, Los Angeles

NORTHERN HORIZON

CEPHEUS

CASSIOPEIA

DRACO

VEGA

PERSEUS

HERCULES

CORONA BOREALIS

POLARIS "NORTH STAR"

CAPELLA

URSA MINOR "LITTLE DIPPER"

AURIGA

TAURUS

EASTERN HORIZON

BOOTES

URSA MAJOR "BIG DIPPER"

WESTERN HORIZON

SERPENS

ARCTURUS

CASTOR

GEMINI

ALDEBARAN

POLLUX

CANCER

BETELGEUSE

RIGEL

LEO

ORION

LIBRA

REGULUS

PROCYON

CANIS MINOR

VIRGO

SPICA

CORVUS

SIRIUS

CANIS MAJOR

HYDRA

SOUTHERN HORIZON

THE NIGHT SKY IN APRIL

Latitude of chart is 34°N, but it is practical throughout the continental United States.

To use: Hold chart vertically and turn it so the direction you are facing shows at the bottom.

Chart time (Local Standard):

10 p.m. First of month

9 p.m. Middle of month

8 p.m. Last of month

Star Chart from *GRIFFITH OBSERVER*, Griffith Observatory, Los Angeles

NORTHERN HORIZON

CASSIOPEIA

CEPHEUS

DENEB

CYGNUS

DRACO

POLARIS "NORTH STAR"

CAPELLA

AURIGA

LYRA

VEGA

URSA MINOR "LITTLE DIPPER"

GEMINI

CASTOR

HERCULES

CORONA BOREALIS

BOOTES

URSA MAJOR "BIG DIPPER"

POLLUX

CANCER

CANIS MINOR

PROCYON

EASTERN HORIZON

OPHIUCHUS

SERPENS

ARCTURUS

LEO

REGULUS

WESTERN HORIZON

VIRGO

LIBRA

SPICA

CORVUS

HYDRA

SCORPIUS

ANTARES

SOUTHERN HORIZON

THE NIGHT SKY IN MAY

Latitude of chart is 34°N, but it is
practical throughout the continental
United States.

To use: Hold chart vertically and turn
it so the direction you are facing
shows at the bottom.

Chart time (Local Standard):

10 p.m. First of month

9 p.m. Middle of month

8 p.m. Last of month

Star Chart from *GRIFFITH OBSERVER*, Griffith Observatory, Los Angeles

THE NIGHT SKY IN JUNE

Latitude of chart is 34°N, but it is practical throughout the continental United States.

To use: Hold chart vertically and turn it so the direction you are facing shows at the bottom.

Chart time (Local Standard):

10 p.m. First of month

9 p.m. Middle of month

8 p.m. Last of month

Star Chart from *GRIFFITH OBSERVER*, Griffith Observatory, Los Angeles

NORTHERN HORIZON

EASTERN HORIZON

WESTERN HORIZON

SOUTHERN HORIZON

THE NIGHT SKY IN JULY

Latitude of chart is 34° N, but it is
practical throughout the continental
United States.

To use: Hold chart vertically and turn
it so the direction you are facing
shows at the bottom.

Chart time (Local Standard):

10 p.m. First of month

9 p.m. Middle of month

8 p.m. Last of month

Star Chart from *GRIFFITH OBSERVER*, Griffith Observatory, Los Angeles

NORTHERN HORIZON

EASTERN HORIZON

WESTERN HORIZON

SOUTHERN HORIZON

THE NIGHT SKY IN AUGUST

Latitude of chart is 34°N, but it is practical throughout the continental United States.

To use: Hold chart vertically and turn it so the direction you are facing shows at the bottom.

Chart time (Local Standard):

10 p.m. First of month

9 p.m. Middle of month

8 p.m. Last of month

Star Chart from *GRIFFITH OBSERVER*, Griffith Observatory, Los Angeles

THE NIGHT SKY IN NOVEMBER

Latitude of chart is 34° N, but it is practical throughout the continental United States.

To use: Hold chart vertically and turn it so the direction you are facing shows at the bottom.

Chart time (Local Standard):

10 p.m. First of month

9 p.m. Middle of month

8 p.m. Last of month

Star Chart from *GRIFFITH OBSERVER*, Griffith Observatory, Los Angeles

NORTHERN HORIZON

DRACO

VEGA

URSA MAJOR "BIG DIPPER"

LITTLE DIPPER

URSA MINOR

POLARIS NORTH STAR

CYGNUS "NORTHERN CROSS"

DENEB

CEPHEUS

DELPHINUS

CASSIOPEIA

CASTOR

POLLUX

GEMINI

CAPELLA

AURIGA

PERSEUS

ANDROMEDA

PEGASUS

CANIS MINOR

PROCYON

TRIANGULUM

ARIES

TAURUS

PLEIADES

AQUARIUS

BETELGEUSE

ALDEBARAN

PISCES

ORION

CANIS MAJOR

SIRIUS

RIGEL

CETUS

FOMALHAUT

LEPUS

COLUMBA

EASTERN HORIZON

WESTERN HORIZON

SOUTHERN HORIZON

THE NIGHT SKY IN DECEMBER

Latitude of chart is 34° N, but it is
practical throughout the continental
United States.

To use: Hold chart vertically and turn
it so the direction you are facing
shows at the bottom.

Chart time (Local Standard):

10 p.m. First of month

9 p.m. Middle of month

8 p.m. Last of month

Star Chart from *GRIFFITH OBSERVER*, Griffith Observatory, Los Angeles

THE NIGHT SKY IN OCTOBER

Latitude of chart is 34°N, but it is practical throughout the continental United States.

To use: Hold chart vertically and turn it so the direction you are facing shows at the bottom.

Chart time (Local Standard):
10 p.m. First of month
9 p.m. Middle of month
8 p.m. Last of month

Star Chart from *GRIFFITH OBSERVER*, Griffith Observatory, Los Angeles

NORTHERN HORIZON

EASTERN HORIZON

WESTERN HORIZON

SOUTHERN HORIZON

THE NIGHT SKY IN SEPTEMBER

Latitude of chart is 34°N, but it is
practical throughout the continental
United States.

To use: Hold chart vertically and turn
it so the direction you are facing
shows at the bottom.

Chart time (Local Standard):

10 p.m. First of month

9 p.m. Middle of month

8 p.m. Last of month

Star Chart from *GRIFFITH OBSERVER*, Griffith Observatory, Los Angeles